Modern Wireless Communications

Simon Haykin

McMaster University
Hamilton, Ontario, Canada

and

Michael Moher

Space-Time DSP Inc.
Ottawa, Ontario, Canada

PEARSON

Prentice
Hall

Upper Saddle River, NJ 07458

Library of Congress Cataloging-in-Publication

Haykin, Simon S.,
  Modern wireless communications / Simon Haykin and Michael Moher.
    p. cm.
  Includes bibliographical references and index.
  ISBN 0-13-022472-3
    1. Wireless communication systems. 2. Spread spectrum communications. I. Moher,
  Michael. II. Title

TK5103.2.H39 2003
621.382--dc22

2003061139

Vice President and Editorial Director, ECS: *Marcia J. Horton*
Vice President and Director of Production and Manufacturing, ESM: *David W. Riccardi*
Executive Managing Editor: *Vince O'Brien*
Managing Editor: *David A. George*
Production Editor: *Craig Little*
Director of Creative Services: *Paul Belfanti*
Art Director: *Jayne Conte*
Cover Designer: *Bruce Kenselaar*
Art Editor: *Greg Dulles*
Manufacturing Manager: *Trudy Pisciotti*
Manufacturing Buyer: *Lisa McDowell*
Marketing Manager: *Holly Stark*

© 2005 Pearson Education, Inc.
Pearson Prentice Hall
Pearson Education, Inc.
Upper Saddle River, NJ 07458

Printed in the United States of America
10 9 8 7 6 5 4 3 2 1

ISBN 0-13-022472-3

Pearson Education Ltd., *London*
Pearson Education Australia Pty. Ltd., *Sydney*
Pearson Education Singapore, Pte. Ltd.
Pearson Education North Asia Ltd., *Hong Kong*
Pearson Education Canada, Inc., *Toronto*
Pearson Educación de Mexico, S.A. de C.V.
Pearson Education—Japan, *Tokyo*
Pearson Education Malaysia, Pte. Ltd.
Pearson Education, Inc., *Upper Saddle River, New Jersey*

In Memory of William S. Hart

and

To Dianne

Contents

Preface

The rapid growth of wireless communications and its pervasive use in all walks of life are changing the way we communicate in some fundamental ways. Most important, reliance on radio propagation as the physical mechanism responsible for the transport of information-bearing signals from the transmitter to the receiver has endowed communications with a distinctive feature, namely, mobility.

Modern Wireless Communications is a new book aimed at the teaching of a course that could follow a traditional course on communication systems, as an integral part of an undergraduate program in electrical engineering or as the first graduate course on wireless communications. The primary focus of the book is on the physical layer, emphasizing the fundamentals of radio propagation and communication-theoretic aspects of multiple-access techniques. Many aspects of wireless communications are covered in an introductory level and book form for the first time.

1. ORGANIZATION OF THE BOOK

The book is organized in seven chapters, nine appendices, and a bibliography.

Chapter 1 motivates the study of wireless communications. It begins with a brief historical account of wireless communications, and then goes on to describe the OSI model of communication networks. The discussion, however, focuses on the issues that arise in the study of the physical layer, which is the mainstay of the book.

Chapter 2 on radio propagation starts with an explanation of the physical mechanisms of the propagation process, including free-space propagation, reflection, and diffraction. These physical mechanisms provide insight into the statistical models that are employed for terrestrial and indoor propagation effects that follow. The small-scale effects of fading and uncorrelated scattering are discussed, leading up to a careful classification of the different wireless channel types. The second half of the chapter describes noise and interference, and how combined with propagation, we may determine wireless communication system performance through a link-budget analysis.

Chapter 3 reviews the modulation process with emphasis on digital transmission techniques. This introductory treatment of modulation paves the way for discussions of the following issues:

- Complex baseband representation of linear modulated signals, and the corresponding input/output descriptions of linear wireless communication channels and linear band-pass filters.
- Practical problems concerning adjacent channel interference and nonlinearities in transmit power amplifiers.

The stage is then set for comparative evaluation of various modulation strategies for wireless communications, discussion of receiver performance in the presence of channel noise and Rayleigh fading, and discussion of frequency-division multiple-access (FDMA).

Chapter 4 focuses on coding techniques and time-division multiple-access (TDMA). After a brief review of Shannon's classical information theory, the source coding of speech signals is discussed, which is then followed by fundamental aspects of convolutional codes, interleavers, and turbo codes. The relative merits of convolutional codes and turbo codes are discussed in the context of wireless communications. The various aspects of channel-estimation, tracking, and channel equalization are treated in detail. The discussion then moves onto TDMA and the advantages it offers over FDMA.

Chapter 5 discusses spread spectrum, code-division multiple-access (CDMA), and cellular systems. It first presents the basics of spread-spectrum systems, namely, direct-sequence, and frequency-hopped systems, and their tolerance to interference. A fundamental component of spread-spectrum systems is the spreading code: a section of the chapter is devoted to explaining Walsh-Hadamard, maximal-length sequences, Gold codes, and random sequences. This discussion is then followed with a description of RAKE receivers, channel estimation, code synchronization, and the multipath performance of direct-sequence systems. This leads naturally to a discussion of how direct-sequence systems perform in a cellular environment.

Chapter 6 is devoted to the notion of space diversity and related topics. It starts with diversity on receive, which represents the traditional technique for mitigating the fading problem that plagues wireless communications. Then the chapter introduces the powerful notion of multiple-input, multiple-output (MIMO) wireless communications, which includes space diversity on receive and space diversity on transmit as special cases. Most important, the use of MIMO communications represents the "spatial frontier" of wireless communications in that, for prescribed communication resources in the form of fixed transmit power and channel bandwidth, it provides the practical means for significant increases in the spectral efficiency of wireless communications at the expense of increased computational complexity. The discussion of MIMO wireless communications also includes orthogonal space-time block codes (STBC), best exemplified by the Alamouti code and its differential form. The discussion then moves onto space-division multiple access (SDMA), and smart antennas.

Chapter 7 links the physical layer and multiple-access topics of the previous chapters with the higher layers of the communications network. This final chapter of the book begins with a comparison of the different multiple-access strategies. The discussion then leads to a consideration of various link-management functions associated with wireless systems, namely, signaling, power control, and handover. The differences between systems used for telephony and those used for data transmission are clearly delineated. This is then followed by a discussion of wireless network architectures, both for telephony and data applications.

1.1 Theme Examples

An enriching feature of the book is the inclusion of Theme Examples within each of the chapters in the book, except for Chapter 1. In a loose sense, they may be viewed as

"Chapters within Chapters" that show the practical applications of the topics discussed in the pertinent chapters. Specifically, the following Theme Examples are discussed:

Chapter 2:   Empirical propagation model, wireless local area networks (LANs), and impulse radio and ultra-wideband

Chapter 4:   Global system for mobile (GSM) communications, joint equalization and decoding, and random-access techniques

Chapter 5:   Code-division multiple access (CDMA) Standard IS-95, GPSS, bluetooth, wideband CDMA and WiFi

Chapter 6:   BLAST architectures, diversity, space-time block codes, and V-BLAST, and keyhole channels

Chapter 7:   Wireless telephone network standards, wireless data network standards, and IEEE 801.11 MAC

1.2 Appendices

To provide supplementary material for the book, nine appendices are included:

- Fourier theory
- Bessel functions
- Random variables and random processes
- Matched filters
- Error function
- Maximum a posteriori probability (MAP) decoding
- Capacity of MIMO links
- Eigendecomposition
- Adaptive antenna array

The inclusion of these appendices is intended to make the book essentially self-sufficient.

1.3 Other Features of the Book

Each chapter includes "within-text" problems that are intended to help the reader develop an improved understanding of the issues being discussed in the text. "End-of-chapter" problems provide an abundance of additional problems, whose solutions will further help the reader develop a deeper understanding of the material covered in the pertinent chapter.

Moreover, each chapter includes examples with detailed solutions covering different aspects of the subject matter.

"Notes and References" included at the end of the chapter provide explanatory notes, and they guide the reader to related references for further reading. All the references so made are assembled in the Bibliography placed at the end of the book.

2. SUPPLEMENTARY MATERIAL ON THE BOOK

The "within-text" problems are provided with answers alongside the problems. The solutions to the additional "end-of-chapter" problems are assembled in the Instructor's Manual, copies of which are available to qualified instructors. Contact your Pearson Prentice Hall representative.

The MATLAB codes used to plot results of the experiments and the graphs in the book are also included in the Instructor's Manual. M files are available on the Companion Website at

http://www.prenhall.com/haykin

Electronic copies in JPG format of selected figures from the book can also be accessed on the website. This material can be used for PowerPoint™ presentations of lectures based on the book.

3. ACKNOWLEDGMENTS

We are indebted to many colleagues for background material and insightful inputs, which, in one way or another, have helped us write certain parts of the book. In this context, the inputs of the following contributors (in alphabetical order) are gratefully acknowledged:

Dr. Claude Berrou, *ENST Bretagne, Brest, France*

Dr. Stewart Crozier, *Communications Research Centre, Ottawa, Ontario*

Dr. Suhas Diggavi, *Swiss Federal Institute of Technology (EPFL), Lausanne, Switzerland*

Dr. David Gesbert, *University of Oslo, Sweden*

Dr. Lajos Hanzo, *University of Southampton, United Kingdom*

Dr. John Lodge, *Communications Research Centre, Ottawa, Ontario*

Dr. Mathini Sellathurai, *Communications Research Centre, Ottawa, Ontario*

Dr. Abbas Yongaçoglu, *University of Ottawa, Ottawa, Canada*

Dr. Mansoor Shafi, *Telecom, Wellington, New Zealand*

The critical inputs and recommendations made by several anonymous reviewers of early versions of the manuscript are deeply appreciated. They provided us with a great deal of food for thought.

Moreover, we are grateful to Kris Huber, McMaster University, and Blair Simpson, Space-Time DSP Inc., for their assistance in generating many of the figures in the book.

We thank the Institute of Electrical and Electronics Engineers, Inc. and the European Telecommunications Standards Institute, for giving us the permissions to produce certain figures in the book.

We are particularly grateful to Tom Robbins, Craig Little, and the production staff at Prentice Hall, Pearson Education for their contributions, which, individually and collectively, have indeed made the production of the book possible.

The continued support and encouragement of our respective wives, Nancy and Dianne, and our families were an essential ingredient in the completion of the book. We are grateful to them all.

Last but by no means least, we are deeply grateful to Lola Brooks, McMaster University, for her tireless effort in all matters relating to the preparation and production of many different versions of the manuscript for the book.

SIMON HAYKIN
Ancaster, Ontario

MICHAEL MOHER
Ottawa, Ontario

CHAPTER 1

Introduction

You see, wire telegraph is a kind of a very, very long cat. You pull his tail in New York and his head is meowing in Los Angeles. Do you understand this? And radio operates exactly the same way: you send signals here, they receive them there. The only difference is that there is no cat.   *Albert Einstein (1879–1955)*

1.1 BACKGROUND

The understanding of radio waves is fundamental to wireless communications, but simply knowing that electromagnetic waves exist is a relatively recent historical event. In the short period since that time, there have been numerous milestones in the development of radio communications. Some of these milestones are the following:

- In 1864, James Clerk Maxwell formulated the electromagnetic theory of light and *predicted the existence of radio waves*. In his honor, the set of equations basic to the propagation of electromagnetic waves is known as *Maxwell's equations*.

- The physical existence of radio waves was first demonstrated by Heinrich Hertz in 1887.

- In 1894, building on the pioneering works of Maxwell and Hertz, Oliver Lodge demonstrated wireless communications, albeit over the relatively short distance of 150 meters.

- During the period from 1895 to 1901, Guglielmo Marconi developed an apparatus for transmitting radio waves over longer distances, culminating in a transmission across the Atlantic Ocean on December 12, 1901, from Cornwall, England, to Signal Hill in Newfoundland, Canada. Similar work was being done by A.S. Popoff of Russia during this time period.

- In 1902, the first point-to-point radio link in the United States was established from California to Catalina Island. These first radio systems were often referred to as wireless telegraphy.

- In 1906, Reginald Fessenden made history by conducting the first radio broadcast, transmitting music and voice using a technique that came to be known as *amplitude modulation* (AM) radio.

- In those early days, the military and merchant marine were quick to adapt wireless techniques. Wireless communications is often given credit for saving over 700 lives during the sinking of the *Titanic* in 1912.

- The early history of *land-mobile wireless* communication is a history of police pioneering. In 1921, the Detroit Police Department made the earliest significant use of wireless communications in a vehicle, operating a radio system at a carrier frequency close to 2 MHz. By 1934, 194 municipal and 58 state police forces were using AM radio for mobile voice communications.

- Parallel work performed on both sides of the Atlantic resulted in the first television broadcasts in 1927. Bell Labs demonstrated television broadcasts in the New York area, and John Baird made similar demonstrations in the United Kingdom.

- *Spread spectrum* techniques made their first appearance just before and during World War II. There were two applications in particular: encryption and ranging. Spread spectrum encryption techniques were often analog in nature with a noiselike signal multiplying a voice signal. The noise signal or information characterizing it was often sent on a separate channel to allow decryption at the receiver.

- In 1946, the first *public mobile telephone systems* were introduced in five American cities.

- In 1947, the first microwave relay system consisting of seven towers connecting New York and Boston became operational. This relay system was capable of carrying 2400 simultaneous conversations between the two cities.

- In 1958, a new era in wireless communications was initiated with the launch of the SCORE (Signal Communication by Orbital Relay Equipment) satellite. This satellite had only the capacity of one voice channel, but its success initiated the start of a new area of radio communications.

- In 1981, the first analog cellular system, known as the Nordic Mobile Telephone (NMT), was introduced in Scandinavia. This was soon followed by the Advanced Mobile Phone Service (AMPS) in North America in 1983.

- In 1988, the first digital cellular system was introduced into Europe. It was known as the *Global System for Mobile (GSM) Communications*. Originally intended to provide a pan-European standard to replace the myriad of incompatible analog systems in operation at the time, GSM was soon followed by the North American IS-54 digital standard.

These are just a few of the accomplishments in wireless communications over the past 150 years. Today, wireless devices are everywhere. Cellular telephones are commonplace. Satellites broadcast television direct to the home. Offices are replacing Ethernet cables with wireless networks. The introduction of these wireless services has increased the mobility and service area for many existing applications and numerous unthought-of applications. Wireless is a growing area of public networks and it plays an equally important role in private and dedicated communications systems. It is an exciting time in radio communications.

1.2   COMMUNICATION SYSTEMS

Today's public communications networks are complicated systems. Networks such as the public switched telephone network (PSTN) and the Internet provide seamless connections between cities, across oceans, and between different countries, languages, and cultures. Wireless is only one component of these complex systems, but there are three areas or layers where it can affect the design of such systems:

1. *Physical layer.* This layer provides the physical mechanism for transmitting *bits* (i.e., binary digits) between any pair of nodes. In wireless systems, it performs the modulation and demodulation of the electromagnetic waves used for transmission, and it includes the transmission medium as well. The module for performing *mo*dulation and *dem*odulation is often called a *modem.*

2. *Data-link layer.* Wireless links can often be unreliable. One purpose of the data-link layer is to perform error correction or detection, although this function is shared with the physical layer. Often, the data-link layer will retransmit packets that are received in error but, for some applications, it discards them. This layer is also responsible for the way in which different users share the transmission medium. In wireless systems, the transmission medium is the radio spectrum. A portion of the data-link layer called the *Medium Access Control (MAC)* sublayer is responsible for allowing frames to be sent over the shared media without undue interference with other nodes. This aspect is referred to as *multiple-access* communications.

3. *Network layer.* This layer has several functions, one of which is to determine the *routing* of the information, to get it from the source to its ultimate destination. A second function is to determine the *quality of service.* A third function is *flow control*, to ensure that the network does not become congested. Wireless systems, in which some of the nodes can be mobile, place greater demands on the network layer, since the associations among nodes are continually changing.

These are three layers of a seven-layer model for the functions that occur in the communications process. The model is called the *open system interconnection* (OSI) reference model[1] and we will discuss it in greater detail later in the book. The physical layer is the lowest layer of this model and is the one directly connected to the transmission medium. Higher layers are proportionately less affected by the transmission medium.

The book will concentrate mainly on the physical and data-link layers. We will finish by illustrating how these are integrated with wireless networks.

1.3   THE PHYSICAL LAYER

The physical layer provides the equivalent of a communications pipe between the source and destination of the information. In wireless and other communications systems, the physical layer has three basic components; the transmitter, the channel, and the receiver, as depicted in Fig. 1.1. *How are these different in a wireless system?*

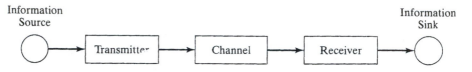

FIGURE 1.1   Block diagram of a communication system linking a source of information to a user of the information.

1.   *Transmitter.* The basic function of the transmitter is to take the information-bearing signal produced by the source of information and modify it into a form suitable for transmission over the channel. In wireless systems,

  (a) the transmitter shapes the signal so that it can pass reliably through the channel, while efficiently using the limited *transmission medium resources* (i.e., *the radio spectrum*)

  (b) the terminals are often mobile and limited by battery power, so the transmitter must use modulation techniques that are both robust and power-efficient; and

  (c) because the medium is shared with other users, the design should minimize interference with other users.

2.   *Channel.* The channel provides the physical means for transporting the signal produced by the transmitter and delivering it to the receiver. In wireless systems, the channel impairments

  (a) include *channel distortion* that may take the form of multipath—the constructive and destructive interference between many copies of the same signal received;

  (b) often have a *time-varying nature,* due to either the terminal mobility or a change in conditions along the propagation path;

  (c) include *interference,* which is produced (accidentally or intentionally) by other sources whose output signals occupy the same frequency band as the transmitted signal; and

  (d) include *receiver noise,* which is produced by electronic devices at the front end of the receiver. Although produced at the receiver, this noise is often considered a channel effect and therefore also referred to as *channel noise.* The effect of receiver noise will depend on the received signal strength, which in turn will depend significantly on the propagation path between the transmitter and receiver.

3.   *Receiver.* The receiver operates on the received signal to produce an *estimate* of the original information-bearing signal. We say "estimate" rather than an exact reproduction of the original information-bearing signal because of the unavoidable presence of *impairments* in the channel. In wireless systems, the receiver

  (a) frequently must *estimate the time-varying nature of the channel* in order to implement compensation techniques;

(b) implements *error-correction techniques* to improve the often poor reliability of wireless channels relative to other media; and

(c) must *maintain synchronization* under rapidly varying channel conditions.

All of these facets make the physical layer very challenging in wireless systems.

1.4   THE DATA-LINK LAYER

Conceptually, at the highest level of the data link is the *multiple access strategy*. This describes the general approach to sharing the physical resources among the different users. For wireless systems, the key physical resource is the radio spectrum. In general, for any wireless service, only a fixed amount of spectrum is allotted. We will consider four multiple access strategies for sharing this spectrum:

1. FDMA – *Frequency-division multiple access* refers to sharing the spectrum by assigning specific frequency channels to specific users on either a permanent or temporary basis.

2. TDMA – *Time-division multiple access* refers to allowing all users access to all of the available spectrum, but users are assigned specific time intervals during which they can access it, on either a temporary or permanent basis.

3. CDMA – *Code-division multiple access* is a form of spread spectrum modulation in which users are allowed to use the available spectrum, but their signal must be spread ("encrypted") with a specific code to distinguish it from other signals.

4. SDMA – *Space-division multiple access* refers to sharing the spectrum between the users by exploiting the spatial distribution of user terminals through the use of "smart" directional antennas that minimize the interference between the terminals.

The objective of all these strategies is to maximize the spectrum utilization—that is, to provide service to more users with the same amount of spectrum. In practice, most wireless systems are a combination of one or more of these multiple access strategies. Historically, due to the existing technology and network topologies, these multiple access strategies developed approximately in the order just presented.

1.4.1   FDMA

Early wireless systems tended to be point-to-point systems with one transmitter communicating with only one receiver. The natural approach was to assign each pair of *user terminals* (UTs) a frequency channel, a form of frequency-division multiple access (FDMA). In the frequency domain, FDMA can be visualized as shown in Fig. 1.2. The radio spectrum is divided into a number of channels, and each pair of users is assigned a different channel. In some cases, a different channel is assigned for each direction of transmission. From an interference viewpoint, it is important that each user signal be kept confined to the assigned band.

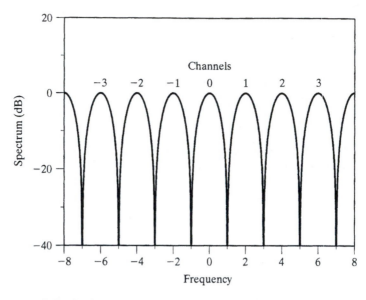

FIGURE 1.2 A frequency-domain representation of FDMA.

1.4.2 TDMA

Later systems were designed to operate in a point-to-multipoint manner, with many users sharing a central base station. The *point-to-multipoint* architecture shown in Fig. 1.3 consists of multiple UTs communicating with a single *base station* (BS). The multiple UTs time-share one or a small number of radio channels. These early time-division multiple access (TDMA) systems used analog modulation with a simple *push-to-talk* protocol. A user waited until the channel seemed unoccupied and then pushed a button on his or her microphone to talk. An example of this type of system would be a taxi-dispatch system.

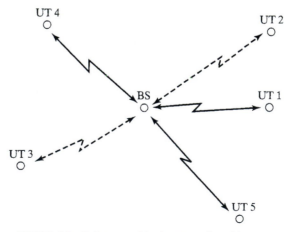

FIGURE 1.3 Point-to-multipoint network architecture.

...	Header	Slot 1	Slot 2	...	Slot N	Header	...

FIGURE 1.4   TDMA frame structure with N slots per frame.

With the advent of digital technology, TDMA systems appeared with more complex and efficient sharing strategies. An example of this is shown in Fig. 1.4, in which time is divided up into frames and each frame is divided into slots. Each active UT is assigned one or more slots from each frame.

From the structure of Fig. 1.4, it would appear that TDMA is suited only to data applications. In fact, many initial applications were voice applications, and the key to such an application is recognizing that the transmission rate for a digital TDMA channel is typically N times faster than that required for a single channel.

1.4.3   CDMA

The object of a *cellular system* is to reuse the radio spectrum over a large area as many times as possible. In a cellular system, the cells are modeled with a hexagon pattern as shown in Fig. 1.5. User terminals (UTs) in each cell communicate with a base station located at the center of the cell as in a point-to-multipoint arrangement, but the UTs can freely roam between cells, changing their base station as they move.

Within a cell, either FDMA or TDMA techniques can be used for sharing the transmission medium. Distinct sets of channel frequencies are assigned to each cell, with the channels being reused in sufficiently separated cells; an example of these *cochannel cells* are shaded in Fig. 1.5.

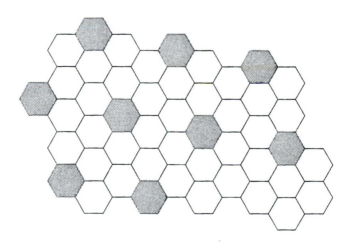

FIGURE 1.5   Hexagonal pattern of cells used in a cellular telephone system.

A drawback of using FDMA and TDMA in cellular systems is that the *reuse distance* is typically limited by worst-case interference. Cochannel cells must be assigned so that the worst-case interference is at an acceptable level. Intuitively, we would expect that reuse could be improved if cochannel cells were assigned on the *basis of average interference*. *Spread spectrum* is a modulation strategy that is quite tolerant of interference, and it forms the basis for the access technique known as code-division multiple access (CDMA). Spread spectrum spreads the information-bearing data over a large bandwidth, achieving the desired averaging effect. As a result, it allows this same spectrum to be used simultaneously by many UTs in adjacent cells, under appropriate conditions.

1.4.4 SDMA

The three multiple access techniques of FDMA, TDMA, and CDMA have increased spectral efficiency by increasing reuse in frequency, time, and space. So far, we implicitly have assumed antennas that are *omnidirectional* (i.e., antennas that operate essentially in a uniform manner along all directions). If the transmit and receive antenna could be focused directly at the other end of the link, then this would provide a number of improvements:

- It would reduce the total power needed to be transmitted, as all power would be flowing in the right direction.
- It would reduce the amount of interference generated by each transmitter, because total transmit power is reduced and localized.
- The receiver would receive a stronger signal due to both antenna gain and less interference.

The development of such "smart" antennas could allow even greater reuse of the radio spectrum. Space-division multiple access (SDMA) improvements in spectral efficiency are thus achieved by exploiting the angular separation of the individual UTs. In particular, *multibeam antennas* are used to separate radio signals by pointing them along different directions. Thus, different users are able to reuse the same radio spectrum, as long as they are separated in angle.

1.5   OVERVIEW OF THE BOOK

The fundamental physical resource in wireless systems is the radio spectrum. *The development of wireless communications has often been a search for more efficient ways of using the radio spectrum.* This evolution is a consequence of both increasing demand for a limited spectral resource and new technologies that have made spectrally efficient techniques a practical reality. This book follows this theme of evolving spectral efficiency.

The fundamental constraints on this evolutionary process are due to the transmission media—the *physical properties of the wireless channel*. These properties shape and influence the transmission techniques that are both practical and reliable. For that reason, understanding the physical media is a key to understanding wireless communications, and this is the topic of Chapter 2.

Early wireless systems were point-to-point systems using FDMA. As the radio spectrum became more crowded, there was a great interest in improving the efficiency of these individual channels. The role that *modulation* plays in the spectral efficiency of individual channels is investigated in Chapter 3. For mobile UTs, the modulation options are constrained by power limitations of the terminal and the need for the signal to be robust and easily detectable in fading.

The introduction of digital technology resulted in large improvements in efficiency for strategies such as TDMA. Similar technology can be applied to reduce the amount of information that needs to be transmitted; in particular, *digital speech-coding* techniques, which encode speech signals intelligently to minimize the number of bits required, are another way of increasing spectral efficiency. Once wireless networks started being used for digitized voice, it was a small step to also use them for data traffic. This created a problem, since existing networks and mobile terminals were designed for voice applications. Voice has a relatively high tolerance for bit errors, whereas data can be highly sensitive to bit errors—just imagine if someone moved the decimal point in your bank account! Data transmission led to the use of *forward error-correction coding* in wireless systems. TDMA and these pertinent technologies are investigated in Chapter 4.

After the time and frequency dimensions, the next area for further improvements in spectral efficiency is the spatial dimension—that is, the reuse of the same spectrum, but in different localities. This leads to the introduction of cellular systems in Chapter 5. The key factor limiting spectrum reuse is interference between user signals, and one approach to managing the effects of interference to optimize spectral efficiency is the use of a *spread spectrum*. This is why Chapter 5 begins with a discussion of spread spectrum modulation and CDMA before introducing the idea behind cellular systems.

While cellular systems produced significant increases in spectral efficiency, further increases were still possible through the *reuse of the spectrum in different angular directions*. This angular reuse requires "smart antenna" technology and is the motivation for SDMA—the topic of Chapter 6. One of the other challenges of wireless communications is the unpredictable and potentially unreliable nature of the propagation path. One well-known method of increasing the reliability of communications is *diversity*. Diversity is the transmission of the same information-bearing data by different methods in order to improve reliability. One of the many forms of diversity is spatial diversity, which usually implies the transmission or reception of the signal through the use of multiple antennas, forcing the signal to travel different paths in space. For this reason, spatial diversity is closely related to SDMA techniques—hence the discussion of both diversity and SDMA in Chapter 6.

The focus of this book is the physical-layer and multiple-access techniques associated with wireless communications. It is important that the reader understand how these technologies fit into the wireless networks that are currently evolving. Toward that end, Chapter 7 includes a comparison of the different multiple-access strategies and link-management functions and discusses how they fit into the architecture of wireless networks.

Notes and References

[1] The OSI reference model was developed as a paradigm for computer-to-computer communication. It has since been applied in a qualitative manner to many other networks. The first chapter of the book by Bertsekas and Gallagher (1992) provides an overview of the OSI model, and later chapters describe its application to data networks.

C H A P T E R   2

Propagation and Noise

2.1 INTRODUCTION

The study of propagation is important to wireless communications because it provides prediction models for estimating the power required *to close a communications link* and provide reliable communications. The study of propagation also provides clues to receiver techniques for compensating the impairments introduced through wireless transmission. In this chapter, we will look at physical models for various propagation phenomena and use these models to predict their effects on transmission. For the most part, the physical models are limited to simple scenarios, but they do provide a basic understanding of how signals propagate. Although the major emphasis of this chapter will be on terrestrial wireless communications, we will consider satellite communications as a baseline because it often provides close to ideal propagation conditions.

The propagation effects and other signal impairments are often collected and categorically referred to as the *channel*. The additive white Gaussian noise (AWGN) channel is the channel model most often used in communications theory development. With this model, zero-mean noise having a Gaussian distribution is added to the signal. The noise is usually assumed to be *white* over the bandwidth of interest; that is, samples of the noise process are uncorrelated with each other. This implies that the noise has an autocorrelation function given by $(N_0/2)\, \delta(t)$, where $\delta(t)$, is the Dirac delta function, or equivalently, it has a flat two-sided power spectral density $N_0/2$ (watts/Hz) across the frequency range $-\infty < f < \infty$. The AWGN channel model is shown in Fig. 2.1.

For fixed satellite communications in which there is a direct line-of-sight path between the transmitter and satellite, and the satellite and receiver, the additive white Gaussian noise (AWGN) channel often provides a reasonably good model[1]. In general, channel models for wireless communications take one of two forms:

1. *Physical models* take into account the exact physics of the propagation environment. Consequently, for any particular situation, site geometry has to be taken into consideration. This method provides the most reliable estimates of propagation behavior but it is computationally intensive.

2. *Statistical models* take an empirical approach, measuring propagation characteristics in a variety of environments and then developing a model based on the

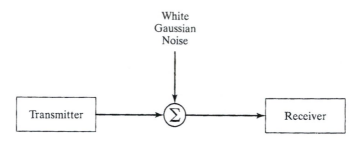

FIGURE 2.1   The Gaussian noise channel.

measured statistics for a particular class of environments. These models are easier to describe and use than the physical models, but do not provide the same accuracy.

We will provide examples of both types of models.

When considering physical models, we will consider the following three basic modes of propagation:

1. Free-space or line-of-sight transmission, as the name implies, corresponds to a clear transmission path between the transmitter and the receiver. Satellite communications generally rely on line-of-sight paths between the transmitter and satellite, and between the satellite and receiver.

2. *Reflection* refers to the bouncing of electromagnetic waves from surrounding objects such as buildings, mountains, and passing vehicles. In terrestrial wireless communications, there is often no direct line-of-sight path between the transmitter and receiver, and communications rely on reflection and diffraction, described next.

3. *Diffraction* refers to the bending of electromagnetic waves around objects such as buildings, or terrain such as hills, and through objects such as trees and other forms of vegetation.

As a result of another propagation phenomenon called *refraction*, electromagnetic waves are bent as they move from one medium to another. An example is the bending of light when it enters water from the air. The occurrence of refraction with wireless communications is limited to special circumstances, such as ionospheric communications.

Often, the received signal is the combination of many of these modes of propagation. That is, the transmitted signal may arrive at the receiver over many paths. The signals on these different paths can constructively or destructively interfere with each other. This is referred to as *multipath*. If either the transmitter or receiver is moving, then these propagation phenomena will be time varying, and *fading* occurs. One advantage of statistical models is their succinct way of describing these complicated situations.

In addition to propagation impairments, the other phenomena that limit communication are *noise* and *interference*. There are many sources for these impairments, including the receiver, human-made sources, and other transmitters.

This chapter has two objectives. The first is to provide a basic understanding of the physics behind various propagation phenomena and noise and how they affect communications. To that end, Section 2.2 explains free-space propagation, Sections 2.3 and 2.4 respectively present physical and statistical models for terrestrial wireless propagation, and Section 2.5 discusses indoor propagation. These explanations are followed by a discussion of local propagation effects such as multipath and Rayleigh fading in Section 2.6 and statistical scattering models in Section 2.7.

The second objective is to provide the reader with a system-level appreciation of the importance of propagation and noise effects in wireless communications. To achieve this, noise and interference effects are described in Section 2.8, and link budgets are presented in Section 2.9. The chapter ends with three theme examples—one describing an empirical model for land mobile propagation, the second one describing how all the facets examined in this chapter fit together in the design of a wireless local area network, and the third describing propagation advantages of ultra-wideband radio.

2.2   FREE-SPACE PROPAGATION

Wireless transmission is characterized by the generation, in the transmitter, of an electric signal representing the desired information, the propagation of corresponding radio waves through space and a receiver that estimates the transmitted information from the recovered electrical signal. The transmission system is characterized by the antennas that convert between electrical signals and radio waves, and the propagation of the radio waves through space. The transmission effects are most completely described by Maxwell's equations; however, in this text we will model the propagation method solely on the basis of the electrical field, and we shall *assume a linear medium in which all distortions can be characterized by attenuation or superposition of different signals.*

2.2.1 Isotropic Radiation

A fundamental concept in the understanding of radio transmission systems is that of an *isotropic antenna*. This is an antenna that transmits equally in all directions. In reality, an isotropic antenna does not exist, and all antennas have some *directivity* associated with them. Still, the isotropic antenna is a reference to which other antennas are compared, and it is useful for explaining fundamentals.

Consider an isotropic source that radiates power P_T watts equally in all directions, as illustrated in Fig. 2.2. The power per unit area, or *power flux density*, on the surface of a sphere of radius R centered on the source is given by

$$\Phi_R = \frac{P_T}{4\pi R^2} \tag{2.1}$$

where $4\pi R^2$ is the surface area of the sphere. Typical units for power flux density are watts per square meter. Equation (2.1) follows directly from the conservation of energy; that is, the energy radiated per unit time by the source must appear at the surface of sphere (or any closed surface) enclosing that source.

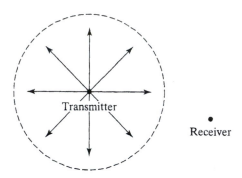

FIGURE 2.2   Illustration of isotropic radiation.

The power P_R that is captured by the receiving antenna depends on the size and orientation of the antenna with respect to the transmitter. The power received by an antenna of *effective area* or *absorption cross section A_e* is given by

$$P_R = \Phi_R A_e = \frac{P_T}{4\pi R^2} A_e \qquad (2.2)$$

An antenna's physical area A and its effective area A_e are related by the *antenna efficiency*

$$\eta = \frac{A_e}{A} \qquad (2.3)$$

This parameter indicates how well the antenna converts incident electromagnetic energy into usable electrical energy. For parabolic antennas, the efficiency typically ranges from 45% to 75%; for horn antennas, the efficiency can range from 50% to 80%.

It is intuitively clear that an isotropic transmit antenna can be considered a point source. What is the effective area of an isotropic receive antenna? If it is considered a *point sink*, then the answer is not obvious. However, from electromagnetic theory, we note that the *effective area of an isotropic antenna in any direction* is given by[2]

$$A_{\text{iso}} = \frac{\lambda^2}{4\pi} \qquad (2.4)$$

where λ is the wavelength of radiation. Substituting Eq. (2.4) into (2.2), we obtain the following relationship between the transmitted and received power for isotropic antennas:

$$P_R = \frac{P_T}{(4\pi R/\lambda)^2} = \frac{P_T}{L_p} \qquad (2.5)$$

In this equation, the path loss is defined as

$$L_p = \left(\frac{4\pi R}{\lambda}\right)^2 \tag{2.6}$$

The quantity L_p is the *free-space path loss* between two isotropic antennas. This definition of path loss depends, somewhat surprisingly, on the wavelength of transmission— a consequence of the dependence of the effective area of an isotropic antenna on the wavelength.

EXAMPLE 2.1 Receiver Sensitivity

Sensitivity is a receiver parameter that indicates the minimum signal level required at the antenna terminals in order to provide reliable communications. The factors it depends on include the receiver design, modulation format, and transmission rate. Receiver sensitivity is often expressed in dBm—that is, the power in milliwatts, expressed in decibels. For example, a commercial mobile receiver for data transmission may be specified with a sensitivity of –90 dBm. Assuming a 100-milliwatt transmitter and free-space path loss between the transmitting and receiving isotropic antennas, what is the radius of the service area of this receiver at a transmission frequency of 800 MHz?

To answer this question, we first note that –90dBm is equivalent to 10^{-9} milliwatts of power. From Eq. (2.5), the maximum tolerable path loss is

$$L_p = \frac{P_T(\text{mW})}{P_R(\text{mW})} = \frac{100}{10^{-9}} = 10^{11}$$

Consequently, observing that $\lambda = c/f$, where c is the speed of light, and using Eq. (2.6), we find that the maximum range is given by

$$R = \frac{\lambda}{4\pi}\sqrt{L_p} = \frac{c}{4\pi f}\sqrt{L_p} = \frac{3 \times 10^8 \text{m/s}}{4\pi \times 800 \times 10^6 \text{s}^{-1}}\sqrt{10^{11}} = 9.2\text{km}$$

The conclusion drawn from this example is that it takes very little transmit power to provide a large service area under free-space propagation conditions. ∎

2.2.2 Directional Radiation

While the isotropic antenna is a useful illustrative device, most antennas are not isotropic. Instead, they have *gain* or *directivity* $G(\theta,\phi)$ that is a function of the *azimuth* angle θ and the *elevation* angle ϕ:

1. The azimuth angle θ is the look angle in the horizontal plane of the antenna relative to a reference horizontal direction. The reference direction could, for example, be due north.

2. The elevation angle ϕ is the look angle of the antenna above the horizontal plane.

With these definitions, the transmit gain of an antenna is defined as

$$G_T(\theta, \phi) = \frac{\text{Power flux density in direction } (\theta, \phi)}{\text{Power flux density of an isotropic antenna}} \tag{2.7}$$

for the same transmit power. Clearly, the transmit gain of an isotropic antenna is unity.

The corresponding definition for the receive antenna gain is

$$G_R(\theta, \phi) = \frac{\text{Effective area in direction } (\theta, \phi)}{\text{Effective area of an isotropic antenna}} \qquad (2.8)$$

Equations (2.7) and (2.8) define the gain for the transmit and receive antennas, respectively. With bidirectional communications, the same antenna is often used to both transmit and receive. *How are the two gains related?* To answer this question, we introduce what is known as the *principle of reciprocity*, which may be stated as follows:[3]

> *Signal transmission over a radio path is reciprocal in the sense that the locations of the transmitter and receiver can be interchanged without changing the transmission characteristics.*

The principle of reciprocity is based on Maxwell's equations of electromagnetic theory. What it implies is that, if the direction of propagation is reversed, the energy of the signal would follow exactly the same paths and suffer the same effects, but in the reverse direction. This implies that the *transmit and receive antenna gains are equal.* The intuitive explanation of why this is the case is the observation that the transmit gain is a measure of how well an antenna emits the radiated energy in a certain area (defined by the azimuth and elevation), while the receive gain is a measure of how well the antenna collects the radiated energy in that area.

Consequently, from the principle of reciprocity and Eq. (2.8), the *maximum transmit or receive gain*, of an antenna, in any direction is given by

$$G = \frac{4\pi}{\lambda^2} A_e \qquad (2.9)$$

The antenna pattern for a parabolic (dish-shaped) antenna is shown in Fig. 2.3. The maximum gain of the antenna is along the axis of the parabola and is given by Eq. (2.9). The gain decreases in the off-axis direction. The *beamwidth* of the antenna is defined as the angular width of the antenna beam at the 3-dB points, as shown in Fig. 2.3. The illustration shows the main lobe of the antenna pattern plus two *sidelobes.* An antenna may also have a *backlobe.* The sidelobes and backlobe are typically not considered for use in the communications link, but they often have to be considered when analyzing interference.

EXAMPLE 2.2 Parabolic Antenna Gain

For a parabolic antenna, looking directly at the antenna boresight, we find that the area is simply $\pi D^2/4$, where D is the diameter of the dish. Consequently, the effective area of the antenna is given by

$$A_e = \eta \frac{\pi D^2}{4}$$

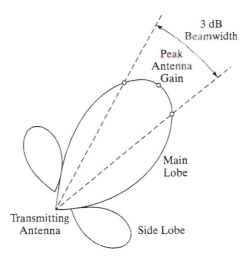

FIGURE 2.3 Illustration of gain pattern of a parabolic antenna.

and the corresponding antenna gain (both transmit and receive) is

$$G = \frac{A_e}{A_{\text{isotropic}}} = \frac{\eta \pi D^2 / 4}{\lambda^2 / 4\pi} = \eta \left(\frac{\pi D}{\lambda}\right)^2$$

This equation illustrates that the antenna gain depends on the wavelength of transmission. For example, a 0.6-m parabolic dish used for receiving a direct broadcast satellite television signal at 12 GHz has a gain of

$$G = \eta \left(\frac{\pi D}{\lambda}\right)^2 = 0.5 \left(\frac{\pi \times 0.6 \times f}{c}\right)^2 = 2842.4$$

If we express this gain in decibels, it corresponds to a gain of 34.5 dB. We have assumed an antenna efficiency of 50%. ∎

The previous example described how to calculate the maximum gain for a parabolic antenna when one is looking directly at the center of the antenna. For such an antenna, the theoretical gain G varies with the off-axis angle ψ, as illustrated by Fig. 2.3. Normalized to unity on-axis gain, the theoretical gain of a parabolic antenna is

$$G(\psi) = \left(\frac{2J_1((\pi D / \lambda)\sin\psi)}{\sin\psi}\right)^2 \left(\frac{\pi D}{\lambda}\right)^{-2} \tag{2.10}$$

where $J_1(x)$ is the Bessel function of the first kind and D is the diameter of the antenna. (The Bessel function is discussed in Appendix D.) This antenna gain characteristic is shown in Fig. 2.4, plotted as a function of the off-axis angle ψ.

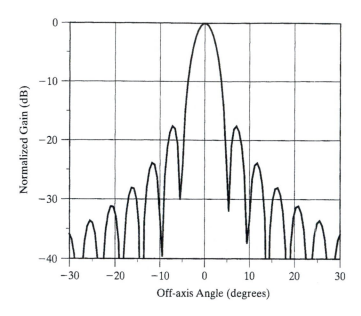

FIGURE 2.4   Off-axis gain characteristic of a 1-meter parabolic antenna at 4 GHz.

The *3-dB beamwidth* is a measure of the size of the antenna beam. For the parabolic antenna, it is the off-axis angle for which the gain falls to 3 dB below its peak value. For a parabolic antenna with an ideal gain characteristic as described by Eq. (2.10), the 3-dB beamwidth[4] is $58.4° \times (\lambda/D)$. From Eq. (2.9), it should be clear that the larger the antenna, the greater the gain it provides. One important further result inferred from Eq. (2.10) is that the larger the antenna, the narrower is the beamwidth of the antenna. That is, large antennas have high gain, but they must be pointed accurately to make use of the gain. This result illustrates a general rule of thumb that *the 3-dB beamwidth is linearly proportional to the wavelength and inversely proportional to the antenna diameter.*

Problem 2.1   Early satellite communications systems often used large 20-m-diameter parabolic dishes with an efficiency of approximately 60% to receive a signal at 4 GHz. What is the gain of one of these dishes in dB?
Ans. 56.2 dB.   ■

2.2.3 The Friis Equation

When nonisotropic antennas are used, the free-space loss relating the received and transmitted power for general antennas is determined by direct substitution of the gain definitions Eqs. (2.7) and (2.8) into Eq. (2.5) to obtain

$$P_R = \frac{P_T G_T G_R}{L_p} \tag{2.11}$$

Equation (2.11) is referred to as the *Friis equation*. To simplify its evaluation, we write it as the decibel relation

$$P_R(\text{dB}) = P_T(\text{dB}) + G_R(\text{dB}) + G_T(\text{dB}) - L_P(\text{dB}) \qquad (2.12)$$

where $X(\text{dB}) = 10 \log_{10}(X)$. The Friis equation is the fundamental *link budget* equation. It relates power received to power transmitted, taking into account transmission characteristics of the radio link. *Closing the link* refers to the requirement that the right-hand side provide enough power at the receiver to detect the transmitted information reliably. That is, the right-hand side must be greater than the receiver sensitivity. In a later section, a more formal link budget will be presented that takes into account a greater number of transmission effects.

Problem 2.2   In terrestrial microwave links, line-of-sight transmission limits the separation of transmitters and receivers to about 40 km. If a 100-milliwatt transmitter at 4 GHz is used with transmitting and receiving antennas of 0.5-m^2 effective area, what is the received power level in dBm? If the receiving antenna terminals are matched to a 50-ohm impedance, what voltage would be induced across these terminals by the transmitted signal?
Ans. $P_R = -55.5$ dBm and $V_{\text{rms}} = 0.37$ millivolt. ∎

2.2.4 Polarization

Electromagnetic waves are transmitted in two orthogonal dimensions, referred to as *polarizations*. There are two commonly used orthogonal sets of polarizations:

1. *Horizontal and vertical polarization.* Vertical polarization is generally used for terrestrial mobile radio communications. At frequencies in the VHF band, vertical polarization is better than horizontal polarization because it produces a higher field strength near the ground. Furthermore, mobile antennas for vertical polarization are more robust and convenient to implement.

2. *Left-hand and right-hand circular polarizations.* These are often used in satellite communications. For well-designed fixed communication links, we can use the two orthogonal polarizations to double the transmission capacity in a given frequency band.

In mobile communications, there is often little frequency reuse advantage to employing both polarizations. *Scattering effects*, described later in the chapter, tend to create a cross-polar component that causes interference between the original orthogonal components and that limits the benefit of polarization.

2.3 TERRESTRIAL PROPAGATION: PHYSICAL MODELS

The free-space propagation equation assumes that the effect of the Earth's surface can be neglected and that there is a clear line of sight between the transmitting and receiving antennas. This is usually the situation with satellite communications. With terrestrial communications, however, buildings, terrain, or vegetation may obstruct the line-of-sight path between the transmit and receive antennas. In this case, communication relies on either the *reflection* of electromagnetic waves from various surfaces or *diffraction* around various objects. With these additional modes of propagation, a multitude of possible paths arise between the transmitter and receiver, and the receiver often receives a significant signal from more than one path. This phenomenon is referred to as *multipath* propagation. With multiple waves arriving at the same location, there is the possibility of either *destructive* or *constructive interference*, and consequently, the propagation properties can be quite different from those obtained with free-space propagation. In the next two subsections, we will look at the physical mechanisms of reflection and diffraction.

Other propagation modes, such as *ducting*, occur when the physical characteristics of the environment create a waveguidelike effect. One example of ducting occurs in high-frequency (HF) radio, for which the carrier frequencies range from 3 to 30 MHz. At these carrier frequencies, differences between the layers of the atmosphere are most pronounced, and the layers act like different media to the transmission. The HF signal can be trapped within a layer of the ionosphere, reflecting off the layers above and below in a manner similar to a stone skipping across a lake. This allows the signal to travel long distances with very little attenuation and is sometimes called *skywave*. However, the skywave form of wireless communications tends to be unreliable, because the boundaries between the atmospheric layers do not have long-term stability.

2.3.1 Reflection and the Plane-Earth Model[5]

Except for very short distances, the presence of the Earth plays an important role in terrestrial propagation. For distances less than a few tens of kilometers, we may often neglect the curvature of the earth and use a flat-Earth model, as illustrated in Fig. 2.5.

The flat-Earth model shows a fixed transmitter with an antenna of height h_T transmitting to a fixed receiving antenna of height h_R. It also shows a two-ray propagation model between the transmitter and the receiver, consisting of a direct path and a path reflected by the Earth's surface. The objective is to determine the power level at the receiving antenna terminals. First, we note the following relationship between the

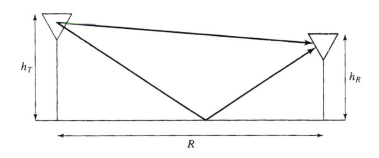

FIGURE 2.5   Plane-Earth reflection model.

power flux density and the *field strength E*:

$$\Phi = \frac{|E|^2}{\eta_0} \tag{2.13}$$

The field strength is measured in volts per meter (or an equivalent), and η_0 is the *characteristic wave impedance of free space.*[6] This characteristic impedance, 120π ohms, is the ratio of the complex amplitude of the electric field to that of the magnetic field in free space. The antenna acts as an impedance transformer to convert the free-space impedance to the impedance seen at the antenna terminals. We shall assume that the electric field is generated by a continuous wave (CW) signal with a transmission frequency f; that is, at a given point in space,

$$E(t) = \sqrt{2}E_0\cos(2\pi f t + \theta)$$
$$= \sqrt{2}\,\mathrm{Re}\left\{E_0 e^{j(2\pi f t + \theta)}\right\} \tag{2.14}$$

where the field strength E_0 and the phase θ depend on the location in space, and Re { } denotes the real part of the quantity inside the parentheses. To simplify the notation, we define the complex phasor

$$\tilde{E} = E_0 e^{j\theta} \tag{2.15}$$

and work extensively with this complex phasor notation in the discussion that follows.

In the case of the reflection of a single ray, if $\tilde{E}_d$ is the incident electric field and $\tilde{E}_r$ is the reflected field, then the relationship between the two fields is given by

$$\tilde{E}_r = \tilde{E}_d \rho e^{j\psi} \tag{2.16}$$

where ρ is the attenuation of the electric field and ψ is the phase change caused by the reflection. These parameters depend on both the angle of incidence of the field and the nature of the reflecting surface, including its smoothness and absorption properties.

Differences between the direct and reflected paths will depend on path length differences. From Fig. 2.5, by the Pythagorean theorem, the length of the direct path is

$$R_d = \sqrt{R^2 + (h_T - h_R)^2} \tag{2.17}$$

and, similarly, the length of the reflected path is

$$R_r = \sqrt{R^2 + (h_T + h_R)^2} \tag{2.18}$$

(This is most easily observed by reflecting the receiving antenna into the Earth.) To calculate the field strength at the receiving antenna, we assume that the difference in attenuation caused by different path lengths between the direct and ground-reflected waves is negligible; that is,

$$|\tilde{E}_r| \approx |\tilde{E}_d| \tag{2.19}$$

However, the phase difference between the two paths is much more sensitive to the path length, and it cannot be neglected. The path difference between the reflected and incident rays is

$$\Delta R = R_r - R_d \tag{2.20}$$

We note that if R is large compared with both h_T and h_R, then the length of the direct path can be approximated by

$$
\begin{aligned}
R_d &= R \sqrt{1 + \frac{(h_T - h_R)^2}{R^2}} \\
&\approx R \left(1 + \frac{(h_T - h_R)^2}{2R^2} \right)
\end{aligned}
\tag{2.21}
$$

In Eq. (2.21), we have used the approximation $\sqrt{1+x} \approx (1+x/2)$ for $x \ll 1$. If the same approximation technique is applied to R_r, then it follows by substituting Eq. (2.21) and its equivalent for R_r into Eq.(2.20) that

$$\Delta R \approx \frac{(h_T + h_R)^2 - (h_T - h_R)^2}{2R} = 2\frac{h_T h_R}{R} \tag{2.22}$$

The corresponding phase difference between the two paths is proportional to the transmission wavelength and is given by

$$\Delta\phi = \frac{2\pi}{\lambda}\Delta R = \frac{4\pi h_T h_R}{\lambda R} \tag{2.23}$$

Since $\tilde{E}_d$ represents the field strength at the receiving antenna due to the direct wave, it follows from using Eq. (2.16) that the total received field is

$$
\begin{aligned}
\tilde{E} &= \tilde{E}_d + \tilde{E}_r \\
&= \tilde{E}_d(1 + \rho \exp(j\psi)\exp(-j\Delta\phi))
\end{aligned}
\tag{2.24}
$$

If the Earth's surface is assumed to be smooth and flat, and the reflected ray is at grazing incidence so that the reflection coefficient is $\rho \exp(j\psi) = -1$, then the total received electric field is

$$\tilde{E} = \tilde{E}_d(1 - \exp(-j\Delta\phi)) \tag{2.25}$$

The magnitude of the combined electric field is given by

$$
\begin{aligned}
\left|\tilde{E}\right| &= \left|\tilde{E}_d\right| \left|1 - \exp(j\Delta\phi)\right| \\
&= \left|\tilde{E}_d\right| [1 + \cos^2\Delta\phi - 2\cos\Delta\phi + \sin^2\Delta\phi]^{1/2} \\
&= \left|\tilde{E}_d\right| [2 - 2\cos\Delta\phi]^{1/2} \\
&= 2\left|\tilde{E}_d\right| \sin\left(\frac{\Delta\phi}{2}\right)
\end{aligned}
\tag{2.26}
$$

Substituting Eq. (2.23) into (2.26) yields

$$
\left|\tilde{E}\right| = 2\left|\tilde{E}_d\right| \sin\left(\frac{2\pi h_T h_R}{\lambda R}\right)
\tag{2.27}
$$

From Eqs. (2.2) and (2.13), the received power is given by

$$
\begin{aligned}
P_R &= \Phi A_e \\
&= \frac{\left|\tilde{E}\right|^2}{\eta_0} A_e \\
&= 4\frac{\left|\tilde{E}_d\right|^2}{\eta_0} A_e \sin^2\left(\frac{2\pi h_T h_R}{\lambda R}\right)
\end{aligned}
\tag{2.28}
$$

The factor $\left|\tilde{E}_d\right|^2 A_e/\eta_0$ in Eq. (2.28) is the power received via the direct path. This direct-path power is governed by free-space propagation conditions and thus is equivalent to the right-hand side of Eq. (2.11). Substituting Eq. (2.6) into (2.11), and using the result to replace the factor $\left|\tilde{E}_d\right|^2 A_e/\eta_0$ in Eq. (2.28), we have the modified Friis' equation

$$
P_R = 4P_T\left(\frac{\lambda}{4\pi R}\right)^2 G_T G_R \sin^2\left(\frac{2\pi h_T h_R}{\lambda R}\right)
\tag{2.29}
$$

If the product λR is much greater than the product $h_T h_R$, we can approximate $\sin\theta$ by θ and Eq. (2.29) becomes

$$
P_R \approx P_T G_T G_R \left(\frac{h_T h_R}{R^2}\right)^2
\tag{2.30}
$$

Equation (2.30) is the *plane-Earth propagation equation*, which differs from the free-space equation in three ways:

1. As a consequence of the assumption that $R \gg h_T, h_R$, the angle $\Delta\phi$ is small, and λ cancels out of the equation, leaving it to be essentially *frequency independent*.

2. It shows an *inverse fourth-power law*, rather than the inverse-square law of free-space propagation. This points to a far more rapid attenuation of the power received.

3. It shows the effect of the transmit and receive antenna heights on propagation losses. The *dependence on antenna height* makes intuitive sense.

The conclusion is that an apparently minor difference in the model can make a significant change in the behavior of radio propagation.

Problem 2.3   Plot and compare the path loss (dB) for the free-space and plane-Earth models at 800 MHz versus distance on a logarithmic scale for distances from 1m to 40 km. Assume that the antennas are isotropic and have a height of 10 m.   ∎

2.3.2 Diffraction[7]

In the previous section, we saw how reflections can significantly affect the propagation characteristics of electromagnetic waves. In this section, we consider a second significant propagation effect for terrestrial propagation, namely, *diffraction*.

In physics, *diffraction* refers to the phenomenon whereby, when electromagnetic waves are forced to travel through a small slit, they tend to spread out on the far end of the slit, as illustrated in Fig. 2.6. The diffraction mechanism is often explained by *Huygens's principle*, which may be stated as follows:

> *Each point on a wave front acts as a point source for further propagation. However, the point source does not radiate equally in all directions, but favors the forward direction, of the wave front.*

Huygens's principle can be derived from Maxwell's equations. Most important, the principle can be used to explain why electromagnetic waves bend over hills and around buildings, as illustrated in Fig. 2.7, to provide communications where we may have thought that there should have been an *electromagnetic shadow*. One portion of the wave front directly hitting the obstacle is blocked (absorbed or perfectly reflected), while the remaining portion tends to bend and illuminate the shadow region.

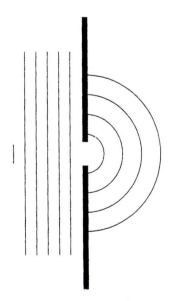

FIGURE 2.6   Behavior of plane wave passing through a slit from left to right.

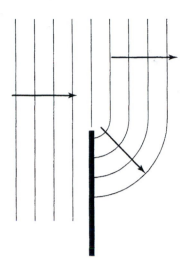

FIGURE 2.7 Illustration of knife-edge diffraction.

To introduce the concepts associated with diffraction, consider a transmitter **T** and receiver **R** in free space, as shown in Fig. 2.8. We also consider a plane normal to the line of sight path at a point **O** between **T** and **R**. On the plane, we construct circles of arbitrary radii. Any wave that propagates from **T** to **R** via a point on any of these circles traverses a longer path than the path **TOR**.

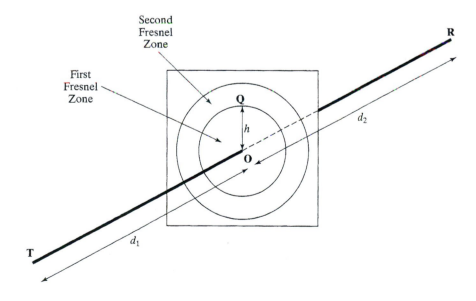

FIGURE 2.8 Fresnel zones of transmitted signal.

In terms of the geometry of Fig. 2.8, the excess path length between the direct path and the nondirect path through the point $\mathbf{Q}$ on the circle of radius h is given by

$$\Delta R = |\mathbf{TQR}| - |\mathbf{TOR}|$$

$$= \sqrt{d_1^2 + h^2} + \sqrt{d_2^2 + h^2} - (d_1 + d_2) \tag{2.31}$$

$$= d_1 \sqrt{1 + \left(\frac{h}{d_1}\right)^2} + d_2 \sqrt{1 + \left(\frac{h}{d_2}\right)^2} - (d_1 + d_2)$$

Assuming that $h << d_1, d_2$ and using the approximation $\sqrt{1+x} \approx 1 + x/2$ for $x<<1$, we find that Eq. (2.31) reduces to

$$\Delta R \approx \left(\frac{h^2}{2} \frac{d_1 + d_2}{d_1 d_2}\right) \tag{2.32}$$

As with the plane-Earth model, the phase difference between the two paths is the critical parameter. This phase difference, corresponding to a transmission wavelength λ, is given by

$$\Delta\phi = \frac{2\pi}{\lambda}\Delta R$$

$$\approx \frac{2\pi h^2}{\lambda} \frac{1}{2}\left(\frac{d_1 + d_2}{d_1 d_2}\right) \tag{2.33}$$

$$= \frac{\pi}{2}h^2\left(\frac{2(d_1 + d_2)}{\lambda d_1 d_2}\right)$$

We next define the *Fresnel–Kirchhoff diffraction parameter*

$$v = h\sqrt{\frac{2(d_1 + d_2)}{\lambda d_1 d_2}} \tag{2.34}$$

The phase difference of Eq. (2.33) can then be written as

$$\Delta\phi \approx \frac{\pi}{2}v^2 \tag{2.35}$$

The Fresnel–Kirchhoff diffraction parameter is a dimensionless quantity that characterizes the phase difference between two propagation paths. It is used in the discussion that follows to characterize diffraction losses in a general situation.

Returning to Fig. 2.8, we construct a family of hypothetical circles having the property that the total path length from $\mathbf{T}$ to $\mathbf{R}$ via each circle is $\Delta R = q\lambda/2$ longer than the path $\mathbf{TOR}$, where q is an integer. The radii of these circles depend on the location of this imaginary plane along the $\mathbf{TR}$-axis. The set of points for which the excess path length is an integer number of half wavelengths defines a family of ellipsoids, with $\mathbf{TOR}$ being the axis of revolution. If we slice a specific ellipsoid perpendicular to $\mathbf{TOR}$, then, with $\Delta R = q\lambda/2$, it follows from Eq. (2.32) that the radius of the resulting circle is

$$h = r_q = \sqrt{\frac{q\lambda d_1 d_2}{d_1 + d_2}} \tag{2.36}$$

and the corresponding diffraction parameter is $v_q = \sqrt{2q}$. From Eq. (2.32), this approximation is valid for $d_1, d_2 >> r_q$. In words, *a particular Fresnel–Kirchhoff parameter v_q defines a corresponding ellipsoid of constant excess path.*

The volume enclosed by the first ellipsoid, $q = 1$, is known as the *first Fresnel zone.* The volume between the first ellipsoid and the second one, $q = 2$, is known as the *second Fresnel zone*, and so on.

Since the qth Fresnel zone is defined as a differential path length

$$\frac{(q-1)\lambda}{2} \le \Delta R \le \frac{q\lambda}{2} \tag{2.37}$$

relative to the line-of-sight path, the corresponding phase difference Eq. (2.33) pertaining to the qth Fresnel zone is $(q-1)\pi \le \Delta\phi \le q\pi$. Consequently, the contributions to the electric field at the receiver from successive Fresnel zones tend to be in phase opposition and therefore interfere destructively rather than constructively. Accordingly, we may state the following:

> *As a general rule of thumb, we must keep the "first Fresnel zone" free of obstructions in order to obtain transmission under free-space conditions.*

EXAMPLE 2.3 Fresnel Zones

Consider the case shown in Fig. 2.9, where the transmitter and receiver are at a height of 10 m and separated from each other by 500 m. From Eq.(2.34), at a point midway between the transmitter and receiver the radius of the first Fresnel zone is given by

$$r_1 = \sqrt{\frac{\lambda d_1 d_2}{d_1 + d_2}} = \sqrt{\frac{\lambda(250)(250)}{500}} = 11.18\sqrt{\lambda} \ \text{m} \tag{2.38}$$

At 800 MHz, $\lambda = 0.375$ m, and from Eq. (2.38), the first Fresnel zone has a radius of 6.84 m at the midway point. At a distance of 100 m from either the transmitter or receiver, the same calculation indicates that the first Fresnel zone has a radius of 5.48 m. Consequently, all objects between the transmitter and receiver must be shorter than 3 to 4 meters in order to have the approximate equivalent of free-space propagation over the 500-m path at 800 MHz. ∎

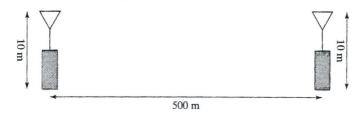

FIGURE 2.9 Scenario for Example 2.3.

Problem 2.4   A company owns two office towers in a city and wants to set up a 4-GHz microwave link between the towers. The two towers have heights of 100 m and 50 m, respectively, and are separated by 3 km. In the line of sight (LOS) and midway between the two towers is a third tower of height 70 meters. Will line-of-sight transmission be possible between the two towers? Justify your answer. Describe an engineering solution to obtain line-of-sight transmission.
Ans. No.   ■

2.3.3 Diffraction Losses

If an ideal, straight, perfectly absorbing screen is placed between **T** and **R** in Fig. 2.8, it will have little effect if the top of the screen is well below the line-of-sight path. The field at **R** will then be the "free-space" value of $\tilde{E}_d$. As the height of the screen is increased, the field strength will vary up and down as the screen blocks more of the Fresnel zones below the path. The amplitude of the oscillation increases until the screen is just in line with **T** and **R** and the field strength is exactly one-half of the unobstructed value. When an obstruction is present, the Fresnel–Kirchhoff parameter is still given by

$$v = h\sqrt{\frac{2(d_1 + d_2)}{\lambda d_1 d_2}} \tag{2.39}$$

but in this case *h* refers to the height of the obstruction. *The height h, and thus also v, is considered positive if the obstruction extends above the line of sight and negative if it does not.*

From advanced diffraction theory,[8] the field strength at the point **R** is determined by the sum of all secondary Huygens's sources in the plane above the obstruction and can be expressed as

$$E = E_d\frac{1+j}{2}\int_v^\infty \exp\left(-j\frac{\pi}{2}t^2\right)dt \tag{2.40}$$

where *v* is the Fresnel–Kirchhoff parameter and $j = \sqrt{-1}$. Equation (2.40) is known as the *complex Fresnel integral*. There is no known closed-form solution to this equation, but it can be evaluated numerically. In Fig. 2.10, we plot the diffraction loss over a knife edge as a function of the Fresnel–Kirchhoff parameter *v*. Note how the loss oscillates around 0 dB when the knife edge is below the first Fresnel zone. Once the screen obstructs that zone and beyond, there is a steady decrease in field strength. When the knife edge is even with the line-of-sight path ($h = 0$), the electric field is reduced by one-half and there is a 6-dB loss in signal power.

For a fixed obstruction at a distance d_1 from the transmitter and a transmission wavelength λ, the diffraction parameter

$$v = h\sqrt{\frac{2(d_1 + d_2)}{\lambda d_1 d_2}} \tag{2.41}$$

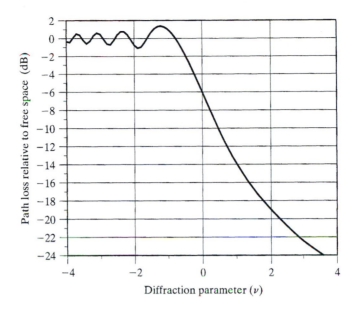

FIGURE 2.10   Diffraction loss over a single knife edge as a function of the Fresnel–Kirchhoff parameter (v).

increases toward infinity as the distance d_2 from the obstruction to the receiver shrinks. This relationship implies a loss becoming infinitely large as the receiver moves more and more into the shadow of the obstruction. The free-space loss referred to in Fig. 2.10 is that corresponding to the straight-line distance between the transmitter and receiver in the absence of the obstruction. Note that the diffraction loss depends on frequency through the wavelength λ that appears in the definition of the diffraction parameter. In addition to the loss introduced relative to the direct path, there will be a phase rotation, as suggested by the presence of the imaginary unit j in Eq.(2.40).

This subsection has described the diffraction loss due to an abrupt "knife-edge" obstruction. The loss provides a second physical model as to why losses may be greater than those predicted by the free-space model. In practice, there will often be obstructions of other shapes and multiple obstructions along the transmission path. The principles for calculating the diffraction losses are the same as described herein, although the calculations involved are more difficult and are frequently done by computer simulations.

Problem 2.5   In Problem 2.4, suppose the middle tower was 80 m and the shorter tower was only 30 m. The separation between the two communicating towers is 2 km. What would the increase in path loss be in this case relative to the free-space loss? How would the diffraction loss be affected if the transmission frequency is decreased from 4 GHz to 400 MHz?
Ans. The losses are approximately 22 dB for 4 GHz and 13 dB for 400 MHz.   ■

2.4   TERRESTRIAL PROPAGATION: STATISTICAL MODELS[9]

The previous sections have identified three modes of propagation:

1. *Free-space propagation,* in which the received power decreases as the square of the distance from the transmitter.

2. *Reflection,* wherein, for the plane-Earth model, the received power decreases as the fourth power of distance. For other reflection models, we may expect power decreases at other powers, typically greater than two.

3. *Diffraction,* which, for a knife-edge model, introduces a constant attenuation that depends on the proportion of the direct path that is blocked. For a terrestrial radio link, the signal may be diffracted a number of times along its path; consequently, we would expect diffraction losses to also exhibit some dependence on distance.

We have seen that, even with simple physical models for these propagation modes, the analysis can be quite complicated. With the *physical approach,* we build up a model of the environment, including the terrain, buildings, and other features that may affect propagation. With this physical model, the possible propagation paths are determined; the process is often referred to as *ray tracing.* For the different paths, the techniques described in the previous sections are used to estimate the path losses between the transmitter and the receiver. In some cases, signals may undergo multiple reflections, multiple diffractions, or a combination of both. This complicated calculation is often programmed and determined by computer. The results, however, apply only to the particular physical model being analyzed.

Alternatively, we may use a *statistical approach* in which the propagation characteristics are empirically approximated on the basis of measurements in certain general types of environments, such as urban, suburban, and rural. The statistical approach is broken down into two components: an estimate of the *median path loss* and a component representing *local variations.* These issues are discussed in the subsections that follow.

2.4.1   Median Path Loss

In the general case, the received signal is the sum of several versions of the transmitted signal received over different transmissions paths. That is, if $\tilde{E}$ is the total received electrical field and $\tilde{E}_d$ is the electrical field of an equivalent direct path, then, assuming that there are N different paths between the transmitter and receiver due to different reflections, etc., the total electric field is

$$\tilde{E} = \tilde{E}_d \sum_{k=1}^{N} L_k e^{j\phi_k} \tag{2.42}$$

The $\{L_k\}$ represent the relative losses for the different paths, and the $\{\phi_k\}$ represent the relative phase rotations. If a direct path exists, then it would nominally be characterized by $L_0 = 1$ and $\phi_0 = 0$.

Empiricists have performed measurements of the field strength in various environments as a function of the distance from the transmitter to the receiver. Together, these investigations motivate a general propagation model for median path loss having the form

$$\frac{P_R}{P_T} = \frac{\beta}{r^n} \tag{2.43}$$

where the *path-loss exponent* n typically ranges from 2 to 5, depending on the propagation environment. The parameter β represents a loss that is related to frequency and that may also be related to antenna heights and other factors. The right-hand side of Eq. (2.43) is sometimes written in the equivalent logarithmic form

$$L_p = \beta_0(\text{dB}) - 10n \log_{10}(r/r_0) \tag{2.44}$$

In this representation, β_0 *represents the measured path loss at the reference distance* r_0, typically 1 meter. *In the absence of other information,* β_0 *is often taken to be the free-space path loss at a distance of 1 meter.*

Numerous propagation studies have been carried out in an attempt to identify the different environmental effects. The conclusion drawn from these studies is that most of these effects are locally dependent and difficult, if not impossible, to characterize in general. Some sample values for n in the model of Eq. (2.43) are shown in Table 2.1. An example of a more complex model, called the *Okumura–Hata model*, is provided in Section 2.10.

While this table indicates general trends, there are exceptions. For example, if the propagation path is along a straight street with skyscrapers on either side, there may be a ducting (waveguide) effect, and propagation losses may be similar to those occurring in free space or be even less. In general, performance in open spaces is not as poor as is predicted by the plane-Earth model.

This model for median path loss is quite flexible, and it is intended for the analytical study of problems, as it allows us to parameterize the performance of various system-related factors. For commercial applications in a terrestrial environment, either a field measurement campaign or detailed modeling of the environment is preferred.

TABLE 2.1   Sample path-loss exponents.

Environment	n
Free space	2
Flat rural	3
Rolling rural	3.5
Suburban, low rise	4
Dense urban, skyscrapers	4.5

2.4.2 Local Propagation Loss

In the previous section, a model for predicting the median path loss was presented. For any particular site, there will be a variation from this median value that will depend on the local characteristics. Several investigators have measured the variation about the median and have suggested that it can be modeled as a *lognormal distribution*. In particular, if μ_{50} is the median value of the path loss (in dB) at a specified distance r from the transmitter, then the distribution x_{dB} of observed path losses at this distance has the probability density function

$$f(x_{dB}) = \frac{1}{\sqrt{2\pi}\sigma_{dB}} e^{-(x_{dB} - \mu_{50}/2\sigma_{dB}^2)} \tag{2.45}$$

Equation (2.45) is known as the *lognormal model* for local *shadowing*. It is the Gaussian distribution in which all quantities are measured in decibels. (See Appendix C for a description of the Gaussian distribution.) Typical values for the standard deviation σ_{dB} range from 5 to 12 dB. In Fig. 2.11, the lognormal distribution

$$\text{Prob}(x_{dB} > x) = \int_{x}^{\infty} f(x_{dB}) dx_{dB} \tag{2.46}$$

is plotted. For a σ_{dB} of 6 dB, the figure indicates that the loss relative to the median path is greater than 10 dB only 1% of the time. Since, by definition of the median, 50% of the samples are expected to be less than the median and 50% greater, regardless of the shadowing, we find that all curves intersect at the median point.

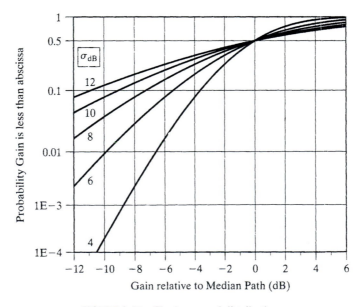

FIGURE 2.11   The lognormal distribution.

We say that a service has 90% *availability* if the received signal level is above receiver sensitivity over 90% of the region of coverage. The availability requirement will depend on the kind of service being offered. For example, safety services require a higher availability than non-safety services.

EXAMPLE 2.4 Availability

A measurement campaign in a large city indicates that the propagation can be reasonably well modeled with a loss exponent of $n = 2.9$. The shadowing deviation about this loss is 6 dB. What is the range of coverage if 99% availability is required for a public-safety radio application? Assume that the receiver sensitivity is –100 dBm and the measured power at 10 meters is 2 milliwatts.

If 99% availability is required with a 6-dB shadowing deviation, then Fig. 2.11 indicates that an extra 10 dB of margin on top of the median path loss is required. Consequently, the median signal power at the edge of coverage must be –90 dBm. Assuming the use of isotropic antennas, the statistical model for median path loss is

$$P_R(\text{dBm}) = P_T(\text{dBm}) + L_p(\text{dB})$$
$$= P_T + \beta_0 - 10n\log_{10}(r/r_0)$$
$$= 3 - 10n\log 10(r/10)$$

where we have used $P_T + \beta_0 = 3$ dBm at a distance $r_0 = 10$ meters.
Solving for r, we obtain

$$10n\log_{10}(r/10) = 3 - P_R(\text{dBm})$$
$$= 3 - (-90)$$
$$= 93 \text{ dB}$$

or

$$r = 10\times10^{93/10n} = 16.1 \text{ km.} \qquad \blacksquare$$

Problem 2.6   A brief measurement campaign indicates that the median propagation loss at 420 MHz in a midsize North American city can be modeled with $n = 2.8$ and a fixed loss (β) of 25 dB; that is,

$$L_p = 25\text{dB} + 10\log_{10}(r^{2.8})$$

Assuming a cell phone receiver sensitivity of –95 dBm, what transmitter power is required to service a circular area of radius 10 km? Suppose the measurements were optimistic and $n = 3.1$ is more appropriate. What is the corresponding increase in transmit power that would be required?
Ans. A transmit power of 12 dBW is required for $n = 2.8$ and 24 dBW for $n = 3.1$. $\blacksquare$

2.5 INDOOR PROPAGATION[10]

With the growth of cellular telephone usage, these appliances are being used more and more in indoor locations such as shopping malls, office buildings, train stations, and airports. It is important that wireless design take into account the propagation characteristics in these high-density locations.

In other applications, wireless local area networks (LANs) are being implemented in buildings and on campuses so as to eliminate the cost of wiring or rewiring buildings to upgrade or provide new services. For these applications, understanding the physics of indoor applications is a key to determining how the locations and numbers of transmitters and receivers will affect the desired quality of service to the user.

In parallel with previous sections, there are two common methods of analyzing indoor propagation: ray tracing, described in Section 2.4, and modifying the free-space model to fit empirical/statistical data and then applying the model for prediction purposes.

The statistical approach provides a more easily understood model. A simple model for the indoor path loss is given by

$$L_p(\text{dB}) = \beta(\text{dB}) + 10\log_{10}\left(\frac{r}{r_0}\right)^n + \sum_{p=1}^{P} \text{WAF}(p) + \sum_{q=1}^{Q} \text{FAF}(q) \qquad (2.47)$$

where r is the distance separating the transmitter from the receiver, r_0 is the nominal reference distance (typically, 1m), n is the path-loss exponent, $\text{WAF}(p)$ is the wall attenuation factor, $\text{FAF}(q)$ is the floor attenuation factor, and P and Q are the number of walls and floors, respectively, separating the transmitter and the receiver.

With indoor propagation, the path-loss exponent n is often chosen to be 2 for small separations, and higher values are selected for greater separations. A typical wall attenuation factor for hollow plaster walls is about 5 dB at 900 MHz; for concrete walls, the attenuation factor is typically 10 dB at the same frequency. The attenuation of concrete floors is usually slightly higher than that of concrete walls, most likely due to the presence of steel beams and re-enforcing rods. These attenuation factors increase as the frequency increases. Some measurements at 1700 MHz indicate a 6-dB increase in the attenuation due to floors and walls over the 900-MHz case. This path-loss model is a relatively simple extension of the models described in earlier sections. It tends to be most accurate when the number of walls and floors is small (less than five).

The shortcomings of the statistical model are that it does not take into account such matters as the dependence of the path-loss exponent on distance and the dependence of the WAF on the angle of incidence. If the transmitter is outside the building, then there is an additional building penetration loss, the size of which depends on the frequency of operation and what floor of the building the receiver is located on.

EXAMPLE 2.5 Indoor Propagation

Suppose, in an office building, a 2.4-GHz transmitter located at a workstation is separated from the network access node (receiver) by a distance of 35 m. The transmission must pass through 5 m of an office, through a plasterboard wall, and then through a large open area. The propagation is modeled as free space for the first 5 m and with a loss exponent of 3.1 for the remainder of the distance. The plasterboard wall causes 6-dB attenuation of the signal. The isotropic transmitter radiates 20 dBm. Can the link be closed if the receiver has a sensitivity of –75 dBm?

TABLE 2.2   Link budget for Example 2.5.

Parameter	Value	Comments
Transmit Power	20 dBm	
Free-space loss	54 dB	$L_p = \left(\dfrac{4\pi R}{\lambda}\right)^2$
Wall attenuation	6 dB	
Open-area loss	26.2 dB	$L_p = (r/r_0)^{3.1}$ with $r/r_0 = 35/5$
Received power	-66.2 dBm	assuming 0-dB receive antenna gain

The propagation path is summarized in Table 2.2. The line items in the table are as follows:

1. The first line of the table is the transmitted power, in dBm.
2. The second line is the free-space loss in the 5 m surrounding the transmitter, using the formula for free-space path loss as shown.
3. The third line is the loss through the plasterboard wall.
4. The fourth line is the relative loss in the open area. To calculate this loss, we assume that the reference distance for the path-loss calculation in the open area is $r_0 = 5$ m. The fourth line is the relative loss between a distance of 5 m and a distance of 35 m.
5. The fifth line is the received power level, obtained by subtracting the previous three lines from the first.

The answer is yes, the link can be closed.   ∎

Problem 2.7   Using the same model as in Example 2.5, predict the path loss for the site geometry shown in Fig. 2.12. Assume that the walls cause an attenuation of 5dB and the floors 10 dB.
Ans. −67.2 dBm.   ∎

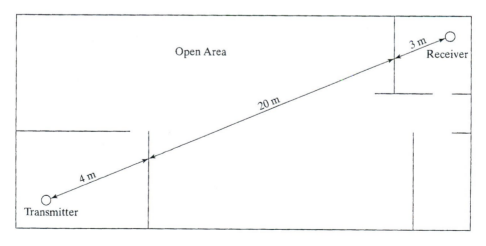

FIGURE 2.12   Site geometry for Problem 2.7.

2.6   LOCAL PROPAGATION EFFECTS WITH MOBILE RADIO[11]

Most mobile communication systems are used in and around centers of population. The major difficulties are caused by the fact that the mobile antenna is well below the surrounding buildings; thus, most communication is via scattering of electromagnetic waves from surfaces or diffraction over and around buildings. These multiple propagation paths, or *multipaths*, have both slow and fast aspects:

1. *Slow fading* arises from the fact that most of the large reflectors and diffracting objects along the transmission path are distant from the terminal. The motion of the terminal relative to these distant objects is small; consequently, the corresponding propagation changes are slow. These factors contribute to the median path losses between a fixed transmitter and a fixed receiver that were described in the previous sections. The statistical variation of these mean losses due to variation of the intervening terrain, vegetation, etc., was modeled as a lognormal distribution for terrestrial applications. The slow-fading process is also referred to as *shadowing* or *lognormal fading*.

2. *Fast fading* is the rapid variation of signal levels when the user terminal moves short distances. Fast fading is due to reflections of local objects and the motion of the terminal relative to those objects. That is, the received signal is the sum of a number of signals reflected from local surfaces, and these signals sum in a constructive or destructive manner, depending on their relative phase relationships, as illustrated in Fig. 2.13. The resulting phase relationships are dependent on relative path lengths to the local objects, and they can change significantly over short distances. In particular, the phase relationships depend on the speed of motion and the frequency of transmission.

In this section, we begin with a statistical analysis of the effects of local reflectors and then consider the case in which the terminal is moving in the environment.

2.6.1   Rayleigh Fading

In wireless communications, a distinction is made between portable terminals and mobile terminals. *Portable terminals* can be moved easily from one place to another, but communications occur when the terminal is stationary. *Mobile terminals* can also be easily moved, but, with them, communications may occur while the terminal is moving. In this section, we consider portable terminals and how local propagation effects may differ at different positions. Subsequently, we will consider mobile terminals.

Consider a stationary receiver that might be at position I_1 or I_2, as shown in Fig. 2.13. We wish to characterize the amplitude distribution of the received signal over a variety of such positions. We model the case in which the transmitted signal reaches a stationary receiver via multiple paths, but the path differences are due only to local reflections. The complex phasor of the N signal rays (reflections) is given by

$$\tilde{E} = \sum_{n=1}^{N} E_n e^{j\theta_n} \tag{2.48}$$

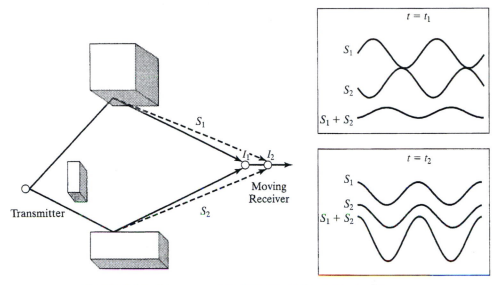

FIGURE 2.13   Illustration of constructive and destructive interference.

where E_n is the electric field strength of the nth path and θ_n is its relative phase. The phasor $\tilde{E}$ is a random variable representing the multiplicative effects of the multipath channel. As seen previously, small differences in path length can make large differences in phase. Since the reflections can arrive from any direction, we assume that the relative phases are independent and uniformly distributed over $[0, 2\pi]$.

With Eq. (2.48) as the model for the complex channel envelope, we see that it is a sum of independent and identically distributed (i.i.d.) complex random variables. From the *central limit theorem* of statistics, the sum of a number of i.i.d. random variables approaches the Gaussian distribution as the number of random variables gets large. (See Appendix C for a review of probability theory and random processes.) That is,

$$\sum_{n=1}^{N} E_n e^{j\theta_n} \; \rightarrow \; Z_r + jZ_i \text{ as } N \rightarrow \infty \tag{2.49}$$

where Z_r and Z_i are real Gaussian random variables. If we consider one of the components of the sum, then we find that the expectation of each component is

$$E[E_n e^{j\phi_n}] = E[E_n]E[e^{j\theta_n}]$$

$$= E[E_n]\frac{1}{2\pi}\int_{0}^{2\pi} e^{j\theta} d\theta \tag{2.50}$$

$$= 0$$

since the mean of a complex phasor with a uniformly distributed phase is zero; **E** denotes the *statistical expectation operator.* Consequently, the mean of the complex envelope is given by

$$\mathbf{E}[\tilde{E}] = \mathbf{E}\left[\sum_{n=1}^{N} E_n e^{j\theta_n}\right]$$

$$= \sum_{n=1}^{N} E_n \mathbf{E}[e^{j\theta_n}] \tag{2.51}$$

$$= 0$$

The variance (power) in the complex envelope is given by the mean-square value

$$\mathbf{E}[|\tilde{E}|^2] = \mathbf{E}\left[\sum_{n=1}^{N} E_n e^{j\theta_n} \sum_{m=1}^{N} E_m e^{-j\theta_m}\right]$$

$$= \sum_{n=1}^{N}\sum_{m=1}^{N} E_n E_m \mathbf{E}[e^{j(\theta_n - \theta_m)}] \tag{2.52}$$

$$= \sum_{n=1}^{N} E_n^2$$

$$= P_0$$

where we have used the fact that the difference of two random phases is also a random phase, unless the two phases are equal. By symmetry, this power is equally distributed between the real and imaginary parts of the complex envelope.

Since the complex envelope has zero mean, the probability density function of Z_r in Eq. (2.49) is given by the Gaussian density function

$$f_{z_r}(z_r) = \frac{1}{\sqrt{2\pi}\sigma} e^{-z_r^2/2\sigma^2} \tag{2.53}$$

where $\sigma^2 = P_0/2$, and similarly for Z_i. We define the amplitude of the complex envelope as

$$R = \sqrt{Z_i^2 + Z_r^2} \tag{2.54}$$

The probability density function of the amplitude is determined from the density function of Eq. (2.53) by performing the appropriate change of variables to obtain

$$f_R(r) = \frac{r}{\sigma^2} e^{-r^2/2\sigma^2} \tag{2.55}$$

(See Appendix C for details.) Equation (2.55) is known as the *Rayleigh probability density function*. Integrating this density function yields the corresponding cumulative probability distribution function:

$$\text{Prob}(r < R) = \int_0^R f_R(r)\,dr$$

$$= 1 - e^{-R^2/2\sigma^2}$$

(2.56)

The mean value of the Rayleigh distribution (the mean value of the absolute envelope) is given by

$$E[R] = \int_0^\infty r f_R(r)\,dr$$

$$= \sigma\sqrt{\frac{\pi}{2}}$$

(2.57)

and the mean-square value is given by

$$E[R^2] = \int_0^\infty r^2 f_R(r)\,dr$$

$$= 2\sigma^2$$

$$= R_{\text{rms}}^2$$

(2.58)

That is, the root-mean-square (rms) amplitude is $R_{\text{rms}} = \sqrt{2}\,\sigma$.

The Rayleigh distribution is plotted in Fig. 2.14. The median of the distribution is $R = R_{\text{rms}}$, which implies that there is constructive interference ($R > R_{\text{rms}}$) for 50% of locations and destructive interference ($R < R_{\text{rms}}$) for 50% of locations. Conceptually, each location corresponds to a different set of $\{\phi_n\}$. Deep fades of 20 dB or more ($R < 0.1 R_{\text{rms}}$) occur only rarely (with a probability of 1%). However, Fig. 2.14 indicates that there can be a wide variation in received signal strength due to local reflections.

These results are for portable receivers. In any one position, the received power will be constant. If we take measurements at a number of random locations, then the received power measurements would show a Rayleigh distribution. For a truly stationary receiver, we would choose a location that minimizes the local reflections and provides the maximum received signal strength.

EXAMPLE 2.6 Margin for Rayleigh Fading

Recall Example 2.4, in which we showed that the system would need 10 dB extra power (*margin*) to provide 99% availability when 6-dB lognormal fading (slow fading) was taken into account. Suppose this same system also undergoes fast fading. What additional margin would be required to maintain the same availability?

From Fig. 2.14, we see that, if the receiver is to operate above threshold for 99% of the locations, then it must be able to tolerate the amplitude dropping to as low as 10% of the rms value, or a fade depth of 20 dB. That is, a 20-dB additional margin would be required for Rayleigh fading in order to maintain 99% availability.

In practice, a well-designed receiver that uses forward error-correction coding (discussed in Chapter 4) can usually significantly reduce the margin required to compensate Rayleigh fading. ∎

Problem 2.8   What are the required margins for lognormal and Rayleigh fading in Example 2.6 if the availability requirement is only 90%?
Ans. The required margins are 5.5 dB for shadowing and 10 dB for Rayleigh fading. ∎

2.6.2 Rician Fading

The Rayleigh-fading model assumes that all paths are relatively equal—that is, that there is no dominant path. Despite the fact that Rayleigh fading is the most popular model, occasionally there is a direct line-of-sight path in mobile radio channels and in indoor wireless as well. The presence of a direct path is usually required to close the link in satellite communications. In this case, the reflected paths tend to be weaker than the direct path, and we model the complex envelope as

$$\tilde{E} = E_0 + \sum_{n=1}^{N} E_n e^{j\theta_n} \tag{2.59}$$

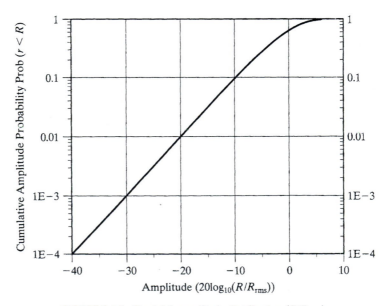

FIGURE 2.14   Rayleigh amplitude distribution (R/R_{rms}).

where the constant term represents the direct path and the summation represents the collection of reflected paths. This model is referred to as a *Rician fading* model. The analysis proceeds in a manner similar to that of the Rayleigh fading case, but with the addition of a constant term. A key factor in the analysis is the ratio of the power in the direct path to the power in the reflected paths. This ratio is referred to as the *Rician K-factor*, defined as

$$K = \frac{s^2}{\displaystyle\sum_{n=1}^{N} |E_n|^2} \tag{2.60}$$

where $s^2 = |E_0|^2$. The Rican K-factor is often expressed in dB.

The calculation of the amplitude density function in the Rician fading case is more involved than with Rayleigh fading, so we merely give the result here. We have

$$f_R(r) = \frac{r}{\sigma^2} e^{-(r^2 + s^2)/2\sigma^2} I_0\left(\frac{rs}{\sigma^2}\right) \quad r \geq 0 \tag{2.61}$$

where $I_0(.)$ is the *modified Bessel function* of zeroth order.[8] (See Appendix C for more details on the Rician distribution.)

The amplitude distributions of the Rician fading channel for different K factors are shown in Fig. 2.15. Deep fades are clearly less probable than with the Rayleigh channel, and the probability of their occurrence decreases as the K factor increases.

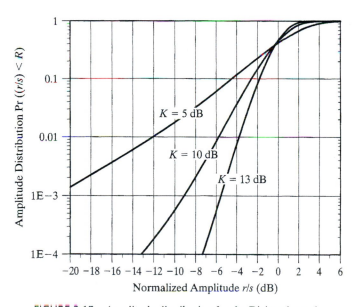

FIGURE 2.15   Amplitude distribution for the Rician channel.

2.6.3 Doppler

Just as a train whistle or car horn appears to have a different pitch, depending on whether it is moving toward or away from one's location, radio waves demonstrate the same phenomenon. If a receiver is moving toward the source, then the zero crossings of the signal appear faster, and consequently, the received frequency is higher. The opposite effect occurs if the receiver is moving away from the source. The resulting change in frequency is known as the *Doppler shift*.

To model the Doppler shift phenomenon, consider Fig. 2.16, which shows a fixed transmitter and a receiver moving at a constant velocity away from the transmitter. If the complex envelope of the signal emitted by the transmitter is $Ae^{j2\pi f_0 t}$, then the signal at a point along the x-axis is given by

$$\tilde{r}(t, x) = A(x)e^{j2\pi f_0(t - x/c)} \tag{2.62}$$

where $A(x)$ represents the amplitude as a function of distance and c is the speed of light. Equation (2.62) shows the signal has a phase rotation that depends on the distance of the signal from the source. If x represents the position of the constant-velocity receiver, then we may write

$$x = x_0 + vt \tag{2.63}$$

where x_0 is the initial position of the receiver and v is its velocity away from the source. Substituting Eq. (2.63) into (2.62), we find that the signal at the receiver is

$$\tilde{r}(t) = A(x_0 + vt)e^{j2\pi f_0\left(t - \frac{x_0 + vt}{c}\right)} \tag{2.64}$$

$$= A(x_0 + vt)e^{-j2\pi f_0 x_0/c}e^{j2\pi f_0(1 - v/c)t}$$

If we focus on the frequency term in the last exponent of Eq. (2.64), we discover that the received frequency is given by

$$f_r = f_0\left(1 - \frac{v}{c}\right) \tag{2.65}$$

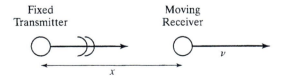

Fixed   Moving
Transmitter   Receiver

FIGURE 2.16   Illustration of Doppler effect.

where f_0 is the carrier transmission frequency. The *Doppler shift* is given by

$$f_D = f_r - f_0 = -f_0 \frac{v}{c} \tag{2.66}$$

Consequently, the relationship between the observed Doppler frequency shift and the velocity away from the source is

$$\frac{v}{c} = -\frac{f_D}{f_0} \tag{2.67}$$

Equation (2.67) describes the case when the direction of terminal motion and the radiation are collinear. This is actually the maximum Doppler shift that can be observed. If the terminal motion and the direction of radiation are at an angle ψ, the more general expression for the shift is

$$f_D = -\frac{f_0}{c} v \cos \psi \tag{2.68}$$

For operating frequencies between 100 MHz and 2 GHz, and for speeds up to 100 km/hr, the Doppler shift can be as large as 185 Hz. This Doppler shift appears as a carrier offset that varies with the speed and direction of the vehicle. Consequently, any mobile receiver must have the capability of both receiving a signal with frequency offsets and tracking fast changes in the frequency offset as the receiver terminal changes velocity.

EXAMPLE 2.7 Aircraft Doppler

An aircraft is headed toward an airport control tower with a speed of 500 km/hr at an elevation of 20°. Safety communications between the aircraft tower and the plane occur at a frequency of approximately 128 MHz. What is the expected Doppler shift of the received signal?

The velocity of the aircraft in the direction away from the tower is

$$v = -500 \text{ km/hr} \times \cos 20° = -130 \text{m/s}$$

The corresponding Doppler shift is

$$f_D = -\frac{f_0}{c} v \cos \alpha$$

$$= -\frac{128 \times 10^6}{3 \times 10^8} \times (-130)$$

$$= 55 \text{ Hz}$$

If the plane banks suddenly and heads in the other direction, the Doppler shift will change from 55 Hz to −55 Hz. This rapid change in frequency, df_D/dt, is sometimes referred to as *frequency slewing*. A good mobile receiver must be capable of tracking the frequency slew rates that can be generated by the receiver's motion.   ∎

Problem 2.9   Suppose that the aircraft in Example 2.7 has a satellite receiver operating in the aeronautical mobile–satellite band at 1.5 GHz. What is the Doppler shift observed at this receiver? Assume that the geostationary satellite has a 45° elevation with respect to the airport. *Ans. The Doppler shift for this receiver is 491 Hz.*   ∎

2.6.4 Fast Fading

If we measure the received signal strength as a function of time for a mobile terminal, we may observe a rapid variation in the signal strength about a median value. This phenomenon is known as *fast fading*. From the previous section on stationary receivers, we would expect the signal power at different locations to have a Rayleigh distribution. In Fig. 2.17, we show a sample trace of the received signal power for a terminal moving at 100 km/hr in a Rayleigh fading environment. The frequency of transmission is 900 MHz.

Over short periods, the received power has a relatively smooth behavior, due to the correlation of the fading at adjacent positions. However, there are the occasional deep fades in the level of the received signal, as would be expected with a Rayleigh distribution. It is the correlation of the fading process that we will investigate in this section.

To model fading with a moving receiver, we expand the model used for stationary fading, given by Eq.(2.48), to include the effects of the Doppler shift. Let

$$\tilde{E}(t) = \sum_{n=1}^{N} E_n e^{j(2\pi f_n t + \phi_n)} \tag{2.69}$$

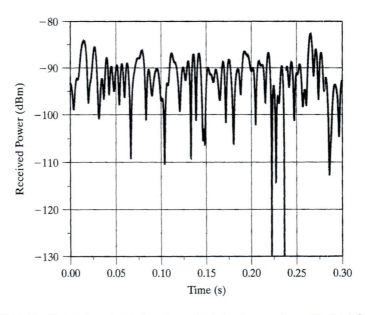

FIGURE 2.17   Illustration of variations in received signal power due to Rayleigh fading.

represent the *complex envelope* of N signal rays. The complex envelope of the channel $\tilde{E}(t)$ is time varying. Since the receiver is moving, each ray has a different Doppler shift $\{f_n\}$ because the relative directions of motion of the rays will be different. The rays are assumed to come from arbitrary angles ψ_n surrounding the receiver, as shown in Fig. 2.18. We assume that all rays are arriving from a horizontal direction; this is known as the *Clarke model.*[12] As before, we assume that the relative phase of each ray is i.i.d.

To study the time dependency of the fading phenomenon, we look at the autocorrelation of the complex envelope and so write

$$R_E(\tau) = \mathbf{E}[\tilde{E}(t)\tilde{E}^*(t+\tau)] \tag{2.70}$$

where the asterisk denotes complex conjugation and τ is a time offset. We begin by computing the cross-correlation between a pair of signal rays. This correlation is given by

$$\mathbf{E}\left[E_n e^{j(2\pi f_c t + \phi_n)} E_m e^{-j(2\pi f_m(t+\tau) + \phi_m)} \right] = \mathbf{E}[E_n e^{j2\pi f_n t} E_m e^{-j2\pi f_m(t+\tau)}]\mathbf{E}[e^{j\theta_n} e^{j\theta_m}]$$

$$= \begin{cases} 0 & \text{if } m \neq n \\ \mathbf{E}[E_n^2 e^{-j2\pi f_n \tau}] & \text{if } m = n \end{cases} \tag{2.71}$$

due to the i.i.d. assumption. Using the result of Eq.(2.71) with Eq. (2.69) in Eq. (2.70),

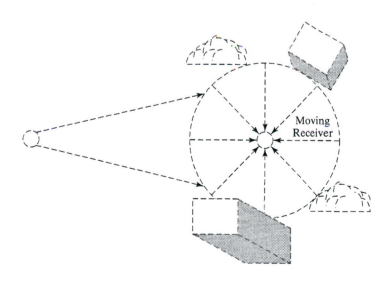

FIGURE 2.18   Illustration of local multipath arriving at a receiver.

we obtain the *autocorrelation of the complex envelope*, given by

$$R_E(\tau) = \mathbf{E}\left[\sum_{n=1}^{N} E_n e^{j(2\pi f_n t + \theta_n)} \sum_{m=1}^{N} E_m e^{-j(2\pi f_m(t+\tau)+\theta_m)}\right]$$

$$= \sum_{n=1}^{N} \mathbf{E}[E_n^2 e^{-j2\pi f_n \tau}] \tag{2.72}$$

$$= \sum_{n=1}^{N} \mathbf{E}[E_n^2]\mathbf{E}[e^{-j2\pi f_n \tau}]$$

where we assume that the effects of small changes in distance on amplitude $\{E_n\}$ are negligible. Recall from Section 2.6.3 that the Doppler shift is proportional to the angle of the radiation relative to the direction of motion; that is,

$$f_n = f_D \cos \psi_n \tag{2.73}$$

where f_D is the *maximum Doppler shift*. The maximum shift occurs for signal rays that are in the same direction as the motion of the terminal. Under the assumption that the multipath is uniformly distributed over the range $[-\pi, \pi]$, the expectation of Eq. (2.72) is independent of n, in which case we have

$$R_E(\tau) = P_0 \mathbf{E}[e^{-j2\pi f \tau}] \tag{2.74}$$

where $P_0 = \sum_{n=1}^{N} \mathbf{E}[E_n^2]$ is the average received power. Substituting Eq. (2.73) into (2.74), we finally obtain

$$R_E(\tau) = P_0 \mathbf{E}[e^{-j2\pi f_D \tau \cos \psi}]$$

$$= P_0\left(\frac{1}{2\pi}\int_{-\pi}^{\pi} e^{-j2\pi f_D \tau \cos \psi}\,d\psi\right) \tag{2.75}$$

$$= P_0 J_0(2\pi f_D \tau)$$

where $J_0(x)$ is the *zeroth-order Bessel function* of the first kind, defined by

$$J_0(x) = \frac{1}{\pi}\int_0^{\pi} e^{-jx\cos\theta}\,d\theta \tag{2.76}$$

In summary, the correlation of the fading with time (or position) is described by Eq. (2.75). Figure 2.19 plots this autocorrelation function of the complex envelope as a function of the normalized parameter $2\pi f_D \tau$. As expected, the autocorrelation peaks at $\tau = 0$, but there is strong correlation for normalized delays as large as $2\pi f_D \tau = 1$. The autocorrelation is symmetric in τ.

The power spectrum of the fading process is determined by the Fourier transform of the autocorrelation function of the envelope and is given by

$$S_E(f) = \mathbf{F}[R_E(\tau)]$$

$$= \mathbf{F}[P_0 J_0(2\pi f_D \tau)]$$

$$= \begin{cases} \dfrac{P_0}{\sqrt{1-(f/f_D)^2}} & f < f_D \\ 0 & f > f_D \end{cases} \tag{2.77}$$

where $\mathbf{F}$ is the Fourier transform operator. Figure 2.20 plots Eq. (2.77) for $P_0 = 1$. The power spectrum is zero at frequencies greater than the Doppler shift f_D and has discontinuities at the edge of the nonzero band. This spectrum is quite similar to what is actually observed, although, in practice, the discontinuities are not present.

Implicit in the Clarke model for fast fading in mobile terrestrial communications is the assumption that all rays are arriving from a horizontal direction. In practice, however, there may be a vertical component to the waves. More sophisticated models for mobile terrestrial communications have taken this component into account, but their conclusions are similar to those presented here. Empirical measurements tend to support the Clarke model, with a Doppler bandwidth related to the frequency of transmission and the mobile velocity. The measured spectra usually show the peaks near the Doppler frequencies, as depicted in Fig. 2.20.

Empirical investigations have indicated that the Clarke model is not always a valid description of the fading process between a mobile terminal and satellite. For example, with aeronautical terminals and satellites, it has been found that a Gaussian

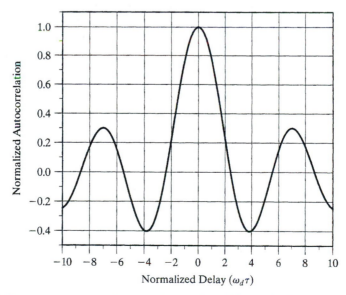

FIGURE 2.19   Autocorrelation of the complex envelope of the received signal according to Clarke's model.

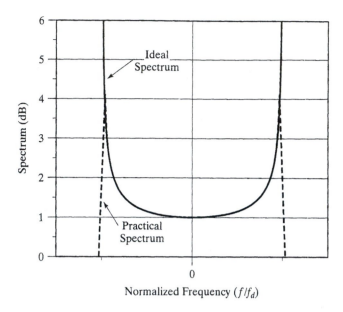

FIGURE 2.20   Power spectrum of the fading process for Clarke's model.

spectrum is a better model of the spectrum of the fading process and that the maximum Doppler frequency of the fading process is *not* proportional to the aircraft's speed, but is typically between 20 Hz and 100 Hz. There are a couple of factors causing this slower rate: (i) Nearby reflections come from the aircraft's fuselage and wings, which are moving (vibrating) at a much slower rate relative to the motion of the antenna, and (ii) more distant reflections from the sea surface tend to be directional rather than omnidirectional. In the case of maritime mobile terminal–satellite communications, a Gaussian fading spectrum with Doppler bandwidths of 1 Hz and less is a more accurate model of the empirical results. This is due not only to the slower motion of the ship through the water, but also to the effects of ocean waves and their movement as reflective surfaces relative to the motion of the ship.

Problem 2.10   A data signal with a bandwidth of 100 Hz is transmitted at a carrier frequency of 800 MHz. The signal is to be reliably received in vehicles travelling at speeds of up to 100 km/hour. What can we say about the minimum bandwidth of the filter at the receiver input?
Ans. This filter should be at least 175 Hz wide.   ∎

2.7   CHANNEL CLASSIFICATION

In previous sections, we investigated the effect that various propagation phenomena have on the received electric field. Some of the observations we made were as follows:

- The destructive interference caused by a single reflection can cause a significant reduction in the received power. This interference causes the power loss as a function of distance to change from a square-law loss, characteristic of free-space propagation, to a fourth-power attenuation.

- The presence of large objects such as hills, buildings, and so on does not prevent a signal from being propagated. Rather, diffraction effects allow the signal to propagate around these large objects, albeit at a reduced level.

- A large number of reflections from objects near the receiver can cause large variations in the received electric field that are highly sensitive to position. Again, this is due to the superposition of multiple electric fields associated with the different reflections.

- Movement of the receiver causes a variation in the received electric field, due to changes in constructive and destructive interference of the various reflections. The rate of variation of this electric field depends on the speed of motion of the receiver.

The first two of these observations can be categorized as large-scale propagation effects, the last two as small-scale effects:

1. *Large-scale effects* are due to the general terrain and the density and height of buildings and vegetation. These effects are characterized statistically by the median path loss and lognormal shadowing. Both of these phenomena have a behavior that varies relatively slowly with time. Large-scale effects are important for predicting the coverage and availability of a particular service.

2. *Small-scale effects* are due to the local environment, nearby trees, buildings, etc., and the movement of the radio terminal through that environment. These effects have a much shorter timescale. They have been characterized statistically as fast Rayleigh fading. A consideration of small-scale effects is important for the design of the modulation format and for general transmitter and receiver design.

When we investigated the behavior of these different physical phenomena, the analysis was based on the transmitted electric field, which was modeled at a point in space as

$$E(t) = E_0 \cos(2\pi f t + \theta) \tag{2.78}$$

and by its equivalent complex phasor notation, $\tilde{E}$. Consequently, the analysis describes the response of the channel at a single frequency f. In practice, the transmitted signal has a nonzero bandwidth and can be represented as

$$s(t) = m(t)\cos(2\pi f t + \phi(t))$$
$$= \text{Re}\{\tilde{s}(t)e^{j2\pi f t}\} \tag{2.79}$$

where $m(t)$ is the amplitude and $\phi(t)$ is the phase. As Eq. (2.79) indicates, the transmitted signal also has a complex phasor representation, $\tilde{s}(t) = m(t)e^{j\phi(t)}$. We shall refer to the complex phasor as the modulating signal. In this section, we consider both the combination of large- and small-scale propagation effects and the effect of nonzero signal bandwidth.

In Section 2.2, we made the assumption that the channel was linear. According to this assumption, all distortions can be characterized by the attenuation or superposition of different signals. In addition, we allow the possibility that the propagation channel may be time varying. As a consequence of these assumptions, the channel can be presented by a dual-time time-varying impulse response $h(t,\tau)$, defined as the response of the system measured at time t due to a unit impulse applied at time $t - \tau$. Hence, unlike the impulse response of a time-invariant system, the impulse response $h(t,\tau)$ changes shape with τ. For an input $s(t)$, the channel output is given by the convolution integral

$$x(t) = \int_{-\infty}^{\infty} h(t, \tau)s(t - \tau)d\tau \tag{2.80}$$

It can be shown (see Chapter 3) that this expression can also be written in complex phasor notation as

$$\tilde{x}(t) = \int_{-\infty}^{\infty} \tilde{h}(t, \tau)\tilde{s}(t - \tau)d\tau \tag{2.81}$$

Channels are classified on the basis of the properties of the time-varying impulse response $\tilde{h}(t, \tau)$. The effects of noise are not considered in classifying channels.

2.7.1 Time-Selective Channels

We begin by extending our model of a fast-fading channel to the case in which the electric field is modulated by the complex phasor $s(t)$. When the signal is modulated, the received signal is given by

$$
\begin{aligned}
x(t) &= \sum_{n=1}^{N} \alpha_n(t)m(t)\cos(2\pi ft + \phi(t) + \theta_n(t)) \\
&= \sum_{n=1}^{N} \mathrm{Re}\left\{ \alpha_n(t)e^{j\theta_n(t)}\tilde{s}(t)e^{j2\pi ft} \right\}
\end{aligned}
\tag{2.82}
$$

where $\alpha_n(t)$ is the attenuation and $\theta_n(t)$ is the phase rotation of the nth reflected path. The channel model for fast fading assumes that the different paths of Eq. (2.82) are all of the same length and thus also have the same delay. Equation (2.82) can be written in complex phasor notation as

$$
\begin{aligned}
\tilde{x}(t) &= \sum_{n=1}^{N} \tilde{\alpha}_n(t)s(t) \\
&= \tilde{\alpha}(t)\tilde{s}(t)
\end{aligned}
\tag{2.83}
$$

where we have made an equivalence between the received signal $\tilde{x}(t)$ and the electric field and where $\tilde{\alpha}_n(t) = \alpha_n(t)e^{j\theta_n(t)}$. That is, the complex gain $\tilde{\alpha}(t)$ (often assumed to have a Rayleigh-distributed magnitude) is the result of the constructive and destructive interference of multiple *local reflections*. It is straightforward to show that $\tilde{\alpha}(t)$ has the same statistical properties as the electric field phasor developed in Section 2.6.4.

Comparing Eq. (2.83) with Eq. (2.81), we can identify the time-varying impulse response. For the model given by Eq. (2.83), it can be verified that the channel impulse response is

$$\tilde{h}(t, \tau) = \tilde{\alpha}(t)\delta(\tau) \tag{2.84}$$

where $\delta(t)$ is the *Dirac delta function* or *unit-impulse function*. Since $\tilde{\alpha}(t)$ in Eq. (2.83) is time varying, the received signal strength is also time varying. For this reason, the channel described by Eq. (2.84) is referred to as a *time-selective* channel, in the sense that the channel is better at selected times than at other times.

From Fourier-transform theory (see Appendix A), we know that multiplication in the time domain is equivalent to convolution in the frequency domain. Accordingly, the power spectrum of the received signal of Eq. (2.83) is the convolution of the power spectra of the transmitted signal and the fading process; that is,

$$S_r(f) = S_\alpha(f) \otimes S_s(f) \tag{2.85}$$

where $\otimes$ is the *convolution operator*. Figure 2.21 depicts the result of convolving the power spectrum of the fading process given by Eq. (2.82) with a nominal transmit spectrum, also shown in the figure. The 3-dB bandwidth of the nominal spectrum is approximately five times the maximum Doppler shift f_D. The conclusion to be drawn from the figure is that, when the Doppler shift is a small fraction of the total signal bandwidth, there is very little effect on the received spectrum. In particular, each frequency component is received at approximately the same relative level as that at which it was transmitted. This type of channel is referred to as a *frequency-flat* channel; that is, the frequency response of the channel is approximately constant and does not change the spectrum of the transmitted signal much.

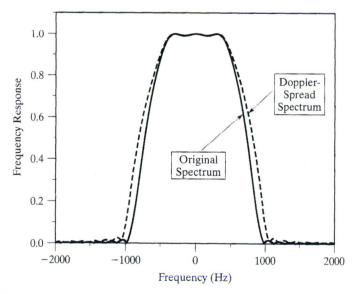

FIGURE 2.21   Comparison of original spectrum with Doppler-spread spectrum. Doppler bandwidth is 20% of signal bandwidth.

As we shall see, a frequency-flat channel is not necessarily time selective; and similarly, a time-selective channel is not necessarily frequency flat. Nevertheless, the combination of a frequency-flat and time-selective channel is common in practice, and we shall refer to it as a *flat-fading* channel.

2.7.2 Frequency-Selective Channels

With large-scale effects, the signal may arrive at the receiver via paths with different path lengths. Consequently, in complex phasor notation, a model for these large-scale effects is

$$\tilde{x}(t) = \sum_{l=1}^{L} \tilde{\alpha}_l \tilde{s}(t - \tau_l) \tag{2.86}$$

where τ_l represents the relative delay associated with the lth path and, for the moment, the complex gains $\{\tilde{\alpha}_l\}$ are assumed constant. (In practice, the complex gains could be constant due to the transmitter and receiver being physically stationary.) With this transmission model, the channel impulse response can be represented as

$$\tilde{h}(t, \tau) = \sum_{l=1}^{L} \tilde{\alpha}_l \delta(\tau - \tau_l) \tag{2.87}$$

This channel is time invariant, but it does show a frequency-dependent response, as illustrated by Example 2.8.

EXAMPLE 2.8 Frequency-Selective Channel

Consider the two-ray channel with the impulse response

$$\tilde{h}(t, \tau) = \delta(\tau) + \tilde{\alpha}_2 \delta(\tau - \tau_2) \tag{2.88}$$

where τ_2 is the relative delay of the second path. If we simulate this channel digitally, with τ_2 equal to the sample period, then the frequency response of the channel for various values of $\tilde{\alpha}_2$ is shown in Fig. 2.22.

For $\tilde{\alpha}_2 = 0.5$, there is attenuation at the edges of the channel. When $\tilde{\alpha}_2 = j/2$, the attenuation is on the left-hand side, and when $\tilde{\alpha}_2 = -j$, there is a null at $+0.25 f_s$, where f_s is the sampling rate. (Recall from Section 2.3 that complex values for α imply a phase rotation.) ∎

In general, channels with an impulse response characterized by Eq. (2.87) show a frequency dependence. For this reason, these channels are called *frequency selective*. By contrast, the channel of Eq. (2.88) is time invariant, and is thus also referred to as a *time-flat* channel.

2.7.3 General Channels

In the preceding two sections, we have described time-selective and frequency-selective channels. A simpler case occurs when the channel is neither time varying nor frequency varying; this channel is referred to as *flat–flat*, since the response is flat in both the time and frequency domains.

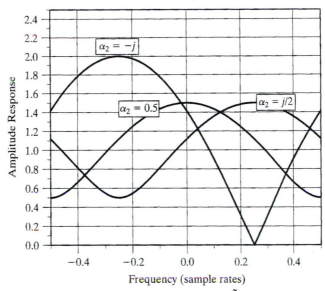

FIGURE 2.22 Frequency response of static two-ray channel $\tilde{h}(t,\ \tau) = \delta(\tau) + \alpha_2\delta(\tau - T_s)$ where T_s is the sample period.

A more interesting case is characterized by the situation in which the received signal can be represented as

$$\tilde{x}(t) = \sum_{l=1}^{L} \tilde{\alpha}_l(t)\tilde{s}(t - \tau_l(t)) \tag{2.89}$$

That is, the signal arrives at the receiver via paths having different lengths, and each of these paths undergoes time-varying fading. The path lengths can also be slowly time varying. Under these conditions, the channel is characterized by the time-varying impulse response

$$\tilde{h}(t,\ \tau) = \sum_{l=1}^{L} \tilde{\alpha}_l(t)\delta(\tau - \tau_l(t)) \tag{2.90}$$

The channel model of Eq. (2.89) includes both large- and small-scale effects. The expression for $h(t,\tau)$ given in Eq.(2.90) is for a finite number of signal paths; in the general case, *we may consider $h(t,\tau)$ as representing a continuum of signal paths with arbitrarily small differences*. A channel characterized by Eq. (2.90) is both *time selective* and *frequency selective*.

In the case of either a continuum or a discrete number of signal paths, the received signal is equal to the convolution of the channel impulse response $\tilde{h}(t,\ \tau)$ with the transmitted signal; that is,

$$\tilde{x}(t) = \int_{-\infty}^{\infty} \tilde{h}(t,\ \tau)\tilde{s}(t - \tau)d\tau \tag{2.91}$$

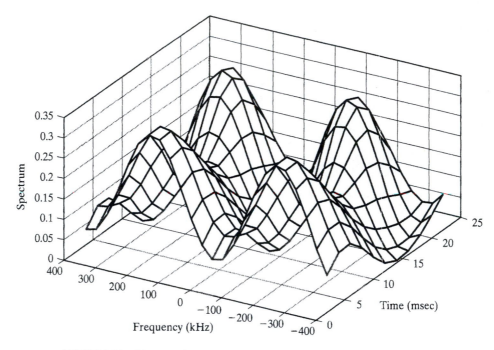

FIGURE 2.23   Time-varying transfer function of a frequency-selective channel.

and related to the time-varying impulse response $h(t,\tau)$, we may define the time-varying frequency response or transfer function

$$H(t,f) = \mathbf{F}[h(t,\tau)] \tag{2.92}$$

where $\mathbf{F}$ is the Fourier transform operator. The quantity $H(t,f)$ represents the time-varying frequency response of the channel, as illustrated in Example 2.9.

EXAMPLE 2.9 Time-Varying Impulse Responses

Consider a channel with the time-varying impulse response

$$\tilde{h}(t,\tau) = \tilde{\alpha}_1(t)\delta(\tau) + \tilde{\alpha}_2(t)\delta(\tau-\tau_2) \tag{2.93}$$

where the $\{\tilde{\alpha}_i(t)\}$ are independent Rayleigh processes. At each time $t = t_0$, we can Fourier transform the impulse response and determine the power spectrum of the result. In Fig. 2.23, we show the progression in time of the transfer function of this fading channel. The channel shows some frequencies with strong responses and other frequencies that correspond to nulls, both of which change with time.   ∎

2.7.4 WSSUS Channels[13]

Two conclusions may be drawn from the channel models presented in the previous sections:

- The mobile channel introduces delay spread into the received signal. That is, the received signal has a longer duration than that of the transmitted signal, due to the different delays of the signal paths. This phenomenon is referred to as *time dispersion*.
- The mobile channel introduces Doppler spread into the received signal. That is, the received signal has a larger bandwidth than that of the transmitted signal, due to the different Doppler shifts introduced by the components of the multipath. This second phenomenon is referred to as *frequency dispersion*.

Both time dispersion and frequency dispersion introduce distortion into the received signal. The amount of degradation caused by this distortion depends on how the signal is designed.

Doppler spreading or frequency dispersion causes variations of the received signal in the time domain. Of interest for the design of data transmission systems is the maximum duration for which the channel can be assumed to be approximately constant. A transmitted data symbol that has a duration less than this time should suffer little distortion from the effects of frequency dispersion. However, it will still suffer the effects of reduced signal levels.

Just as fading does in the time domain, time dispersion causes slow variations in the received signal in the frequency domain. (See Example 2.8.) Of interest in the design of digital transmissions systems is the maximum transmission bandwidth over which there is little variation. A signal contained within that bandwidth should suffer little distortion from the effects of time dispersion.

For design purposes, two parameters characterize the effects of frequency and time dispersion:

- *Coherence time* refers to the time separation at which the amplitudes of two time-domain samples of the channel become decorrelated.
- *Coherence bandwidth* refers to the frequency separation at which the attenuation of two frequency-domain samples of the channel becomes decorrelated.

In other words, the coherence time is a measure of the length of time for which the channel can be assumed to be approximately constant in the time domain. The coherence bandwidth is a measure of the approximate bandwidth within which the channel can be assumed to be nearly constant. The parameters of coherence time and coherence bandwidth are useful for assessing both the performance of various modulation techniques and the distortion that will be caused by the channel. The threshold for declaring decorrelation in the definitions of coherence time and coherence bandwidth is a matter for debate. Although not definitive, a threshold of 0.5 is used by Jakes (1974) and Lee (1982), and we shall use this definition in the discussion that follows.

The task ahead of us is to develop expressions for estimating the coherence time and coherence bandwidth for a general channel impulse response $\tilde{h}(t, \tau)$. We begin the task by assuming that $\tilde{h}(t, \tau)$ is *wide-sense stationary (WSS)*. As explained in Appendix C, a random process is wide-sense stationary if it has a mean that is time independent and a correlation function $R(t_1, t_2) = R(t_1 - t_2)$. What does this mean for the time-varying impulse response? It means that the autocorrelation function of the channel response takes the form

$$\mathbf{E}[\tilde{h}(t_1, \tau_1)\tilde{h}^*(t_2, \tau_2)] = R_h(t_1 - t_2; \tau_1, \tau_2) \tag{2.94}$$

That is, the channel autocorrelation function depends only on the time difference, $t_1 - t_2$.

It is also commonly assumed that, in multipath channels, the gain and phase shift at one delay are uncorrelated with the gain and phase shift at another delay. This type of behavior is referred to as *uncorrelated scattering (US)*, a phenomenon in which the $\{\tilde{\alpha}_1(t)\}$ are uncorrelated random processes. Under the uncorrelated-scattering assumption, we find that

$$R_h(t_1 - t_2; \tau_1, \tau_2) = R_h^w(t_1 - t_2; \tau_1)\delta(\tau_1 - \tau_2) \tag{2.95}$$

That is, the autocorrelation is nonzero only when $\tau_1 = \tau_2$. The combination of a wide-sense stationary signal and uncorrelated scattering is referred to as a *wide-sense stationary uncorrelated scattering (WSSUS) channel*.

EXAMPLE 2.10 Uncorrelated Scattering

Suppose that the channel impulse response is given by

$$\tilde{h}(t, \tau) = \sum_{l=1}^{L} \tilde{\alpha}_l(t)\delta(\tau - \xi_1) \tag{2.96}$$

and the scattering processes are uncorrelated. Determine the autocorrelation function of the channel.

The autocorrelation function of the channel is given by

$$
\begin{aligned}
R_h(t_1, t_2; \tau_1, \tau_2) &= \mathbf{E}[h(t_1, \tau_1)h^*(t_2, \tau_2)] \\
&= \mathbf{E}\left[\sum_{l=1}^{L} \tilde{\alpha}_l(t_1)\delta(\tau_1 - \xi_l)\sum_{k=1}^{L} \tilde{\alpha}_k^*(t_2)\delta(\tau_2 - \xi_k)\right] \\
&= \sum_{l=1}^{L}\sum_{k=1}^{L} \mathbf{E}[\tilde{\alpha}_l(t_1)\tilde{\alpha}_k^*(t_2)]\delta(\tau_1 - \xi_l)\delta(\tau_2 - \xi_k) \\
&= \sum_{l=1}^{L} \mathbf{E}[\tilde{\alpha}_l(t_1)\tilde{\alpha}_l^*(t_2)]\delta(\tau_1 - \xi_l)\delta(\tau_2 - \xi_l)
\end{aligned}
\tag{2.97}
$$

where the last equality follows from the third by the uncorrelated-scattering assumption. Then, with a Clarke model for the scattering process described in Eq. (2.75), the autocorrelation of the fading for the lth path is given by

$$\mathbf{E}[\tilde{\alpha}_l(t_1)\tilde{\alpha}_l^*(t_2)] = P_l J_0(2\pi f_D(t_1 - t_2)) \tag{2.98}$$

where P_l is the average power of the lth path. Consequently, the autocorrelation function of the channel reduces to

$$R_h(t_1, t_2; \tau_1, \tau_2) = \left[\sum_{l=1}^{L} P_l J_0(2\pi f_D(t_1 - t_2))\delta(\tau_1 - \xi_l) \right] \delta(\tau_1 - \tau_2) \qquad (2.99)$$

Equation (2.99) has the form indicated by Eq. (2.95). ∎

2.7.5 Coherence Time

In Section 2.7.4, the coherence time is defined as the period over which there is a strong correlation of the channel time response. For a WSSUS process, this is the range of Δt over which $R_h^w(\Delta t; \tau_1) \equiv R_h(t, t + \Delta t; \tau_1, \tau_2)$ of Eq. (2.95) is significant. If we set $\tau_1 = 0$ in $R_h^w(\Delta t; \tau_1)$, then we find that the Fourier transform of the resulting autocorrelation function is given by

$$S_D(f) = \mathbf{F}[R_h^w(\Delta t; 0)] \qquad (2.100)$$

The function $S_D(f)$ is referred to as the *Doppler power spectrum* of the channel. A general property of the Fourier transform is that the width of a signal in the time domain is inversely proportional to the width of the signal in the frequency domain; see Appendix A. Consequently, if the Doppler spread $2f_D$, where f_D is the maximum Doppler frequency, is the approximate width of the Doppler power spectrum, then the coherence time is given approximately by

$$T_{\text{coherence}} \approx \frac{1}{2f_D} \qquad (2.101)$$

That is, the coherence time of the channel is approximately the inverse of the Doppler spread of the channel.

EXAMPLE 2.11 Coherence Time for a Flat-Fading Channel

For the Clarke model for fast fading described in Section 2.6.4, what is the coherence time of the channel?

From Eq. (2.75), the normalized autocorrelation of the fading process of a flat-fading channel described by the Clarke model is given by

$$R_h^w(\Delta t, 0) = R_h(\Delta t) = J_0(2\pi f_D \Delta t) \qquad (2.102)$$

If we define the coherence time as that range of values, Δt, over which the correlation is greater than 0.5, then substituting $R_h^w(T_{\text{coherence}}; 0) = 0.5$ into the left-hand side of Eq. (2.102) implies that the coherence time is given by

$$T_{\text{coherence}} = \frac{1}{2\pi f_D} J_0^{-1}(0.5)$$

$$= \frac{0.30}{2f_D}$$

where $J_0^{-1}(.)$ is the inverse Bessel function of order zero. As predicted by the preceding analysis, the coherence time is inversely related to the Doppler spread.   ■

The two inversely related parameters of coherence time and Doppler spread provide one measure of the channel characteristics. If the coherence time is finite, then the channel is *time selective*, that is, *time varying*. If the coherence time is infinite, then the channel is *time flat*, that is, *time invariant*.

2.7.6   Power-Delay Profile

With a WSSUS channel, we found that the autocorrelation of the channel impulse response is given by Eq. (2.95); that is, the wide-sense stationary assumption implies that there is a dependence only on Δt and not on the pair (t_1, t_2), while the uncorrelated scattering assumption implies that τ_1 and τ_2 must align for nonzero correlation. We define the function

$$P_h(\tau) = R_h^w(0; \tau) \qquad (2.103)$$

based on $R_h^w(\Delta t; \tau)$ with $\Delta t = 0$. This can be shown to be equivalent to $\mathbf{E}[|h(t, \tau)|^2]$ for the WSSUS channel. The function $P_h(\tau)$, which is known as the *multipath intensity profile* or *power-delay profile* of the channel, provides an estimate of the average multipath power as a function of the relative delay τ.

EXAMPLE 2.12   Discrete Power-Delay Profile

Consider the situation in which the channel impulse response is given by

$$\tilde{h}(t, \tau) = \sum_{l=1}^{L} \tilde{\alpha}_l(t) \delta(\tau - \xi_l) \qquad (2.104)$$

Then, from Example 2.10, the corresponding autocorrelation function of $\tilde{h}(t, \tau)$ is given by Eq. (2.99); that is,

$$R_h(t_1, t_2; \xi_1, \xi_2) = \left[\sum_{l=1}^{L} P_l J_0(2\pi f_D \Delta t) \delta(\tau_1 - \xi_l) \right] \delta(\tau_1 - \tau_2) \qquad (2.105)$$

and the power-delay profile is given by

$$P_h(\tau) = \sum_{l=1}^{L} P_l \delta(\tau - \xi_l) \qquad (2.106)$$

That is, at the delay ξ_l, the power-delay profile is the average power of the *l*th path, as would be expected from the definition.   ■

The different rays that compose a multipath signal will travel different paths between the transmitter and the receiver. These paths will often have different lengths, and there will be correspondingly different transmission delays. Intuitively, we would

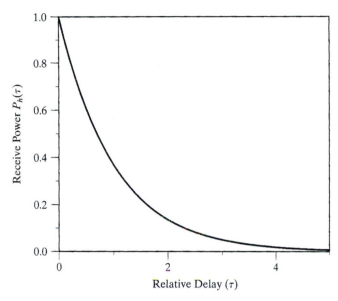

FIGURE 2.24   Sample average power-delay profile.

expect signals at greater delays to be weaker simply because they have traveled greater distances. This is often the case, but not always: the actual propagation losses depend on the reflection coefficients and diffraction effects. The multipath intensity is sometimes modeled with an exponential power-delay profile, as shown in Fig. 2.24. That is, the average power decreases exponentially with relative delay. Sample channels may vary significantly, but the power-delay profile refers to the power as a function of delay, averaged over a large number of channels.

The power-delay profile may be characterized by two statistical parameters. With $P_h(\tau)$ representing the profile, the *average delay* is given by

$$T_D = \frac{1}{P_m} \int_0^\infty \tau P_h(\tau) d\tau \tag{2.107}$$

where the *average power* is

$$P_m = \int_0^\infty P_h(\tau) d\tau \tag{2.108}$$

The second central moment of the average power-delay profile is given by

$$\mu_2 = \frac{1}{P_m} \int_0^\infty (\tau - T_D)^2 P_h(\tau) d\tau$$

$$\tag{2.109}$$

$$= \frac{1}{P_m} \int_0^\infty \tau^2 P_h(\tau) d\tau - T_D^2$$

There are two parameters in common use to characterize this time dispersion or excess delay. The first is *delay spread,* defined by

$$S = \sqrt{\mu_2} \qquad (2.110)$$

which is simply the root-mean-square (rms) delay. The second parameter is the *multipath spread,* given by

$$T_M = 2\sqrt{\mu_2} = 2S \qquad (2.111)$$

The multipath spread is the two-sided rms delay, or the approximate width of the excess delay. The multipath spread is effectively the time interval over which the power-delay profile is nonzero.

2.7.7 Coherence Bandwidth

Just as the coherence time of a channel is related to the autocorrelation function of the channel impulse response, the coherence bandwidth of a channel is related to the autocorrelation function of the time-varying frequency response. Recall from Section 2.7.3 that the time-varying frequency response of the channel is given by the Fourier transform

$$H(t, f) = \mathbf{F}[\tilde{h}(t, \tau)] \qquad (2.112)$$

With this definition of the time-varying frequency response, we may define the autocorrelation function

$$R_H(f_1, f_2; \Delta t) = \mathbf{E}[H(t, f_1)H^*(t + \Delta t, f_2)] \qquad (2.113)$$

This is the *frequency-spaced, time-spaced correlation function;* it depends only on the time difference Δt due to the assumed WSS properties of the channel. It can be shown that the frequency-spaced, time-spaced correlation also depends only on Δf for a WSSUS channel, not on the individual f_1 and f_2, so we can define

$$P_H(\Delta f, \Delta t) = R_H(f, f + \Delta f; \Delta t) \qquad (2.114)$$

It can also be shown that there is a Fourier transform relationship between the two quantities $[P_h(\tau)]$ and $(P_H(\Delta f, 0))$, as shown by

$$P_H(\Delta f, 0) = \mathbf{F}[P_h(\tau)] \qquad (2.115)$$

The function on the left is the *coherence spectrum* of the channel. The function on the right is the Fourier transform of the power-delay profile. The coherence spectrum represents the average correlation between frequencies separated by Δf at any particular instant in time ($\Delta t = 0$). Consequently, the coherence spectrum is closely related to the coherence bandwidth of the signal. In particular, because of the inverse relationship between the time and frequency domains, we have the relationship

$$BW_{coh} \approx \frac{1}{T_M} \qquad (2.116)$$

That is, *the coherence bandwidth is inversely proportional to the multipath spread of the channel.*

The coherence bandwidth of the channel is the bandwidth over which the frequency response is strongly correlated—that is, relatively flat. If the coherence bandwidth is small with respect to the bandwidth of the transmitted signal, then the channel is said to be *frequency selective.* That is, insofar as the transmitted signal is concerned, the channel has a nonflat frequency response. When the coherence bandwidth is large with respect to the bandwidth of the transmitted signal, the channel is said to be *frequency nonselective* or *frequency-flat.*

2.7.8 Stationary and Nonstationary Channels

The key feature of wide-sense stationary uncorrelated-scattering (WSSUS) channels is that correlation of the channel response depends only on the time difference and not on the absolute time. These stationary models for channel characteristics are convenient for analysis, but often, except for short time intervals, they are not an accurate description of reality. For example, terrestrial mobile channels are usually highly nonstationary, for the following reasons, among others:

1. The propagation path often consists of several discontinuities, such as buildings, that can cause significant changes in the propagation characteristics (e.g., when a terminal moves in and out of the electromagnetic shadow of a building).
2. The environment itself is physically nonstationary. There may be moving trucks, moving people, or other elements of the environment that can significantly affect propagation. An example is the effect of a person's head on the propagation path for a cell phone user as the user nods or changes position.
3. The interference caused by other users sharing the same frequency channel will vary dynamically as these other users come onto and leave the system.

All of these factors contribute to the nonstationarity of the link. Different types of radio links have different degrees of nonstationarity; nevertheless, some wireless links, such as fixed satellite links, can often be considered essentially stationary.

Stationary models still have their application. Even though many communications links are highly nonstationary, the WSSUS model provides a reasonably accurate account of the propagation characteristics for short periods of time. For many applications, this is sufficient. For example, with digitized voice transmissions, the size of a voice frame is often on the order of 20 milliseconds. Since communications link can frequently be considered stationary over many voice frames, the WSSUS model is an important tool for analyzing system performance.

2.7.9 Summary of Channel Classification

This section introduced several classifications of wireless channels:

- *Time-flat channels* are time-invariant channels. An example is a transmitter and receiver that are both physically stationary, with the propagation environment unchanging.
- *Frequency-flat channels* have a frequency response that is approximately flat over a bandwidth greater than or equal to the bandwidth of the transmitted signal.
- *Time-selective channels* are time-varying channels. An example is a wireless terminal moving through the environment and undergoing Rayleigh fading.
- *Frequency-selective channels* have a frequency response that cannot be assumed to be flat over the bandwidth of the signal. Frequency selectivity is due to a multipath that has significant delay spread relative to the symbol period of the transmission.

The classification of a channel depends on the signal that is to be transmitted. Narrowband mobile channels are often flat-fading (flat frequency, time selective), because the coherence bandwidths of most radio channels at transmission frequencies less than 1 gigahertz are typically tens of kilohertz. The exception to this is HF radio, wherein due to propagation modes such as ducting through the ionosphere, even very narrow channels can be frequency selective. Wideband channels, in which either the transmitter or the receiver is moving, are often both time selective and frequency selective.

For data transmission systems, there are two parameters of interest:

- *Coherence time* is the maximum duration for which the channel can be assumed to be approximately constant. If a transmitted data symbol has a duration that is less than the coherence time, it should suffer little distortion from the effects of frequency dispersion.
- *Coherence bandwidth* is the maximum transmission bandwidth over which there is little variation. A signal contained within the coherence bandwidth should suffer little distortion from the effects of time dispersion.

If we assume that the channel is wide-sense stationary with uncorrelated scattering (WSSUS), then the coherence time is inversely proportional to the Doppler spread of the channel and the coherence bandwidth is inversely proportional to the delay spread of the channel.

Problem 2.11   Measurements of a radio channel in the 800 MHz frequency band indicate that the coherence bandwidth is approximately 100 kHz. What is the maximum symbol rate that can be transmitted over this channel that will suffer minimal intersymbol interference?

Ans. The multipath spread of the channel is approximately 10 microseconds. If we assume that spreading of the symbol by less than 10% causes negligible interference, then maximum symbol period is 100 microseconds and the corresponding symbol rate is 10 kHz. ∎

Problem 2.12   Calculate the *rms* delay spread for an HF radio channel for which

$$P(\tau) = 0.6\delta(\tau) + 0.3\delta(\tau - 0.2) + 0.1\delta(\tau - 0.4)$$

where τ is measured in milliseconds. Assume that signaling with a 5-kHz bandwidth is to use the channel. Will delay spread be a problem? That is, is it likely that some form of compensation (an equalizer) will be necessary?

Ans. The rms delay spread is 0.29 millisecond and the coherence bandwidth is 1720 Hz, so an equalizer is likely necessary. ∎

Problem 2.13   Show that the time-varying impulse response of Eq.(2.84) and a time-invariant impulse response are related by

$$\tilde{h}_{\text{time–invariant}}(t) = \tilde{h}_{\text{time–varying}}(t, t)$$

Explain, in words, what the preceding equation means. ∎

2.8   NOISE AND INTERFERENCE

The primary objective of the previous sections is to estimate the received power level and characterize its behavior as a function of the environment. *What sets the minimum acceptable received power level is the receiver noise and interference levels.* In general, noise can be defined as unwanted (and usually uncontrollable) electrical signals interfering with the desired signal. Unwanted signals arise from a variety of sources, both natural and artificial. Artificial sources include noise from automobile ignition circuits, commutator sparking in electric motors, 60-cycle hum, and signals from other communications systems. Numerous artificial sources result from harmonics of the natural frequency. For example, spark plug firings in car engines have a frequency on the order of thousands of rpm; consequently, although the fundamental frequency is less than 1 kHz, the energy emitted by this excitation is often so strong that high-order harmonics can cause significant interference in radio systems. Natural sources of noise include circuit noise, atmospheric disturbances, and extraterrestrial radiation. In this section, we will characterize various sources of noise. Later sections will show how they influence communication system design.

2.8.1   Thermal Noise[14]

Thermal noise can be considered a fundamental property of matter above the absolute temperature of 0°K. In an insulator, electrons are bound to the atom. In a conductor, such as a wire or a resistor, the attraction between the electrons and the nucleus is not as strong, and electrons are free to move from one atom to another. In any conductor at a temperature above 0°K, these free electrons are never really stationary, but move about with random velocities in all directions. These random motions have an average velocity that is zero in any direction over a long period of time. Over short intervals, there are statistical fluctuations from this zero average—equivalent to intermittent currents. We call these spontaneous fluctuations *thermal noise.*

Consider the voltage that is observed between the two terminals of a metallic resistor. The average voltage is zero. The mean-square voltage is nonzero, due to thermal noise. Quantum physics provides an expression for this mean-square voltage namely,

$$\bar{v}^2 = \frac{2\pi^2 k^2 T^2}{3h} R \tag{2.117}$$

where

R is the resistance in ohms

$k = 1.37 \times 10^{-23}$ W-s/°K is Boltzman's constant

$h = 6.62 \times 10^{-34}$ W-s^2 is Planck's constant

T is the absolute temperature of the resistor in degrees Kelvin

Consequently, the resistor can be modeled as equivalent to the Thévenin circuit shown in Fig. 2.25, in which the resistor is ideal (noiseless) and the voltage source has zero mean and a mean-square voltage given by Eq. (2.117). Since the voltage is the sum of the motions of a large number of electrons, it is accurately modeled as a *Gaussian distribution* with zero mean. The mean-square (m.s.) voltage across the resistor terminals is finite and depends on the resistance R. The maximum power delivered by the circuit of Fig. 2.25 occurs when the load resistance is matched to R. By dividing both sides of Eq. (2.117) by R and scaling by a factor of 1/4, we obtain an expression for the maximum available noise power.

Any communications link occupies only a very narrow portion of the whole electromagnetic spectrum. What is of interest to communications engineers is the distribution of thermal noise power across frequency, or $S_n(f)$—that is, the *noise spectral density*. Quantum physics provides an expression for this as well. The important result is that, *for frequencies up to 10^{12} Hz, the available thermal noise spectral density (watts/Hz) is approximately constant* and is given by

$$S_n(f) = kT/2$$
$$\equiv N_0/2, \qquad -\infty < f < \infty \tag{2.118}$$

Equation (2.118) describes the noise power delivered to a matched load; it is independent of the resistance. This upper limit of 10^{12} Hz falls in the near-infrared portion of the electromagnetic spectrum, far above where conventional electrical components have ceased to respond. Above these frequencies, the noise power decreases exponentially as a function of frequency.

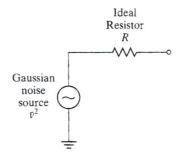

FIGURE 2.25   Thévenin model of a resistor.

On the basis of Eq. (2.118), we use the *two-sided spectral model* for thermal noise depicted in Fig. 2.26. This model shows that thermal noise has a flat spectrum across all frequencies (positive as well as negative), with density $N_0/2$. This type of noise is referred to as *white noise*, since it contains all frequencies at an equal level. The amplitude distribution at any frequency is Gaussian, and there is no correlation between samples of the equivalent time-domain noise process $n(t)$. That is, for a *real* white-noise process, the autocorrelation function is given by

$$R_n(\tau) = E[n(t)n(t+\tau)]$$

$$= \frac{N_0}{2}\delta(\tau) \tag{2.119}$$

The noise spectral density is conventionally defined as a two-sided quantity $N_0/2$ applicable over the range $-\infty < f < \infty$. In practice, the single-sided value N_0 is often used, for two reasons:

- Conventional instrumentation measures only positive frequencies.
- In Chapter 3, it will be shown that noise has a complex envelope representation similar to that of any other signal. In the case of noise, both the in-phase and quadrature components are white, Gaussian, and independent of each other. The corresponding complex baseband process $\tilde{n}(t)$ has a two-sided noise spectral density of N_0, and autocorrelation is $R_{\tilde{n}}(\tau) = N_0\delta(\tau)$.

The noise power in a bandwidth B is

$$P = (N_0/2)(2B)$$

$$= N_0 B \tag{2.120}$$

where the first equality corresponds to the two-sided representation and the second to the single-sided representation. If a filter has a frequency response $H(f)$, then we define the *noise-equivalent bandwidth*, or *noise bandwidth*, of the filter as

$$B_{eq} = \frac{\displaystyle\int_{-\infty}^{\infty} |H(f)|^2 (N_0/2)df}{N_0/2} \tag{2.121}$$

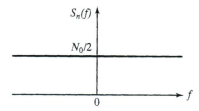

FIGURE 2.26   Spectral model for white noise.

The noise bandwidth is equivalent to the bandwidth of an ideal rectangular filter that would pass the same amount of noise power. For many filters, the noise bandwidth is approximately equal to their *3-dB bandwidth*. We may define the noise-equivalent bandwidth of a signal in a similar manner.

Problem 2.14 What would be the rms voltage observed across a 10-$M\Omega$ metallic resistor at room temperature? Suppose the measuring apparatus has a bandwidth of 1GHz, with an input impedance of 10 $M\Omega$. What voltage would be measured then? Compare your answer with the voltage generated across the antenna terminals by the signal defined in Problem 2.5. Why is the avoidance of large resistors recommended for circuit design? What is the maximum power density (W/Hz) that a thermal resistor therefore delivers to a load?

Ans. The maximum rms voltage is 1.6 volts. Over 1 GHz into a 10-$M\Omega$ load, the voltage is approximately 6 millivolts. The maximum power density delivered to a matched load is $kT = 4 \times 10^{-21}$ watt/Hz at room temperature. ∎

2.8.2 Equivalent Noise Temperature and Noise Figure

There are other sources of noise that exhibit behavior similar to that of thermal noise. That is, the noise produced by these sources is Gaussian in distribution and white over the range of frequencies of interest. For this reason, the concept of an *equivalent noise temperature* is often assigned to these sources. The equivalent noise temperature of a device is not necessarily its physical temperature, but rather a measure of the noise that the device produces.

EXAMPLE 2.13 Earth's Radiation

The antenna of a satellite looks directly at the Earth in order to receive signals emitted by ground-based terminals. The Earth has an average temperature of approximately $T_0 = 290°K$ and, like any warm body, radiates energy that is proportional to this temperature. The radiated energy is effectively white across the frequency band. Any of this energy that falls within the antenna aperture is collected and added to the received signal, in which case we say the antenna has an *equivalent noise temperature* of T_0. ∎

Amplifiers add noise to the signal in much the same way that passive devices such as resistors do. The noise added depends on the amplifier design and must be calibrated for each device. Although amplifiers could be specified in terms of their equivalent noise temperature, they are conventionally specified in terms of their *noise figure*. One reason for this alternative representation is that amplifiers usually have a finite operational bandwidth; consequently, the noise effects may be frequency dependent. For applications considered in this book, we will assume that noise effects are white.

The *noise figure* represents the increase in noise at the output of the amplifier, referenced to the input. Let $G(f)$ be the available power gain of the device as a function of frequency. G is defined as the ratio of the available signal power at the output of the device to the available signal power of the source when the source is a sinusoid of frequency f. The noise figure is then defined as

$$F = \frac{S_{no}(f)}{G(f)S_{ni}(f)} \qquad (2.122)$$

where $S_{no}(f)$ is the spectrum of the output noise power and $S_{ni}(f)$ is the spectrum of the input noise power. For an ideal noiseless amplifier, the noise figure is unity; that is, the amplifier simply amplifies the input noise, but adds no noise itself. Equation (2.122) permits the calculation of the noise figure dependence on the frequency f if necessary.

The noise figure is usually expressed in decibels, but for very low-noise amplifiers such as those used as receiver front ends in satellite communications, the noise figure is often very close to 0 dB. The decibel scale makes it difficult to distinguish between the performance of the different amplifiers in this case. In these situations, it is often preferable to use the equivalent noise temperature of the amplifier. That is, equivalent noise temperature and noise figure are equivalent methods of specifying the noise contribution of a device.

To show the relationship between noise figure and equivalent noise temperature, assume that the amplifier has a power gain G that is constant; this is referred to as a *flat spectral response*. A metallic resistor is connected across the input terminals. The available noise power at the output due to the input noise is GkT_0. The total output noise power is defined as $G(kT_e + kT_0)$, where T_e is the equivalent noise temperature of the device. This noise model is shown in Fig. 2.27. Consequently, the noise figure can be written as

$$F = \frac{G(kT_e + kT_0)}{GkT_0} = \frac{T_e + T_0}{T_0} \qquad (2.123)$$

where T_0 is the standard temperature of 290°K. The relationship between the equivalent noise temperature T_e and the noise figure is given by

$$T_e = (F - 1)T_0 \qquad (2.124)$$

EXAMPLE 2.14 Noise Figure and Receiver Sensitivity

Suppose a wireless receiver has a noise figure of 8 dB and includes a modem that requires an SNR of 12 dB for proper operation in a 5-kHz bandwidth. What is the receiver sensitivity?

The noise power due to the receiver in a bandwidth B is given by

$$N = F(kT_0)B$$

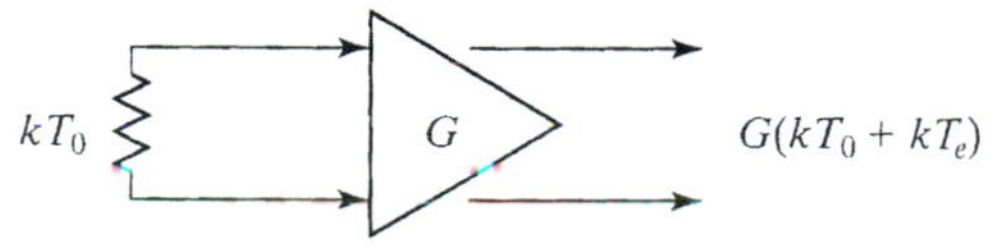

FIGURE 2.27   Noise model of amplifier.

where F is the noise figure of the receiver. In the decibel representation, this equation becomes

$$N(\text{dBm}) = k(\text{dBm} - \text{sK}^{-1}) + T_0(\text{dBK}) + F(\text{dB}) + B(\text{dBHz})$$
$$= -198.6 + 10\log_{10}290 + 8 + 10\log_{10}5000$$
$$= -129.0 \text{ dBm}$$

where we have used the following equivalences for Boltzmann's constant:

$$k = 1.37\times10^{-23}\text{Watt} - \text{sec}°\text{K}$$
$$= -228.6\text{dBW} - \text{s}/\text{K}$$
$$= -198.6\text{dBm} - \text{s}/\text{K}$$

The corresponding receiver sensitivity is given by

$$S = \text{SNR} \times N$$

So, in a decibel representation, the sensitivity (i.e., the minimum signal level that the receiver can reliably detect) is

$$S(\text{dB}) = \text{SNR}(\text{dB}) + N(\text{dBm})$$
$$= 12 + (-129)$$
$$= -117 \text{ dBm}$$ ∎

Problem 2.15   The noise figure of a cell phone receiver is specified as 16 dB. What is the equivalent noise temperature? Assume that reliable detection of a 30-kHz FM signal by this receiver requires an SNR of 13 dB. What is the receiver sensitivity in dBm?
Ans. The equivalent noise temperature is 11,255°K and the receiver sensitivity is –100 dBm.   ∎

2.8.3   Noise in Cascaded Systems

A communications receiver is composed of a number of components, including an antenna, amplifiers, filters, and mixers, and these components may be connected by transmission lines. The overall *system temperature* involves noise contributions from these components in a weighted manner. In this section, we show how to combine the noise contributions of these different portions of the system.

We model each of the aforementioned devices with the two-port model shown in Fig. 2.28, where each device has a noise factor F and a gain (or loss) G. With this model, the input to the device is summed with an internal noise source. The spectral density of the internal noise source is $(F - 1)kT_0$. The combined signal is then processed

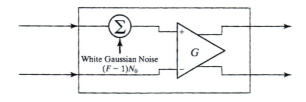

FIGURE 2.28   Two-port model for a communications system component with $N_0 = kT_0$.

by an ideal (noiseless) amplifier with gain G. If $S(f)$ is the device input, then the output of the circuit is given by

$$Y(f) = GS(f) + G(F-1)kT_0 \qquad (2.125)$$

That is, the output has both a signal component and a noise component.

Now consider a system with two such devices, having noise factors F_1 and F_2 and gains G_1 and G_2, respectively, as shown in Fig. 2.29. To determine the noise figure of the cascaded system, we assume that the input is connected to a metallic resistor with noise density N_0.

Under these assumptions, the output of the first stage and input to the second stage is $F_1G_1N_0$. The output of the second stage is $G_2\{(F_2-1)N_0 + F_1G_1N_0\}$. In accordance with the definition of noise figure given in Eq. (2.122), if we compare the input and output of the cascaded devices, the overall noise figure is given by

$$F = \frac{G_2\{(F_2-1)N_0 + F_1G_1N_0\}}{G_1G_2N_0}$$

$$= F_1 + \frac{F_2-1}{G_1} \qquad (2.126)$$

That is, *the combined noise figure is the noise figure of the first device, plus the noise figure of the second device attenuated by the gain of the first device.* In the general case, the noise figure for a multistage system is given by

$$F = F_1 + \frac{F_2-1}{G_1} + \frac{F_3-1}{G_1G_2} + \cdots, \qquad (2.127)$$

Correspondingly, the equivalent noise temperature of the system is

$$T_{\text{sys}} = T_1 + \frac{T_2}{G_1} + \frac{T_3}{G_1G_2} + \cdots. \qquad (2.128)$$

where T_k and G_k are the noise temperature and gain of the kth stage of the receiver amplification chain, with $k = 1,2,\ldots$. If we include the equivalent noise temperature of the antenna in Eq. (2.128), the equation becomes

$$T_{\text{sys}} = T_A + T_1 + \frac{T_2}{G_1} + \frac{T_3}{G_1G_2} + \cdots. \qquad (2.129)$$

where T_A is the equivalent noise temperature of the antenna.

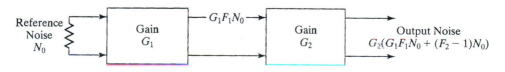

FIGURE 2.29   Illustration of cascaded noise calculation for two-device system.

EXAMPLE 2.15 System Noise Figure Calculation

Suppose the antenna of Example 2.13 is connected to an amplifier stage with a gain of 40 dB and an equivalent noise temperature of 50°K. A subsequent amplifier stage has a gain of 30 dB and noise figure of 5. What is the system noise figure?

From Example 2.13, the antenna has an equivalent noise temperature of T_0 or, equivalently, a noise figure of 2 and a gain of 1. For the first amplifier, the noise figure is

$$F_1 = 1 + T_e/T_0 = 1.17$$

The system noise figure is then

$$F_{sys} = F_{ant} + (F_1 - 1) + \frac{(F_2 - 1)}{G_1}$$

$$= 2 + (1.17 - 1) + \frac{5}{10^4} = 2.17$$

From this example, it should be clear that it is the noise performance of the first amplifier stage, often referred to as the *low-noise amplifier* (LNA), that is critical to communications performance. ∎

Problem 2.16   Show that the system temperature, in general, is given by Eq. (2.129). ∎

2.8.4 Man-Made Noise[15]

Natural background and internal noise sources may be minimized with good system and antenna design. Beyond this, artificial noise levels may set the sensitivity limit of a receiver. Some artificial sources of noise may be other communications systems operating in the same frequency band; other sources may be systems that operate in other frequency bands, but unintentionally generate RF signals in the band of interest. In this section, we will consider three such sources.

Devices that produce large electrical discharges often cause *impulse noise*. The electromagnetic pulses may have durations of 5 to 10 nanoseconds and may occur at rates up to several hundred pulses per second. Due to the short pulse duration, impulse noise is often modeled as a broadband type of noise. The impulses vary in their strength and frequency. Some impulses have power levels well above the receiver noise floor, but may occur infrequently. Impulses of weaker strength occur more frequently. The spectral density of the impulse power decreases with frequency, being greater in frequency bands below 500 MHz and decreasing rapidly at high frequencies.

The sources of impulse noise include the following:

- Noise from electrical machinery (particularly commutating motors)
- Noise from spark ignition systems in automobile or other internal combustion engines
- Switching transients
- Discharge lighting.

A general characterization of impulse noise is difficult to come by, but impulse noise is a nonstationary phenomenon that depends on the frequency of operation of the source, the time of day, the surrounding environment, and various other factors. Accurate modeling requires measurements in the frequency band of interest, in differing environments, and at different times of day, week, and year.

The impact of impulse noise on digital communications systems depends on the impulse characteristics, such as pulse power, duration, and repetition rate. The impulse noise can affect both the front-end and intermediate stages of a receiver and can cause variations in signal strength, synchronization loss, and message errors.

A second type of man-made noise or interference is *out-of-band transmissions* from other communicationslike services. These out-of-band transmissions include the harmonics of high-power radars and television signals, services that may be using a highly separate frequency band, but that have a second or third harmonic which may fall directly into the communications band of interest. Of particular concern are examples where one service transmits at much higher levels than another. For example, the Global Positioning Satellite Service (GPSS) operates at very low power levels in the frequency band centered on 1.476 GHz. The band at 738 MHz is assigned for wideband public safety messages, with base station transmit powers of up to 30W permitted. Clearly, the second harmonic of the latter signal must be sufficiently attenuated to prevent interference with the former. (See Problem 2.44.)

A third type of interference related to system design, *multiple access interference*, is discussed in the next section.

Problem 2.17   Microwave ovens operate at a natural frequency of the water molecule at approximately 2.45 GHz. This frequency falls in the middle of a band from 2.41 to 2.48 GHz that has been allocated for low-power unlicensed radio use, including wireless local area networks (sometimes referred to as Wi-Fi—see Section 5.16). The oscillators used in some microwave ovens have poor stability and have been observed to vary ±10 MHz around their nominal frequency. Discuss how such variability would affect the signal design for the use of this band.   ∎

2.8.5   Multiple-Access Interference

Due to a limited frequency spectrum, communication frequencies are reused the world over. Often, the rules for frequency reuse are subject to international agreements to minimize and control the effects of cross-border interference. An example is the assignment of television broadcast frequencies along international borders. In other cases, international agreements are in place for the use of common frequencies. For example, the VHF radio band, 118 MHz to 130 MHz, is used for aeronautical safety communication the world over. The channel frequencies and permitted modulation types within that band are part of the agreement.

For wireless communications, frequencies are often assigned by federal authorities on a regional basis. These same frequencies can then be assigned to another user or service provider in a different region, sufficiently distant that there is no interference. To maximize frequency reuse, reuse distances are often made as small as possible. As a consequence, a service is not interference free, but rather must have a certain interference tolerance. A typical example is an FM receiver that must tolerate a signal-to-interference ratio (SIR) of 20 dB at its edge of coverage, such as that illustrated in Fig. 2.30.

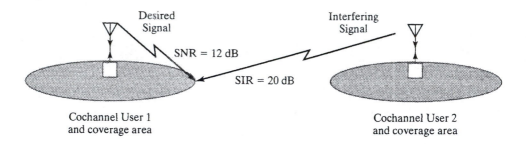

FIGURE 2.30   Example of traditional frequency reuse.

Frequency reuse policies have made significant advances in recent years with the introduction of *cellular systems*, wherein a single service provider is given a band of frequencies for a given service area. The service provider tries to reuse the frequencies as many times as possible within the given service area in order to meet customer demands. In a cellular system, the cells are often modeled with a hexagon pattern, as shown in Fig. 2.31, with users in each cell communicating with a base station located at the center of the cell. Cellular telephones connect the mobile user wirelessly to the base station, which acts as an interface to the wired telephone network; or it could potentially connect the user to another mobile user. The term *cellular* is usually applied to terrestrial systems, but similar considerations apply to multibeam satellites. For nongeostationary satellite systems, there is the added issue that the cells move with respect to the earth as the satellite moves.

The hexagonal cell distribution is an approximation of practical cellular patterns that is useful for illustration because of its ease of analysis. With a hexagon geometry,

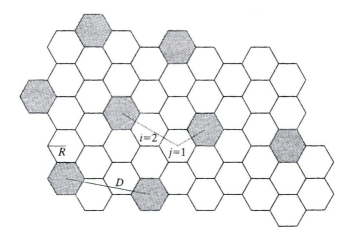

FIGURE 2.31   Hexagonal pattern of cells used in cellular telephone system.

the reuse pattern can be defined relative to a given reference cell as follows: *Move i cells along any chain of hexagons, and turn clockwise 60 degrees; then move j cells along the chain that lies on this new heading,* as shown in Fig. 2.31. The *j*th cell and the reference cell are *cochannel cells.* That is, the same frequency channels are used in each of these cells. With this hexagonal geometry, the cells form natural *clusters* around the reference cell and each of its cochannel cells. Each cell in a cluster uses different frequencies, but all available frequencies are used in each cluster. The number of cells per cluster is given by

$$N = i^2 + ij + j^2 \qquad (2.130)$$

For the geometry shown in Fig. 2.31, with $i = 2$ and $j = 1$, there are seven cells per reuse cluster. The ratio of D, the distance between the centers of nearest-neighboring cochannel cells, to R, the cell radius, is the *normalized reuse distance* and is given by

$$\frac{D}{R} = \sqrt{3N} \qquad (2.131)$$

With different choices of i and j, we obtain *reuse factors* of 1-in-3, 1-in-4, 1-in-7, 1-in-9, 1-in-12, etc. In terrestrial systems, cochannel isolation is determined by propagation losses; consequently, the reuse distance is a function of the propagation loss. (For satellite systems, the reuse distance is a function of the isolation between the spot beams that form the cells.) In the preceding expression, the reuse distance D is a function of both the cell radius and N, the number of cells per cluster.

Recall our model for the median propagation loss between a transmitter and a receiver from Section 2.4.1, which is given by

$$P_R = \frac{\beta P_T}{(r/r_0)^n} \qquad (2.132)$$

where β is the loss at distance r_0 and n is a path-loss exponent that typically ranges from two to five, depending on the propagation environment (see Table 2.1). For users with similar modulation and transmit power, and with similar propagation environments, the mean carrier-to-interference ratio can be approximated by

$$
\begin{aligned}
\frac{C}{I} &= \frac{\text{Received power of desired user}}{\sum \text{Received powers of interfering users}} \\[2mm]
&= \frac{\beta P_T / (r_d / r_0)^{n_d}}{\sum_{k \neq d} \beta P_T / (r_d / r_0)^{n_k}} \\[2mm]
&= \frac{r_d^{-n_d}}{\sum_{k \neq d} r_k^{-n_k}}
\end{aligned}
\qquad (2.133)
$$

where the subscript d corresponds to the desired user and $k \ne d$ corresponds to interfering cochannel users. Equation (2.133) assumes that βP_T is the same for all users. The range r_k is the range of the kth transmitter to the receiver of interest.

With frequency-division multiple access (FDMA) strategies, frequencies are not reused in each cell, since that would cause excessive *cochannel interference*. (FDMA was discussed briefly in Chapter 1, and will be discussed in greater detail in Chapter 3.)

EXAMPLE 2.16 Minimum Reuse Pattern

Consider a cellular system with a one-in-N reuse pattern, where the base station receivers are at the center of each cell. The desired user is at the edge of the cell. Assume that the six closest interferers in a cellular system cause most of the interference and that these interferers are at the centers of their cells. What is the best reuse pattern that achieves a carrier-to-interference ratio of $(C/I)_{min}$?

From Eq. (2.133), for the desired user at the edge of the cell, $r_d = R$, and for all cochannel users, $r_k \approx D$. So

$$\frac{R^{-n}}{6D^{-n}} \ge \left(\frac{C}{I}\right)_{min} \tag{2.134}$$

where R is the cell radius, D is the reuse distance, and n is the path-loss exponent. It follows from Eqs. (2.131) and (2.134) that the number of cells per cluster (the inverse of the *frequency reuse factor*) has a lower bound of

$$N \ge \frac{1}{3}\left[6\left(\frac{C}{I}\right)_{min}\right]^{2/n} \tag{2.135}$$

If the minimum tolerable carrier-to-interference ratio for an analog communication system is 18 dB and the path-loss exponent is 3.1, then

$$N \ge \frac{1}{3}[6\times10^{18/10}]^{2/3.1} = 15,$$

The smallest value of N satisfying both this equation and Eq. (2.130) is $N = 19$, with $i = 2$ and $j = 3$. That is, to achieve the desired C/I ratio, the smallest cluster size and reuse factor is 19. ∎

Problem 2.18   The minimum tolerable C/I ratio depends on the modulation and coding strategy and the quality of service required. Suppose a narrowband digital system requires a $C/I = 12$ dB. What would be the maximum frequency reuse factor? If the addition of forward error-correction coding would reduce this to 9 dB without increasing the signal bandwidth, what would be the relative improvement in reuse factor? Assume that the propagation loss exponent is 2.6.
Ans. For a C/I ratio of 12 dB, the minimum reuse is 12. For a C/I ratio of 9 dB, the minimum reuse is 7. ∎

2.9   LINK CALCULATIONS

2.9.1   Free-Space Link Budget

In designing a system for reliable communications, we must perform a link budget calculation to ensure that sufficient power is available at the receiver to close the link and meet the SNR requirement.

The basis for the link budget is the Friis equation (2.11) described in Section 2.2.3. This equation describes the relationship between the received power and the transmitted power under given assumptions about propagation. The equation does not include the effects of noise. As we have seen, there are many sources of noise in wireless communications: receiver noise, antenna noise, artificial noise, and multiple access interference. In what follows, we assume that the dominant effect is receiver noise and that this noise is characterized by the single-sided noise spectral density N_0.

To include the effects of noise, we divide each side of the Friis equation by the noise density N_0. For free-space propagation, the basic link budget equation is given by

$$\frac{P_R}{N_0} = \frac{P_T G_R G_T}{L_p k T_e} \tag{2.136}$$

where $N_0 = kT_e$ and T_e is the equivalent noise temperature of the system. For satellite applications, this equation is often written as

$$\frac{C}{N_0} = \text{EIRP} - L_p + (G/T) - k \tag{2.137}$$

where all quantities are expressed in decibels and

$C/N_0 = P_R/N_0$	is the received carrier-to-noise density ratio (dB-Hz);
$\text{EIRP} = G_T P_T$	is the *equivalent isotropic radiated power* of the transmitter (dBW);
L_p	is the path loss (dB);
$G/T = G_R/T_e$	is the ratio of receiver antenna gain to noise temperature (dB-K^{-1});
k	is Boltzmann's constant (-228.6 dBW-sK^{-1}).

The C/N_0 ratio is one of a number of equivalent ways of expressing the SNR. Indeed, in many cases it is the preferred choice, because it makes no assumption about the underlying modulation strategy.

EXAMPLE 2.17   *G/T* of a Satellite

Consider a geosynchronous satellite whose global beam covers all of the visible Earth's surface. The radius of the Earth is about 6400 km, and the altitude of the satellite is 36,000 km.

Consequently, the satellite antenna gain, relative to the isotropic situation, is equivalent to the inverse ratio of the cross-sectional area of the earth to the surface area of a sphere at 36,000 km, assuming 100 percent efficiency; that is,

$$G_{sat} \approx \frac{4\pi R_{alt}^2}{\pi R_{earth}^2} = \frac{4\pi \times 36,000^2}{\pi \times 6400^2} = 126.6 \tag{2.138}$$

Expressed in decibels, the satellite antenna (power) gain is 21.0 dB. The satellite antenna is looking directly at the Earth and thus has an equivalent noise temperature of approximately 290°K. Usually, the satellite receiver noise can be considered small relative to the antenna noise. Consequently, the G/T ratio of the satellite receiving system is

$$
\begin{aligned}
(G/T) &= G_{sat}(dB) - 10\log_{10}T_{ant} \\
&= 21.0 - 24.6 \\
&= -3.6 \text{ dBK}^{-1}
\end{aligned}
$$

The system has a G/T ratio of -3.6 dBK^{-1}. ∎

Problem 2.19 A fixed satellite terminal has a 10-m parabolic dish with 60% efficiency and a system noise temperature of 70°K. Find the G/T ratio of this terminal at 4 GHz. Suppose it was a terrestrial mobile radio using an omnidirectional antenna. What would you expect the equivalent noise temperature of the mobile antenna to be?
Ans. The satellite terminal G/T is 31.7 dBK^{-1}. The mobile antenna has a noise temperature of approximately 290°K. ∎

EXAMPLE 2.18 Link Budget from Earth Station to Satellite.

A simple link budget for transmitting from a ground Earth station to a satellite is provided in Table 2.3. The table is broken down into four subsections that can be loosely identified with the transmitter, the transmission path, the receiver, and the combined system. We break down the line items of the table according to these divisions:

1. *Earth station transmitter*, which identifies:
   - the transmit frequency from the ground Earth station to the satellite.
   - the transmitted power transmitted by the Earth station (as measured at the transmit antenna terminals).
   - the antenna gain. If the antenna is a parabolic dish, then the power gain can be determined from the diameter and the transmit frequency.
   - the transmit EIRP is the dB sum of the transmit power and the antenna gain.

2. *Losses*, which identify:
   - the satellite elevation. For a geostationary satellite, this is the only geometric information required to compute the range of the satellite.
   - the satellite range, as computed from geometrical considerations. (See Problem 2.20.)
   - the free-space loss, which is determined from the satellite range and transmit frequency.

TABLE 2.3   Sample link budget from Earth station to satellite.

Parameter	Units		Comments
Earth station transmitter			
Transmit frequency	GHz	6.0	
Transmit power	dBW	−10	100-mW transmitter
Transmit antenna gain	dBi	30	60% efficiency for a 10-m parabolic dish
Transmit EIRP	dBW	20	
Losses			
Satellite elevation	°	45	
Satellite range	km	37,630	
Free-space loss	dB	193.5	$=20 \times \log_{10}(2\pi r/\lambda)$
Satellite receiver			
Antenna gain	dB	23.0	
Received power	dBm	−150.5	
Receive G/T	dB-K	−2.5	Global beam looking at Earth (290°K)
System			
Boltzmann's constant	dBW-K	−228.6	
Uplink C/N_0	dB-Hz	52.6	From link equation (2.137)

3. *The satellite receiver*, which identifies:
   - the satellite antenna gain. This also depends on the transmit frequency.
   - the received power level. This is the dB sum of the transmit EIRP, the free-space loss, and the receive antenna gain.
   - the satellite G/T. This empirically provided parameter characterizes the satellite design. In satellite systems, a terminal is usually specified in terms of its G/T rather than in terms of separate gain and system noise components.

4. *System*, which identifies
   - Boltzmann's constant, which is necessary to convert a system temperature into an equivalent noise density.
   - the uplink C/N_0, as determined by Eq. (2.137).

The conclusion from this table is that the link from Earth station to satellite will support any signaling system that requires a C/N_0 ratio of 52.6 dB-Hz or less.   ∎

Problem 2.20   For a geostationary satellite at altitude h (36,000 km), determine a formula relating the range r from the satellite to an earth station to the satellite elevation ϕ relative to the earth station. (Let $R_e = 6400$ km be the radius of the Earth.)
Ans. The range is

$$r = R_e \sin\phi + \sqrt{(R_e + h)^2 + R_e \cos^2\phi}$$ ∎

EXAMPLE 2.19 Satellite-to-Mobile Terminal Link Budget

In Table 2.4, we show a more detailed link budget for the link from the satellite to a mobile Earth station and combine it with the link results of the previous example. In this case, the satellite transmit power is proportional to the received power obtained from Table 2.3. Table 2.4 is broken down into subsections similar to those in Table 2.3, but we have included a number of additional effects corresponding to a more realistic link budget:

1. *Satellite transmitter*, which includes
   - the transmit frequency.
   - the receive power. Since this is a two-hop system, the transmit power is determined by the power received at the satellite from the earth station (Example 2.18).
   - the satellite gain. This depends on the satellite design and is an empirically provided parameter.
   - the transmit antenna gain. This again depends on the antenna size and the transmit frequency.
   - transmit EIRP. For the satellite, this is the dB sum of the satellite-received power, the satellite amplifier gain, and the satellite transmit antenna gain.

2. *Losses*, which include
   - the elevation of the satellite with respect to the mobile terminal. This is typically different from the elevation of the satellite with respect to the (base) Earth station.
   - the range of the downlink. For mobile terminals, the link budget must consider the worst-case path loss. This is equivalent to the largest range or smallest elevation angle.
   - the free-space loss.
   - the absorption loss due to the atmosphere. There are losses in RF signal strength due to atmospheric absorption, which will be largest at the lowest elevation angles, because, at those angles, the signal passes through more of the atmosphere.

3. *Earth station receiver*, which includes
   - the receive G/T.
   - implementation losses. These losses, which are due to nonideal implementation of the receiver/modem, can include filtering losses, synchronization losses, and distortion losses due to slight nonlinearities in receiver processing, finite precision effects, etc. They could also include losses due to interference, or the latter could be included as a separate line item.

4. *System*, which includes
   - Boltzmann's constant.
   - the downlink C/N_0 calculated with the use of Eq. (2.137).
   - the uplink C/N_0, which is included for reference purposes.
   - the combined C/N_0 for the two links. This calculation is described next.

The overall link has two components in this example: Earth station–to-satellite (uplink) and then satellite-to–mobile terminal (downlink). *The use of the terms* uplink *and* downlink *in satellite systems differs from those used in terrestrial systems.* We assume the satellite acts as a "bent pipe"; that is, it simply retransmits whatever it receives, and there is no onboard processing for regenerating the signal. If it is a bent-pipe satellite, the overall link carrier-to-noise ratio is given by

$$\frac{C}{N_0} = \left(\left(\frac{C}{N_0} \right)_{uplink}^{-1} + \left(\frac{C}{N_0} \right)_{downlink}^{-1} \right)^{-1} \tag{2.139}$$

TABLE 2.4   Downlink budget from satellite to mobile Earth station.

Parameter	Units		Comments
Satellite transmitter			
Transmit frequency	GHz	1.5	
Receive power	dBW	−150.5	From uplink budget
Satellite gain	dB	150	Internal satellite amplifiers
Transmit Antenna gain	dBi	26	Corresponds to 10-m parabolic dish on the satellite
Transmit EIRP	dBW	25.5	
Losses			
Elevation	°	20	Worst-case elevation angle
Range	km	39,809	
Free-space loss	dB	181.9	
Absorption loss	dB	0.2	Due to atmosphere
Earth station receiver			
Receive G/T	dBi	−22.6	Approximately omnidirectional in upper hemisphere
Receive implementation loss	dB	1.0	Loss due to nonideal receiver
System			
Boltzmann's constant	dBW-sK^{-1}	−228.6	
Downlink C/N_0	dB-Hz	48.6	From link equation
Uplink C/N_0	dB-Hz	52.6	From uplink budget
Overall C/N_0	dB-Hz	47.1	

This calculation sums the noise of both the uplink and the downlink, weighted according to the received power level. The calculation is done in absolute terms, not in decibels. The result is shown in the system section of Table 2.4. In this case, the C/N_0 ratio of the downlink is dominant, and we say that the *channel is downlink limited*.

The final step is to compare the overall C/N_0 ratio with that required by the modem delivering the service. If the C/N_0 ratio is less than the required one, then reliable communication is not possible. If the C/N_0 ratio is greater than the required one, we say that the link has a *margin*. Margin is desirable for a variety of reasons. Transmit power that varies with temperature and losses that vary with weather conditions are examples of other considerations that have not been included in the link budget. In the absence of further information, the margin is a tolerance available for uncharacterized transmission impairments. If further information were available about any of these effects, it could be included as part of the link budget. ∎

Problem 2.21   Numerous other quantities may be included in the satellite link budget. For example, if the satellite amplifier (which is typically nonlinear) is shared with a number of carriers, then *intermodulation distortion* will be generated. When there is a large number of equal-power carriers present, intermodulation distortion can be modeled as white noise with spectral density I_0. This will produce a term C/I_0 at the satellite that must be combined with the uplink and downlink C/N_0 to produce the overall $C/(N_0 + I_0)$. Given the uplink and downlink C/N_0 of Examples 2.18 and 2.19, what is the overall $C/(N_0 + I_0)$ if $C/I_0 = 50$ dB-Hz?

Ans. The $C/(N_0 + I_0)$ is 45.3 dB-Hz. ∎

2.9.2 Terrestrial Link Budget

For a terrestrial link budget, many of the line items are similar to those in the free-space link budget examples, but their calculations are different.

EXAMPLE 2.20 Terrestrial Link Budget

The link budget in Table 2.5 is modeled after a police radio service that is intended to service the city core and the surrounding suburban and rural areas. The base station is assumed to be located close to the city core, with an antenna mounted on a high tower that provides a relatively unobstructed view of the surrounding rural area. In this case, although we would expect large path-loss exponents in the built-up areas, the median path losses in the outlying rural areas are assumed to be close to free-space losses. Shadowing losses are still expected, due to small undulations of the terrain, trees, bushes, and other forms of vegetation. Table 2.5 is broken down into the sections described on the next page.

TABLE 2.5   Terrestrial link budget from base to mobile.

Parameter	Units		Comments
Base station transmitter			
Transmit frequency	MHz	705	Mobile public safety band
Transmit power	dBW	15	
Transmit Antenna gain	dBi	2	Uniform radiation in azimuth
Transmit EIRP	dBW	17	Maximum EIRP of 30 dBW
Power at 1 m	dBm	17.6	$P_0 = P_T/(4\pi/\lambda)^2$
Losses			
Path-loss exponent		2.4	Applicable at edge of coverage
Range	km	10	Range at edge of coverage
Median path loss	dB	96	$2.4 \times 10\log_{10}(r/r_0)$
Lognormal shadowing	dB	8	standard deviation of log-normal shadowing
Shadowing margin	dB	13.2	for 95% availability ($1.65\sigma_{dB}$)
Received Signal			
Receive antenna gain	dBi	1.5	Vertically polarized whip antenna
Received signal strength	dBm	−90.1	$P_R = P_0 + G_R - 21\log_{10}(r/r_0) - M_{shadow}$
Receiver characteristics			
Required C/N_0	dB-Hz	69.8	from modem characteristics
Boltzmann's constant	dBm-K	−198.6	
Receiver noise figure	dB	6.0	provided
Receiver sensitivity	dBm	−98.2	$S = C/N_0 + NF + kT_0$
Margin	dB	8.1	Margin $= P_R - S$

1. Base station transmitter, which includes
   - the transmit frequency.
   - the transmit power at the antenna terminals.
   - the transmit antenna gain. We assume a donut-shaped radiation pattern that is approximately uniform in azimuth.
   - the transmit EIRP is the dB sum of the previous two quantities.

2. Losses, which include:
   - the path-loss exponent. Slightly worse than free-space propagation is assumed for the rural areas at the edge of coverage.
   - the range. This is the expected radius of the service area.
   - the median-path loss. This is obtained from the general statistical model, assuming a path-loss exponent of 2.4.
   - lognormal shadowing. This describes the standard deviation of the shadowing expected at the edge of coverage.
   - shadowing margin. This is the margin required to maintain 95% availability. The shadowing margin is obtained from the lognormal distribution with the preceding deviation. Since police radio is a safety service, it typically has higher availability requirements than nonsafety services.

3. *Received Signal*, which includes
   - the receive antenna gain. Here, we assume a whip antenna; a *whip antenna* is a flexible rod or wire antenna, such as commonly used for cellular telephones or car radios.
   - the received signal strength. This is the reference power at 1 meter, minus the path losses, minus the shadowing margin, plus the receive antenna gain.

4. *Receiver characteristics*

   - computes the receiver sensitivity from modem characteristics and the receiver noise figure.
   - compares the receiver sensitivity with the received signal level.

The receiver noise has a spectral density that is characterized by the product of the receiver noise factor F and thermal noise at the nominal temperature, namely, kT. Consequently, noting that $N_0 = F(kT)$, we observe that the received signal strength S and the carrier-to-noise ratio are related by

$$S(\text{dBm}) = \frac{C}{N_0}(\text{dBHz}) + kT(\text{dBHz}) + F(\text{dB}) \qquad (2.140)$$

where T is the nominal temperature (290°). Equation (2.140) is used for the calculation in the receiver characteristics section of Table 2.5. Equation (2.140) also defines the relationship between the receiver sensitivity and the threshold C/N_0 of the receiver. ■

Problem 2.22 Repeat the link budget of Table 2.5, analyzing the performance in the city core. Assume that the maximum range within the city core is 2 km, but that the path-loss exponent is 3.5 and the lognormal shadowing deviation is 10 dB. Is this service limited by the receiver sensitivity? What is the expected service availability for the city core?

Ans. Yes, it is limited by the receiver sensitivity; only 11.5 dB of margin is available to compensate shadowing, and this implies an availability of approximately 75%. ■

2.10   THEME EXAMPLE 1: OKUMURA–HATA EMPIRICAL MODEL[16]

We now discuss an example of a land mobile propagation model that is based on empirical measurements. This model applies to propagation in the UHF/VHF frequency band from 150 MHz to 1 GHz. The original data were collected by Okumura and others in several areas of Japan. Numerous charts were provided illustrating the many factors that affect land mobile propagation, including building characteristics and antenna height. Hata later provided analytical approximations to these data that captured most of the major effects. This analytical representation is described here.

This *Okumura–Hata model* predicts the median-path loss in three types of environment: urban, suburban, and open. The median-path losses, in dB, for the three environments are given by the equations

$$L_p = A + B\log_{10}r \qquad \text{Urban}$$
$$L_p = A + B\log_{10}r - C \qquad \text{Suburban} \qquad (2.141)$$
$$L_p = A + B\log_{10}r - D \qquad \text{Open}$$

where r is the range in kilometers. The parameters in these equations depend on the frequency of operation, f_c, the height of the transmitting station, h_b, and the height of the receiving station, h_m. These parameters are given by the empirical formulas

$$A = 69.55 + 26.16\log_{10}f_c - 13.82\log_{10}h_b - a(h_m)$$
$$B = 44.9 - 6.55\log_{10}h_b$$
$$C = 5.4 + 2[\log_{10}(f_c/28)]^2 \qquad (2.142)$$
$$D = 40.94 + 4.78(\log_{10}f_c)^2 - 18.33\log_{10}f_c$$

where f_c is measured in MHz, h_b and h_m are in meters, and $a(h_m)$ is a correction factor that is defined in what follows. This model is valid for the following range of parameter values:

$$150 \text{ MHz} < f_c < 1000 \text{ MHz}$$
$$30 \text{ m} < h_b < 200 \text{ m}$$
$$1 \text{ m} < h_m < 10 \text{ m} \qquad (2.143)$$
$$1 \text{ km} < r < 20 \text{ km}$$

The parameter A represents a fixed loss that depends on the frequency of the signal and the heights of the base station and mobile station antennas. Its dependence on frequency is approximately a 2.6 power law, slightly higher than the square-law dependency we would expect for free-space propagation and certainly not independent of frequency as predicted by the plane-Earth model. The dependence on antenna heights has a term proportional to $h_b^{1.382}$ that is similar to the plane-Earth model, which has a quadratic dependence on h_b. The term $a(h_m)$ is a correction factor based on the mobile antenna height and is a function of the environment. For a large city, it is given by

$$a(h_m) = 8.29(\log 1.54h_m)^2 - 1.1\text{dB} \qquad \text{for } f_c \leq 300 \text{ MHz}$$
$$a(h_m) = 3.2(\log 11.75h_m)^2 - 4.97\text{dB} \qquad \text{for } f_c > 300 \text{ MHz} \qquad (2.144)$$

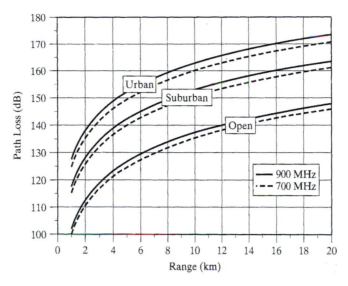

FIGURE 2.32   Performance loss predictions with Okumura–Hata propagation model with 30-m base antenna and 1-m mobile antenna for a midsized city.

For small and medium-sized cities, the correction factor is given by

$$a(h_m) = (1.1\log f_c - 0.7)h_m - (1.56\log f_c - 0.8)dB \qquad (2.145)$$

for all frequencies. The dependence on the mobile antenna height as described by $a(h_m)$ differs significantly from the quadratic dependence of the plane-Earth model.

From Eq. (2.141), the parameter B represents the path-loss exponent. The worst-case path-loss exponent is approximately 4.5, improving with higher base station antennas, as we would intuitively expect.

The parameters C and D represent reductions in the fixed losses for the less demanding suburban and open propagation environments. These improvements are frequency dependent.

Although this model is mathematically complicated, it is straightforward to evaluate on a computer and estimate path losses. In Fig. 2.32, the computed path loss for frequencies of 700 and 900 MHz is plotted for the three different environments of the Okumura–Hata model. We see two expected trends:

1. Propagation losses increase with frequency.
2. Propagation losses increase in built-up areas.

There have been enhancements and modifications to the Okumura–Hata model to extend the frequencies, the distances, and the environments over which it is valid.

EXAMPLE 2.21 Base Station Antenna Height

For public safety applications in the 700-MHz band, the maximum transmitter power is 30 W. What base station height would be needed to communicate with a patrol car at a distance of 10 km in a suburban environment? Assume that the mobile station antenna height is 1 m.

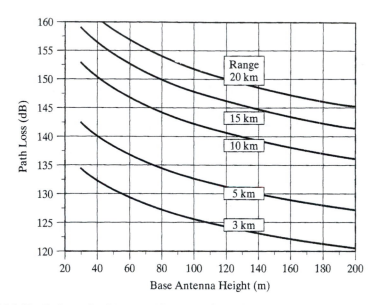

FIGURE 2.33   Path loss for Okamura–Hata model in a midsized suburban environment at 700 MHz as a function of base antenna height.

In Fig. 2.33, we plot the path loss from Eq. (2.141) of the Okumura–Hata model as a function of the base station antenna height for the suburban portion of a midsized city. Assuming a receiver sensitivity of –105 dBm and a fading margin of 10 dB, the allowable path loss would be

$$L_p(\text{dB}) = P_T(\text{dBm}) - P_R(\text{dBm}) - M_f(\text{dB})$$
$$= 45 - (-105) - 10$$
$$= 140\text{dB}$$

From Fig. 2.33, we find that the base station height would need to be at least 130 meters.   ∎

EXAMPLE 2.22 Receiver Range

A mobile phone operating at 900 MHz has a receiver sensitivity of –95 dBm and an antenna with an effective antenna height of 1 m and gain of 0 dBi. If the base antenna transmits 2 watts with an antenna height of 60 m, what is the range of the cell phone in an urban environment?

At the edge of coverage, the path loss (assuming no shadowing margin) is

$$L_p(\text{dB}) = P_T(\text{dBm}) - P_R(\text{dBm})$$
$$= 33 - (-95)$$
$$= 128\text{dB}$$

Using the Okumura–Hata model and, in particular, the results of Fig. 2.33, we find that the expected range is slightly less than 3 km.   ∎

Problem 2.23   Evaluate the path loss at 900 MHz, using the Okumura–Hata model for the suburban environment. Assume that the base and mobile station antenna heights are 30 m and 1 m, respectively.   ∎

2.11 THEME EXAMPLE 2: WIRELESS LOCAL AREA NETWORKS[17]

Wireless local area networks (LANs) are being implemented in offices and plants to reduce the cost and inconvenience of wiring network connections. There are also applications, such as airport lounges, libraries, etc., in which the users are transient and a wireless connection to a local node may be convenient. Of the various approaches to wireless LANs, many fall under the umbrella of the IEEE 802.11 standard. The one in most common use is IEEE 802.11b: however, we will discuss the more future-oriented IEEE 802.11a. This standard operates in indoor environments in the 5-GHz band and offers information rates ranging from 6 Mbps up to 54 Mbps. In this section, we will consider propagation and noise issues related to this service.

2.11.1 Propagation Model[18]

The propagation model that the International Telecommunications Union (ITU) has recommended for evaluating indoor systems in the 5-GHz band is given by

$$P_R = P_T - 41\text{dB} - 31\log_{10}r - \sum_q \text{WAF}(q) - \sum_q \text{FAF}(q) \qquad (2.146)$$

where the wall attenuation factor (WAF) and floor attenuation factor (FAF) were described in Section 2.5. Let us consider each of the terms on the right-hand side of Eq. (2.146) in turn. The second term is a fixed loss of 41 dB. The third term indicates a path-loss exponent of 3.1, where r is the distance in meters. The fourth and fifth terms represent the sum of losses the signal encounters, depending on how many walls and floors are in the path, as described in Section 2.5.

2.11.2 Receiver Sensitivity

The IEEE 802.11a wireless LAN standard recommends that the receiver have a noise figure of 10 dB or better. On the basis of this figure, we can calculate a nominal value for the noise floor of the receiver of

$$N_0 = FkT_0 \qquad (2.147)$$

or, expressed in decibels,

$$
\begin{aligned}
N_0 &= 10\log_{10}F + k(\text{dBm} - sK^{-1}) + T(\text{dBK}) \\
&= 10 + (-198.6) + 24.6 \\
&= -164 \text{ dBm/Hz}
\end{aligned}
\qquad (2.148)
$$

The 6-Mbps service in this standard uses binary phase-shift keying (BPSK) modulation. We shall discuss BPSK modulation in Chapter 3—but for now, we note that, for digital services, the SNR is often expressed as E_b/N_0, where E_b is the energy transmitted per bit. Theoretically, BPSK modulation requires an E_b/N_0 of 11 dB to achieve a bit error rate of 10^{-6}. In practice, however, with a low-cost solution, there will be implementation losses, which we will assume are no larger than 3 dB. Consequently, the practical requirement for E_b/N_0 is 14 dB.

With the foregoing information, we can estimate the receiver sensitivity. Equation (2.140) defines the sensitivity, and we note that the carrier power is related to the energy per bit by $C = E_b R_b$, where R_b is the bit rate. Substituting the expression for C into Eq. (2.140), we can relate the sensitivity to the required E_b/N_0 by

$$S(\text{dBm}) = E_b/N_0(\text{dB}) + R_b(\text{dBHz}) + N_0(\text{dBm/Hz}) \qquad (2.149)$$

That is, the sensitivity is equal to the required E_b/N_0 times the data rate R_b and multiplied by the noise density N_0. For this example, the required receiver sensitivity is

$$S = 14 + 10\log_{10} 6 \times 10^6 + (-164)$$
$$= -82 \text{ dBm} \qquad (2.150)$$

Although we have made some simplifying assumptions, the result given in Eq. (2.150) agrees with the required sensitivity specified in the standard for reception under AWGN conditions.

Problem 2.24   The 54-Mbps service of IEEE 802.11a uses 64-quadrature-amplitude-modulation (QAM). (This method of modulation is considered in Chapter 3) Suppose the practical E_b/N_0 required to achieve a bit error rate of 10^{-6} with 64 QAM is 30 dB. For this data rate, what is the sensitivity of the receiver just discussed?
Ans. The required sensitivity is –53 dBm, including a 3-dB implementation loss.   ∎

2.11.3 Range

Radios for 5-GHz wireless LAN applications must transmit 200 milliwatts (23 dBm) or less. What is the expected range of this service? If we assume an open office environment with no walls, we suppose that and the transmitter and receiver are on the same floor, then the propagation model of Eq. (2.144) simplifies to

$$P_R = P_T - 41(\text{dB}) - 31\log_{10} r \qquad (2.151)$$

To close the link with the 6-Mbps service, the received power must be at least –82 dBm from Section 2.11.2, so

$$31\log_{10} r(m) = P_T(\text{dBm}) - P_R(\text{dBm}) - 41\text{dB}$$
$$= 23 - (-82) - 41 \qquad (2.152)$$
$$= 64 \text{ dB}$$

Solving this equation for r gives a maximum range of 116 meters.

Problem 2.25   From the results of Problem 2.24, what is the maximum range expected with the 54-Mbps service?
Ans. The expected range is 13.5 m.   ∎

2.11.4 Power-Delay Profile

Empirical measurements of the multipath intensity profile indicate that the rms delay spread for indoor applications in the 5-GHz band is typically between 40 and

60 nanoseconds. A discrete-time model that has been suggested for the multipath power-delay profile is depicted in Fig. 2.34. The model shows an exponential decrease in the average power of a multipath ray as a function of the delay.

With this model, the impulse response $\{\tilde{h}_k\}$ of the channel is composed of complex samples with a random uniformly distributed phase and Rayleigh-distributed magnitude, with average power calculable from the power-delay profile shown in the figure. Mathematically, the samples of the impulse response are given by

$$\tilde{h}_k = N(0, \sigma_k^2/2) + jN(0, \sigma_k^2/2) \tag{2.153}$$

where $N(\mu, \sigma^2)$ is a Gaussian random variable with mean μ and variance σ^2. The variances in Eq. (2.153) are given by

$$\sigma_k^2 = \sigma_0^2 e^{-kT_s/T_{\text{rms}}} \tag{2.154}$$

where T_s is the sampling period of the impulse response and T_{rms} is the rms delay spread. The constant $\sigma_0^2 = 1 - e^{-T_s/T_{\text{rms}}}$ is chosen to normalize the average power; that is,

$$\sum_{k=0}^{\infty} \sigma_k^2 = 1 \tag{2.155}$$

For a sampling period of 12.5 nanoseconds and T_{rms} of 50 nanoseconds, an example of the multipath spectrum for this application is shown in Fig. 2.35. The spectrum is

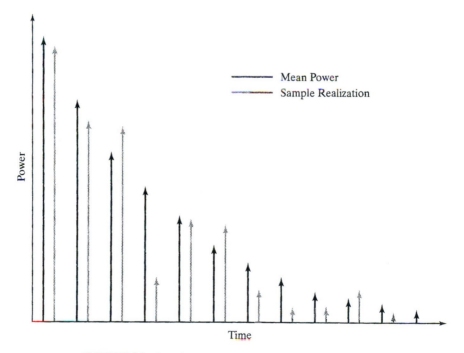

FIGURE 2.34   Impulse response model for multipath channel.

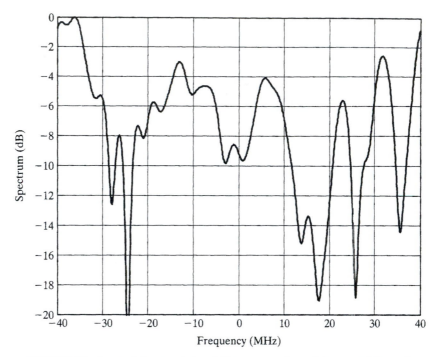

FIGURE 2.35   Spectral characteristics of 50-nanosecond rms delay spread channel.

depicted at complex baseband and ranges from –40 MHz to +40 MHz. The signal bandwidth for Fig. 2.35, IEEE 802.11a is approximately 20 MHz for all data rates. We see that there is significant variation over any 20-MHz band in this spectrum. The channel is frequency selective and, from Fig. 2.35, the coherence bandwidth is on the order of 1 Megahertz.

2.11.5 Modulation

As the results in the previous section indicate, the wireless LAN application has significant *frequency-selective characteristics.* One method of compensating for these characteristics is to include in the receiver a device known as an *equalizer,* which effectively estimates the channel and tries to invert its effects at the receiver. Unfortunately, these devices are difficult to implement for this application. The alternative approach is to design the modulation so that it is easier to compensate for the channel characteristics. This is what has been done in the IEEE 802.11a standard.

The standard uses a form of *multicarrier modulation* known as *orthogonal frequency division multiplex (OFDM).* This technique, which will be described in Chapter 3, consists of a number of narrowband carriers transmitted in a synchronous fashion. When all the details in the IEEE 802.11a standard are accounted for, it is seen that there are 52 carriers transmitted, each of which has a nominal bandwidth of 312.5 kHz.

By inspection of the spectrum plot of Fig. 2.34, we see that 312.5 kHz is significantly less than the coherence bandwidth of the channel. That is, over most 312.5-kHz pieces of the spectrum, the channel is approximately constant. This means that the channel appears like a flat-fading channel to an individual carrier. As we shall see in Chapter 3, it is easier to track and compensate for a flat-fading channel than for a frequency-selective channel. Theme Example 2 is a precursor of the discussions in the chapters that follow, where we will see a number of cases in which the modulation strategy has been selected to take advantage of, or to compensate for, the wireless channel characteristics.

2.12 THEME EXAMPLE 3: IMPULSE RADIO AND ULTRA-WIDEBAND[19]

Traditional radio transmission strategies impress information on an RF carrier, a process known as modulation, for propagation from the transmitter to the receiver. The bandwidth of the resulting signal is typically much less than the carrier frequency. The result is known as a bandpass signal. As we shall see in the following chapters, the analysis of bandpass signals and systems is fundamental to much of communications theory. However, there are other methods of transmitting information via radio waves.

One method that has captured attention recently is known as *impulse radio*. With this technique, information is sent by means of very narrow pulses that are widely separated in time. This is reminiscent of the "spark-gap" transmitter used by Marconi when he made the first radio transmission across the Atlantic Ocean. Since the pulse widths are very narrow, the spectrum of the resulting signal is very broad, and consequently, this technique is a form of *ultra-wideband (UWB) radio transmission*.

There are several possible definitions for ultra-wideband transmission, but two popular ones are:

- any transmission where the RF bandwidth exceeds 1 GHz, or
- any transmission where the RF bandwidth at the −10 dB points of the spectrum exceeds 25% of the center frequency.

Ultra-wideband transmission is considered best for low-power indoor applications where there is high clutter, that is, the surrounding environment causes significant amounts of multipath.

Impulse radio is a form of ultra-wideband radio that is based on the use of very short pulses. One type of pulse used for this application is the *Gaussian monocycle*, which is based on the first derivative of the Gaussian function. The waveform of the Gaussian monocycle is given by

$$v(t) = A\frac{t}{\tau}\exp\left\{-6\pi\left(\frac{t}{\tau}\right)^2\right\} \tag{2.156}$$

where A is an amplitude scale factor and τ is the time constant of the pulse. This signal is depicted in Fig. 2.36. It consists of a positive lobe followed by a negative lobe, with a total pulse width of approximately τ. For impulse radio applications, the pulse width τ is typically between 0.20 and 1.50 nanoseconds.

The spectrum of a sequence of these pulses can be obtained from the Fourier transform of an individual pulse and this spectrum is shown in Fig. 2.37. (Fourier theory is reviewed in Appendix A.) The frequency axis in Fig. 2.37 has been normalized in

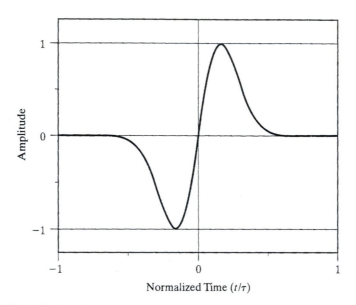

FIGURE 2.36   Illustration of a Gaussian monopulse used for impulse radio.

terms of the time constant τ; for $\tau = 1.0$ nanosecond, this frequency axis ranges from 0 to 5 GHz. In Fig. 2.37, the center frequency of the signal is approximately $1/\tau$, and at the -10 dB points the bandwidth of the signal is approximately $2/\tau$. Thus we see that the bandwidth of the UWB signal is greater than its center frequency, which certainly violates a fundamental requirement for a bandpass signal.

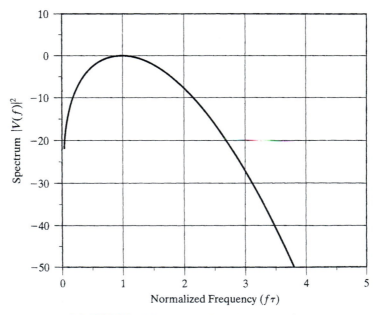

FIGURE 2.37   Spectrum of a Gaussian monopulse.

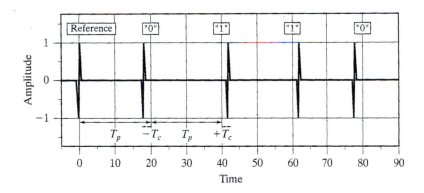

FIGURE 2.38   Pulse-position modulation of impulse radio.

There are several methods for modulating such an impulse wave. One method, known as *pulse-position modulation*, is illustrated in Fig. 2.38. With this method, there is a nominal time separation, T_p, between successive pulses. To transmit a binary signal "0," the pulse is transmitted slightly early $(-T_c)$. To transmit a binary signal "1", the pulse is transmitted slightly late $(+T_c)$. The receiver detects this *early/late timing* and demodulates the data accordingly. Typical separations between pulses (T_p) range from 25 nanoseconds to 1000 nanoseconds, resulting in a range of data rates from 40 Mbits/s to 1 Mbits/s.

The ultra-wideband nature of the modulated signal has both good aspects and bad. Since the signal power is spread over such a large bandwidth, the amount of power that falls in any particular narrowband channel is small. However, such power falls in all such narrowband channels. Consequently, there is a concern that ultra-wideband radios will cause harmful interference into existing narrowband radio services occupying the same radio spectrum. As a consequence, although ultra-wideband radio has been allowed in various jurisdictions, there are strict limits on the power spectra that may be transmitted. Due to this limitation on transmit power, ultra-wideband radio is limited to short-range applications, typically less than a few hundred meters.

EXAMPLE 2.23 Spectral Density of UWB Compared to Noise Floor

Compare the interference caused by a one-milliwatt impulse radio $(\tau = 1 \text{ ns})$ into an 800 MHz radio to the noise floor of the narrowband radio. Assume the 800 MHz radio has a noise figure of 10 dB and the separation between the impulse and narrowband radios is 10 meters.

In the problem statement, one milliwatt refers to the average power transmitted by the radio. Since the impulse radio has a low duty cycle, the peak power is much higher than this average. However, the switching rate of the impulse radio is much greater than the bandwidth of most narrowband signals. This has a smoothing effect, implying that we can use the average power for analysis.

From Fig. 2.37, the average power of the impulse radio is spread over a bandwidth of between $1/\tau$ to $2/\tau$. If we use the lower of these two bandwidths, the effective power spectrum is one milliwatt spread over 1 GHz. Assuming an approximate uniform distribution for this power, this is equivalent to one picowatt per hertz, or -90 dBm/Hz. Assuming free-space propagation from an isotropic antenna, the path loss between the impulse radio and the narrowband receiver at 800 MHz is, from Eq. (2.6),

$$L_p = \left(\frac{4\pi R}{\lambda}\right)^2$$

$$= \left(\frac{4 \times 10}{c/800 \times 10^6}\right)^2 \tag{2.157}$$

$$= 112,212$$

$$= 50.5 \text{ dB}$$

The resulting interference density at the narrowband receiver input is then

$$J_0 = -90 \text{ dBm/Hz} - 50.5 \text{ dB}$$

$$= -(140.5 \text{ dBm/Hz}) \tag{2.158}$$

We compare this interference density to the noise floor of the 800 MHz radio, which is given by

$$N_0 = kTF$$

$$\sim -174 \text{ dBm/Hz} + 10 \text{ dB} \tag{2.159}$$

$$= -164 \text{ dBm/Hz}$$

where F is the noise figure of the receiver. Since the interference density is approximately 30 dB higher than the noise density of the receiver, the 800 MHz radio will see a significant degradation of performance in the presence of the impulse radio, unless it has at least 24 dB margin. ∎

This example illustrates a number of interesting points. First, even though impulse radio spreads power over a large bandwidth, the transmit powers must still be very low to be comparable to the noise floor, typically in the microwatt range. Second, it reminds us that free-space path loss depends on frequency, which means it varies considerably over a wideband signal. However, the receiver antenna gain usually has an inverse dependency on frequency that compensates for this loss.

The advantage of impulse radio is that it has a very simple transmitter and receiver design. There is no up-conversion to, and down-conversion from, carrier frequencies; transmission is performed at baseband. In addition, impulse radio has the capability to provide very high data rates, albeit over short distances.

As mentioned previously, the major application of ultra-wideband radio is in indoor environments where there is significant multipath. In this application, impulse radio has a significant advantage in its ability to mitigate the effects of multipath. Multipath is due to echoes of the transmitted signal arriving slightly later than the transmitted signal. In indoor environments, the multipath differential delay, or delay spread, tends to be small, often on the order of nanoseconds.[20] Since most of the time an impulse radio is not transmitting, by proper design, these echoes will arrive at the receiver in between the signaling epochs, as illustrated in Fig. 2.39. Consequently, for a receiver that is attuned to the timing of the main signal, the echoes will have no effect.[21] Note that the speed of light is approximately 0.3 meters per nanosecond. Hence, reflections, even with a path difference as little as one meter, can be resolved by the impulse radio receiver. The upper limit on allowable path differences is determined by the pulse interval T_p.

In summary, impulse radio is a reincarnation of very early radio transmission techniques that did not use an RF carrier. For this reason, it results in a simple implementation for the transmitter and receiver. It also permits the transmission of very high data rates over short distances in dense multipath environments. There are a number of issues related to ultra-wideband radio. A practical issue is the interference that impulse radio may cause into existing narrowband radio services. Impulse radio also poses a modeling challenge. The time resolution of impulse radio is such that it can distinguish individual multipath components, and thus the Rayleigh model for amplitude distribution of the multipath is no longer valid.

Impulse radio is not the only ultra-wideband transmission technique. Other methods, based on an extension of the direct-sequence spread spectrum (to be discussed in Chapter 5), have also been proposed. Other modifications of the basic technique move the signal spectrum to frequencies above 2 GHz to reduce some of the concerns with interference.

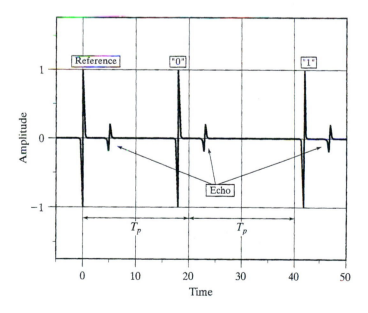

FIGURE 2.39   Illustration of multipath effects on ultra-wideband radio.

2.13 SUMMARY AND DISCUSSION[22]

In this chapter, we have looked at propagation and noise phenomena and their effect on a wireless communications link.

(a) Three principal *propagation modes* were identified:

- Free-space propagation, which corresponds to the ideal situation and a path loss proportional to the square of the distance between transmitter and receiver.

- Reflection, which is common in terrestrial propagation and shows significantly greater path loss than that of free-space propagation.

- Diffraction, which is also common in terrestrial propagation and explains why links may be closed in apparent electromagnetic shadows, albeit with some significant path losses.

(b) *Statistical propagation models* were then presented, always with the recognition that the computation of the exact physical behavior of a signal is too complex in most scenarios. Statistical models were discussed for

- Median-path loss that depends upon the range, with a path-loss exponent that is related to the environment in which propagation is occuring.

- Shadowing losses that are variations about the median-path loss due to the particular propagation path. These shadowing losses are often modeled with a log-normal distribution.

- Fast fading due to local reflections that quickly change their relative phase as the user terminal moves. Fast fading is often modeled with a Rayleigh distribution.

(c) *Noise* and *interference* were characterized as constraints on the performance of a communications system. Various forms of noise and their sources were examined, including

- Receiver noise that is due to the electronics of the receiver and the fundamental properties of circuits. This is ordinarily modeled as white noise.

- Antenna noise due to the natural radiation from the local environment. This noise contribution depends upon the antenna pattern, its height, and the environment.

- Artificial noise due to electrical machinery and discharges that produce harmonics or impulse noise in the radio frequencies used for communications. The effects of artificial noise decrease as the frequency of a signal increases.

- Multiple-access noise is due to other terminals accessing the same radio frequencies as those used by a signal, either in an adjacent channel or even in the same channel. Multiple-access noise is a system issue that relates the individual performance of a particular terminal to the number of user terminals that are allowed to access the system.

(d) The final topic of the chapter dealt with combining all of the previous elements in a *link budget* describing the communications link and its performance. The link budget depends on the following parameters:

- Propagation characteristics of the link, which can vary widely, depending upon the environment. The propagation determines the average signal power that the receive terminal can expect at its antenna port for a given transmit power.
- Noise characteristics determined by the receiver design and also by the environment in which the receiver operates.
- The SNR required by the modem to provide reliable communications. The combination of the received signal strength and the noise determines whether the communications link can be closed.

NOTES AND REFERENCES

[1] The additive white Gaussian model is often an excellent model for fixed satellite channels. More details on satellite channels and the factors involved in satellite link budgets can be found in Chapter 3 and 10 of Roddy (1996) and Chapter 6 of Spilker (1977).

[2] The derivation of the effective area of an isotropic antenna is provided in Chapter 2 of Stutzman (1998).

[3] There are various reciprocity theorems in electromagnetics, and a description of the theorems applicable to antennas may be found in Chapter 12 of Ramo (1968) and Chapter 9 of Stutzman(1998).

[4] The characteristics of parabolic antennas are discussed in Chapter 6 of Spilker (1977).

[5] More detailed analysis of the effects of ground-reflected waves and diffraction may be found in Chapters 2 and 3 of Parsons (1992).

[6] The characteristic impedance of free space, or wave impedance, is defined in analogy to the characteristic impedance of an infinite transmission line. The characteristic impedance of free space is the ratio of the total electric field to the total magnetic field in a plane perpendicular to the direction of propagation of a wave. For any such plane, the impedance is $\eta_0 = 120\pi$ ohms, as described in Chapter 6 of Ramo (1966) and Chapter 1 of Stutzman (1998).

[7] Further explanation of diffraction effects may be found in Chapter 3 of Rapport (1999) and in Chapter 3 of Parsons (1992). Further information on Fresnel integrals may be found in Chapter 7 of Abramowitz and Stegun (1964).

[8] A more advanced discussion of diffraction and Fresnel integrals may be found in Chapter 12 of Stutzman (1998).

[9] Mobile terrestrial channel propagation characteristics are discussed at a similar level to that of ours in the text by Rappaport (1999).

[10] The International Telecommunications Union (ITU) is an international body that makes recommendations regarding communications, both wired and wireless. The purpose of the recommendations is to promote common standards and interoperability among the communications systems of different nations. The ITU recommendations are published and are often a good source for propagation and interference models. ITU Recommendation P.1238-2 (2001) describes models for indoor propagation at various transmission frequencies. Detailed

information on indoor propagation models may be found in the papers by Saleh and Valen-zuela (1987) and Cheung et al. (1998). The latter paper discusses extensions to the statistical model proposed in the text to more accurately model the effects of in-building diffraction.

[11] Further discussion of local propagation effects can be found in Chapter 1 of the classic book by Jakes (1974) that has been reprinted by the IEEE Press. Further information on Bessel functions may be found in Appendix B and in Chapter 9 of Abramowitz and Stegun (1964).

[12] Chapter 1 of Jakes (1974) provides a description of the Clarke model. The original presentation may be found in Clarke (1968).

[13] A more detailed description of WSSUS channels may be found in Chapter 7 of Proakis (1995). This book describes Bello's system functions (including the spaced-time spaced-frequency correlation and coherence spectrum functions) and the Fourier transform relationships between them. Similar material may be found in Chapter 6 of Parsons (1992).

[14] A more detailed description of the physics behind thermal noise may be found in Carlson (1975).

[15] More details on the analysis, characteristics, and measurement of artificial noise—particularly, impulse noise—can be found in Chapter 9 of Parsons (1992).

[16] The Okumura–Hata model for land-mobile propagation is often considered the "Bible" in this area. The original empirical investigations behind the model are described in the paper by Okumura et al. (1968), while the analysis and model fitting are described in the paper by Hata (1980).

[17] The Institute of Electrical and Electronic Engineers (IEEE) publishes a number of standards related to communications. IEEE Standard 802 was initially developed to specify the interchange of data on Ethernet networks. It has since been expanded manyfold to describe the interchange of data on many types of networks. IEEE Std. 802.11 (1999), in particular, describes the specifications for a number of wireless local area networks operating at different frequencies.

[18] This propagation model is part of ITU Recommendation P.1238-2 (2001) for indoor applications.

[19] The details of mathematically modeling impulse radio are described in the paper by Win and Scholtz (1998).

[20] Channel models for the ultra-wideband transmission are described in the paper by Cassioli et al (2002).

[21] Alternatively, one could use a device known as a Rake receiver to collect the energy in the echoes to improve performance. Rake receivers will be discussed in Section 5.6.

[22] Much of the material presented in this chapter can be found at a more advanced level in the book by Stüber (2001), which also considers in detail the effects on propagation of different modulation strategies for mobile radio.

ADDITIONAL PROBLEMS

Propagation

Problem 2.26 Consider a communications link with a geostationary satellite such that the transmitter–receiver separation is 40,000 km. Assume the same transmitter and receiver characteristics as described in Problem 2.2. What is the received power level in dBm? What implications does this power level have on the receiver design? With a land-mobile satellite terminal, the typical antenna gain is 10 dB or less. What does this imply about the data rates that may be supported by such a link?

Problem 2.27   Consider a 10-watt transmitter communicating with a mobile receiver having a sensitivity of –100 dBm. Assume that the receiver antenna height is 2 m, and the transmitter and receiver antenna gains are 1 dB. What height of base station antenna would be necessary to provide a service area of radius 10 km? If the receiver is mobile, and the maximum radiated power is restricted by regulation to be 10 watts or less, what realistic options are there for increasing the service area?

Problem 2.28   In Problem 2.26, the satellite–receiver separation was 40,000 km. Assume the altitude of a geostationary satellite was said to be 36,000 km. What is the elevation angle from the receiver to the satellite in Problem 2.26? What was the increased path loss, in decibels, relative to a receiver when the satellite is in the *zenith* position (directly overhead)? If the transmitter–receiver separation in Problem 2.2 had been 20 km, what would the path loss have been? What can be said about comparing the dB path losses in satellite and terrestrial scenarios as a function of absolute distance?

Problem 2.29   Suppose that, by law, a service operator is not allowed to radiate more than 30 watts of power. From the plane-Earth model, what antenna height is required for a service radius of 1 km? 10 km? Assume that the receiver sensitivity is –100 dBm.

Problem 2.30   A satellite is in a geosynchronous orbit (i.e., an orbit in which the satellite appears to be fixed relative to the Earth). The satellite must be at an altitude of 36,000 km above the equator to achieve this synchrony. Such a satellite may have a global beam that illuminates all of the Earth in its view. What is the approximate 3-dB beamwidth of this global beam if the radius of the earth is 6400 km. What is the antenna gain?

Problem 2.31   A 100-m base station tower is located on a plateau as depicted in Fig. 2.40. In one directions there is a valley bounded by two (nonreflective) mountain ranges. If the transmit power is 10 watts, then, assuming ideal diffraction losses, what is the expected received signal strength. Assume free-space loss over the plains and a transmission frequency of 400 MHz.

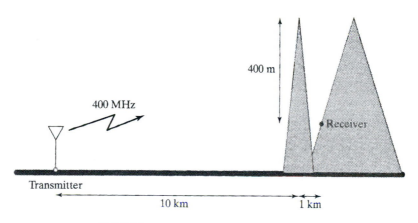

FIGURE 2.40   Site geometry for Problem 2.31.

Problem 2.32   In Problem 2.6, the estimates were based on median path loss. In the same city, the deviation about the median-path loss was estimated to be $\sigma_{dB} = 8$ dB. Assuming a log-normal model, how much additional power must be transmitted to cover the same service area with 90% *availability* at the edge of coverage when local shadowing is taken into account?

Problem 2.33   In Problem 2.7, we assume that the path-loss exponent depends on distance. For the first 10 m the path-loss exponent is 2, while beyond that distance this it is 3.5. What is the expected path loss in this case?

Problem 2.34   Suppose satellites are placed in geosynchronous orbit around the equator at 3° intervals. An Earth station with a 1-m parabolic antenna transmits 100 watts of power at 4 GHz toward the intended satellite. How much power is radiated toward the adjacent satellites on either side of the intended satellite? What implications does this amount of power have about low-power and high-power signals on adjacent satellites? What does it imply about the diameter of Earth station antennas?

Multipath

Problem 2.35   Prove that the Rayleigh density function is given by Eq. (2.55). (*Hint*: Let $Z_i = R\cos\theta$ and $Z_j = R\sin\theta$.)

Problem 2.36   Suppose the aircraft in Example 2.7 were accelerating at a rate of 0.25 g (i.e., approximately 2.5 m/s^2). What would the frequency slew rate be?

Problem 2.37   Suppose an aircraft is at an altitude of 4000 m in a circular holding pattern above an airport. The radius of the holding pattern is 4 km. If the aircraft speed is 400 km/hr, what is the frequency slew rate, df_r/dt (Hz/s), that an aircraft receiver must be capable of tracking. Assume that the transmitter is located at ground level at the center of the holding pattern. Aircraft safety communications use the VHF frequency band from 118 to 130 MHz.

Problem 2.38   A common device used in digital communications over fading channels is an *interleaver*, which is discussed in detail in Chapter 4. It boosts the performance of many forward error-correction codes. An interleaver takes a block of data from the encoder and permutes the order of the bits before transmitting them. At the receiver, the inverse operation, de-interleaving of the bits, is applied before the bits are decoded. The objective of the de-interleaver is to make the fading on adjacent bits (as seen by the decoder) appear uncorrelated. What is the ideal (minimum) spacing of bits in an interleaver to achieve this objective for a terminal moving at 100 km/hr and transmitting at 800 MHz?

Problem 2.39   From the development of Section 2.6, we could construct a model to simulate a Rayleigh fading process and the effects of fast fading. This model is illustrated in Fig. 2.41. The outputs of two independent white Gaussian noise generators are added in quadrature and then processed with a filter that approximates the desired fading spectrum. In practice, these operations are often done digitally. Since the fading process is frequently much slower than desired signalling rates, the spectrum-shaping filter may be very narrowband, requiring many interpolation stages to model accurately in the digital domain.

Develop a MATLAB program to generate a Rayleigh fast-fading signal.

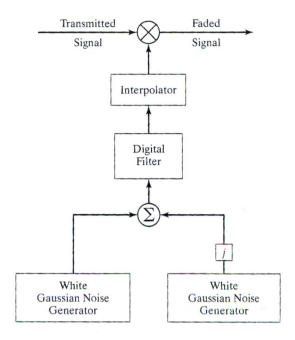

FIGURE 2.41   Illustration of digital model for simulating Rayleigh fading in Problem 2.39.

Problem 2.40   Generate a Rayleigh distribution with a 100-Hz fading bandwidth. A suggested approximation is to filter white Gaussian noise samples at 1 kHz with the one-pole filter $y_n = 0.7y_{n-1} + 0.3u_n$. For the output samples,

(a) plot the distribution of the output samples. Does this change with fading rate?
(b) plot the spectrum of the output samples. What is the 3-dB fading bandwidth?
(c) how much of the time is the signal −4dB or lower? How does this time compare with theoretical values?
(d) determine the average 3-dB fade duration. That is, each time the signal drops 3 dB below average, how long does it stay 3 dB or more below average?

Problem 2.41   Plot the frequency response of the channel $\tilde{h}(t) = \delta(t) - \alpha_2\delta(t - \tau)$ for $\alpha_2 = 0.3j$ and $\tau = 10$ microseconds. From the results obtained, what is the approximate maximum bandwidth over which the channel could be considered frequency flat?

Noise

Problem 2.42   Assume that a terrestrial radio receiver has a sensitivity of −100 dBm. What other information is necessary to compare this receiver to a satellite receiver that has a G/T of −22 dB/K?

Problem 2.43   Suppose the receiver in Problem 2.42 has a noise figure of 6 dB and an IF bandwidth of 30 kHz. What is its equivalent G/T ratio?

Problem 2.44   The sensitivity of a commercial GPS receiver is –130 dBm. If the receiver should function within 100 m of a 738–MHz base station transmitting 30 watts of power, how much attenuation of the second harmonic of the transmitted signal is required? Assume free-space loss between the transmitter and the GPS receiver, and that the GPS receiver requires a carrier-to-interference ratio, C/I, of 10 dB.

Problem 2.45   Suppose a laboratory spectrum analyzer has a noise figure of 25 dB. What N_0 would you expect to see when measuring a noiseless signal?

Problem 2.46   A satellite antenna is installed on the tail of an aircraft and connected to the receiver in an equipment bay located behind the cockpit. The antenna has a noise temperature of 50°K, and the transmission cable connecting the antenna to the low-noise amplifier (LNA) in the bay has a loss of 7 dB. The LNA has a gain of 60 dB and an equivalent noise temperature of 70°K. A secondary gain stage with 40 dB gain and a noise temperature of 1500°K follows. What is the noise temperature of the overall system? Where would be a better location for the LNA? What is the noise temperature of the new system?

Problem 2.47   A satellite is in a geosynchronous orbit at an altitude of 36,000 km. The satellite has three spot beams that illuminate approximately one-third of the visible Earth each. What is the approximate 3-dB beamwidth of the satellite spotbeam if the radius of the Earth is 6400 km. What is the antenna gain? Since the average temperature of the Earth is 290°K, what is the satellite G/T ratio?

Problem 2.48   A handheld radio has a small omnidirectional antenna. The first amplifier stage of the radio provides a 20-dB gain and has a noise temperature of 1200°K. The second amplifier stage provides another 20-dB gain and has a noise temperature of 3500°K. What is the combined noise figure of the antenna and first two stages of the radio? If the baseband processing of the radio requires an SNR of 9 dB in 5 kHz, what is the receiver sensitivity?

Problem 2.49   When a directional antenna is pointing toward the empty sky, the noise temperature falls to about 3°K at frequencies between 1 GHz and 10 GHz. The 3°K temperature represents the residual background radiation of the universe. Cosmological theory suggests that it is a remnant of the initial Big Bang. If the antenna is connected directly to a low-noise amplifier with a gain of 60 dB and a noise temperature of 100°K, what is the system noise temperature? What is the equivalent noise figure for the system?

Problem 2.50   Prove that, with a 1-in-N reuse pattern, the number of cochannel cells at the minimum distance is six.

Problem 2.51   Consider a dense urban environment in which the path-loss exponent is 4.5 and the required C/I is 13 dB. What reuse factor is permissible? If the cell size is reduced to increase capacity, what other changes would also be required?

Problem 2.52   Consider the previous problem, 2.51, but in a rolling rural environment with a path-loss exponent of 3. What reuse factor is permissible? What are the disadvantages and advantages related to the path-loss exponent?

Link Budgets

Problem 2.53 Suppose that a particular 100-kbps service requires a bit error rate (BER) of 10^{-5} in AWGN. Could this service be supported by the link budget described in Table 2.3? Justify your answer.

Problem 2.54 What is the power flux density on the surface of the Earth for Example 2.17, assuming that the satellite delivers 50 watts to the antenna terminals. How much power is collected by a 1.8-meter antenna with 50% efficiency? What voltage level would be generated across a 50-ohm resistor by that amount of power? What is the equivalent rms thermal noise voltage of this resistor in a 10-kHz bandwidth?

Problem 2.55 Construct a link budget for the return link from mobile to base of a commercial cell phone that transmits at most 600 milliwatts. Find the range of the service for the propagation models shown in Table 2.6, assuming that a service availability of 95% is required. Assume that the base station receiver noise figure is 3 dB and the required SNR is 8 dB for a 9.6-kbps service.

How do these results change if the required availability is 90%? What additional things could be done at the base station to help close the return link at greater distances?

Problem 2.56 Consider the design of a radio-controlled model airplane with a maximum range of 300 m. The receiver requires a C/N_0 ratio of 47 dB-Hz. Due to poor isolation from the aircraft engine, the receiver has a noise figure of 22 dB. What EIRP would have to be transmitted to achieve the maximum range? Assume line-of-sight transmission at 45 MHz, and assume that transmit and receive antennas have gains of −3 dB relative to an isotropic antenna.

Problem 2.57 A typical 1.5-volt "AA" cell has 1.5 ampere-hours (Ah) of energy. Assuming that a 3-V transmitter in the model airplane of Problem 2.56 has a 50% efficiency in converting input energy into EIRP, how long would a pair of batteries last? How long would they last if the transmitter had a 10% duty cycle?

Problem 2.58 You are a delegate to an ITU-R meeting representing your country. Your country's objective is to have the frequency band from 1910 to 1930 MHz designated worldwide for personal communication services (PCS). You have prepared submissions detailing the suitability of this band for this application, bearing in mind propagation characteristics, technology, and the need for the designation of this band, based on commercial projections of growth in the PCS market. The proposal has support from a number of participants at the meeting, but a number of countries that have existing fixed microwave services operating in the band from 1910 to 1920 MHz object to the proposal. Suggest a compromise proposal that may achieve your long-term objectives.

TABLE 2.6 Propagation models for Problem 2.55.

n	β(dB)	σ_{dB}
2	0	8
3	20	8
3.5	10	10
4	6	12

Problem 2.59   When G. Marconi made the first radio transmission in 1899 across the Atlantic Ocean, he used all of the spectrum available worldwide to transmit a few bits per second. It has been suggested that, in the period since then, spectrum usage (bits/s/Hz worldwide) has increased by a factor of a million. List the factors that have resulted in this substantial increase. Which factor will likely result in the largest increase in the future?

Problem 2.60   In Example 2.23, to what would the transmit power need to be reduced in order to produce an interference level equivalent to the noise floor of the 800 MHz receiver? If the transmit power remains at one milliwatt, what is the minimum separation of the impulse radio from the 800 MHz radio to produce equivalent interference and noise floor densities? Assume free-space propagation conditions.

CHAPTER 3

Modulation and Frequency-Division Multiple Access

3.1 INTRODUCTION

As mentioned in Chapter 1, *multiple access* is a technique that permits the communication resources of a wireless channel to be shared by a large number of users in different locations. There are four basic forms of multiple access applied to wireless communications, depending on which particular resource is exploited:

1. Frequency-division multiple access (FDMA)
2. Time-division multiple access (TDMA)
3. Code-division multiple access (CDMA)
4. Space-division multiple access (SDMA)

In a way, this list also orders the evolution of the spectral efficiency of wireless communication systems through the years, with the motivation for change being driven by improved utilization of the available spectrum through the use of increasingly sophisticated modulation and coding techniques. In this chapter, the focus is on FDMA.

As the name implies, FDMA operates by dividing the bandwidth of a wireless channel equally among a number of users wanting to access the channel. Specifically, FDMA can be visualized as shown in Fig. 1.2, reproduced here as Fig. 3.1 for convenience of presentation. To facilitate the kind of frequency division portrayed in this chapter, we resort to the use of *modulation*, which is a process of transforming the frequency content of a particular user's information-bearing signal so as to lie inside the frequency band allotted to that user. The block diagram of Fig. 3.2 presents the basic functional blocks that constitute the transmitter and receiver of the wireless communication system serving a single user; the figure also includes the idealized power spectra at different points in the system. The baseband processor performs operations such as filtering on the speech signal and thereby prepares it for modulation. These operations are followed by transmission over the wireless channel via a pair of antennas, one at the transmitter and the other at the receiver. The receiver performs inverse operations on the received signal, with the objective of delivering an estimate of the speech signal to a user.

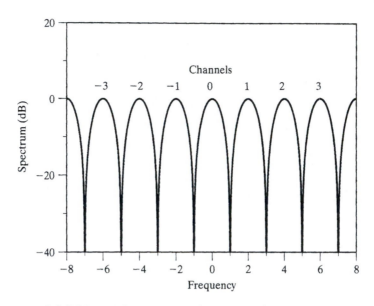

FIGURE 3.1 A frequency-domain representation of FDMA.

Section 3.2 reviews the essentials of the modulation process and the different approaches to impressing an information-bearing signal onto a carrier. A discussion of linear modulation techniques follows in Section 3.3, with emphasis on binary data transmission. Section 3.4 describes the raised cosine (RC) spectrum for pulse shaping so as to mitigate the so-called intersymbol interference problem. The material presented sets the stage for the complex representation of linear modulated signals and linear bandpass systems in Section 3.5. Section 3.6 discusses issues relating to the geometric representation of digitally modulated signals. Nonlinear modulation techniques are examined in Section 3.7, again with emphasis on binary data transmission. Section 3.8 presents two practical issues, adjacent channel interference and nonlinearity,

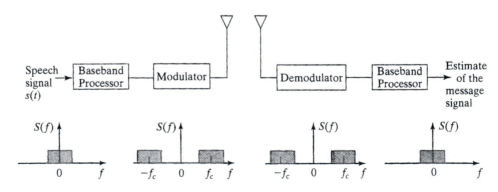

FIGURE 3.2 Illustration of the bandpass characteristic of a wireless system.

that are of practical concern in wireless communications. With this material at hand, the stage is set for a comparison of modulation techniques in the context of wireless communications, which is presented in Section 3.9. Section 3.10 takes up the issue of channel estimation and tracking at the receiver. The discussion of modulation techniques is completed in Section 3.11 with a look at the noise performance of digital modulation schemes.

Section 3.12 discusses the frequency-division multiple access technique. As with other chapters in the book, Sections 3.13 and 3.14 discuss two theme examples. The first, on orthogonal frequency-division multiplexing (OFDM), builds on a computationally efficient algorithm, namely, the fast Fourier transform, that is widely used in digital signal processing. The second theme example discusses the ubiquitous cordless telephone.

3.2   MODULATION

Modulation is formally defined as *the process by which some characteristic of a carrier wave is varied in accordance with an information-bearing signal.* In this context, the information-bearing signal is referred to as the *modulating signal*, and the output of the modulation process is referred to as the *modulated signal*. The device that performs the modulation process in the transmitter is referred to as a *modulator*, and the device used to recover the information-bearing signal in the receiver is referred to as a *demodulator*.

We can identify three practical benefits that result from the use of modulation in a wireless communication system:

1. *Modulation is used to shift the spectral content of a message signal so that it lies inside the operating frequency band of the wireless communication channel.*
   Consider, for example, telephonic communication over a cellular radio channel. In such an application, the frequency components of a speech signal from about 300 to 3100 Hz are considered adequate for the purpose of communication. In North America, one band of frequencies assigned to cellular radio systems is 800–900 MHz. The subband of 824–849 MHz is used to receive signals from mobile users, and the subband of 869–894 MHz is used for transmitting signals to mobile users. For this form of telephonic communication to be feasible, we clearly need to do two things: shift the essential spectral content of a speech signal so that it lies inside the prescribed subband for transmission, and shift it back to its original frequency band on reception. The first of these two operations is one of modulation, and the second is one of demodulation.

2. *Modulation provides a mechanism for putting the information content of a message signal into a form that may be less vulnerable to noise or interference.*
   In a wireless communication system, the received signal is ordinarily corrupted by noise generated at the front end of the receiver or by interference picked up in the course of transmission. Some specific forms of modulation (e.g., frequency modulation, considered in Section 3.4) have the inherent ability to trade

off increased transmission bandwidth for improved system performance in the presence of noise.

3. *Modulation permits the use of multiple-access techniques.*
A cellular radio channel represents a major capital investment and must therefore be deployed in a cost-effective manner, permitting mobile users access to the channel. *Multiple access* is a signal-processing operation that makes this possible. In particular, it permits the simultaneous transmission of information-bearing signals from a number of independent users over the channel and on to their respective destinations.

In wireless communications, the carrier, denoted by $c(t)$, is typically sinusoidal and is written as

$$c(t) = A_c \cos(2\pi f_c t + \theta) \tag{3.1}$$

where A_c is the amplitude, f_c is the frequency, and θ is the phase. With these three carrier parameters individually available for modulation, we have three basic methods of modulation whose specific descriptions depend on whether the information-bearing signal is of an analog or a digital nature. These two families of modulation are discussed in what follows.

3.2.1 Linear and Nonlinear Modulation Processes

Figure 3.3 shows the block diagram of a modulator supplied with a sinusoidal carrier $c(t)$. The modulating signal, acting as input, is denoted by $m(t)$. The modulated signal, acting as output, is denoted by $s(t)$. The input–output relation of the modulator is governed by the manner in which the output $s(t)$ depends on the input $m(t)$. On this basis, we may classify the modulation process as one of two basic types: linear and nonlinear.

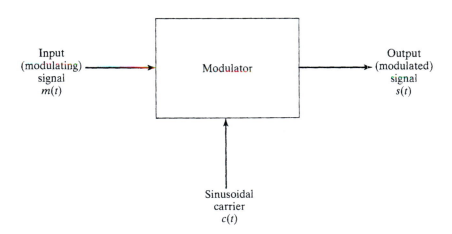

FIGURE 3.3   Block diagram of modulator.

The modulation process is said to be *linear* if the input-output relation of the modulator satisfies the *principle of superposition*. According to this principle, the modulation process, or equivalently, the modulator, satisfies two conditions:

1. The output of the modulator produced by a number of inputs applied simultaneously is equal to the sum of the outputs that result when the inputs are applied one at a time.

2. If the input is scaled by a certain factor, the output of the modulator is scaled by exactly the same factor.

The modulation process, or equivalently, the modulator, is said to be *nonlinear* if the principle of superposition is violated in part or in full.

The linearity or nonlinearity of a modulation process has important consequences, in both theoretical as well as practical terms, as we shall see in the remainder of the chapter.

3.2.2 Analog and Digital Modulation Techniques

Another way of classifying the modulation process is on the basis of whether the message signal $m(t)$ is derived from an analog or a digital source of information. In the analog case, the message signal $m(t)$ is a continuous function of time t. Consequently, the modulated signal $s(t)$ is, likewise, a continuous function of time t. It is for this reason that a modulation process of the analog kind is commonly referred to as *continuous-wave (CW) modulation*.

In the digital case, by contrast, the modulated signal $s(t)$ may exhibit discontinuities at the instants of time at which the message signal $m(t)$ switches from symbol 1 to symbol 0 or vice versa. Note, however (as we will find out later on in the chapter), that under certain conditions it is possible for the modulated signal to maintain continuity even at the instants of switching.

In other words, we may distinguish between analog and digital modulation processes as follows:

- All analog modulated signals are continuous functions of time.

- Digital modulated signals can be continuous or discontinuous functions of time, depending on how the modulation process is performed.

3.2.3 Amplitude and Angle Modulation Processes

Yet another way of classifying modulation processes is on the basis of which parameter of the sinusoidal carrier $c(t)$ is varied in accordance with the message signal $m(t)$. Accordingly, we speak of two kinds of modulation:

1. *Amplitude modulation*, in which the amplitude of the carrier, A_c, is varied linearly with the message signal $m(t)$.

2. *Angle modulation*, in which the angle of the carrier, namely,

$$\psi(t) = 2\pi f_c t + \theta \qquad (3.2)$$

is varied linearly with the message signal $m(t)$.

Angle modulation may itself be classified into two kinds:

2.1. *Frequency modulation*, in which the frequency of the carrier, f_c, is varied linearly with the message signal $m(t)$.

2.2. *Phase modulation*, in which the phase of the carrier, θ, is varied linearly with the message signal $m(t)$.

In historical terms, the design of communication systems was dominated by analog modulation techniques. Nowadays, however, we find that the use of digital modulation techniques is the method of choice, due to the pervasive use of silicon chips and digital signal-processing techniques. For this reason, the focus in what follows is on digital modulation techniques.

3.3   LINEAR MODULATION TECHNIQUES

3.3.1   Amplitude Modulation[1]

By definition, amplitude modulation (AM), produced by an analog message signal $m(t)$, is described by

$$s(t) = A_c(1 + k_a m(t))\cos(2\pi f_c t) \qquad (3.3)$$

where k_a is the *sensitivity* of the amplitude modulator. For convenience of presentation, we have set the carrier phase θ equal to zero, as it has no bearing whatsoever on the transmission of information.

Appendix A briefly reviews *Fourier theory*, which is basic to the spectral analysis of signals. In light of the theory presented therein, we may portray the spectral characteristics of amplitude modulation as illustrated in Fig. 3.4. The figure clearly shows that the bandwidth of the AM signal $s(t)$ is $2W$, where W is the bandwidth of $m(t)$ itself. Most important, except for the frequency shift in the spectrum of the message signal $m(t)$, denoted by $M(f)$, and the retention of the carrier, exemplified by the impulses at $\pm f_c$, amplitude modulation has no other effect on the spectrum $S(f)$ of $s(t)$.

Problem 3.1   Show that amplitude modulation is a nonlinear process, as it violates the principle of superposition.   ∎

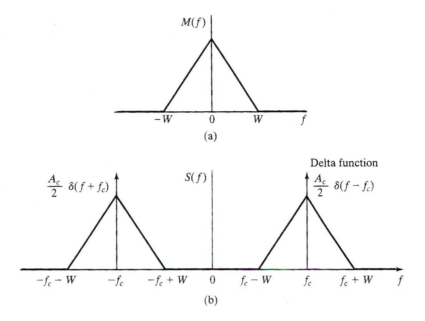

FIGURE 3.4   (a) Message spectrum. (b) Spectrum of corresponding AM signal.

Note, however, that the violation of the principle of superposition described in Problem 3.1 is of a mild sort which permits the application of the Fourier transform to an AM signal, as described in Fig. 3.4.

Another important point to note is that, insofar as information transmission is concerned, retention of the carrier in the composition of the AM signal represents a loss of transmitted signal power. To mitigate this shortcoming of amplitude modulation, the carrier is suppressed, in which case the process is referred to as *double sideband–suppressed carrier (DSB–SC) modulation*. Correspondingly, the DSB–SC modulated signal is defined simply as the product of the message signal $m(t)$ and the carrier $c(t)$; that is,

$$\begin{aligned} s(t) &= c(t)m(t) \\ &= A_c m(t) \cos(2\pi f_c t) \end{aligned} \tag{3.4}$$

Figure 3.5 depicts the spectrum of the new $s(t)$. Comparing this spectrum with that of Fig. 3.4, we see clearly that the absence of the delta functions at $\pm f_c$ is testimony to the suppression of the carrier in Eq. (3.4). Nevertheless, the AM signal of Eq. (3.3) and the DSB–SC modulated signal of Eq. (3.4) do share a common feature: They both require the use of a transmission bandwidth equal to twice the message bandwidth, namely, $2W$.

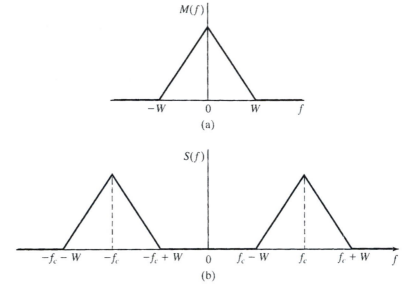

FIGURE 3.5 (a) Message spectrum. (b) Spectrum of corresponding DSB–SC modulated signal.

Problem 3.2 Consider the sinusoidal modulating signal

$$m(t) = A_m \cos(2\pi f_m t)$$

Show that the use of DSB-SC modulation produces a pair of side frequencies, one at $f_c + f_m$ and the other at $f_c - f_m$, where f_c is the carrier frequency. What is the condition that the modulator has to satisfy in order to make sure that the two side-frequencies do not overlap?
Ans. $f_c > f_m$. ∎

3.3.2 Binary Phase-Shift Keying

Consider next the case of digital modulation in which the modulating signal is in the form of a binary data stream. Let $p(t)$ denote the *basic pulse* used in the construction of this stream. Let T denote the *bit duration* (i.e., the duration of binary symbol 1 or 0). Then the binary data stream, consisting of a sequence of 1's and 0's, is described by

$$m(t) = \sum_k b_k p(t - kT) \tag{3.5}$$

where

$$b_k = \begin{cases} +1 & \text{for binary symbol 1} \\ -1 & \text{for binary symbol 0} \end{cases} \tag{3.6}$$

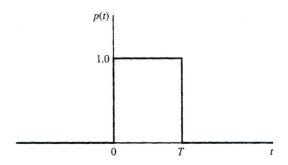

FIGURE 3.6   Rectangular pulse.

For example, in the case of a rectangular pulse we have

$$p(t) = \begin{cases} +1 & \text{for } 0 \le t \le T \\ 0 & \text{otherwise} \end{cases} \tag{3.7}$$

which is depicted in Fig. 3.6.

In binary phase-shift keying (BPSK), the simplest form of digital phase modulation, the binary symbol 1 is represented by setting the carrier phase $\theta(t) = 0$ radians, and the binary symbol 0 is represented by setting $\theta(t) = \pi$ radians. Correspondingly,

$$s(t) = \begin{cases} A_c \cos(2\pi f_c t) & \text{for binary symbol 1} \\ A_c \cos(2\pi f_c t + \pi) & \text{for binary symbol 0} \end{cases} \tag{3.8}$$

Recognizing that

$$\cos(\theta(t) + \pi) = -\cos(\theta(t)) \text{ for all time } t,$$

we may rewrite Eq. (3.8) as

$$s(t) = \begin{cases} A_c \cos(2\pi f_c t) & \text{for symbol 1} \\ -A_c \cos(2\pi f_c t) & \text{for symbol 0} \end{cases} \tag{3.9}$$

In light of Eqs. (3.6), (3.7), and (3.9), we may express the BPSK signal in the compact form

$$s(t) = c(t)m(t) \tag{3.10}$$

where $m(t)$ is itself defined by Eq. (3.5). Most important, Eq. (3.10) shows that BPSK is another example of linear modulation.

Problem 3.3   Consider a binary data stream $m(t)$ in the form of a square wave with amplitudes ± 1, centered on the origin. Determine the spectrum of the BPSK signal obtained by multiplying $m(t)$ by a sinusoidal carrier whose frequency is ten times that of the fundamental frequency of the square wave.
Ans.

$$s(t) = \sum_{k = 1, 3, 5, \ldots} \frac{\sin(k\pi/2)}{(k\pi/2)} \cos\left(2\pi\frac{kf_c}{10}t\right)\cos(2\pi f_c t)$$

$$= \frac{1}{2}\sum_{k = 1, 3, 5, \ldots} \frac{\sin(k\pi/2)}{(k\pi/2)}\left[\cos\left(2\pi t\left(f_c + \frac{kf_c}{10}\right)\right)\cos\left(2\pi t\left(f_c - \frac{kf_c}{10}\right)\right)\right]$$

The spectrum of the BPSK signal consists of side-frequencies at $f_c \pm \dfrac{kf_c}{10}$ with decreasing amplitude in accordance with $\dfrac{1}{2}\sin(k\pi/2)/(k\pi/2)$, where $k = 1, 3, 5 \ldots$ ∎

3.3.3 Quadriphase-Shift Keying

As with DSB–SC modulation, BPSK requires a transmission bandwidth twice the message bandwidth. Now, channel bandwidth is a primary resource that should be conserved, particularly in wireless communications. How then can we retain the property of linearity that characterizes BPSK, yet accommodate the transmission of a digital phase-modulated signal over a channel whose bandwidth is equal to the bandwidth of the incoming binary data stream? The answer to this fundamental question lies in the use of a digital modulation technique known as *quadriphase-shift keying* (QPSK).

  To proceed with a description of QPSK, suppose the incoming binary data stream is first demultiplexed into two substreams, $m_1(t)$ and $m_2(t)$. Next, note that, as the name implies, the phase of the carrier in QPSK assumes one of four equally spaced values, depending on the composition of each *dibit*, or group of two adjacent bits, in the original binary data stream. For example, we may use 0, $\pi/2$, π, and $3\pi/2$ radians as the set of four values available for phase-shift keying the carrier. Specifically, the values 0 and π radians are used to phase-shift key one of the two substreams, $m_1(t)$, and the remaining values, $\pi/2$ and $3\pi/2$ radians, are used to phase-shift key the other substream, $m_2(t)$.

  Accordingly, we can formulate the block diagram for the QPSK modulator as shown in Fig. 3.7. On the basis of this structure, we can then describe the QPSK modulator as *the parallel combination of two BPSK modulators that operate in phase quadrature with respect to each other.* By phase quadrature, we mean the arrangement of phases such that the phase of the carrier in the lower path of the modulator is 90° out of phase with respect to the carrier in the upper path.

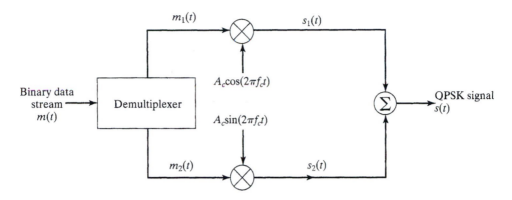

FIGURE 3.7   Block diagram of a QPSK generator, using a phase-quadrature pair of carriers $A_c\cos(2\pi f_c t)$ and $A_c\sin(2\pi f_c t)$.

As remarked previously, $m_1(t)$ and $m_2(t)$ denote the two binary substreams that result from demultiplexing of the binary data stream $m(t)$. Extending the mathematical description of Eq. (3.5) to the situation at hand, we may express the corresponding descriptions of binary substreams $m_1(t)$ and $m_2(t)$ as follows:

$$m_i(t) = \sum_k b_{k,i} p(t - kT) \quad \text{for } i = 1, 2 \tag{3.11}$$

For $i = 1,2$ we have

$$b_{k,i} = \begin{cases} +1 & \text{for symbol 1} \\ -1 & \text{for symbol 0} \end{cases} \tag{3.12}$$

and for the case of a rectangular pulse,

$$p(t) = \begin{cases} +1 & \text{for } 0 \le t \le 2T \\ 0 & \text{otherwise} \end{cases} \tag{3.13}$$

Then the BPSK signal produced in the upper path of Fig. 3.7 is described by

$$s_1(t) = A_c m_1(t) \cos(2\pi f_c t) \tag{3.14}$$

The BPSK signal produced in the lower path of Fig. 3.7 is described by

$$s_2(t) = A_c m_2(t) \sin(2\pi f_c t) \tag{3.15}$$

The QPSK signal is obtained by adding these two BPSK signals:

$$\begin{aligned} s(t) &= s_1(t) + s_2(t) \\ &= A_c m_1(t) \cos(2\pi f_c t) + A_c m_2(t) \sin(2\pi f_c t) \end{aligned} \tag{3.16}$$

Since both BPSK signals $s_1(t)$ and $s_2(t)$ are linear, the QPSK signal $s(t)$ is likewise linear.

The transmission bandwidth requirement of the QPSK signal $s(t)$ is the same as that of the original binary data stream $m(t)$. We justify this important property of QPSK signals as follows:

- The original binary data stream $m(t)$ is based on bits, whereas the substreams $m_1(t)$ and $m_2(t)$ are based on dibits. The symbol duration of both $m_1(t)$ and $m_2(t)$ is therefore twice the symbol duration of $m(t)$.
- The bandwidth of a rectangular pulse is inversely proportional to the duration of the pulse. Hence, the bandwidth of both $m_1(t)$ and $m_2(t)$ is one-half that of $m(t)$.
- The BPSK signals $s_1(t)$ and $s_2(t)$ have a common transmission bandwidth equal to twice that of $m_1(t)$ or $m_2(t)$.
- The QPSK signal $s(t)$ has the same transmission bandwidth as $s_1(t)$ or $s_2(t)$.
- Hence, the transmission bandwidth of the QPSK signal is the same as that of the original binary data stream $m(t)$.

EXAMPLE 3.1 QPSK Waveform

Figure 3.8(a) depicts the waveform of a QPSK signal for which the carrier phase assumes one of the four possible values $0°$, $90°$, $180°$, and $270°$. Moreover, the waveform is the result of transmitting a binary data stream with the following composition over the interval $0 \leq t \leq 10T$:

- The input dibit (i.e., the pair of adjacent bits in $m(t)$) changes in going from the interval $0 \leq t \leq 2T$ to the next interval $2T \leq t \leq 4T$.
- In going from the interval $2T \leq t \leq 4T$ to the next interval $4T \leq t \leq 6T$, there is no change in the input dibit.
- The input dibit changes again in going from the interval $4T \leq t \leq 6T$ to $6T \leq t \leq 8T$.
- The input dibit is unchanged in going from the interval $6T \leq t \leq 8T$ to the next interval $8T \leq t \leq 10T$. ∎

Examining the waveform of Fig. 3.8(a), we see that QPSK signals exhibit two unique properties:

1. The carrier amplitude is maintained constant.
2. The carrier phase undergoes jumps of $0°$, $\pm90°$ or $\pm180°$ every $2T$ seconds, where T is the bit duration of the incoming binary data stream.

3.3.4 Offset Quadriphase-Shift Keying

Property 2 of the conventional QPSK signal, namely, the fact that the carrier phase may jump by $\pm90°$ or $\pm180°$ every two bit durations can be of particular concern when the QPSK signal is filtered during the course of transmission over a wireless channel.

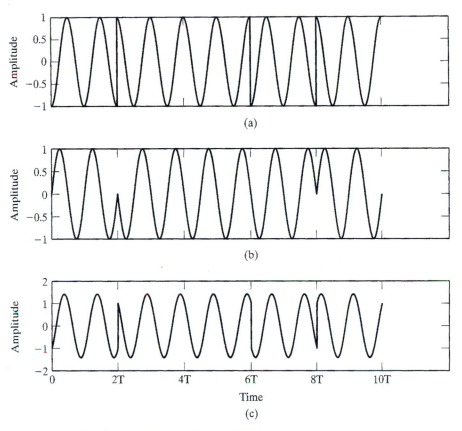

FIGURE 3.8   Waveforms of (a) conventional QPSK (b) offset QPSK, and (c) $\pi/4$-shifted QPSK.

The filtering action can, in turn, cause the carrier amplitude (i.e., the envelope of the QPSK signal) to fluctuate, thereby making the receiver produce additional symbol errors over and above those due to channel noise.

The extent of amplitude fluctuations exhibited by conventional QPSK signals may be reduced by using *offset quadriphase-shift keying (OQPSK)*, which is also referred to as *staggered QPSK*. In this variant of QPSK, the second substream $m_2(t)$, multiplied by the 90° phase-shifted carrier $A_c\sin(2\pi f_c t)$, is delayed (i.e., offset) by a bit duration T with respect to the first substream $m_1(t)$, multiplied by the carrier $A_c\cos(2\pi f_c t)$. Accordingly, unlike the phase transitions in conventional QPSK, the phase transitions likely to occur in offset QPSK are confined to 0°, ±90°, as illustrated in the next example. However, the ±90° phase jumps in OQPSK occur twice as frequently, but with a reduced range of amplitude fluctuations, compared with those of conventional QPSK. Since, in addition to ±90° phase jumps, there are ±180° phase jumps in conventional QPSK, we usually find that amplitude fluctuations in OQPSK due to filtering have a smaller amplitude than in conventional QPSK.

EXAMPLE 3.2 OQPSK Waveform

Part (b) of Fig. 3.8 depicts the waveform of the OQPSK for the same binary data stream responsible for generating the conventional QPSK waveform depicted in part (a) of the figure. Here again, we see that the carrier amplitude of OQPSK is maintained constant. However, unlike the carrier phase in the conventional QPSK of Fig. 3.8(a), the carrier phase of the OQPSK shown in Fig. 3.8(b) has jumps of only $\pm 90°$.   ∎

3.3.5 $\pi/4$-Shifted Quadriphase-Shift Keying

As mentioned previously, ordinarily the carrier phase of a conventional QPSK signal may reside in one of two possible discrete settings:

1. $0, \pi/2, \pi,$ or $3\pi/2$ radians.
2. $\pi/4, 3\pi/4, 5\pi/4,$ or $7\pi/4$ radians.

These two phase settings are shifted by $\pi/4$ radians relative to each other. The QPSK waveform depicted in Fig. 3.8(a) follows setting 1. In another variant of QPSK known as $\pi/4$-*shifted QPSK*, the carrier phase used for the transmission of successive dibits is alternatively picked from settings 1 and 2.

An attractive feature of $\pi/4$-shifted QPSK signals is that amplitude fluctuations due to filtering are significantly reduced, compared with their frequency of occurrence in conventional QPSK signals. Thus, the use of $\pi/4$-shifted QPSK provides the bandwidth efficiency of conventional QPSK, but with a reduced range of amplitude fluctuations. The reduced amplitude fluctuations become important when the transmitter includes a slightly nonlinear amplifier, as we shall see in Section 3.9. Indeed, it is for this reason that $\pi/4$-shifted QPSK has been adopted in the North American digital cellular time-division multiple access (TDMA) standard, IS-54 as well as the Japanese digital cellular standard.[2]

EXAMPLE 3.3 $\pi/4$-Shifted QPSK Waveform

Figure 3.8(c) depicts the $\pi/4$-shifted QPSK waveform produced by the same binary data stream used to generate the conventional QPSK waveform of Fig. 3.8(a). Comparing these two waveforms, we see that (1) the phase jumps in the $\pi/4$-shifted QPSK are restricted to $\pm\pi/4$ and $\pm 3\pi/4$ radians and (2) the $\pm\pi$ phase jumps of QPSK are eliminated—hence the advantage of $\pi/4$-shifted QPSK over conventional QPSK. However, this advantage is attained at the expense of increased complexity.   ∎

3.4 PULSE SHAPING

The pulse defined in Eq. (3.7) for representing binary symbol 1 or 0 is rectangular in shape. From a practical perspective, the use of a rectangular pulse shape is undesirable for two fundamental reasons:

1. The spectrum (i.e., Fourier transform) of a rectangular pulse is *infinite in extent*. Correspondingly, the spectrum of a digitally modulated signal based on the use

of a rectangular pulse is shifted by an amount equal to the carrier frequency but, most important, its frequency content is also infinite in extent. However, a wireless channel is bandlimited, which means that the transmission of such a digitally modulated signal over the channel will introduce *signal distortion* at the receiving end.

2. A wireless channel has *memory* due to the presence of multipath. Consequently, the transmission of a digitally modulated signal over the channel results in a special form of interference called *intersymbol interference* (ISI), which refers to interference between consecutive signaling symbols of the transmitted data sequence.

Now, in a wireless communication system, the goal is to accommodate the largest possible number of users in a prescribed channel bandwidth. To satisfy this important requirement we must use a *premodulation filter*, whose objective is that of *pulse shaping*. Specifically, the shape of the basic pulse used to generate the digitally modulated signal must be designed so as to overcome the signal distortion and ISI problems cited under points 1 and 2.

The design criteria for pulse shaping are covered by the fundamental theoretical work of Nyquist.[3] Let $P(f)$ denote the overall frequency response made up of three components: the transmit filter, the channel, and the receive filter. According to Nyquist, the effect of intersymbol interference can be reduced to zero by shaping the overall frequency response $P(f)$ so as to consist of a *flat* portion and sinusoidal *rolloff* portions, as illustrated in Fig. 3.9(a). Specifically, for a data rate of R bits/second, the channel bandwidth may extend from the minimum value $W = R/2$ to an adjustable value from W to $2W$ by defining $P(f)$ as follows:

$$P(f) = \begin{cases} \dfrac{1}{2W} & 0 \le |f| \le f_1 \\[2mm] \dfrac{1}{4W}\left[1 + \cos\left(\dfrac{\pi}{2W\rho}(|f| - W(1-\rho))\right)\right] & f_1 \le |f| < 2W - f_1 \\[2mm] 0 & |f| \ge 2W - f_1 \end{cases} \tag{3.17}$$

The frequency parameter f_1 and bandwidth W are related by the parameter

$$\rho = 1 - \frac{f_1}{W} \tag{3.18}$$

Called the *roll-off factor*, ρ indicates the excess bandwidth over the *ideal* solution corresponding to $\rho = 0$. An important characteristic of the frequency response $P(f)$ is that its inverse Fourier transform, denoted by $p(t)$ (i.e., the overall impulse response of the transmit filter, the channel, and the receive filter) has the value of unity at the current signaling instant (i.e., $p(0) = 1$) and zero crossings at all other consecutive signaling instants (i.e., $p(nT) = 0$ for nonzero integer n), as shown in Fig. 3.9(b). The zero crossings of the impulse response $p(t)$ ensure that the ISI problem is reduced to zero. The frequency response of Fig. 3.9(a) is called the *raised-cosine (RC) spectrum*, so called because of its trigonometric form as defined in Eq. (3.17).

The variability of the rolloff factor ρ over the range $(0,1)$ allows the designer to tradeoff transmitted signal bandwidth for robustness of the pulse shape.

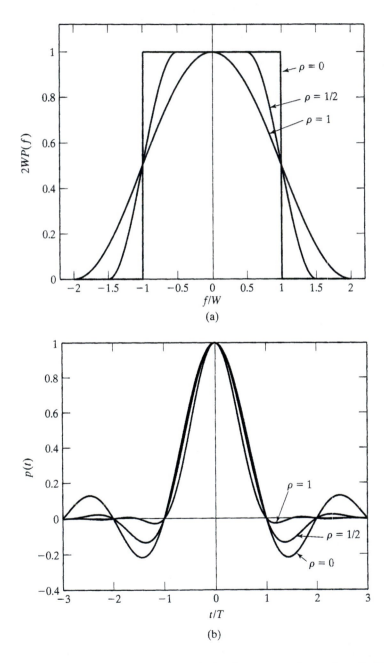

FIGURE 3.9 (a) Frequency response of the raised cosine spectrum for varying roll-off rates. (b) Impulse response of the Nyquist shaping filter (i.e., inverse Fourier transform of the spectrum plotted in part (a) for varying roll-off rates).

Problem 3.4

(a) Starting with the RC spectrum $P(f)$ of Eq. (3.17), evaluate the inverse Fourier transform of $P(f)$ and thus show that.

$$p(t) = \left(\frac{\cos(2\pi\rho Wt)}{1 - 16\rho^2 W^2 t^2} \right) \text{sinc}\,(2Wt) \tag{3.19}$$

(b) Determine $P(f)$ and $p(t)$ for the special case of $\rho = 1$, which is known as the *full-cosine roll-off pluse*.

Ans. (b) $p(t) = \text{sinc}(4Wt)/(1 - 16W^2 t^2)$ ∎

3.4.1   Root Raised-Cosine Pulse Shaping[4]

A more sophisticated form of pulse shaping uses the *root raised-cosine (RC) spectrum* rather than the regular RC spectrum of Eq. (3.17). Specifically, the spectrum of the basic pulse is now defined by the square root of the right-hand side of this equation. Thus, using the trigonometric identity

$$\cos^2\theta = \frac{1}{2}(1 + \cos 2\theta)$$

where, for the problem at hand,

$$\theta = \frac{\pi}{2W\rho}(|f| - W(1 - \rho))$$

and retaining $P(f)$ as the symbol for the root RC spectrum, we may write

$$P(f) = \begin{cases} \dfrac{1}{\sqrt{2W}} & 0 \le |f| \le f_1 \\[2mm] \dfrac{1}{\sqrt{2W}} \cos\left(\dfrac{\pi}{4W\rho}(|f| - W(1 - \rho)) \right) & f_1 \le |f| < 2W - f_1 \\[2mm] 0 & |f| \ge 2W - f_1 \end{cases} \tag{3.20}$$

where, as before, the roll-off factor ρ is defined in terms of the frequency parameter f_1 and the bandwidth W as in Eq. (3.18).

If, now, the transmitter includes a *pre-modulation filter* with the transfer function defined in Eq. (3.20) and the receiver includes an identical filter, then the overall pulse waveform will experience the spectrum $P^2(f)$, which is the regular raised cosine spectrum. In effect, by adopting the root RC spectrum $P(f)$ of Eq. (3.20) for pulse shaping, we would be working with $P^2(f)$ in an overall transmitter-receiver sense. On this basis, we find that in the context of wireless communications, if the channel is affected by both flat fading and additive white Gaussian noise, and the pulse-shape filtering is partitioned equally between the transmitter and the receiver in the manner described herein, then effectively the receiver would maximize the output signal-to-noise ratio at the sampling instants.

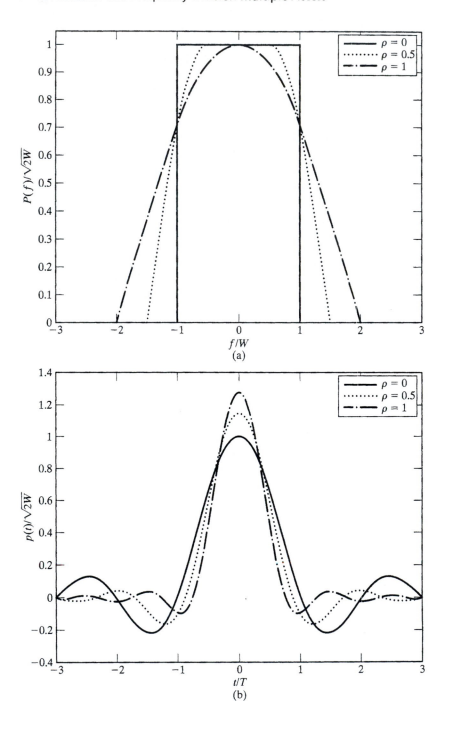

FIGURE 3.10 (a) $P(f)$ for root raised - cosine spectrum. (b) $p(t)$ for root raised - cosine spectrum.

The inverse Fourier transform of Eq. (3.20) defines the root RC shaping pulse

$$p(t) = \frac{\sqrt{2W}}{(1-(8\rho Wt)^2)}\left(\frac{\sin(2\pi W(1-\rho)t)}{2\pi Wt} + \frac{4\rho}{\pi}\cos(2\pi W(1+\rho)t)\right) \quad (3.21)$$

The important point to note here is the fact that the root RC shaping pulse $p(t)$ of Eq. (3.21) is radically different from the standard RC shaping pulse of Eq. (3.19). In particular, the new shaping pulse satisfies an *orthogonality constraint under T-shifts*, as shown by

$$\int_{-\infty}^{\infty} p(t)p(t-nT)dt = 0 \text{ for } n = \pm 1, \pm 2, ... \quad (3.22)$$

where T is the symbol duration. Yet, $p(t)$ has exactly the same excess bandwidth as the standard RC pulse.

It is important to note, however, that despite the added property of orthogonality, the root RC shaping pulse of Eq. (3.21) lacks the zero-crossing property of the regular RC shaping pulse defined in Eq. (3.19).

Figure 3.10(a) plots the root RC spectrum $P(f)$ for roll-off factor $\rho = 0, 0.5, 1$; the corresponding time-domain plots are shown in Fig. 3.10(b). These plots are different from those of Fig. 3.9 for nonzero ρ. The following example contrasts the waveform of a specific binary sequence using the root RC shaping pulse with the corresponding waveform using the regular RC shaping pulse.

EXAMPLE 3.4 Pulse Shaping Comparison

Using the root RC shaping pulse $p(t)$ of Eq. (3.21) with roll-off factor $\rho = 0.5$, plot the waveform for the binary sequence 01100, and compare it with the corresponding waveform obtained by using the regular RC shaping pulse $p(t)$ of Eq. (3.19) with the same roll-off factor.

Using the root RC pulse $p(t)$ of Eq. (3.21) with a multiplying plus sign for binary symbol 1 and multiplying minus sign for binary symbol 0, we get the dashed and dotted pulse train shown in Fig. 3.11 for the sequence 01100. The solid pulse train shown in the figure corresponds to the use of the regular RC pulse $p(t)$ of Eq. (3.15). The figure shows the root RC waveform occupies a larger dynamic range than the regular RC waveform. ∎

Problem 3.5

(a) Starting with Eq. (3.20), derive the root RC pulse shape $p(t)$ of Eq. (3.21).

(b) Evaluate $p(t)$ at (i) $t = 0$, and (ii) $t = \pm 1/(8\rho W)$.

(c) Show that the $p(t)$ derived in part (a) satisfies the orthogonality constraint described in Eq. (3.22)

Ans. (b)   (i) $p(0) = \sqrt{2W}\left(1 - \rho + \frac{4\rho}{\pi}\right)$

(ii) $p\left(\frac{\pm 1}{8\rho W}\right) = \sqrt{2W\rho}\left(\left(1 + \frac{2}{\pi}\right)\sin\left(\frac{\pi}{4\rho}\right) + \left(1 - \frac{2}{\pi}\right)\cos\left(\frac{\pi}{4\rho}\right)\right)$ ∎

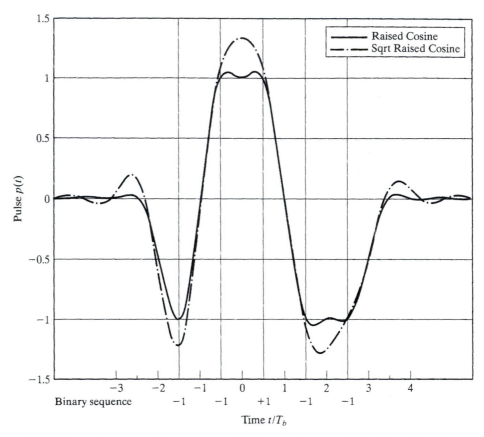

FIGURE 3.11 Two pulse trains for sequence 01100, one using regular RC pulse and the other using root RC pulse.

3.5 COMPLEX REPRESENTATION OF LINEAR MODULATED SIGNALS AND BAND-PASS SYSTEMS[5]

The linear modulation schemes considered in Section 3.3 may be viewed as special cases of the *canonical representation of a band-pass signal*:

$$s(t) = s_I(t)\cos(2\pi f_c t) - s_Q(t)\sin(2\pi f_c t) \tag{3.23}$$

It is customary to refer to $s_I(t)$ as the *in-phase component* of $s(t)$ and to $s_Q(t)$ as the *quadrature component*. This terminology follows from the definition of the sinusoidal carrier $c(t)$ in Eq. (3.1). Table 3.1 summarizes descriptions of AM, DSB–SC, BPSK, and QPSK in terms of the components $s_I(t)$ and $s_Q(t)$.

We may simplify matters further by introducing the complex signal

$$\tilde{s}(t) = s_I(t) + js_Q(t) \tag{3.24}$$

TABLE 3.1   Special Cases of the Canonical Equation (3.23).

	Type of modulation	In-phase component $s_I(t)$	Quadrature component $s_Q(t)$	Defining equation
Analog	Amplitude modulation	$A_c(1 + k_a m(t))$	0	3.3
	Double sideband– suppressed carrier modulation	$A_c m(t)$	0	3.4
Digital	Binary phase-shift keying	$A_c \sum_k b_k p(t - kT)$	0	3.5
	Quadriphase-shift keying	$A_c \sum_k b_{k,1} p(t - 2kT)$	$-A_c \sum_k b_{k,2} p(t - 2kT)$	3.11

where j is the square root of -1. For obvious reasons, the new signal $\tilde{s}(t)$ is referred to as the *complex envelope* of the modulated signal $s(t)$. Next, we invoke *Euler's formula*

$$\exp(j2\pi f_c t) = \cos(2\pi f_c t) + j\sin(2\pi f_c t) \tag{3.25}$$

Hence, in light of Eqs. (3.24) and (3.25), we may considerably simplify the formulation of the modulated signal $s(t)$ of Eq. (3.23) as

$$s(t) = \text{Re}\left\{\tilde{s}(t)\exp(j2\pi f_c t)\right\} \tag{3.26}$$

Equation (3.26) is referred to as a *single-carrier transmission*.

The material presented in this section is of profound theoretical importance in the study of linear modulation theory, be it in the context of analog or digital modulation techniques. Specifically, we may make the following four statements:

1. The in-phase component $s_I(t)$ and the quadrature component $s_Q(t)$ are both real-valued functions of time that are uniquely defined in terms of the baseband (i.e., message) signal $m(t)$. Given the two components $s_I(t)$ and $s_Q(t)$, we may thus use the scheme shown in Fig. 3.12(a) to *synthesize* the modulated signal $s(t)$.
2. Given the modulated signal $s(t)$, we may use the scheme shown in Fig. 3.12(b) to *analyze* the modulated signal $s(t)$ and thereby construct the in-phase component $s_I(t)$ and the quadrature component $s_Q(t)$.
3. The in-phase and quadrature components are *orthogonal* to each other, occupying exactly the same bandwidth as the message signal $m(t)$.
4. The complex envelope $\tilde{s}(t)$ given in Eq. (3.24) completely preserves the information content of the modulated signal $s(t)$ except for the carrier frequency f_c.

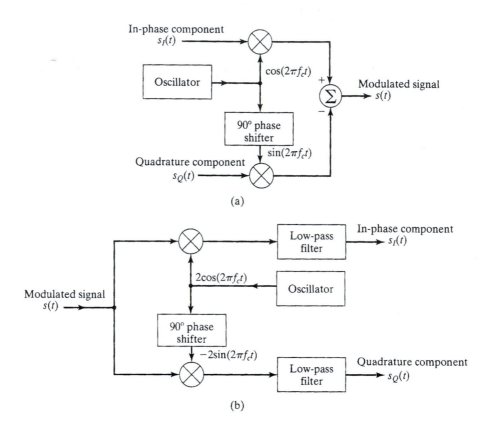

FIGURE 3.12   (a) Synthesizer for constructing a modulated signal from its in-phase and quadrature components. (b) Analyzer for deriving the in-phase and quadrature components of the modulated band-pass signal.

3.5.1 Complex Representation of Linear Band-Pass Systems

In a communication system, modulators exist alongside linear band-pass systems represented by band-pass filters and narrowband communication channels. As with any other linear system, these band-pass systems are uniquely characterized by an impulse response in the time domain and the corresponding transfer function in the frequency domain. From an analytic viewpoint, we find it highly instructive to develop complex representations of linear band-pass systems in a manner analogous to that followed for linear modulated (i.e., band-pass) signals. This form of representation not only simplifies the mathematical analysis of communication systems, but also provides the basis for their simulation on a digital computer.

Consider, then, a linear band-pass system with impulse response $h(t)$ and fed by an input signal $x(t)$ to produce an output signal $y(t)$, as depicted in Fig. 3.13. Two assumptions are made:

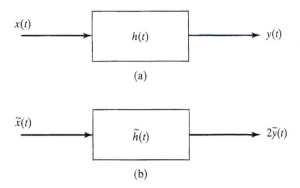

FIGURE 3.13   (a) Block diagram of linear band-pass system driven by a modulated signal $x(t)$ to produce the band-pass signal $y(t)$ as output. (b) Equivalent complex baseband model of the system, where the input signal, the impulse response, and output signal are all complex and in their baseband form.

1. The system is *narrowband*, which means that the system bandwidth is small compared with the midband frequency.
2. The input signal is a modulated signal whose carrier frequency, f_c, is the same as the midband frequency of the system.

The analysis of the system depicted in Fig. 3.13(a) is considerably simplified by replacing it with the *equivalent complex baseband model* of Fig. 3.13(b). This equivalent model is said to be complex in that its impulse response may be expressed as

$$\tilde{h}(t) = h_I(t) + jh_Q(t) \tag{3.27}$$

where $h_I(t)$ is the *in-phase component* of $\tilde{h}(t)$ and $h_Q(t)$ is the *quadrature component*. In a manner similar to the complex representation of modulated signals, the impulse response of the original band-pass system, $h(t)$, is related to the complex impulse response $\tilde{h}(t)$ as

$$h(t) = \text{Re}\left\{\tilde{h}(t)\exp(j2\pi f_c t)\right\} \tag{3.28}$$

The input signal supplied to the equivalent model, namely, $\tilde{x}(t)$, is the complex envelope of the original input signal $x(t)$. Likewise, the output signal of the equivalent model, namely, $\tilde{y}(t)$, is the complex envelope of the original output signal $y(t)$. These two complex envelopes are related by the complex convolutional integral

$$\begin{aligned}
\tilde{y}(t) &= \frac{1}{2}\int_{-\infty}^{\infty}\tilde{x}(\lambda)\tilde{h}(t-\lambda)d\lambda \\
&= \frac{1}{2}\int_{-\infty}^{\infty}\tilde{h}(\lambda)\tilde{x}(t-\lambda)d\lambda
\end{aligned} \tag{3.29}$$

Using shorthand notation, we may simply write

$$
\begin{aligned}
\tilde{y}(t) &= \frac{1}{2}\tilde{x}(t) \otimes \tilde{h}(t) \\
&= \frac{1}{2}(\tilde{h}(t) \otimes \tilde{x}(t))
\end{aligned}
\tag{3.30}
$$

where the symbol $\otimes$ stands for convolution. The two lines of Eq. (3.29) and those of Eq. (3.30) simply emphasize the *commutative property* of convolution. Note the need for adding the multiplying factor 1/2 in these two equations, which is required to maintain the exact equivalence between the real system of Fig. 3.13(a) and its complex equivalent baseband model of Fig. 3.13(b). Note also that this equivalence assumes that the midband frequency of the system and the carrier frequency of the modulated input $x(t)$ are coincident.

The important point to take from the equivalence between the two systems, one real and the other complex, is that the carrier frequency f_c has been eliminated from the equivalent model. In effect, we have *traded complex analysis for the elimination of the carrier frequency*, thereby simplifying the analysis considerably without any loss of information. In particular, having determined the complex envelope $\tilde{y}(t)$ of the output signal by convolving the complex envelope $\tilde{x}(t)$ of the input signal with the complex impulse response $\tilde{h}(t)$, we may readily determine the actual output signal $y(t)$ in Fig. 3.13(a) by using the formula

$$
y(t) = \mathrm{Re}\left\{\tilde{y}(t)\exp(j2\pi f_c t)\right\}
\tag{3.31}
$$

A similar formula holds for the input signal $x(t)$ in terms of its complex envelope $\tilde{x}(t)$.

3.6   SIGNAL-SPACE REPRESENTATION OF DIGITALLY MODULATED SIGNALS[6]

Digital modulation distinguishes itself from analog modulation in that the transmitted signal $s(t)$ assumes one of a discrete set of possible forms. The corresponding mapping of the complex envelope $\tilde{s}(t)$ onto a signal space is referred to as a *signal constellation* or *signal pattern*. Much of the literature on digital modulation techniques focuses on *two-dimensional* signal constellations whose structure naturally depends on the specific form of modulation under study. The traditional approach is to use energy-normalized versions of the in-phase component $s_I(t)$ and quadrature component $s_Q(t)$ of the modulated signal $s(t)$ as the horizontal axis and vertical axis of the two-dimensional signal space, respectively. In what follows, these *normalized coordinates of unit energy* are denoted by $\phi_1(t)$ and $\phi_2(t)$, respectively, defined in turn by

$$
\phi_1(t) = \sqrt{\frac{2}{T}}\cos(2\pi f_c t) \qquad 0 \leq t \leq T
\tag{3.32}
$$

and

$$\phi_2(t) = \sqrt{\frac{2}{T}}\sin(2\pi f_c t) \qquad\qquad 0 \le t \le T \qquad\qquad (3.33)$$

where T is the symbol duration and the carrier frequency is an integer multiple of $1/T$. From these two equations we readily see the following two properties:

1. $\int_0^T \phi_1(t)\phi_2(t)dt = 0$ $\qquad\qquad\qquad\qquad\qquad\qquad\qquad\qquad$ (3.34)

which confirms the orthogonality of $\phi_1(t)$ and $\phi_2(t)$ over the interval $0 \le t \le T$.

2. $\int_0^T \phi_1^2(t)dt = \int_0^T \phi_2^2(t)dt = 1$ $\qquad\qquad\qquad\qquad\qquad\qquad$ (3.35)

which shows that both $\phi_1(t)$ and $\phi_2(t)$ have unit energy.

Thus, $\phi_1(t)$ and $\phi_2(t)$ are said to constitute an *orthonormal set*.

Figure 3.14(a) shows the constellation of binary phase-shift keying (BPSK). The constellation is one dimensional as the coordinate $\phi_1(t)$ is sufficient for the description of BPSK. The constellation has two signal points along the ϕ_1 - axis at $\pm\sqrt{E_b}$, which correspond to the transmission of symbol 1 (the plus sign) and the symbol 0 (the minus sign).

Figure 3.14(b) shows the two-dimensional signal constellation of *quadriphase-shift keying* (QPSK), in which the phase of the transmitted signal takes on one of four possible values: 45°, 135°, 225°, and 315°. The radius of the circle on which the four points of the constellation lie is equal to $\sqrt{2E_b}$, denoting the square root of the transmitted signal energy per symbol. Table 3.2 summarizes the coordinates of the four possible message points characterizing the QPSK signal constellation of Fig. 3.14(b). In terms of the message points (s_{i1}, s_{i2}) defined in the table, we may now express the QPSK signal as

$$s_i(t) = s_{i1}\phi_1(t) + s_{i2}\phi_2(t), \qquad \begin{array}{c} i = 1, 2 \\ 0 \le t \le T \end{array} \qquad (3.36)$$

Each point in the constellation of Fig. 3.14(b) corresponds to a specific dibit made up of a group of two binary symbols; hence, the symbol period T for QPSK is twice the bit duration of the incoming binary stream. Figure 3.14(b) includes the four dibits corresponding to the *message points* of the constellation. Note that as we go around the constellation from one message point to the adjacent one, only one of the two binary symbols in a dibit experiences a change. This form of message representation is referred to as a *Gray code*. Table 3.2 also includes the Gray encoding descriptions of the four input dibits.

As remarked previously, the QPSK signal may also be constructed from the discrete set of phase values: 0°, 90°, 180°, and 270°. Figure 3.14(c) shows the signal constellation for this second construction of QPSK.

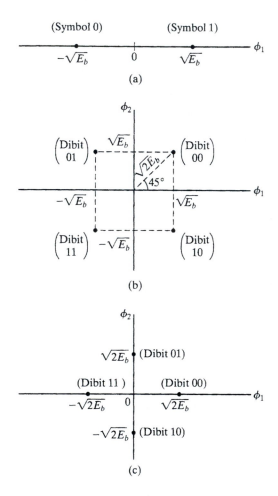

FIGURE 3.14   Signal constellations for (a) BPSK, (b) one version of QPSK, and (c) another version of QPSK.

TABLE 3.2   Signal-space characterization of the QPSK signal constellation described in Fig. 3.14(b).

Input dibit	Gray-encoded phase of QPSK signal (radians)	Coordinates of message points	
		s_{i1}	s_{i2}
10	$7\pi/4$	$+\sqrt{E_b}$	$-\sqrt{E_b}$
11	$5\pi/4$	$-\sqrt{E_b}$	$-\sqrt{E_b}$
01	$3\pi/4$	$-\sqrt{E_b}$	$+\sqrt{E_b}$
00	$\pi/4$	$+\sqrt{E_b}$	$+\sqrt{E_b}$

Problem 3.6 Construct a table summarizing the signal-space characterization of the QPSK signal constellation described in Fig. 3.14(c).
Ans.

Input dibit	Coordinates of message points
00	$(\sqrt{2E_b}, 0)$
01	$(0, \sqrt{2E_b})$
11	$(-\sqrt{2E_b}, 0)$
10	$(0, -\sqrt{2E_b})$

■

Figure 3.15 shows the signal constellation of another commonly used linear form of modulation: *16-quadrature amplitude modulation* (16-QAM). The modulated signal illustrated is hybrid in that it combines amplitude and phase modulations. Each message point of the constellation corresponds to a specific *quadbit*, which is made up of a group of four adjacent bits. Here again, we have used Gray coding in that, as we go from one message point to the adjacent one, only one of the binary symbols in a quadbit experiences a change. However, there is a major difference between the signal constellations of Figs. 3.14 and 3.15: In the QPSK signal represented by the constellation of

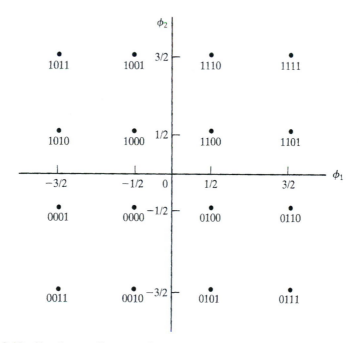

FIGURE 3.15 Signal-space diagram of *M*-ary QAM for *M* = 16; the message points in each quadrant are identified with Gray-encoded quadbits.

Fig. 3.14(a) and (b), the energy transmitted remains fixed; in the 16-QAM signal represented by the constellation of Figs. 3.15, the energy transmitted is variable, depending on the particular quadbit (i.e., 4-bit symbol) chosen for transmission. Note that QPSK signal can be transmitted over a nonlinear channel, whereas the transmission of the 16-QAM signal requires the use of a linear channel.

3.7   NONLINEAR MODULATION TECHNIQUES

The canonical representation of Eq. (3.23) naturally provides the Cartesian basis for the construction of signal constellations of linear modulated signals. Equivalently, we may recast that equation in the *polar* form

$$s(t) = a(t)\cos[2\pi f_c t + \theta(t)] \tag{3.37}$$

where

$$a(t) = \sqrt{s_I^2(t) + s_Q^2(t)} \tag{3.38}$$

is called the *envelope* of the modulated signal $s(t)$ and

$$\theta(t) = \tan^{-1}\left(\frac{s_Q(t)}{s_I(t)}\right) \tag{3.39}$$

is the *phase* of $s(t)$. Equation (3.37) provides the natural basis for the study of *nonlinear* modulation techniques, as discussed next.

3.7.1   Frequency Modulation

In *frequency modulation* (FM), a form of angle modulation, the instantaneous frequency of the sinusoidal carrier, denoted by $\psi(t)$, is varied in accordance with the information-bearing signal $m(t)$. A complete oscillation occurs whenever the phase $\psi_i(t)$ changes by 2π radians. If $\psi_i(t)$ increases monotonically with time, the average frequency in hertz, over an incremental interval from t to $t + \Delta t$, is given by

$$f_{\Delta t}(t) = \frac{\psi(t + \Delta t) - \psi(t)}{2\pi\Delta t} \tag{3.40}$$

Letting Δt approach zero leads to the following definition for the instantaneous frequency of the FM signal:

$$
\begin{aligned}
f(t) &= \lim_{\Delta t \to 0} f_{\Delta t}(t) \\
&= \lim_{\Delta t \to 0}\left[\frac{\psi(t + \Delta t) - \psi(t)}{2\pi\Delta t}\right] \\
&= \frac{1}{2\pi}\frac{d\psi(t)}{dt}
\end{aligned}
\tag{3.41}
$$

Equation (3.41) states that, except for a scaling factor, the instantaneous frequency $f(t)$ is the derivative of the instantaneous phase $\psi(t)$ with respect to time t. Conversely, we may define the instantaneous phase as the integral of the instantaneous frequency with respect to time; that is,

$$\psi(t) = \int_0^t f(\tau)d\tau \qquad (3.42)$$

where, for convenience of presentation, it is assumed that the initial phase

$$\psi(0) = \int_{-\infty}^0 f(t)dt$$

is zero.

To formally describe frequency modulation in mathematical terms, we write

$$f(t) = f_c + k_f m(t) \qquad (3.43)$$

where k_f is the sensitivity of the frequency modulator and f_c is the frequency of the unmodulated carrier. Substituting Eq. (3.43) into Eq. (3.39) and multiplying the result by 2π, we get

$$\psi(t) = 2\pi f_c t + 2\pi k_f \int_0^t m(\tau)d\tau \qquad (3.44)$$

The frequency-modulated signal is thus described in the time domain by

$$\begin{aligned} s(t) &= A_c \cos(\psi(t)) \\ &= A_c \cos\left(2\pi f_c t + 2\pi k_f \int_0^t m(\tau)d\tau\right) \end{aligned} \qquad (3.45)$$

From Eq. (3.45), we see that frequency modulation is a *nonlinear* process, in that the dependence of the modulated signal on the information-bearing signal violates the *principle of superposition*. Consequently, unlike the spectrum of the AM signal of Eq. (3.2), the spectrum of the FM signal of Eq. (3.45) is *not* related to the spectrum of the information-bearing signal $m(t)$ in a simple manner. This means that, in general, the spectral analysis of an FM signal is a more difficult task than that of an AM signal. Elsewhere,[7] it is shown that the transmission bandwidth of the FM signal $s(t)$ is approximately given by *Carson's rule*, which states that

$$B_T \approx 2\Delta f\left(1 + \frac{1}{D}\right) \qquad (3.46)$$

Carson's rule involves two parameters:

1. *The frequency deviation, Δf*, which is defined as the maximum deviation in the instantaneous frequency of the FM signal away from the carrier frequency f_c.

2. *The deviation ratio, D*, which is defined as the ratio of the frequency deviation to the highest frequency component contained in the modulating signal $m(t)$.

Unfortunately, Carson's rule does not always provide a good estimate of the bandwidth requirement of a wireless communication system using frequency modulation. For a more accurate estimate of the bandwidth requirement, we may have to resort to the use of computer simulations or actual measurements.

Unlike amplitude modulation, frequency modulation offers the means for exchanging the increased transmission bandwidth for an improved noise performance. Specifically, under the simplifying assumption of a high carrier-to-noise ratio, we may make the following statement:

An increase in the FM transmission bandwidth B_T produces a quadratic increase in the signal-to-noise ratio at the output of the receiver.

Indeed, it is because of the *bandwidth–noise trade-off* capability that frequency modulation was adopted in the *first generation* of wireless communication systems, based on frequency-division multiple access (FDMA).

However, with the ever-increasing emphasis on the use of digital signal-processing techniques, analog frequency modulation is being superseded by its digital counterparts, discussed in the rest of the section.

3.7.2 Binary Frequency-Shift Keying

In binary frequency-shift keying (BFSK), the symbols 1 and 0 are distinguished from each other by the transmission of one of two sinusoidal waves that differ in frequency by a fixed amount. A typical pair of sinusoidal waves is described by

$$s_i(t) = \begin{cases} \sqrt{\dfrac{2E_b}{T}} \cos(2\pi f_i t) & 0 \le t \le T_b \\ 0 & \text{otherwise} \end{cases} \tag{3.47}$$

where $i = 1,2$, T is the symbol (bit) duration, and E_b is the energy transmitted per bit; the frequency transmitted is

$$f_i = \frac{n_c + i}{T} \quad \text{for some fixed integer } n_c \text{ and } i = 1,2 \tag{3.48}$$

The symbol 1 is represented by $s_1(t)$ and the symbol 0 by $s_2(t)$. The BFSK signal described here is known as *Sunde's FSK*, in recognition of its inventor. It is a *continuous-phase signal* in the sense that phase continuity is maintained everywhere, including the interbit switching times. This form of digital modulation is an example of *continuous-phase frequency-shift keying* (CPFSK), on which we have more to say in the next subsection.

Figure 3.16 shows the signal constellation of BFSK, with the two coordinates of the constellation defined by

$$\phi_i(t) = \begin{cases} \sqrt{\dfrac{2}{T}}\cos(2\pi f_i t) & 0 \le t \le T \\ 0 & \text{otherwise} \end{cases} \qquad (3.49)$$

where $i = 1,2$, and f_i is itself defined by Eq. (3.48).

Problem 3.7   What conclusion can you draw from the signal-space diagram of Fig. 3.16 for BFSK, compared with that of Fig. 3.14(a) for BPSK?
Ans. The signal-space diagram of BPSK in Fig. 3.14(a) is one-dimensional, whereas that of BFSK in Fig. 3.16 has two dimensions. Dimensions as two coordinates, $\phi_1(t)$ and $\phi_2(t)$, are required for the description of BFSK. ∎

Problem 3.8   How is BFSK extended to M-ary FSK?
Ans. The dimensionality of the BFSK signal is two, exactly the same as the number of symbols embodied in the generation of the signal. In the M-ary FSK signal, there are M distinct signals to be accounted for; hence, the dimensionality of the M-ary FSK signal is M. M-ary FSK includes BFSK as a special case. ∎

3.7.3 Continuous-Phase Modulation: Minimum Shift Keying[8]

In constructing the BFSK signal constellation of Fig. 3.16, the only design constraint we imposed on the construction is that the two frequencies f_1 and f_2 (representing symbols 1 and 0, respectively) satisfy the requirement of Eq. (3.48). However, through a more constrained choice of these two frequencies, we can make a significant difference to two important characteristics of this nonlinear method of digital signaling, namely, improved spectral efficiency and improved noise performance.

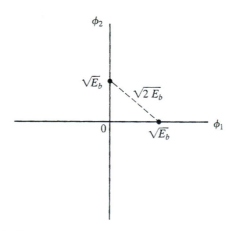

FIGURE 3.16   Signal constellation of binary FSK signal.

Consider, then, a *CPFSK signal*, which, for the interval $0 \le t \le T$, is defined as

$$
s(t) = \begin{cases}
\sqrt{\dfrac{2E_b}{T}} \cos[2\pi f_1 t + \theta(0)] & \text{for symbol 1} \\[4mm]
\sqrt{\dfrac{2E_b}{T}} \cos[2\pi f_2 t + \theta(0)] & \text{for symbol 0}
\end{cases}
\tag{3.50}
$$

where E_b is the energy transmitted per bit and T is the symbol (bit) duration. The phase $\theta(0)$, denoting the value of the phase at time $t = 0$, sums up the past history of the modulation process up to time $t = 0$. The frequencies f_1 and f_2 are sent in response to binary symbols 1 and 0 respectively appearing at the modulator input.

Another useful way of representing the CPFSK signal $s(t)$ is to express it in the conventional form of an angle-modulated signal as

$$
s(t) = \sqrt{\frac{2E_b}{T}} \cos[2\pi f_c t + \theta(t)]
\tag{3.51}
$$

where $\theta(t)$ is the phase of $s(t)$; Eq. (3.51) is a special form of Eq. (3.37) with $a(t) = \sqrt{2E_b/T}$. When the phase $\theta(t)$ is a continuous function of time, we find that the modulated signal $s(t)$ itself is also continuous at all times, including the interbit switching times. The phase $\theta(t)$ of a CPFSK signal increases or decreases linearly with time during each bit duration of T seconds, as shown by the relationship

$$
\theta(t) = \theta(0) \pm \frac{\pi h}{T} t \qquad 0 \le t \le T
\tag{3.52}
$$

where the plus sign corresponds to sending symbol 1 and the minus sign corresponds to sending symbol 0; the parameter h is to be defined. Substituting Eq. (3.52) into Eq. (3.51), and then comparing the angle of the cosine function with that of Eq. (3.50), we deduce the following pair of relations:

$$
f_c + \frac{h}{2T} = f_1
\tag{3.53}
$$

$$
f_c - \frac{h}{2T} = f_2
\tag{3.54}
$$

Solving Eqs. (3.53) and (3.54) for f_c and h, we thus get

$$
f_c = \frac{1}{2}(f_1 + f_2)
\tag{3.55}
$$

and

$$
h = T(f_1 - f_2)
\tag{3.56}
$$

The nominal carrier frequency f_c is therefore the arithmetic mean of the transmit frequencies f_1 and f_2. The difference between the frequencies f_1 and f_2, normalized with respect to the bit rate $1/T$, defines the dimensionless parameter h, which is referred to as the *deviation ratio* of the CPFSK signal. We have purposely used a different symbol for this deviation ratio to distinguish it from the deviation ratio of an analog FM signal.

From Eq. (3.52) we find that at time $t = T$,

$$\theta(T) - \theta(0) = \begin{cases} \pi h & \text{for symbol 1} \\ -\pi h & \text{for symbol 0} \end{cases} \tag{3.57}$$

That is to say, sending of the symbol 1 increases the phase of a CPFSK signal $s(t)$ by πh radians, whereas sending of the symbol 0 reduces the phase by an equal amount.

The variation of phase $\theta(t)$ with time t follows a path consisting of a sequence of straight lines, the slopes of which represent changes in frequency. Figure 3.17 depicts possible paths starting from time $t = 0$. A plot like that shown is called a *phase tree*. The tree makes clear the transitions of phase across interval boundaries of the incoming sequence of data bits. Moreover, it is evident from the figure that the phase of a CPFSK signal is an odd or even multiple of πh radians at odd or even multiples, respectively, of the bit duration T.

The phase tree described in Fig. 3.17 is a manifestation of phase continuity, which is an inherent characteristic of a CPFSK signal. To appreciate the notion of phase continuity, let us go back for a moment to Sunde's FSK, which is the simplest form of a CPFSK scheme. In this case, the deviation ratio h is exactly unity, in accordance with Eq. (3.56). Hence, according to Fig. 3.17 the phase change over one bit interval is $\pm\pi$ radians. But a change of $+\pi$ radians is exactly the same as a change of $-\pi$ radians, modulo 2π. It follows, therefore, that in the case of Sunde's FSK, there is no memory; that is, knowing which particular change occurred in the *previous* bit interval provides no help in the *current* bit interval.

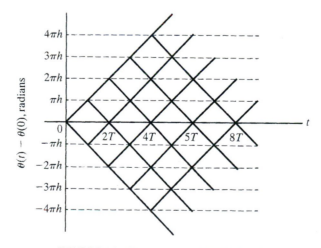

FIGURE 3.17   Phase tree of a CPFSK signal.

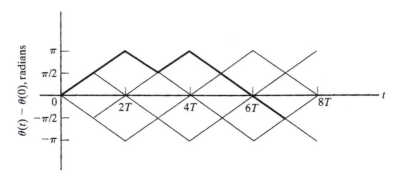

FIGURE 3.18   Phase trellis; boldfaced path represents the sequence 1101000.

In contrast, we have a completely different situation when the deviation ratio h is assigned the special value 1/2. We now find that the phase can take on only the two values $\pm\pi/2$ at odd multiples of T and only the two values 0 and π at even multiples of T, as shown in Fig. 3.18. This second graph is called a *phase trellis*, since a trellis is a treelike structure with reemerging branches. Each path from left to right through the trellis of the figure corresponds to a specific binary sequence input. For example, the path shown in boldface in Fig. 3.18 corresponds to the binary sequence 1101000 with $\theta(t) = 0$. Henceforth, we assume that $h = 1/2$.

With $h = 1/2$, we find from Eq. (3.56) that the frequency deviation (i.e., the difference between the two signaling frequencies f_1 and f_2) equals half the bit rate. This is the minimum frequency spacing that allows the two FSK signals representing the symbols 1 and 0, as in Eq. (3.51), to be coherently orthogonal, in the sense that they do not interfere with one another in the process of detection. It is for this reason that a CPFSK signal with a deviation ratio of one-half is commonly referred to as *minimum shift keying* (MSK).

Table 3.3 delineates how the memory of the MSK generator manifests itself. For example, if $\theta(0) = 0$, then the transmission of the symbol 0 results in $\theta(T) = -\pi/2$. If, however, $\theta(0) = \pi$, then the transmission of the symbol 0 results in $\theta(T) = \pi/2$.

Problem 3.9   Assume that the frequencies f_1 and f_2 are both odd integer multiplies of $1/4T$. Then show that the MSK signal may be expressed in terms of the orthonormal pair of coordinates

$$\phi_1(t) = \sqrt{\frac{2}{T}}\cos\left(\frac{\pi}{2T}t\right)\cos(2\pi f_c t) \qquad 0 \le t \le T \qquad (3.58)$$

and

$$\phi_2(t) = \sqrt{\frac{2}{T}}\sin\left(\frac{\pi}{2T}t\right)\sin(2\pi f_c t) \qquad 0 \le t \le T \qquad (3.59)$$

In particular, show that

$$s(t) = s_1\phi_1(t) + s_2\phi_2(t) \qquad (3.60)$$

TABLE 3.3   Transition characterization of MSK.

Transmitted Binary Symbol, $0 \le t \le T$	Phase States (radians)		Coordinates of Message Points	
	$\theta(0)$	$\theta(T)$	s_1	s_2
0	0	$-\pi/2$	$+\sqrt{E_b}$	$+\sqrt{E_b}$
1	π	$-\pi/2$	$-\sqrt{E_b}$	$+\sqrt{E_b}$
0	π	$+\pi/2$	$-\sqrt{E_b}$	$-\sqrt{E_b}$
1	0	$+\pi/2$	$+\sqrt{E_b}$	$-\sqrt{E_b}$

where

$$s_1 = \int_0^{2T} s(t)\phi_1(t)dt \qquad -T \le t \le T \qquad (3.61)$$
$$= \sqrt{E}\cos(\theta(0))$$

and

$$s_2 = \int_{-T}^{T} s(t)\phi_2(t)dt \qquad 0 \le t \le 2T \qquad (3.62)$$
$$= \sqrt{E}\sin(\theta(T))$$

Hence, using the formulas of Eqs. (3.61) and (3.62), verify the entries for s_1 and s_2 in Table 3.3. ■

Problem 3.10   Using the results of Problem 3.9, construct the waveform of the MSK signal for the binary sequence 1101000, assuming that $f_1 = 5/(4T)$, and $f_2 = 3/(4T)$ by plotting the components $s_1\phi_1(t)$ and $s_2\phi_2(t)$.
Ans. See Fig. 3.19, where the input sequence occupies the time interval $0 \le t/T \le 7$ ■

3.7.4 Power Spectra of MSK Signal

It is informative to formulate the power spectrum of an MSK signal. To that end, we assume that the input binary sequence is random, with the symbols 1 and 0 equally likely and the symbols transmitted during different time slots being statistically independent. We may thus make the following observations:

1. Depending on the value of the phase state $\theta(0)$, the in-phase component equals $+g_I(t)$ or $-g_I(t)$, where

$$g_I(t) = \begin{cases} \sqrt{\dfrac{2E_b}{T}}\cos\left(\dfrac{\pi t}{2T}\right) & -T \le t \le T \\ \\ 0 & \text{otherwise} \end{cases} \qquad (3.63)$$

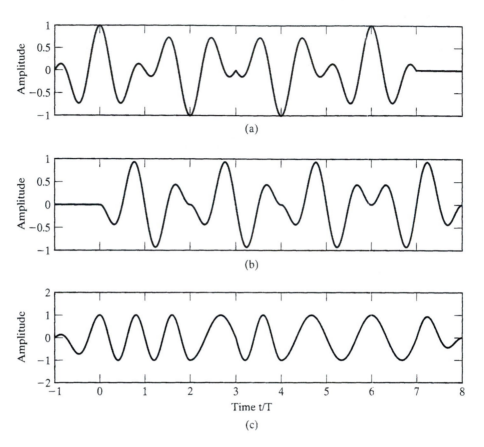

FIGURE 3.19   Waveforms for Problem 3.10: (a) $s_1\phi_1(t)$, (b) $s_2\phi_2(t)$, and (c) MSK signal $s(t)$.

2. Depending on the value of the phase state $\theta(T)$, the quadrature component equals $+g_Q(t)$ or $-g_Q(t)$, where

$$
g_Q(t) = \begin{cases} \sqrt{\dfrac{2E_b}{T}}\sin\left(\dfrac{\pi t}{2T}\right) & -T \leq t \leq T \\[2ex] 0 & \text{otherwise} \end{cases}
\tag{3.64}
$$

3. The in-phase and quadrature components of the MSK signal are statistically independent, yielding the baseband power spectral density of the MSK signal:

$$
\begin{aligned}
S_B(f) &= 2\left[\frac{\Psi_g(f)}{2T}\right] \\[2ex]
&= \frac{32E_b}{\pi^2}\left[\frac{\cos(2\pi Tf)}{16T^2f^2 - 1}\right]^2
\end{aligned}
\tag{3.65}
$$

where $\psi_g(f)$ is the common power spectral density of $g_I(t)$ and $g_Q(t)$. According to Eq. (3.65), the baseband power spectral density of the MSK signal falls off as the inverse fourth power of frequency, compared with the inverse square power of frequency for QPSK. Accordingly, MSK does not produce as much interference outside the signal band of interest as does QPSK. This is a desirable characteristic of MSK, especially when the digital communication system operates with a bandwidth limitation.

Problem 3.11   Find the frequencies at which the baseband power spectrum of the MSK signal attains its minimum value of zero.

Ans. $f = 1/4T$.   ∎

3.7.5 Gaussian-Filtered MSK

From the study of minimum shift keying (MSK) just presented, we may summarize the desirable properties of the MSK signal. The MSK signal should have

1. a constant envelope,
2. a relatively narrow bandwidth, and
3. a coherent detection performance equivalent to that of QPSK.

However, the out-of-band spectral characteristics of MSK signals, as good as they are, still do not satisfy the stringent requirements of wireless communications. To illustrate this limitation, we find from Eq. (3.65) that at $fT = 0.5$, the baseband power spectral density of the MSK signal drops by only $10\log_{10} 9 = 9.54$ dB below its midband value. Hence, when the MSK signal is assigned a transmission bandwidth of $1/T$, the adjacent channel interference of a wireless communication system using MSK is not low enough to satisfy the practical requirements of such a multiuser communications environment.

Recognizing that the MSK signal can be generated by direct frequency modulation of a voltage-controlled oscillator, we may overcome this serious limitation of MSK by modifying the power spectrum of the signal into a *compact* form, while maintaining the constant-envelope property of the signal. This modification can be achieved through the use of a premodulation low-pass filter, hereafter referred to as a baseband *pulse-shaping filter.* Such a filter should have

1. a frequency response with a narrow bandwidth and sharp cutoff characteristics,
2. an impulse response with relatively low overshoot, and
3. a phase trellis which evolves in such a manner that the carrier phase of the modulated signal assumes the two values $\pm\pi/2$ at odd multiples of T and the two values 0 and π at even multiples of T, as in MSK.

Condition 1 is needed to suppress the high-frequency components of the transmitted signal. Condition 2 avoids excessive deviations in the instantaneous frequency of the FM signal. Finally, condition 3 ensures that the modified FM signal can be coherently detected in the same way as the MSK signal is, or else it can be noncoherently detected as a simple BFSK signal.

These desirable properties can be achieved by passing a polar *nonreturn-to-zero (NRZ)* binary data stream through a baseband pulse-shaping filter whose impulse response (and, likewise, frequency response) is defined by a *Gaussian* function. In this line code, the binary symbols 0 and 1 are represented by levels −1 and +1, respectively. The resulting method of binary frequency modulation is naturally referred to as *Gaussian-filtered MSK* or just *GMSK*. The use of filtering has the effect of introducing additional memory into the generation of GMSK.

Let W denote the *3-dB baseband bandwidth* of the pulse-shaping filter. We may then respectively define the transfer function $H(f)$ and impulse response $h(t)$ of the pulse-shaping filter as

$$H(f) = \exp\left(-\frac{\log_e 2}{2}\left(\frac{f}{W}\right)^2\right) \tag{3.66}$$

and

$$h(t) = \sqrt{\frac{2\pi}{\log_e 2}}\, W \exp\left(-\frac{2\pi^2}{\log_e 2}W^2 t^2\right) \tag{3.67}$$

The response of this Gaussian filter to a rectangular pulse of unit amplitude and duration T (centered on the origin) is given by

$$
\begin{aligned}
g(t) &= \int_{-T/2}^{T/2} h(t - \tau)d\tau \\
&= \sqrt{\frac{2\pi}{\log_e 2}}\, W \int_{-T/2}^{T/2} \exp\left(-\frac{2\pi^2}{\log_e 2}W^2(t - \tau)^2\right) d\tau
\end{aligned}
\tag{3.68}
$$

which may be expressed as the difference between two complementary error functions:

$$g(t) = \frac{1}{2}\left[\operatorname{erfc}\left(\pi\sqrt{\frac{2\pi}{\log_e 2}}\,WT\left(\frac{t}{T} - \frac{1}{2}\right)\right) - \operatorname{erfc}\left(\pi\sqrt{\frac{2\pi}{\log_e 2}}\,WT\left(\frac{t}{T} + \frac{1}{2}\right)\right)\right] \tag{3.69}$$

(For a formal definition of the *complementary error function* and its properties, see Appendix E.) The pulse response, $g(t)$ constitutes the *frequency-shaping pulse* of the GMSK modulator, with the dimensionless *time–bandwidth product* WT playing the role of a design parameter.

The frequency-shaping pulse $g(t)$, as defined in Eq. (3.69), is noncausal, in that it is nonzero for $t < -T/2$, where $t = -T/2$ is the time at which the input rectangular pulse (symmetrically positioned around the origin) is applied to the Gaussian filter. For a causal response, $g(t)$ must be truncated and shifted in time. Figure 3.20 presents plots of $g(t)$, which have been truncated at $t = \pm 2.5T$ and then shifted in time by $2.5T$, for $WT = 0.2, 0.25$, and 0.3. Note that as WT is reduced, the time spread of the frequency-shaping pulse is correspondingly increased.

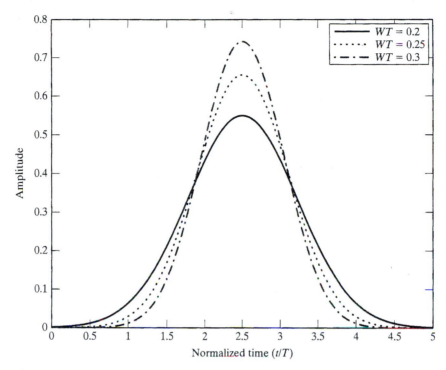

FIGURE 3.20   Frequency-shaping pulse $g(t)$ of Eq. (3.63), shifted in time by $2.5T$ and truncated at $\pm 2.5T$ for varying time–bandwidth product WT.

Figure 3.21 shows some machine-computed power spectra of MSK signals (expressed in decibels) versus the normalized frequency difference $(f - f_c)T$, where f_c is the midband frequency and T is the bit duration. The plots are for varying values of the time–bandwidth product WT. From the figure, we may make the following observations:

- The curve for the limiting condition $WT = \infty$ corresponds to the ordinary MSK.
- When WT is less than unity, increasingly more of the transmit power is concentrated inside the passband of the GMSK signal.

An undesirable feature of GMSK, readily apparent from Fig. 3.20, is that the processing of NRZ binary data by a Gaussian filter generates a modulating signal that is no longer confined to a single bit interval as in ordinary MSK. Stated in another way, the tails of the Gaussian impulse response of the pulse-shaping filter cause the modulating signal to spread out to adjacent symbol intervals. The net result is the generation of a *controlled form of intersymbol interference*, the extent of which increases with decreasing WT. In light of this observation and the observation we made on the basis of Fig. 3.21 about the power spectra of GMSK signals, we may say that the choice of the time–bandwidth product WT offers a trade-off between spectral compactness and a reduction in receiver performance. The compromise value $WT = 0.3$ ensures that the sidelobes of the power spectrum of the GMSK signal drop by an amount larger than 40 dB relative to the midband frequency, which means that the effect of adjacent channel interference is practically negligible. (The issue of adjacent channel interference is discussed in Section 3.9.)

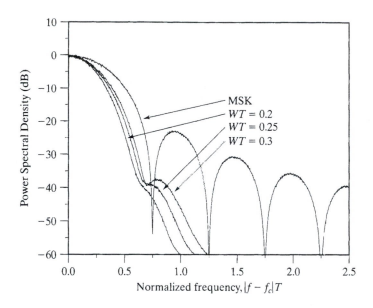

FIGURE 3.21   Power spectra of MSK and GMSK signals for varying time–bandwidth product.

The corresponding degradation in noise performance incurred by the choice of $WT = 0.3$ is about 0.46 dB, compared with the standard MSK, which is a small price to pay for the highly desirable spectral compactness of the GMSK signal; this degradation in performance is justified in Section 3.12.

3.8   FREQUENCY-DIVISION MULTIPLE ACCESS[7]

In *frequency-division multiple access* (FDMA), the available spectrum is divided into a set of continuous frequency bands labeled 1 through N, and the bands are assigned to individual mobile users for the purpose of communications on a continuous-time basis for the duration of a telephone call. An issue of particular concern in the design of FDMA systems is that of adjacent channel interference, which is discussed in Section 3.9 in the context of different modulation techniques. In any event, to reduce this form of interference, two precautionary measures are usually taken:

1. The power spectral density of the modulated signal is carefully *controlled* so that the power radiated into the adjacent band is 60 to 80 dBs below that in the desired band; this system requirement, in turn, requires the use of highly selective filters.

2. *Guard bands* are inserted as buffer zones in the channel assignment, as illustrated in Fig. 3.22. The guard bands provide additional protection because of the impossibility of achieving the ideal (i.e., brick wall) filtering characteristic to separate the different users.

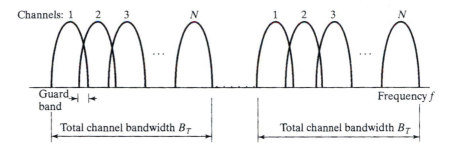

FIGURE 3.22   Channel allocations in FDD/FDMA system.

In a wireless communication system, it is highly desirable for a mobile user to send information-bearing signals to the base station while receiving information-bearing signals from the base station itself on the same antenna. To provide for such a capability, some form of *duplexing* is used, either in the frequency domain or in the time domain. Figure 3.23 shows the block diagram of a *frequency-division diplexer* (FDD), which incorporates a modulator and demodulator, with the modulator acting on input *control* data and the demodulator producing corresponding output data.

For *frequency-division diplex* (FDD) transmissions in an FDMA system accommodating, N users, as illustrated in Fig. 3.22, there are two groups of subbands or channels:

- One group of N contiguous subbands, occupying a total bandwidth of B_T hertz, is used for *forward-link radio transmissions* from a base station to its mobile users; a forward-link is also referred to as a *downlink*.

- A similar group of N subbands also occupying a total bandwidth of B_T hertz. These subbands are used for reverse-link radio transmissions from the mobile users to their base station; a reverse-link is also referred to as an *up-link*.

Separation of these two groups of subbands is facilitated by the FDD, as illustrated in Fig. 3.23. Each mobile user is allocated a subband in both groups of subbands, as well as the FDD band.

The first generation of analog wireless communication systems uses FDMA/FDD, with speech signals being transmitted over the forward or reverse links through frequency modulation (FM). The data control functions are performed digitally by means of frequency-shift keying (FSK) for data transmission, which is the simplest digital form of frequency modulation. Specifically, one frequency is used for the transmission

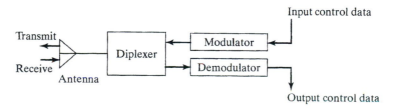

FIGURE 3.23   Frequency-division diplexer.

of binary symbol 0 and a different frequency is used for the transmission of binary symbol 1. (See Subsection 3.7.2.)

A useful feature of FDMA systems is that, for each mobile user, the radio transmission takes place over a narrow channel of bandwidth (B_T/N) hertz, in which case any fading encountered in the course of transmission tends to be flat. From Chapter 2 we recall that flat fading is relatively easy to handle, hence the adoption of FDMA for the first generation of analog wireless communications. However, FDMA systems suffer from a serious practical limitation: A separate transmitter-receiver, termed *transceiver*, is needed at the base station for each of the mobile users actively operating in its coverage area. This limitation is mitigated by using TDMA, which is discussed in Chapter 4.

3.9   TWO PRACTICAL ISSUES OF CONCERN

As already stated, modulation lies at the heart of an FDMA system. Moreover, modulation, in one form or another, features prominently in the other types of multiple-access techniques. It is therefore important that we provide a comparative assessment of the various modulation techniques in the context of wireless communications, keeping in mind two pertinent practical issues:

1. Adjacent channel interference
2. Power amplifier nonlinearity

These two issues are discussed in this section. The comparison of modulation strategies in the context of these issues is deferred to the next section.

3.9.1   Adjacent Channel Interference[10]

A performance measure of profound practical importance in the design of wireless communication systems is that of spectral efficiency. Formally, *spectral efficiency* is defined as the ratio of the permissible source rate (in bits per second) to the available channel bandwidth (in hertz); hence, spectral efficiency is measured in bits/s/Hz. The more spectrally efficient a multiple-access system is, the greater will be the number of mobile users that can operate satisfactorily inside the prescribed channel bandwidth. In the FDMA system discussed in Section 3.8, the spectral efficiency depends on how closely the individual channels (frequency bands) can be spaced. There are several factors that limit this spacing, the most important of which is *adjacent channel interference* (ACI), referred to in the preceding discussions on GMSK and FDMA.

To illustrate the ACI problem, consider a rectangular pulse $p(t)$ of duration T and whose energy spectral density is defined by

$$|P(f)|^2 = \left| \frac{\sin(\pi f T)}{(\pi f T)} \right|^2$$

$$= \operatorname{sinc}^2(f T) \tag{3.70}$$

Equation (3.70) also defines the power spectrum of a random binary data stream in which the symbols 0 and 1, represented by the amplitudes ±1, occur with equal probability. Suppose the pulse $p(t)$ is transmitted simultaneously over two channels labeled 0 and 1. Figure 3.24 presents two plots of Eq. (3.70). One plot, shown as a solid line, pertains to channel 0; the other plot, shown as a dashed line, pertains to channel 1. Pulse transmission over these two channels is accounted for by the main lobes of the respective spectra. The sidelobes account for the ACI problem. Specifically, the first two sidelobes to the left of the main lobe of channel 1 give rise to ACI in channel 0. By the same token, the first two sidelobes to the right of the main lobe of channel 0 give rise to ACI in channel 1. In both cases, however, the levels of the sidelobes are suppressed by only 13 and 17 dB relative to the main lobes (i.e., desired signals). If we combine the interference generated in channel 0 by signal transmission in channel 1 with the interference generated by the next adjacent channels +2 and −1, we see that the transmission of the desired signal in channel 0 is significantly affected by the presence of ACI.

In Fig. 3.24, the received signals are all assumed to be at the same power level. In wireless communications, signals may be transmitted at the same power level. Because of the effects of propagation over different distances to the receivers, the transmitted signals are then received at widely differing power levels with the result that in some systems, the range of received signal levels can vary by as much as 80 dB or more. If, for example, the signal on channel 1 is received 30 dB stronger than the signal on channel 0, it is possible for the ACI produced by channel 1 to completely overwhelm the desired signal in channel 0. This phenomenon is called the *near-far problem.*

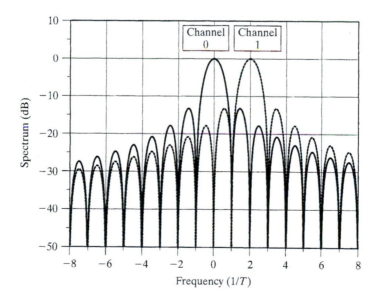

FIGURE 3.24   Adjacent-channel interference problem, illustrated with the spectrum of a rectangular pulse of duration T.

3.9.2 Power Amplifier Nonlinearity

One of the most critical constraints imposed on mobile radio terminals is their *limited battery power*. Terminals are designed for a certain battery life or time between recharges, and the corresponding circuitry must respect the underlying power budget. A significant consumer of power in mobile radios is the *transmit power amplifier*. For this reason, considerable attention is always paid to this component of the wireless system.

There are many amplifier designs, and they have been traditionally categorized in the electronics literature as Class A, Class B, Class AB, Class C, Class D, and so on, where Class A represents a linear amplifier and the amplifier classes that follow are typically increasingly nonlinear. Although Class A is considered to be a linear amplifier, no amplifier is truly linear; what linearity means in this context is that the operating point is chosen such that the amplifier behaves linearly over the signal range. The drawback of a Class A amplifier is that it is power inefficient. Typically, 25 percent or less of the input power is actually converted to radio frequency (RF) power; the power that is left is converted to heat and, therefore, wasted. The remaining amplifier classes are designed to provide increasingly improved power efficiency, but at the expense of making the amplifier increasingly more nonlinear.

In Fig. 3.25, we show the measured gain characteristic of a solid-state power amplifier at two different frequencies: 1626 MHz and 1643 MHz. The curves show that the amplifier gain is approximately constant, that is, the amplifier is linear over a wide range of inputs. However, as the input level increases, the gain decreases, indicating that the amplifier is saturating. It can also be seen that there is a significant difference in

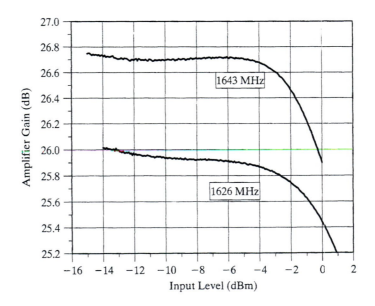

FIGURE 3.25 Gain characteristic of a solid-state amplifier at two different operating frequencies: 1626 MHz and 1643 MHz.

amplifier performance at different frequencies. If this amplifier is operated at an average input level of −10 dBm with an amplitude swing of ±2 dB, then the amplifier would be considered linear. If, however, the signal has an amplitude swing of ±10 dB, the amplifier would be considered nonlinear. The fact that the gain is not constant over all input levels means that the amplifier introduces *amplitude distortion* in the form of amplitude modulation (AM). Since the amplitude distortion depends upon the input level, it is referred to as *AM-to-AM distortion.*

An ideal amplifier does not affect the phase of an input signal, except possibly for a constant phase rotation. Unfortunately, a practical amplifier behaves quite differently, as illustrated in Fig. 3.26, which shows the phase characteristic of the same power amplifier considered in Fig. 3.25. The fact that the phase characteristic is not constant over all input levels means that the amplifier introduces *phase distortion* in the form of phase modulation (PM). Since the phase distortion depends upon the input level, this second form of distortion is called *AM-to-PM distortion.*

An "ideal" amplifier nonlinearity has the AM-to-AM characteristic illustrated in Fig. 3.27. That is, the amplifier acts linearly up to a given point, whereafter it sets a hard limit on the input signal. This can sometimes be achieved by placing appropriate compensation around a nonideal amplifier. With this ideal nonlinearity, the phase distortion is assumed to be zero. The operating point of the amplifier is often specified as the *input back-off,* defined as the root-mean-square (rms) input signal level $V_{\text{in, rms}}$ relative to the saturation input level $V_{\text{in, sat}}$ in dB. That is,

$$\text{Input back-off} = 10\log_{10}\left(\frac{V_{\text{in, rms}}}{V_{\text{out, sat}}}\right)^2 \tag{3.71}$$

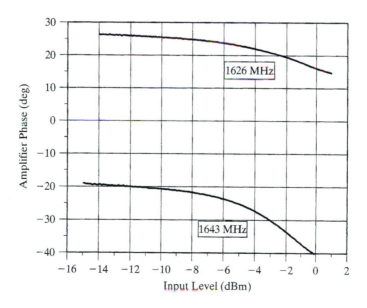

FIGURE 3.26   Phase characteristic of a nonlinear amplifier at two different operating frequencies: 1626 MHz and 1643 MHz.

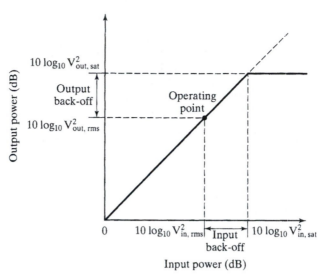

FIGURE 3.27   AM/AM characteristic of an "ideal" form of amplifier nonlinearity.

Alternatively, the operating point can be expressed in terms of the *output back-off*, defined as follows:

$$\text{Output back-off} = 10\log_{10}\left(\frac{V_{\text{out, rms}}}{V_{\text{out, sat}}}\right)^2 \tag{3.72}$$

where $V_{\text{out, rms}}$ is the rms output signal level and $V_{\text{out, sat}}$ is the saturation output level. In both Eqs. (3.71) and (3.72), the closeness to saturation determines the amount of distortion introduced by the amplifier.

3.10 COMPARISON OF MODULATION STRATEGIES FOR WIRELESS COMMUNICATIONS

With the material presented in Section 3.9 at hand, we are now ready to address the issue of comparing the modulation strategies described in Sections 3.3 and 3.7, linear and nonlinear modulation strategies for wireless. The comparison will be made under two headings: linear and nonlinear channels, which is determined by whether the transmit power amplifier is operated in its linear or nonlinear region. In both cases, the discussion will focus on digital modulation techniques because of the ever-increasing pervasiveness of digital wireless communications.

3.10.1 Linear Channels

Consider first the effect of adjacent channel interference, in which case the *transmit spectrum* is clearly an important criterion for selecting between modulation strategies in a wireless system. The transmit spectrum is determined by three factors:

1. *Pulse shaping* that is performed as part of the modulation process, for example, rectangular pulse shaping, Gaussian filtering, and so on.
2. *Other filtering* that may be necessary in the transmitter RF chain, to remove the presence of images that may occur during the up-conversion to an RF frequency.
3. *Presence of nonlinearities* in the transmitter RF chain, for example, power amplifiers that may be operated close to saturation.

In general, the "other filtering" is often the least significant of the three and we shall ignore it in this presentation. This subsection is devoted to linear channels; nonlinear effects will be covered in the next subsection.

In Fig. 3.28, we compare the baseband transmit spectra of the three modulation strategies discussed in this chapter.[11] The baseband spectra shown in the figure are normalized to the same peak spectral density. The three spectra correspond to

1. QPSK with rectangular pulse shaping; this spectrum is characterized as the $\text{sinc}^2(fT)$ function where T is the symbol duration.
2. MSK with its pulse shape corresponding to half a sine wave.
3. QPSK with root raised-cosine (RC) pulse shaping with 50% rolloff as defined in Section 3.4.

Comparing QPSK with rectangular pulse shaping to MSK, we find that QPSK has a narrower mainlobe but the sidelobes decrease much more slowly than MSK. For $f \gg 1/T$, the baseband power spectrum of MSK signal falls off as the inverse fourth power of frequency, whereas in the case of QPSK, it falls off as the inverse square of frequency. Consequently, the ACI produced by MSK is less troublesome than that produced by QPSK.

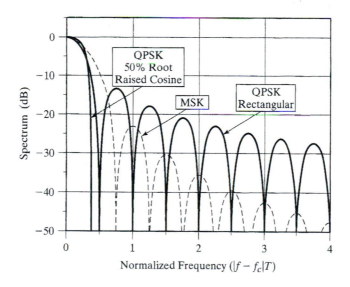

FIGURE 3.28 Comparison of the spectra of QPSK with rectangular pulse shaping, MSK, and QPSK with 50% root-raised-cosine pulse shaping.

Comparing QPSK with root pulse shaping to the other two modulation strategies, the advantages of the former linear method of modulation are clear:

- The main lobe of QPSK with root RC pulse shaping is the narrowest among the three modulation strategies considered herein.
- The QPSK with root RC pulse shaping has negligibly small sidelobes.

In a linear channel, the QPSK with root RC pulse shaping is clearly the superior strategy in terms of minimizing adjacent channel interference. In practical implementations of root RC filters, there are sidelobes; however, by careful filter design, these sidelobes can be easily kept 40 to 50 dB below the level of the main lobe. Note that in a linear channel, the spectra of the three modulations—QPSK, OQPSK, and $\pi/4$-QPSK—are identical when they use the same pulse shaping.

3.10.2 Nonlinear Channels

The use of a nonlinear transmit power amplifier may have a significant effect on system performance, depending upon on how close to saturation the amplifier is operated. The effect of the amplifier also depends on the type of modulation employed. The two examples of Subsection 3.10.1, namely, rectangular QPSK and MSK, are constant-envelope modulations. Since the nonlinear effects depend upon envelope variation, the spectra of these modulations are unaffected by a nonlinear amplifier. The same can be said of the GMSK, the power spectrum, which is illustrated in Fig. 3.21.

However, the same cannot be said of modulations such as QPSK with root RC filtering, which rely on its envelope variations to produce a compact spectrum. In fact, because the envelope characteristics of the three modulation schemes QPSK, OQPSK, and $\pi/4$-shifted QPSK are different, the effect of the nonlinear amplifier on the respective modulated signals will differ. In Fig. 3.23, we compare the spectra of these three modulations after passing through an "ideal" nonlinear amplifier. This ideal nonlinearity has a linear response up to the saturation point, at which point it clips the complex envelope of the signal to a constant level in accordance with Fig. 3.27. For the results shown in Fig. 3.29, the average input power was 1 dB below saturation.

For this ideal nonlinearity, the different responses of the three modulation types are evident but the variation between them is small. Of the three, OQPSK has the better spectral response. Compared to the spectrum experienced in a linear channel, the nonlinear channel has raised the sidelobes up to 30 dB with a slow rolloff. This is still superior to MSK, but not as good as various forms of GMSK, whose power spectra are shown in Fig. 3.21.

In practice, the choice of modulation is a tradeoff between various characteristics. The transmit spectrum is an important consideration in the tradeoff. Other considerations include the simplicity of detection and error-rate performance; the latter issue is discussed in Section 3.12. For many new wireless communication systems, it appears that linear modulation strategies, such as QPSK with root raised-cosine filtering, have emerged as the methods of choice.

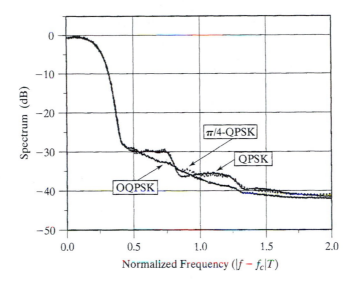

FIGURE 3.29   Comparison of different QPSK spectrum when passed through an ideal nonlinear amplifier with a 1 dB input backoff. All three modulation schemes use root-raised-cosine pulse shaping with 50% rolloff.

One last comment on phase distortion is in order. Just as with AM-to-AM distortion, AM-to-PM distortion varies with the signal level at the power amplifier input. With small signal levels, there is little phase distortion to be concerned with. As the signal level approaches the point of saturating the amplifier, the phase distortion increases. For modulation schemes with a constant envelope, AM-to-PM distortion amounts to a simple phase rotation of the signal and is not really a distortion at all. For nonconstant envelope modulations, the amount of distortion that can be tolerated depends upon the constellation. With BPSK modulation—where there is a shift of 180 degrees between constellation points—significant phase distortion can be tolerated. However, with more complicated modulation schemes such as 64-QAM, where the phase shift between constellation points can be as low as 10 degrees or less, the tolerance to phase distortion is much lower.

3.11   CHANNEL ESTIMATION AND TRACKING

From Chapter 2 we recall that fading is a major source of channel impairment in wireless communications. By definition, however, with slowly fading channels, the phase change over one symbol interval due to fading is *small* compared with the phase change in the transmitted signal. Accordingly, we can exploit this property and use differential detection to track phase variations in the transmitted signal. Differential detection is simple to implement—hence its wide application in practice. But the use of differential detection is usually limited to phase modulation systems, although there are some extensions to systems that combine amplitude and phase modulation.

Another method for tracking a slowly fading channel is based on pilot symbol transmission, which operates by sending known symbols at regular intervals (bursts) throughout the course of data transmission. Pilot symbol transmission schemes offer a potential performance advantage over differential detection in that, in addition to tracking, they can be used to estimate the impulse response of the channel; hence, they can improve the noise performance of the receiver. This performance advantage, however, is attained at the cost of a somewhat reduced spectral efficiency and increased system complexity.

3.11.1 Differential Detection

In pass-band transmission systems, the primary concern is the recovery of the desired signal from its assigned frequency channel. If we assume linear modulation, then in light of the material presented in Section 3.3, the transmitted band-pass signal can be represented as

$$s(t) = A \text{Re} \left\{ m(t) e^{j2\pi f_c t} \right\} \tag{3.73}$$

where $m(t)$ is to be defined, A is the transmit amplitude and f_c is the transmit carrier frequency. In previous sections of this chapter, we have assumed perfect recovery of the carrier frequency and phase in converting to complex baseband. Although the nominal frequency of the channel may be known, there will be some frequency errors due to inaccuracies of the transmit and receive oscillators. In practice, the received complex baseband signal is accurately represented as

$$\tilde{x}(t) = A' m(t) e^{j(2\pi\Delta f t + \phi)} + w(t) \tag{3.74}$$

where A' is the received amplitude, Δf is the residual frequency error after downconversion, ϕ is the residual phase error, and $w(t)$ is the additive channel noise. Equation (3.74) assumes an AWGN channel. It is left to the baseband circuitry to acquire and track this residual frequency and phase error.

In wireless channels, the coherent recovery and tracking of the residual frequency error Δf can be difficult. A simple, robust technique for recovering the carrier frequency is *differential detection.*[12] In this technique, it is assumed that the receiver "knows" the transmit frequency to within a certain accuracy; typically, this error is less than 5% of the data rate.

Differential detection can be illustrated with linear modulation, with the complex baseband signal written as

$$m(t) = \sum_k b_k p(t - kT) \tag{3.75}$$

where $p(t)$ is the pulse shape used to modulate each data symbol b_k. If the residual frequency error is small compared with the signal bandwidth, it is practically transparent to the matched-filtering process, in which case the sampled value of the matched filter output is given by the convolution integral

$$y_k = \int_{nT}^{(n+1)T} p^*(t - kT)\tilde{x}(t)dt$$

which approximates to

$$y_k = A'b_k e^{j(2\pi\Delta fkT + \phi)} + w_k \tag{3.76}$$

where it is assumed that the pulse shape is normalized to unit energy. (The characterization of *matched filters* is discussed in Appendix D.) To simplify the presentation, the received amplitude A' is hereafter ignored without loss of generality.

If the data stream is *differentially encoded* at the transmitter such that $b_k = a_k b_{k-1}$, where a_k is the original data symbol, then the original data can be recovered with the "delay-and-multiply" circuit shown in Fig. 3.30. Mathematically, this circuit performs the following operation:

$$\begin{aligned}
\hat{a}_k &= y_k y_{k-1}^* \\
&= [b_k e^{(j2\pi\Delta fkT+\phi)} + w_k][b_{k-1}e^{(j2\pi\Delta f(k-1)T+\phi)} + w_{k-1}]^* \\
&= b_k b_{k-1}^* e^{j2\pi\Delta fT} + \eta_k \\
&\approx a_k + \eta_k
\end{aligned} \tag{3.77}$$

In the last line of Eq. (3.77), it is assumed that Δf is small enough for $e^{j2\pi\Delta fT} \approx 1$. The noise term η_k represents the sum of three noise-related terms resulting from the multiplication of terms in the second line of the equation. The approach described herein is well suited for both BPSK and QPSK forms of modulation.

To sum up, the practical significance of differential detection is twofold:

1. Differential detection is based on phase differences. It therefore lends itself naturally to linear phase modulation schemes, but it can also be applied to nonlinear phase modulation schemes,[12] so long as the phase is differentially encoded.

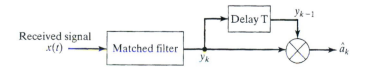

FIGURE 3.30   Delay-and-multiply circuit for differential detection.

2. If amplitude is not important (i.e., the data stream is phase modulated), then the effects of fading are similar to a small time-varying frequency error on the received signal. Provided that the fading is slow relative to the data rate, differential detection may be applied to such wireless systems.

For these reasons, differential detection has established itself as a robust strategy for the detection of phase modulation over flat-fading channels.

3.11.2 Pilot Symbol Transmission[13]

In some applications, it may be necessary to perform *coherent detection* so as to avoid the degradation in receiver performance associated with differential detection; or it could be that some of the information about the data is encoded in amplitude variations, as in QAM for example. In these cases, the insertion of *known pilot symbols* in the transmit data stream is the preferred approach.

If the channel exhibits flat fading, then the complex baseband received signal of Eq. (3.74) may be modeled as

$$\tilde{x}(t) = \tilde{\alpha}(t)m(t) + \tilde{w}(t) \tag{3.78}$$

where $\tilde{\alpha}(t)$ is a complex Rayleigh fading process. In Eq. (3.78), we have included the received amplitude A' as part of the fading process $\tilde{\alpha}(t)$. Also, to simplify matters, we have neglected the small frequency error Δf. In a manner analogous to the development of Eq. (3.76), with linear modulation and Nyquist pulse shaping, the samples at the output of the matched filter can be represented in discrete time as

$$y_k = \alpha_k b_k + w_k \tag{3.79}$$

where b_k is the data symbol in symbol period k, w_k is the corresponding white Gaussian noise sample, and α_k is the complex gain due to the fading process. The model of Eq. (3.79) is based on the assumption that the fading is approximately constant over the significant portion of the pulse length. If we use a Rayleigh-fading channel model, then $\{\alpha_k\}$ are samples of a Rayleigh fading process.

If the pilot symbols are known at regular times—say, at $k = Ki$, where K is the pilot-symbol spacing and i is an integer—(i.e., the pilot symbols are known at the receiver), then we can use Eq. (3.79) to form the equation

$$
\begin{aligned}
h_{Ki} &= b_{Ki}^* y_{Ki} \\
&= \alpha_{Ki} |b_{Ki}|^2 + b_{Ki}^* w_{Ki} \\
&= \alpha_{Ki} + w_{Ki}
\end{aligned}
\tag{3.80}
$$

The development from the second to the third line of Eq. (3.80) assumes that the pilot symbols are BPSK modulated—that is, $b_{Ki} = \pm 1$. (The random nature of the noise term w_{Ki} is unaffected by whether b_{ki} equals −1 or +1). It is straightforward to show that, under these assumptions, the statistical properties of the noise samples are unchanged.

The samples $\{h_{Ki}\}$ can be used as estimates of the fading process, since the known modulation has been removed. However, they are corrupted by noise.

In practice, we usually wish to improve upon this estimate of the fading process by smoothing adjacent estimates. In particular, we make a *linear estimator* of the fading, based on the $2L + 1$ surrounding samples, obtaining

$$\hat{\alpha}_{Ki} = \sum_{m=-L}^{L} a_m h_{K(i+m)} \tag{3.81}$$

where the a_m are complex weighting parameters for smoothing the estimate. These parameters are often chosen to minimize the expected error in the estimate. That is, the $\{a_m\}$ are chosen to minimize the *cost function*

$$\begin{aligned} J &= \mathbf{E}[|\hat{\alpha}_{Ki} - \alpha_{Ki}|^2] \\ &= \mathbf{E}\left[\left|\sum_{m=-L}^{L} a_m h_{K(i+m)} - \alpha_{Ki}\right|^2\right] \end{aligned} \tag{3.82}$$

where $\mathbf{E}$ is the statistical expectation operator. Equation (3.82) can be expanded as follows:

$$\begin{aligned} J = \mathbf{E}\bigg[|\alpha_{Ki}|^2 &+ \sum_{m=-L}^{L} \sum_{n=-L}^{L} a_m h_{K(i+m)} a_n^* h_{K(i+n)}^* \\ &- \sum_{m=-L}^{L} a_m h_{K(i+m)} \alpha_{Ki}^* - \sum_{n=-L}^{L} a_n^* h_{K(i+n)}^* \alpha_{Ki}\bigg] \end{aligned} \tag{3.83}$$

Next, we define the average power

$$P = \mathbf{E}|\alpha_{Ki}|^2 \tag{3.84}$$

and the autocorrelation of the fading-plus-noise samples as

$$\begin{aligned} \mathbf{E}[h_{K(i+m)} h_{K(i+n)}^*] &= \mathbf{E}[(\alpha_{K(i+m)} + w_{K(i+m)})(\alpha_{K(i+n)}^* + w_{K(i+n)}^*)] \\ &= \mathbf{E}[\alpha_{K(i+m)} \alpha_{K(i+n)}^*] + \mathbf{E}[w_{K(i+m)} w_{K(i+n)}^*] \\ &= \begin{cases} r_{K(m-n)} & m \neq n \\ r_0 + \sigma^2 & m = n \end{cases} \end{aligned} \tag{3.85}$$

where $r_m = R(mT)$ is the sampled autocorrelation function of the fading process and T is the symbol period. The second line of Eq.(3.85) follows from the first line due to the statistical independence of the noise and fading processes. The third line follows from the second line because the noise samples are drawn from a white-noise process and are therefore uncorrelated. Similarly, we find that

$$
\begin{aligned}
\mathbf{E}[h_{K(i+m)}\alpha_{Ki}^*] &= \mathbf{E}[(\alpha_{K(i+m)} + w_{K(i+m)})\alpha_{Ki}^*] \\
&= \mathbf{E}[\alpha_{K(i+m)}\alpha_{Ki}^*] + \mathbf{E}[w_{K(i+m)}w_{K(i+n)}^*] \\
&= r_{Km}
\end{aligned}
\tag{3.86}
$$

Now we define the matrix $\mathbf{R}$ with elements $R_{ij} = r_{K(i-j)}$ and the vector $\mathbf{r}$ with elements r_{Ki}. Then, by using Eqs. (3.84) through (3.86), we may rewrite Eq. (3.83) in the compact form

$$
J = P + \mathbf{a}^\dagger \mathbf{R}\mathbf{a} - \mathbf{a}^\dagger \mathbf{r} - \mathbf{r}^\dagger \mathbf{a}
\tag{3.87}
$$

where $\mathbf{a} = [a_{-L}, ..., a_L]^T$, and the superscript denotes Hermitian transposition. By completing the square, Eq.(3.87) can be rewritten as

$$
J = P - \mathbf{r}^\dagger \mathbf{R}^{-1}\mathbf{r} + (\mathbf{a} - \mathbf{R}^{-1}\mathbf{r})^\dagger \mathbf{R}(\mathbf{a} - \mathbf{R}^{-1}\mathbf{r})
\tag{3.88}
$$

Since the vector $\mathbf{a}$ is the only free parameter in this equation, the cost function J is clearly minimized if

$$
\mathbf{a} = \mathbf{R}^{-1}\mathbf{r}
\tag{3.89}
$$

Equation (3.89) is a form of what is known as the *Wiener–Hopf equation* in the signal-processing literature. The weighting parameter vector $\mathbf{a}$ depends only on the correlation properties of the fading process and the noise. It provides a solution for the weighting that is an optimal estimate of the fading in the mean-square error sense. (Appendix I also includes a discussion of adaptive filters in the context of adaptive antennas.)

Problem 3.12   Alternatively, we can derive the Wiener–Hopf equation (3.89) by differentiating the cost function J with respect to the weighting parameter vector $\mathbf{a}$, setting the result equal to zero, and then solving for $\mathbf{a}$. Show that this procedure also leads to Eq. (3.89).   ∎

The combination of data estimation and channel tracking may be implemented as shown in Fig. 3.31. The signal is first matched-filtered, as described by Eq. (3.76), to produce, at its output, $\{y_k\}$. The pilot samples are then demultiplexed from the data,

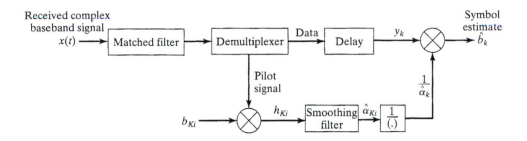

FIGURE 3.31   Pilot scheme for tracking and compensating for channel variations. Note: The b_{Ki} are pilot symbols that, by definition, are known at the receiver.

and the data modulation is removed from the pilot samples according to Eq. (3.80). Then the pilot samples are processed by a smoothing filter, in accordance with Eq. (3.81). The parameter vector of the smoothing filter is given by Eq. (3.89). Since the smoothing filter (i.e., linear estimator) introduces a delay, the data in the upper branch of Fig. 3.31 must be delayed to match the delay introduced by the filter. After this delay, we compensate for the channel by multiplying the received sample stream by the inverse of the estimated channel, $\hat{b}_k = y_k / \hat{\alpha}_k$ to obtain estimates of the data. We note that, since the pilot samples are spaced K symbols apart, the smoothing filter shown in Fig. 3.31 will also need to perform some interpolation to provide estimates of the fading channel on the intervening symbols.

   If we plot the frequency response of the finite-duration impulse-response (FIR) filter defined by the parameter vector **a**, we find that this filter is approximately matched to the fading process. In practice, if the fading bandwidth is unknown, we would typically consider the worst-case fading in designing this filter. The pilot symbols will not only track the fading but will also track any residual frequency and phase errors between the transmitter and receiver, as long as such errors fall within the bandwidth of the tracking filter. Thus, the pilot symbols permit channel estimation, tracking, and coherent detection in a single operation.

   Pilot-symbol techniques do introduce some losses. Energy that could otherwise be devoted to the information symbols must be devoted to the pilot symbols. In addition, the bandwidth of the transmitted signal must be expanded to accommodate the inclusion of the pilot symbols as well as the data. What is the minimum number of pilot symbols required? From Eq. (3.80), we see that the pilot symbols are effectively sampling the fading process. Consequently, if the pilot symbol rate is at least twice the bandwidth of the fading process, then, by Nyquist's theorem, we should be able to estimate the channel reliably. In practice, the pilot symbol rate is typically several times the fading process bandwidth, compensating for the effects of noise on the samples and simplifying the interpolation process for estimating the fading on data symbols between the pilot symbols.

Problem 3.13   If the transmission frequency is 1.9 GHz and the receiver uses a simple quartz crystal for regenerating the channel frequency, what is the range of residual frequency error on the baseband signal? A quartz crystal typically has an accuracy of 10 parts per million (ppm). If the data rate is 9.6 kilobits/s, how much phase rotation will occur during one symbol period with this error? *Ans.* ±19 kHz; ±2π *radians (approximately).*   ∎

Problem 3.14   Show that frequency shifting and phase rotation do not affect the statistical properties of zero-mean white Gaussian noise.   ∎

3.12 RECEIVER PERFORMANCE: BIT ERROR RATE

A study of digitally modulated signals would be incomplete without discussing how the presence of channel noise and channel fading would affect the performance of a wireless receiver. Recognizing the underlying input-output binary operation of a digital wireless communication system varies randomly with time, the performance of the system is measured in terms of the *average probability of symbol error*, P_e. An error is incurred when the symbol 1 in transmitted but the receiver decides in favor of the symbol 0, or vice versa. The P_e is obtained by averaging over these two conditional error probabilities with respect to the prior probabilities of symbols 0 and 1. In such a situation, P_e is often referred to as the *bit error rate* (BER).

In what follows, the receiver performance is considered under the following two conditions:

- The presence of additive channel noise
- The presence of multiplicative noise exemplified by a frequency-flat, slowly fading channel

For coherent receivers operating on simple modulation schemes, such as BPSK, QPSK, BFSK, and MSK, exact formulas are derivable for P_e. On the other hand, in the case of coherent receivers operating on more elaborate modulation schemes, such as Gaussian-filtered MSK, *M*-ary PSK, *M*-ary QAM, and *M*-ary FSK, the analytic approach becomes formidable and we therefore resort to the use of error bounds or computer simulations to derive approximate formulas for P_e.

3.12.1 Channel Noise[14]

In a *coherent receiver*, the locally generated carrier is synchronized with the carrier in the transmitter in both phase and frequency. To achieve synchronization, additional circuitry is required in designing the receiver. The receiver design can be simplified by ignoring phase information, in which case the receiver is said to be *noncoherent*. However, the design simplification is accomplished at the expense of degraded noise performance.

First of all, reference to Table 3.4 presents a summary of the bit error rates for the following receivers:

- Coherent binary phase-shift keying (BPSK)
- Coherent quadriphase-shift keying (QPSK)
- Coherent binary frequency-shift keying (BFSK)

TABLE 3.4   Summary of Formulas for the bit error rate (BER) of coherent and noncoherent digital communication receivers.

Signaling Scheme	BER (Additive white Gaussian noise channel)	BER (Slow Rayleigh fading channel)
(a) Coherent BPSK Coherent QPSK Coherent MSK	$\dfrac{1}{2}\mathrm{erfc}\left(\sqrt{\dfrac{E_b}{N_0}}\right)$	$\dfrac{1}{2}\left(1-\sqrt{\dfrac{\gamma_0}{1+\gamma_0}}\right)$
(b) Coherent BFSK	$\dfrac{1}{2}\mathrm{erfc}\left(\sqrt{\dfrac{E_b}{2N_0}}\right)$	$\dfrac{1}{2}\left(1-\sqrt{\dfrac{\gamma_0}{2+\gamma_0}}\right)$
(c) Binary DPSK	$\dfrac{1}{2}\exp\left(-\dfrac{E_b}{N_0}\right)$	$\dfrac{1}{2(1+\gamma_0)}$
(d) Noncoherent BFSK	$\dfrac{1}{2}\exp\left(-\dfrac{E_b}{2N_0}\right)$	$\dfrac{1}{2+\gamma_0}$

Definitions:

E_b = transmitted energy per bit

N_0 = one-sided power spectral density of channel noise

γ_0 = mean value of the received energy per bit-to-noise spectral density ratio

- Coherent minimum shift keying (MSK)
- Differential phase-shift keying (DPSK)
- Noncoherent binary frequency-shift keying (BFSK)

In noncoherent BFSK, the receiver consists basically of a frequency discriminator that responds only to shifts imposed on the carrier frequency by the incoming binary data. DPSK combines differential detection with binary phase-shift keying; it may, therefore, be viewed as the noncoherent version of BPSK; differential detection was discussed in subsection 3.11.1.

In Fig. 3.32, we have used the formulas summarized in Table 3.4 for an additive white Gaussian noise channel to plot the BER as a function of the signal energy per bit-to-noise spectral density ratio, E_b/N_0, which is also referred to as the signal-to-noise ratio (SNR). On the basis of the noise performance curves shown in the figure, we can make the following observations:

- The BERs for all the receivers listed in the previous paragraph decrease monotonically with increasing E_b/N_0; the curves have a similar shape in the form of a *waterfall*.
- For any prescribed E_b/N_0, coherent versions of BPSK, QPSK, and MSK produce a smaller BER than does any of the other modulation schemes.
- For a prescribed BER, coherent BPSK and DPSK require an E_b/N_0 that is 3 dB less than the corresponding values for coherent BFSK and noncoherent BFSK, respectively.

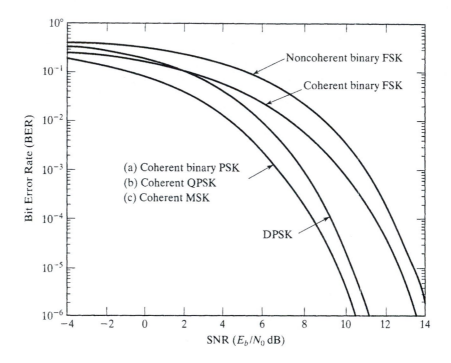

FIGURE 3.32   Comparison of the noise performance of different PSK and FSK schemes.

- At high values of E_b/N_0, DPSK and noncoherent BFSK perform almost as well (to within 1 dB) as coherent BPSK and coherent BFSK, respectively, for the same BER and signal energy per bit.

The noise performance of Gaussian-filtered minimum-shift keying (GMSK) does not lend itself to exact analysis in the way that conventional MSK does. This is rather unfortunate, since GMSK is widely used in digital wireless communications. Nevertheless, the BER of GMSK is reasonably well described by the empirical formula

$$P_e \approx \frac{1}{2}\text{erfc}\left(\sqrt{\frac{\beta E_b}{2N_0}}\right) \tag{3.90}$$

where β is a constant whose value depends on the time–bandwidth product of the MSK. For $\beta = 2$, Eq. (3.90) reduces to the exact formula for the BER of conventional MSK. Thus, we may view $10\log_{10}(\beta/2)$, expressed in decibels, as a measure of the degradation in performance of GMSK compared with that of conventional MSK. For GMSK with a time–bandwidth product $WT = 0.3$, the degradation in noise performance of the receiver is about 0.46, which corresponds to $(\beta/2) = 0.9$. This degradation is a small price to pay for the highly desirable spectral compactness of the GMSK signal.

Frequently Flat, Slowly Fading Channel

Table 3.4 also includes the exact formulas for the bit error rates for a slow Rayleigh fading channel, where

$$\gamma_0 = \frac{E_b}{N_0} \mathbf{E}[\alpha^2] \tag{3.91}$$

is the *mean value of the received signal energy per bit-to-noise spectral density ratio*. In Eq. (3.91), the expectation $\mathbf{E}[\alpha^2]$ is the mean value of the Rayleigh-distributed random variable α characterizing the channel. For the derivations of the fading-channel formulas listed in the last column of Table 3.4, the reader is referred to Problem 3.35. Comparing these formulas with the formulas for their additive white Gaussian noise (i.e., nonfading) channel counterparts, we find that the Rayleigh fading process results in a severe degradation in the noise performance of a digital communication receiver, with the degradation measured in terms of decibels of additional mean signal-to-noise spectral density ratio. In particular, the asymptotic decrease in the bit error rate with γ_0 follows an *inverse law*. This form of asymptotic behavior is dramatically different from the case of a nonfading channel, for which the asymptotic decrease in the bit error rate with γ_0 follows an *exponential law*.

In graphical terms, Fig. 3.33 plots the formulas under part (a) of Table 3.4 to compare the bit error rates of BPSK over the AWGN and Rayleigh fading channels. The

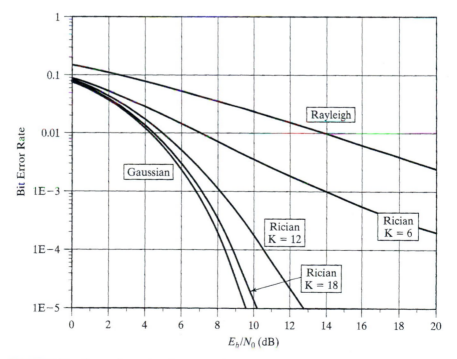

FIGURE 3.33 Comparison of performance of coherently-detected BPSK over different fading channels.

figure also includes corresponding plots for the Rice fading channel with different values of the Rice factor K. We see that as K increases from zero to infinity, the behavior of the receiver varies all the way from the Rayleigh channel to the AWGN channel. Note that the results plotted in Fig. 3.33 for the Rice channel were obtained using a procedure based on computer simulation.[15]

From Fig. 3.33 we see that as matters stand, we have a serious problem caused by channel fading. For example, at a signal-to-noise ratio of 10 dB and the presence of Rayleigh fading, the use of BPSK results in a BER of about 3×10^{-2}, which may not be good enough for the transmission of speech or digital data over the wireless channel. To overcome this problem, the traditional approach is to use diversity-on-receive, which is discussed in Chapter 6.

3.13 THEME EXAMPLE 1: ORTHOGONAL FREQUENCY-DIVISION MULTIPLEXING[16]

In the first theme example of this chapter, we discuss parts of a multicarrier system used for wireless local area networks (LANs) that fall under IEEE802.11a; this standard was discussed in Section 1.22. Recall that the object of the service is to provide wireless data links supporting rates up to 54 megabits/s to link workstations, laptops, printers, and personal digital assistants to a network access node without the expense of cabling. The service illustrates a number of the topics discussed in this chapter.

The transmission bandwidth for the service is constrained to be less than 20 MHz. One of the permitted transmission rates is 36 megabits/s of user information. If we add the overhead due to forward error-correction coding (discussed in Chapter 4), the required throughput is 48 megabits/s. To transmit 48 megabits/s through a channel bandwidth less than 20 MHz means that we need to transmit over 2 bits/Hz. This need demands that an M-ary digital modulation technique be used. The particular technique used for the service is the *16-QAM*, discussed in Section 3.6. With that technique, the in-phase and quadrature channels are independently modulated with a four-level signal derived from the incoming binary data stream. Nominally, the levels are ±1 and ±3. Figure 3.15 plots the pattern constellation of 16-QAM. At each symbol time, one of the 16 points of the constellation is selected for transmission. Since the points are equally likely to be selected, $\log_2 M = 4$ bits are transmitted at each symbol time.

At this point in the discussion, it is instructive to use the complex representation of band-pass signals presented in Section 3.5. Specifically, the complex envelope of the 16-QAM signal is defined by

$$\tilde{s}(t) = b_k\, p(t - kT_d) \qquad (k-1)T_d \le t < kT_d \qquad (3.92)$$

where $p(t)$ is a rectangular pulse, the complex coefficient b_k is selected in accordance with the 16-QAM constellation, and T_d is the symbol period of the 16-QAM signal (i.e., four times the incoming bit duration). At each symbol time kT_d, a different symbol is selected to be transmitted. In a conventional system, the band-pass signal to be transmitted is defined by Eq. (3.26), reproduced here for convenience of presentation:

$$s(t) = \text{Re}\{\tilde{s}(t)\exp(j2\pi f_c t)\} \tag{3.93}$$

For reasons discussed in Chapter 2, it is preferred to use *multicarrier transmission* for this application. In particular, instead of sending 36 megabits/s (effectively 48 mega-bits/s) over one carrier, it is preferred to send 750 kilobits/s over each of 48 distinct *subcarriers.* The term subcarrier is used to distinguish this set of carriers from the RF carrier f_c defined by Eq.(3.93). We denote the subcarrier frequencies as $f_1, f_2, \ldots, f_{48}$ and consider all of the subcarriers in their complex form

$$\tilde{c}_i(t) = \exp(j2\pi f_i t) \qquad i = 1,2,\ldots,48 \tag{3.94}$$

If the data set $\{b_k\}$ is demultiplexed into 48 parallel streams represented by $\{b_{k,i}\}_{i=1}^{48}$, running at 1/48 of the incoming data rate, then the modulation of each subcarrier can be represented in the complex low-pass form as follows:

$$
\begin{aligned}
&\vdots \\
\tilde{s}_i(t) &= b_{k,i}g(t-kT)\exp(j2\pi f_i t) \\
\tilde{s}_{i+1}(t) &= b_{k+1,i}g(t-kT)\exp(j2\pi f_{i+1}t) \\
&\vdots
\end{aligned}
\left.\begin{aligned}\end{aligned}\right\}
\begin{aligned}
&\text{for } (k-1)T \le t < kT \\
&\quad i = 1, 2, \ldots, 48
\end{aligned}
\tag{3.95}
$$

where the new symbol duration $T = 48T_d$ and the new pulse $g(t)$ is 48 times as long as $p(t)$.

Each of the terms in Eq. (3.95) is equivalent to a band-pass signal, and the aggregate of all the band-pass signals is the overall signal to be transmitted over the wireless channel. The complex low-pass representation of the combined modulation for one symbol period is given by

$$\tilde{s}(t) = \sum_{i=1}^{48} \tilde{s}_i(t) \tag{3.96}$$

At first sight, the combination of Eqs. (3.95) and (3.96) appears to be a complicated modulation strategy to implement, even though we can represent it rather simply in its complex low-pass equivalent form. However, analogous to the Fourier transform, briefly reviewed in Appendix A, is a discrete equivalent called the *discrete Fourier transform* (DFT). The DFT transforms a set of samples in the time domain into an equivalent set of samples in the frequency domain. The *inverse discrete Fourier transform* (IDFT)[17] performs the reverse operation. This transform-pair is described mathematically by

$$
\left.
\begin{array}{ll}
\text{DFT:} & b_n = \displaystyle\sum_{m=0}^{M-1} B_m \exp(-j2\pi mn/M) \quad n = 0, 1, \ldots, M-1 \\[1em]
\text{IDFT:} & B_m = \dfrac{1}{M}\displaystyle\sum_{n=0}^{M-1} b_n \exp(j2\pi mn/M) \quad m = 0, 1, \ldots, M-1
\end{array}
\right\}
\tag{3.97}
$$

where the sequences $\{b_n\}$ and $\{B_m\}$ are the frequency-domain and time-domain samples, respectively. Considering Eq. (3.95) in the context of the IDFT and noting that $g(t)$ has a rectangular pulse shape, we observe that, for a fixed k, we may represent samples at each subcarrier frequency f_i, $i = 1, 2, \ldots, 48$. We now make two assumptions:

1. The subcarrier frequencies are selected such that

$$
f_i = \frac{i}{T} \qquad i = 1, 2, \ldots, 48
\tag{3.98}
$$

2. Each subcarrier is sampled M times per symbol interval T.

If we also let the output be sampled at M times per symbol interval T, then the sampled low-pass equivalent of Eq. (3.96) is given by

$$
\tilde{s}(m) = \sum_{n=0}^{M-1} b_n \exp(j2\pi mn/M) \qquad m = 0, 1, \ldots, M-1
\tag{3.99}
$$

That is, the samples of the complex low-pass equivalent of the transmitted signal are given by the IDFT of the subcarriers.

The typical implementation of the transmitter is shown in Fig. 3.34(a). First, the incoming binary data stream is forward error-correction encoded, followed by 16-QAM modulation. (Error-correction encoding is discussed in Chapter 4.) Then it goes through a serial-to-parallel conversion device to create 48 independent data streams. Next, these independent data streams are combined by a computationally efficient implementation of the IDFT known as the *inverse fast Fourier transform (IFFT) algorithm*. The output of the IFFT consists of the time-domain samples to be transmitted over the channel. Besides displaying the 48 data-bearing subcarriers, Fig. 3.34(a) shows additional subcarriers, used by the receiver for synchronization and tracking purposes. Furthermore, the figure indicates the use of a 64-point IFFT. This is because the computationally efficient implementation of the FFT and IFFT algorithms requires that the number of samples be an integer power of 2. With this representation, any unused

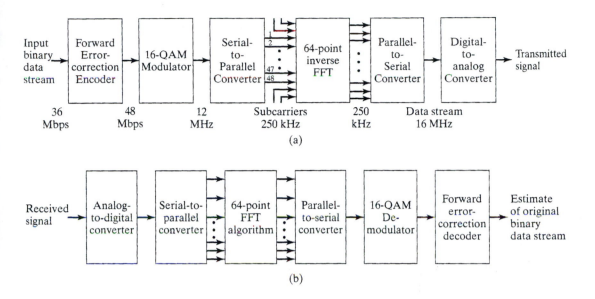

FIGURE 3.34   Block diagram of (a) OFDM transmitter, and (b) OFDM receiver.

subcarriers on the input are set equal to zero. The output of the IFFT consists of 64 of time samples of a complex envelope for each period T, which are parallel-to-serial con-verted and then finally digital-to-analog converted to facilitate transmission of the incoming data stream over the wireless channel.

Figure 3.34(b) presents the corresponding implementation of the receiver, which follows a sequence of operations in the reverse order to those performed in the trans-mitter of Fig. 3.34(a). Specifically, to recover the original input binary data stream, the received signal is passed through the following processors:

- Analog-to-digital converter
- Serial-to-parallel converter
- 64-point FFT algorithm
- Parallel-to-serial converter
- 16-QAM demodulator
- Forward error-correction decoder

The two parts of the system in Fig. 3.34 are instructive from a computationally efficient implementation point of view. To develop insight into the underlying communication-theoretic operations carried out in the system, consider Fig. 3.35 that depicts another viewpoint of what goes on basically in the system. Parts (a) and (b) of the figure per-tain to transmission and reception aspects of the system, respectively. In particular, focusing on the transmission depicted in Fig. 3.35(a), we may make two statements:

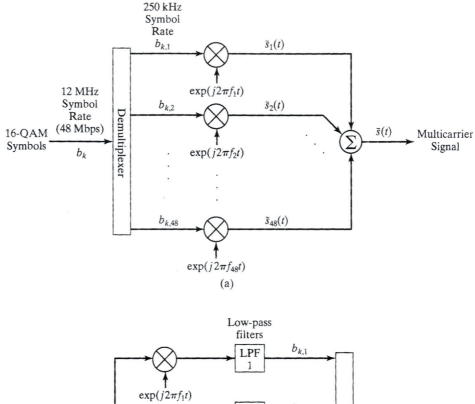

FIGURE 3.35 Communication - theoretic interpretation of the OFDM system: (a) Transmission (b) Reception.

1. *The subcarriers $\tilde{c}_i(t)$ constitute an orthogonal (orthonormal, to be more precise) set.* Given that the frequencies of the subcarriers satisfy Eq. (3.98), it follows that the subcarriers themselves satisfy the conditions of orthonormality over the symbol period T, as the following series of manipulations of the cross-correlation of the complex exponential subcarriers shows:

$$\frac{1}{T}\int_0^T \tilde{c}_i(t)\tilde{c}_k^*(t)dt = \frac{1}{T}\int_0^T \exp(j(2\pi i)t/T)\exp((-j))(2\pi k)t/T)dt$$

$$= \frac{1}{T}\int_0^T \exp(j2\pi(i-k)t/T)dt$$

$$= \frac{1}{T}\int_0^T \cos(2\pi(i-k)t/T)dt + \frac{j}{T}\int_0^T \sin(2\pi(i-k)t/T)dt \quad (3.100)$$

$$= \frac{\sin(2\pi(i-k))}{2\pi(i-k)}$$

$$= \begin{cases} 1 & \text{for all integer } i = k \\ 0 & \text{for all integer } i \neq k \end{cases}$$

2. *The complex modulated (heterodyned) signals are multiplexed in the frequency domain.*

Put together, these two statements therefore justify referring to the communication system of Fig. 3.34 as an *orthogonal frequency-division multiplexing (OFDM) system.*

3.13.1 Cyclic Prefix

Guard intervals are included in the serial data stream of the OFDM transmitter so as to overcome the effect of intersymbol interference (ISI) produced by signal transmission over the wireless channel. To take care of this matter, the pertinent OFDM symbol is cyclically extended in each guard interval. Specifically, the *cyclic extension* of an OFDM symbol is the periodic extension of the DFT output, as shown by

$$s(-k) = s(N-k) \text{ for } k = 1, 2, ..., v \quad (3.101)$$

where N is the number of subchannels in the OFDM system and v is the duration of the baseband impulse response of the wireless channel. In effect, v is the memory length of the channel. The condition described in Eq. (3.101) is called a *cyclic prefix*. Clearly, inclusion of cyclic prefixes results in an increase in the OFDM transmission bandwidth.

Having stuffed the guard intervals in the parallel-to-serial converter in the transmitter with cyclic prefixes, the guard intervals (and with them, the cyclic prefixes) are removed in the serial-to-parallel converter in the receiver *before* the conversion takes place. Then the output of the serial-to-parallel converter is in the correct form for discrete Fourier transformation.

To summarize, the theme example on OFDM has demonstrated the following desirable features:

1. Spectrally efficient digital modulation schemes, such as 16-QAM can be represented simply by their complex low-pass equivalents.

2. Modulations of even greater complexity such as OFDM, can be presented and understood easily in terms of their complex low-pass equivalents.

3. The clear understanding of these modulation schemes allows us to take advantage of digital signal-processing techniques such as the fast Fourier transform algorithm so as to simplify the implementation of some quite complicated modulation schemes with the use of multiple carriers.

3.14 THEME EXAMPLE 2: CORDLESS TELECOMMUNICATIONS

Cordless telephones are almost a ubiquitous consumer item. Although commonplace, they include some of the more advanced concepts, on a very small scale, that we will discuss later in this book. In particular, they are an example of an FDMA system operating with a very small cell size. Accordingly, they usually operate with near *line-of-sight* conditions, with a minor amount of the propagation effects considered in Chapter 2. In this second theme example of the chapter, we will describe a European standard for cordless telephony known as CT-2.

The CT-2 cordless telephones operate in the 4-MHz band of frequencies extending from 864.15 to 868.15 MHz. This band is divided into 40 channels, each with a bandwidth of 100 kHz. Why so many channels? The system is designed to allow operation of many cordless phones in close proximity. This could be for multiple phones in an office environment, or it could be just for phones in separate residences that are closely spaced. The access strategy for using these 40 channels is known as *dynamic channel allocation* (DCA) to assign frequencies. With this strategy, each base and portable unit selects an appropriate channel on the basis of their measurements of traffic conditions and channel quality. This is an example of FDMA in which multiple users share the same 4 MHz of bandwidth by dividing it up into 100-kHz channels and accessing the free channels when needed.

Transmissions from the base to a portable and from the portable to the base use the same frequency channel with a technique known as *time-division duplex* (TDD). The frame structure for this duplex transmission is shown in Fig. 3.36. In a 2-millisecond frame, the base transmits a burst of 66 bits and the portable follows with a burst of 66 bits; the sequence repeats every two milliseconds. The two bursts are separated by *guard intervals* that have a duration equivalent to 5.5 and 6.5 bits, respectively. TDD is a simple form of *time-division multiple access* (TDMA) in which the base station transmitter and mobile transmitter share the same channel by using different time slots. A more sophisticated form of TDMA will be considered in Chapter 4.

With CT-2, speech is encoded digitally, using a method known as *adaptive pulse-coded modulation* (ADPCM).[18] With this technique, the analog voice signal is converted into a bit stream that can be transmitted with digital modulation techniques.

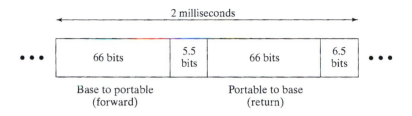

FIGURE 3.36   Frame structure for CT-2 for cordless telecommunication.

The ADPCM standard uses a data rate of 32 kbps to provide a high-quality voice signal. From the frame structure of Fig. 3.36, it is clear that the overall bit rate is 72 kilobits/s; this allows for 32 kilobits/s transmission in both directions, plus the overhead for the guard intervals.

The phones use digital modulation with a variant of FSK signalling that includes Gaussian filtering. This form of digital modulation is closely related to Gaussian minimum-shift keying (GMSK), discussed in Section 3.7. Prior to filtering, the modulated signal at complex baseband is given by

$$\tilde{s}(t) \; = \; \sqrt{\frac{2E_b}{T}} \cos(2\pi f_i t + \phi) \qquad\qquad 0 \le t < T \qquad\qquad (3.102)$$

where the $f_i, i = 1,\dots,M$, are the tones in the M-ary FSK modulation strategy. The transmitted signal is given by (see Eq. (3.26))

$$s(t) \; = \; \mathrm{Re}\{\tilde{s}(t)\exp(j2\pi f_c t)\} \qquad\qquad (3.103)$$

where f_c is the center frequency of one of the 40 RF channels. The separation of the two transmitter operations of modulation in Eq. (3.102) and upconversion in Eq. (3.103) not only is convenient for analysis, but also has its implementation advantages. Many advanced signal-processing techniques can be applied at complex baseband with the use of a microprocessor that would be very difficult to carry out if the modulation process was to be performed at higher frequencies.

The advantage of M-ary FSK for wireless channels is that it is a robust modulation strategy. The demodulator is noncoherent, being implemented as a simple *energy detector* situated on each of the frequency bins with center frequency f_c. The receiver is noncoherent in that it does not require knowledge of the phase of the signal—a property that follows from the discussion of fading channels presented in Chapter 2.

Power control is an effective way of minimizing interference and extending the life of the battery. The CT-2 system uses a simple two-level power control scheme. Normal transmissions occur with a nominal transmit power of 5 milliwatts of power. However, if the received signal strength exceeds a certain level, then inband signalling is used to request the transmitter to reduce its power level by approximately 15 dB.

Problem 3.15   What is the round-trip path delay for a 60-m path with CT-2? Does this amount of delay explain the allocated guard intervals?
Ans. The round trip is 0.4 microseconds. The guard intervals include (1) processing time for the terminals, and (2) allowances for other timing errors.   ■

Problem 3.16   The maximum frequency error allowed in the CT-2 standard is 10 kHz or less. What is the relative frequency error, expressed in parts per million (ppm)?
Ans. The clock error is 22 ppm.   ■

3.15 SUMMARY AND DISCUSSION

In this chapter, we discussed the modulation process, focusing on those techniques which are of particular interest to wireless communications. Among the analog techniques, frequency modulation (FM) stands out by virtue of its inherent capability to trade off channel bandwidth for improved noise performance at the expense of increased system complexity.

Much of the discussion, however, was devoted to digital modulation techniques, which distinguish themselves from their analog modulation counterparts by their signal-space representation, in which the transmitted message points show up as discrete points in the signal space. The discussion also highlighted detailed treatments of quadriphase-shift keying (QPSK), minimum-shift keying (MSK), and its Gaussian-filtered version (GMSK). A variant of QPSK known as the $\pi/4$-shifted QPSK is used in the TDMA standard IS-54. On the other hand, GMSK features prominently in the design of a TDMA system, commonly referred to as GSM, which will be discussed in Chapter 4.

The analysis of modulation systems, be they of an analog or digital nature, is simplified considerably by using the complex baseband representation of band-pass signals and systems with no loss of information. The simplification results from exchanging the use of complex baseband signals and systems for their real counterparts, complicated by the presence of a carrier frequency. Simply put, elimination of the sinusoidal carrier permits us to get rid of tedious trigonometry at the expense of complex signal analysis.

Frequency-division multiple access (FDMA), rooted in frequency-domain concepts, relies on the use of modulation and band-pass filtering to place a user's information-bearing signal inside a permissible subband. It is largely because of its conceptual simplicity and the status of technology that frequency modulation became the technique upon which the early generation of wireless communication systems were based. This analog method of modulation provides protection against channel noise by virtue of its ability to exchange increased transmission bandwidth for improved receiver noise performance in accordance with a square power law. For data control, FDMA uses binary frequency-shift keying, a digital form of frequency modulation.

However, FDMA suffers from the limitation that the system needs a separate transceiver (i.e., transmitter-receiver) for each active user at the base station under its coverage. It is this limitation that motivated the development of time-division multiple access (TDMA), the deployment of which has been made possible by virtue of the availability of digital signal-processing techniques. (TDMA is discussed in Chapter 4.)

NOTES AND REFERENCES

[1] For a detailed treatment of analog modulation techniques including both amplitude modulation and frequency modulation, time-domain and frequency-domain representations, methods of generation and detection, and the evaluation of noise performances, see Chapter 2 of Haykin (2001).

[2] For details of the IS-54 standard, see the Telecommunication Industry Association Report 1992. For details of the Japanese digital cellular standard using the $\pi/4$-shifted QPSK, see Nakahima *et al.* (1990). See also Pahlavan and Levesque (1995), pp. 272–273.

[3] The fundamental work of Nyquist on pulse shaping is described in the classic 1928 paper.

[4] The root raised-cosine pulse shaping is discussed in Chennakesku and Saulnier (1993) in the context of $\pi/4$-shifted differential QPSK for digital cellular radio. It is also discussed in Anderson (1999), pp. 26–27, and Stüber (1996), pp. 169–172.

[5] The complex representation of band-pass signals and systems is discussed in Haykin (2001). The material presented therein involves the use of the *Hilbert transform*, which may be viewed as a device whose frequency response satisfies two conditions:

- The amplitude response is constant at all frequencies.
- The phase response is equal to +90° for negative frequencies and –90° for positive frequencies.

[6] For a detailed exposition of the signal-space analysis of digitally modulated signals, see Chapter 5 of Haykin (2001). The classic book on this topic is that of Wozencroft and Jacobs (1965).

[7] For a discussion of Carson's rule on the bandwidth of FM signals, see Chapter 2 of Haykin (2001).

[8] The invention of minimum shift keying (MSK) may be traced to Doelz and Heald (1961). Independently of this early work, deBuda (1972) derived the fast FSK, another way of viewing MSK. For discussion of GMSK, see Stüber (1996) and Steele and Hanzo (1999). The noise performances of MSK and GMSK are discussed in Chapter 6 of Haykin (2001).

[9] Frequency-division multiple access is discussed in Steele and Hanzo (1999), and Rappaport (2002).

[10] The adjacent channel interference problem is discussed in Shankar (2002).

[11] Spectral analysis of QPSK and MSK signals is presented in Haykin (2001), pp. 360–361, and pp. 394–396, respectively.

[12] Early digital radios were made by modifying FM voice radios. An FM modulator integrates the baseband signal to determine the transmitted phase, analogous to the way a differential encoder works. Since FM is a constant-envelope modulation scheme, FM receivers typically place a hard limit on the received signal to remove the effect of channel gain variations. In addition, the received signal is passed through a discriminator, which, in mathematical terms, is equivalent to a differentiator. The discriminator performs an operation quite similar to differential detection. For that reason, differential detection of digital FM was an important early transmission scheme for mobile radio. Generalizations of this approach are called continuous phase modulation (CPM) and the reader can find more information in Anderson *et al.* (1986).

[13] Moher and Lodge (1989) and earlier papers were the first to propose the use of pilot symbols for estimating and compensating the variations of flat-fading in mobile radio. Pilot symbol techniques are used in a variety of forms, not necessarily just for uniform spacing, in many current radio systems.

[14] For a detailed treatment of the noise performance of modulation schemes for passband data transmission, see Chapter 6 of Haykin (2001).

[15] Benedetto and Biglieri (1999) describe a general technique for computing bit error rates over a fading channel. The technique proceeds from a bound known as the union bound. Specifically, it is shown that the calculation of the bit error rate can be reduced to the calculation of the probability that a random variable takes on a negative value. This probability is in the form of an integral that lends itself to numerical computation, which can be made arbitrarily accurate.

[16] OFDM theory is closely related to the theory of discrete multitone (DMT) modulation, which is discussed in Starr *et al.* (1999). Bahai and Saltzberg (1999) is devoted to OFDM theory and applications; this book also discusses wireless local area networks and future trends using OFDM.

[17] The Discrete Fourier Transform and its inverse are discussed in Oppenheim *et al.* (1999), which also describes the Fast Fourier Transform (FFT) algorithm.

[18] Pulse-code modulation (PCM) provides a technique for the conversion of an analog signal into a binary stream. The conversion relies on three basic operations:

- Sampling
- Quantization
- Coding

PCM provides a significant improvement over FM in terms of the bandwidth-noise trade-off issue. Whereas this trade-off follows a square-power law in FM, the law in PCM is of an exponential form. Adaptive pulse-code modulation provides an improvement over the standard PCM by making the quantizer, the encoder, or both adaptive. For a discussion of ordinary PCM and adaptive PCM, see Chapter 3 of Haykin (2001).

ADDITIONAL PROBLEMS

Modulation Techniques

Problem 3.17   Figure 3.37(a) presents the block diagram of a passband digital phase modulator, which lends itself to VLSI implementation (Steele and Hanzo, 1999). The pre-modulation filter of impulse response $h(t)$, as configured in Fig. 3.37(b), is designed to produce a data-dependent phase signal $\theta(t)$, which addresses two read-only-memory (ROM) units to yield values of the trigonometric terms $\cos(\theta(t))$ and $\sin(\theta(t))$. The resulting digital signals are converted into analog form.

(a) In effect, the baseband model of Fig. 3.12(a) is being extended to deal with nonlinear phase modulation. This extension is however subject to the assumption that the essentially highest frequency component of both $\cos(\theta(t))$ and $\sin(\theta(t))$ is less than the carrier frequency f_c. Justify the need for this assumption.

(b) Under this assumption, show that the radiated output of Fig. 3.37(a) can be formulated as the phase-modulated signal

$$s(t) = A\cos(2\pi f_c t + \theta(t))$$

where A is a constant amplitude.

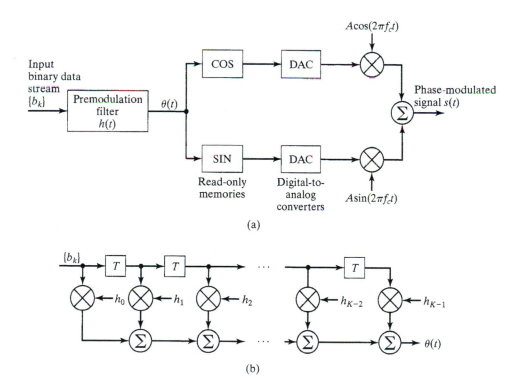

FIGURE 3.37   Block diagrams for Problem 3.17.

(c) The premodulation filter can be implemented in the form of a tapped-delay-line filter, as in Fig. 3.37(b), where T is the symbol duration. Justify this method of implementation.

Problem 3.18

(a) Construct and label the constellations of M-ary PSK for (i) $M = 8$, and (ii) $M = 16$.

(b) Discuss the differences that distinguish 16-PSK considered in part (a) and 16-QAM described in Fig. 3.15, doing so in the context of information transmission over a wireless channel.

Problem 3.19 Quadriphase-shift keying and minimum-shift keying provide two spectrally efficient methods for the digital transmission of binary data over a wireless channel. List the advantages and disadvantages of these two methods of digital modulation.

Problem 3.20 The $\pi/4$-*shifted DQPSK* is characterized by two combined features:

- the use of 8 carrier-phase states, and
- the transmission of an information-bearing signal in the differential carrier phase.

Specifically, the differential carrier phase $\Delta\theta_k$ is governed by the mapping

$$
\Delta\theta_k = \begin{cases}
-3\pi/4 & \text{for } b_k = -3 \\
-\pi/4 & \text{for } b_k = -1 \\
+\pi/4 & \text{for } b_k = +1 \\
+3\pi/4 & \text{for } b_k = +3
\end{cases}
$$

where $\{b_k\}$ denotes the incoming data stream.

Formulate the expression or the complex envelope of the $\pi/4$-shifted DQPSK signal.

Problem 3.21 Continuing with the $\pi/4$-shifted DQPSK described in Problem 3.20, suppose that the incoming pulse amplitude is shaped in accordance with the *square root raised cosine spectrum* discussed in Section 3.4.1.

(a) Using simulation, compute the phase trajectory of the $\pi/4$-shifted DQPSK signal by plotting its quadrature component versus the in-phase component for an incoming random quaternary sequence.

(b) Demonstrate that the phase trajectory does not pass through the origin. What are the practical implications of this property in the context of information transmission over a wireless channel?

Problem 3.22 In this problem, we explore the effect of a multipath channel on the waveform of three digitally modulated signals. The multipath channel is represented by the simple tapped-delay model of Fig. 3.38, where T denotes the symbol duration. Values of the tap-weights are

$$w_0 = 0.25$$
$$w_1 = 0.25$$
$$w_2 = 0.25$$

The incoming binary sequence is ...0000110101110101...

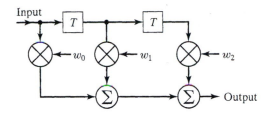

FIGURE 3.38   Tapped-delay-line model of multipath channel.

(a) Plot the waveforms of the modulated signals at the model input, using the following three methods:

    (i) QPSK

    (ii) OQPSK

    (iii) $\pi/4$ -shifted QPSK

(b) For each of these methods, plot the waveform produced at the model output.

(c) What conclusions can you draw from the results of parts (a) and (b)?

Assume the use of non-return-to-zero signaling.

Problem 3.23   Repeat Problem 3.22, this time using the more difficult set of tap-weights:

$$w_0 = 0.5$$
$$w_1 = \sqrt{0.5}$$
$$w_2 = 0.5$$

Why is this set of tap-weights more difficult to deal with than those of Problem 3.22?

Problem 3.24
Equations (3.58) and (3.59) define the two coordinates for signal-space analysis of MSK. Given these two orthonormal coordinates, depict the signal-space representation of the MSK signal.

Problem 3.25   Equation (3.67) defines the impulse response of the pulse-shaping filter $h(t)$ used to generate a GMSK signal.

(a) Show that the time function $h(t)$ satisfies all the properties of a probability density function.

(b) Expanding on the interpretation of $h(t)$ as a probability density function, determine the variance of the distribution. What is the significance of this interpretation?

Problem 3.26   The *tamed frequency modulation (TFM)*, due to deJager and Dekker (1978), is designed to provide a frequency-modulated signal whose power spectrum is compact without sidelobes. This desirable spectral characteristic is achieved by careful control of the phase transitions of the frequency-modulated signal. Figure 3.39 depicts the TFM modulator that consists of a premodulation filter feeding a voltage-controlled oscillator. The premodulation filter itself

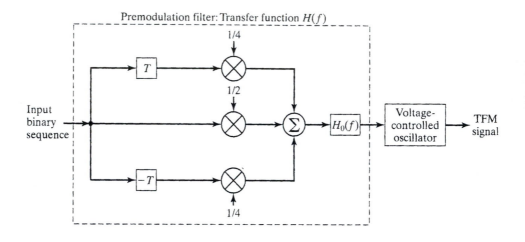

FIGURE 3.39   Block diagram of TFM modulator.

consists of the tapped-delay-line filter cascaded with a low-pass filter whose impulse response is the inverse Fourier transform of the example transfer function

$$H_0(f) = \begin{cases} (\pi f T)/\sin(\pi f T) & \text{for } 0 \le |f| \le 1/2T \\ 0 & \text{otherwise} \end{cases}$$

where T is the symbol duration.

(a) Show that the overall transfer function of the corresponding premodulation filter is given by

$$H(f) = \begin{cases} \dfrac{(\pi f T)}{\sin(\pi f T)} \cos^2(\pi f T) & \text{for } 0 \le |f| \le 1/2T \\ 0 & \text{otherwise} \end{cases}$$

(b) Using computer simulation, plot the overall impulse response of the premodulation filter, denoted by $h(t)$.

(c) The filter defined in part (b) is noncausal. Propose the use of a delay that would make the filter causal for all practical purposes.

(d) How does the impulse response of the premodulation filter computed in part (b) compare with that of the premodulation filter used in the GMSK modulator?

(e) Discuss the practical benefit that could be gained by using TFM for a FDMA system.

Frequency-Division Multiple Access

Problem 3.27   An FDMA system using frequency modulation accommodates a total of $N = 100$ mobile users assigned to a particular cell. The largest frequency component of the speech signal is $W = 3.4$ kHz. Using Carson's rule, determine the bandwidth of the uplink and downlink of the system for each of the following frequency deviations:

(a) $D = 1$

(b) $D = 2$

(c) $D = 3$

Problem 3.28   The use of frequency hopping makes it possible for the carrier frequency to hop randomly from one frequency to another. Discuss how the use of frequency hopping can improve the performance of an FDMA system operating in a wireless communication environment.

Adjacent Channel Interference

Problem 3.29   As a measure of the adjacent channel interference problem illustrated in Fig. 3.24, consider the following index of performance:

$$\text{ACI} = \frac{P_{SL}}{P_{ML}}$$

where P_{SL} is the power spilling over into a channel of interest due to the sidelobes of an adjacent channel, and P_{ML} is the power produced in that channel due to its own main lobe.

(a) Using Eq. (3.70), derive a formula for ACI.

(b) Calculate the index of performance for the example illustrated in Fig. 3.24.

Problem 3.30   A general formula for assessing the adjacent-channel interference problem is

$$\text{ACI}(\delta f) = \frac{\int_{-\infty}^{\infty} G(f)|H(f - \delta f)|^2 df}{\int_{-\infty}^{\infty} G(f)|H(f)|^2 df}$$

where $G(f)$ is the power spectral density of the input signal, $H(f)$ is the frequency response of the band-pass filter used to separate adjacent channels, and δf is the frequency separation between the two channels. Justify the validity of this formula.

Amplifier Nonlinearities

Problem 3.31   One way of linearizing a nonlinear power amplifier is to predistort the input signal. In effect, the cascade connection of two nonlinear components behaves like a linear memoryless system. Discuss the rationale of how such a scheme can be implemented.

Channel Estimation and Tracking

Problem 3.32   Plot the loss in dB as a function of residual phase error using BPSK modulation. For typical modem implementations, there is an allocated implementation margin that may range from 0.5 dB to 2 dB, depending upon the application. This implementation margin includes all losses due to nonideal implementation of the modem. If the portion of the implementation margin allocated for phase errors is 0.25dB, what is the maximum phase error allowed if the target BER is 10^{-5}?

Problem 3.33   The statement "The residual frequency error is small compared to the signal bandwidth" implies that there is minimal phase rotation over between two successive symbols. If the maximum phase rotation permitted is $10°$, what is the maximum frequency error that would be permitted as a fraction of the symbol rate? What does this imply about the accuracy the local oscillator for down-converting the received signal?

Problem 3.34   Show that pilot symbols can be used to track both fading and small residual frequency errors in the receiver. What are the constraints on this residual frequency error?

Receiver Noise Performance

Problem 3.35   The last column of Table 3.4 lists the exact formulas for the bit error rates of different digital modulation schemes operating over a slow Rayleigh fading channel. The parameter γ_0 denotes the mean value of the received signal-energy-to-noise spectral density ratio.

   (a) Derive these exact formulas.
   (b) Assuming that γ_0 is large compared to unity, find the approximate forms of these formulas.

Problem 3.36   A digital communication system uses MSK for information transmission. The requirement is to do the transmission with a bit error rate that must not exceed 10^{-4}. Calculate the minimum signal-to-noise ratio needed to meet this requirement for the following two scenarios:

   (a) Additive white Gaussian noise channel
   (b) Rayleigh fading channel

Orthogonal Frequency-Division Multiplexing

Problem 3.37   In this problem, we address the issue of evaluating the power spectrum of an OFDM signal. From the discussion presented in Section 3.13, we may treat the OFDM signal as a modulated set of orthogonal subcarriers whose frequencies are separated by the reciprocal of the symbol duration T. Consider an incoming signal constellation characterized by two features:

   - zero mean, and
   - amplitude-shaping pulse $P(f)$.

   (a) Derive the expression for the power spectrum of the complex envelope of the OFDM signal.
   (b) Plot the power spectrum derived in part (a) for the following specifications:

      (i) Number of subcarriers, $N = 16$.
      (ii) Pulse-amplitude shaping pulse in the form of a rectangular function of time t.

   (c) Repeat the power spectrum computation of part (b) for $N = 48$.

C H A P T E R   4

Coding and Time-Division Multiple Access

4.1 INTRODUCTION

In conceptual terms, FDMA operates by partitioning the prescribed radio spectrum (i.e., prescribed for wireless communications by regulatory agencies) among potential users. Hence, FDMA belongs to the *analog world*. In contrast, TDMA operates by partitioning prescribed *time intervals* among potential users. This alternative to multiple access on a frequency-division basis belongs to the *ever-expanding digital world* that continues to improve over time in terms of both computing power and the cost of fabricating equipment. Thus, although the many-faceted implementation of TDMA requires the use of sophisticated digital signal-processing techniques, thanks to breakthroughs in digital signal-processing theory, groundbreaking discoveries in solid-state physics, and major industrial refinements of microfabrication machinery, we find the following

- For the same level of performance, TDMA systems are cheaper to build than FDMA systems.
- For the same cost, TDMA systems deliver a superior performance compared with FDMA systems.

Later in the chapter, we present a detailed account of the advantages of TDMA over FDMA. For the present, it suffices to say that the increase in system complexity of TDMA systems (compared with FDMA systems) comes about from a sequence of operations, each of which is designed for a specific purpose:

Coding. For digital signal processing, the speech signal is sampled at a uniform rate, and the speech samples are subsequently encoded into a digital sequence through two operations:

1. Source encoding for effective utilization of channel bandwidth
2. Channel encoding for protection against channel noise

An attractive feature of TDMA is that it does accommodate the transmission of source-channel encoded digital data alongside digitized speech in a straightforward manner.

Equalization. With TDMA, the channel bandwidths are wider than with FDMA, and the fading is no longer frequency flat, but rather frequency selective. This form of modification introduces a type of distortion known as *intersymbol interference* (ISI). Channel equalization provides a practical means to mitigate the ISI problem.

Modulation. For the transmission of digitized speech and data over a wireless channel, we naturally require the use of passband modulation techniques, which, in turn, mandates the use of synchronization so as to establish a strict one-to-one correspondence between the locally generated carrier frequency, carrier phase, and symbol timing at the receiver, on the one hand, and their corresponding counterparts at the transmitter, on the other. Modulation was discussed in Chapter 3.

Figure 4.1 shows the block diagram of a *basic TDMA link*. The source signal (e.g., speech signal) is known to contain redundant information. With the efficient utilization of channel bandwidth as a primary objective of wireless communications, the transmitter begins by sampling the incoming speech signal, followed by encoding the signal through a process designed to remove as much of the *natural redundancy* in the speech signal as possible without compromising the ability of the receiver to provide a high-quality reproduction of the original signal. The next functional block in the transmitter is a *channel encoder*, whose function is to introduce *controlled redundancy* into the speech-encoded signal to provide protection against channel noise. The channel encoder fulfills this function by making the controlled redundancy known to the receiver. One other important point to take into account is that a wireless channel distinguishes itself from other communication channels (e.g., an ordinary telephone channel) by its tendency to produce errors in the form of *bursts*, attributed to deep fades, which degrade receiver performance. To mitigate this particular channel impairment, an *interleaver* is included in the transmitter for the purpose of pseudorandomizing the order of the binary symbols in the channel-encoded signal in a deterministic manner. As with the controlled redundancy introduced by the channel encoder, the pseudorandomization introduced by the interleaver is also known to the receiver. The next functional block in the transmitter is a *packetizer*, whose function is to convert the encoded and interleaved sequence of digitized speech data into successive *packets*, each of which occupies a significant portion of a *frame*. Each frame also includes a *synchronization information-bearing signal*, the purpose of which is to synchronize the timing operations in the receiver with the corresponding ones in the transmitter. The remainder of the frame is occupied by a *probing signal*, whose function, as the name implies, is to probe the channel and thereby make it possible to estimate the unknown impulse response of the channel on a frame-by-frame basis. With the estimate of the channel impulse response at hand, equalization of the channel at the receiving end of the link is made possible. The final functional block of the transmitter is used to *modulate* the packetized speech data onto a sinusoidal carrier for transmission over the channel.

As indicated in the figure, the receiver consists of a cascade of the following blocks: a quadrature demodulator, a baseband processor for channel estimation and equalization, a deinterleaver, a channel decoder, a speech decoder, and a reconstruction (low-pass) filter. The individual functions of these blocks are to reverse the corresponding operations performed by the channel and the transmitter. The *quadrature demodulator* converts the received RF signal into its baseband form without any loss of information; this conversion permits the use of digital signal-processing tools.

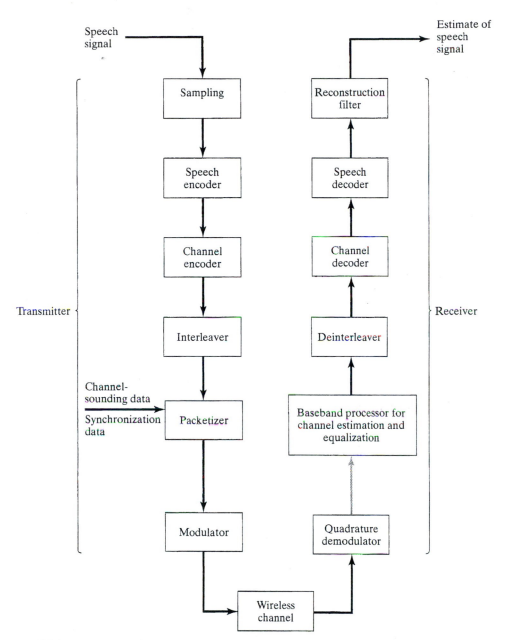

FIGURE 4.1   Block diagram of a basic TDMA link; the shaded arrow indicates a complex signal with real and imaginary parts.

The baseband processor operates on the resulting *complex baseband signal* to perform two functions: estimating the unknown channel impulse response, and using the estimate obtained to reverse the convolution performed on the transmitted signal by the channel (i.e., channel equalization). The resulting output is then *deinterleaved, channel* decoded, *source decoded*, and, finally, low-pass filtered. In this way, an *estimate* of the original speech signal is delivered to a mobile user at the receiver output. Note that the shaded arrow connecting the baseband processor and the quadrature demodulator indicates the transmission of complex baseband signals (i.e., signals with real and imaginary parts).

From the block diagram of Fig. 4.1 for a basic TDMA link, it is apparent that the use of digital wireless communications requires a considerable amount of electronic circuitry. Fortunately, nowadays electronics are inexpensive, due to the ever-increasing availability of very large-scale integrated (VLSI) circuits in the form of silicon chips. Thus, although cost considerations used to be a factor in selecting analog wireless communications over digital wireless communications, that is no longer the case today.

The chapter is organized as follows. Section 4.2 reviews the sampling process, which is the first step in the digitization of an analog signal (e.g., speech signal), followed by rationale for the use of source and channel coding techniques in Section 4.3. The stage is then set for a review of Shannon's information theory, which provides the mathematical foundation for the various signal-processing operations embodied in the TDMA link of Fig. 4.1. Section 4.5 discusses speech-coding techniques that are of particular interest in wireless communications. Sections 4.6 through 4.13 examine channel-coding and Turbo-coding techniques, including several related issues that are pertinent to wireless communications. Section 4.14 discusses partial-response modulation. This is followed by the treatment of channel estimation/equalization in Section 4.15. At that point in the chapter, the stage is set for a description of TDMA in Section 4.16. Finally, Sections 4.17 through 4.19 discuss three theme examples: the global system of mobile (GSM) communications; joint equalization-decoding in the receiver; and random access techniques, in that order. The chapter draws to a conclusion in Section 4.17.

4.2 SAMPLING

We started our introductory discussion of the basic TDMA link of Fig. 4.1 by considering the sampling process, which is basic to the digital representation of all analog signals, including speech. The purpose of sampling is to convert an analog information-bearing signal, without significant loss of information, into a corresponding sequence of samples that are usually spaced uniformly in time. For such a process to have practical utility, it is necessary that we choose the sampling rate properly, so that the sequence of samples uniquely defines the original signal, which, in turn, enables the original signal to be reconstructed from the sequence of samples. This is the essence of the sampling theorem.

A derivation of the sampling theorem follows from the Fourier transform that provides a mathematical link between the time- and frequency-domain descriptions of an analog signal. In light of the derivation, presented in Appendix A, we may state the *sampling theorem* for strictly band-limited signals of finite energy in two equivalent parts:

1. *A band-limited signal of finite energy that has no frequency components greater than W hertz is completely described by specifying the values of the signal at instants of time separated by 1/2W seconds.*

2. *A band-limited signal of finite energy that has no frequency components higher than W hertz may be completely recovered from a knowledge of its samples taken at the rate of 1/2W samples per second.*

Part 1 of the sampling theorem applies to the transmitter, while Part 2 applies to the receiver. The sampling rate of $2W$ samples per second for a signal bandwidth of W hertz is called the *Nyquist rate*; its reciprocal, $1/2W$ (measured in seconds), is called the *Nyquist interval*.

The statement of the sampling theorem, as presented herein, is based on the assumption that the information-bearing signal $m(t)$ is strictly band limited. In practice, however, $m(t)$ is *not* strictly band limited, with the result that some degree of undersampling is encountered. Consequently, some *aliasing* is produced by the sampling process, resulting in (hopefully) only a minor loss of information. *Aliasing* refers to the phenomenon of a high-frequency component in the spectrum of the signal $m(t)$ seemingly taking on the identity of a lower frequency in the spectrum of the sampled version of the signal, as illustrated in Fig. 4.2. The aliased spectrum, shown by the solid curve in Fig. 4.2(b), pertains to an undersampled version of the signal represented by the spectrum of Fig. 4.2(a).

To combat the effects of aliasing in practice, we use two corrective measures:

1. Prior to sampling, a low-pass *antialiasing filter* is used to attenuate high-frequency components of the signal $m(t)$ that are not essential to the information being conveyed.

2. The output of the low-pass filter is sampled at a rate slightly higher than the Nyquist rate, which has the beneficial effect of easing the design of the reconstruction filter used to recover the original signal in the receiver.

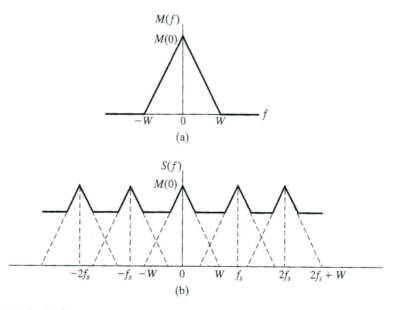

FIGURE 4.2 (a) Spectrum of a message signal band limited to $-W \leq f \leq W$.
(b) Spectrum of the corresponding sampled version of the signal for a sampling rate $f_s < 2W$.

For example, the spectrum of a speech signal extends well into the kilohertz region. However, insofar as telephonic communication is concerned, limiting the spectrum of the speech signal to about 3.1 kHz is adequate for such an application. Thus, to cater to corrective measure 1, the customary practice is to pass the speech signal through a low-pass filter with a cutoff frequency equal to 3.1 kHz. To cater to corrective measure 2, the filtered speech signal is sampled at a rate equal to 8 kHz, which is slightly larger than the Nyquist rate of $2 \times 3.1 = 6.2$ kHz. Indeed, the sampling rate of 8 kHz is the international standard for the sampling of speech signals.

4.3   WHY FOLLOW SAMPLING WITH CODING?

An analog signal (e.g., a speech signal) has a continuous range of amplitudes; therefore, its samples will have a continuous amplitude range, too. In other words, within the amplitude range of the signal, we find an infinite number of possible amplitude levels. However, recognizing that any human sense (e.g., the ear) can detect only finite intensity differences, we may *approximate* the analog signal by its quantized version, which is constructed of discrete amplitudes selected from an available set on a minimum-error basis. This approximation process is called *quantization*, and, unlike the sampling process, it is not reversible. In any event, with the combined use of sampling and quantization at our disposal, the specification of the analog signal becomes limited to a *discrete set of values*—discrete in both time and amplitude. Unfortunately, such a set is not in the form best suited for transmission over a wireless channel. Rather, to facilitate the transmission in an efficient manner, we require the use of an *encoding process* that translates the discrete set of sample values to a more appropriate form. In this context, the use of a *binary code* (with two symbols, namely, 0 and 1) offers the maximum benefit over the effects of channel noise on the signal. Thus, by sampling, quantizing, and encoding the analog signal, in that order, we end up with a *digital* representation of the signal in binary form.

However, with a code word of n binary symbols used to represent each sample of the analog signal, the digital representation of the analog signal requires an expansion of the channel bandwidth by a factor of n. This requirement imposes a new burden on the design of the wireless communication system. To lessen the burden, we recognize that when a speech signal, for example, is sampled at a rate slightly higher than the Nyquist rate, the resulting sampled signal is found to exhibit a high degree of *correlation* between adjacent samples. The meaning of this correlation is that the speech signal does not change rapidly from one sample to the next; the result is that the encoded version of the signal contains *redundant information*. Accordingly, not all of the symbols resulting from the encoding process are essential to the transmission of information over the wireless channel. By removing the redundant elements before encoding, we obtain a more efficient coded signal. The operation of redundancy removal, called *source coding*, has the net effect of reducing the channel bandwidth required to transmit the speech signal over the wireless channel.

The use of an encoded version of the speech signal has another beneficial effect: It offers the potential for mitigating the effects of channel noise on the signal. In particular, through the use of *channel coding* as a follow-up to source coding, a wireless communication system is capable of correcting transmission errors incurred in transporting the information-bearing signal from the source at the transmitter input to the sink (user) at the receiver output.

The efficient implementation of coding systems relies on digital signal-processing technology in the form of silicon chips. Interestingly enough, it is this same enabling technology that has made it possible to build time-division multiple-access systems and incorporate source-coding and channel-coding techniques so as to realize the following two practical objectives of wireless communications in a cost-effective manner:

- Efficient transmission of an information-bearing signal across a wireless channel
- Reliable delivery of the signal to its destination

4.4   SHANNON'S INFORMATION THEORY

In a classic paper published in 1948, Claude Shannon laid down the mathematical foundation of communication, which has survived the test of time. In basic terms, Shannon's information theory addresses two issues of practical importance: the efficient encoding of a source signal and its reliable transmission over a noisy channel. Both issues are of fundamental importance to the study of wireless communications. In what follows, we briefly review the underpinnings of information theory.

4.4.1   Source-Coding Theorem[1]

The *source-coding theorem* is motivated by two facts:

- A common characteristic of information-bearing signals generated by physical sources (e.g., speech signals) is that, in their natural form, they contain a certain amount of information that is *redundant*, the transmission of which is wasteful of primary communication resources, namely, transmit power and channel bandwidth.
- For *efficient* signal transmission, the redundant information should be removed from the information-bearing signal prior to transmission.

In its simplest form, the source-coding theorem may be stated as follows:

> *Given a discrete memoryless source characterized by a certain amount of entropy, the average code-word length for a distortionless source-encoding scheme is upper bounded by the entropy.*

In information theory, *entropy* is a measure of the average information content per symbol emitted by the source. According to the source-coding theorem, entropy represents a fundamental limit on the average number of bits per source symbol necessary to represent a discrete memoryless source, in that that number can be made as small as, but no smaller than, the entropy of the source. We may thus express the *efficiency* of a source encoder as

$$\eta = \frac{H(S)}{\bar{L}} \tag{4.1}$$

where $H(S)$ is the entropy of the source with source alphabet S and $\bar{L}$ is the average number of bits per symbol used in the source-encoding process. Entropy is itself defined by

$$H(S) = \sum_{k=0}^{K-1} p_k \log_2\left(\frac{1}{p_k}\right) \tag{4.2}$$

where p_k is the probability that a certain symbol s_k is emitted by the source. With the base of the logarithm in this definition equal to 2, the entropy is measured in *bits*, a basic unit of information.

4.4.2 Channel-Coding Theorem

Another practical reality is the inevitable presence of channel noise, which produces errors between the output and input data sequences of a digital communication system. For a wireless communication channel, the probability of error may exceed 10^{-1}, which means that (on the average) only 9 out of 10 transmitted symbols are received correctly. For many applications, this level of reliability is unacceptable. To achieve *reliable* communication over a noisy wireless channel, we resort to the use of *channel coding*, which consists of mapping an incoming data sequence into an output data sequence in such a way that the overall effect of channel noise on the system is minimized. The mapping operation performed in the transmitter is accomplished by a *channel encoder*, and the inverse mapping operation performed in the receiver is accomplished by a *channel decoder*. The channel encoder and channel decoder are both under the designer's control, with the objective of optimizing the overall reliability of the wireless communication system. The approach taken to realize this objective is to purposely introduce *redundancy* in the channel encoder so as to reconstruct the original data sequence as accurately as possible. Thus, in a rather loose sense, we may view channel coding as the *dual* of source coding, in that the former introduces controlled redundancy to improve data transmission reliability, whereas the latter reduces natural redundancy to improve data transmission efficiency.

In a theoretical context, a concept basic to channel coding is that of *channel capacity*, which, for a discrete memoryless channel, represents the maximum amount of information that can be transmitted per channel use. The *channel-coding theorem* may be stated as follows:

> If a discrete memoryless channel has capacity C and a source generates information at a rate less than C, then there exists a coding technique such that the output of the source may be transmitted over the channel with an arbitrarily low probability of symbol error.

The channel-coding theorem thus specifies the channel capacity C as a *fundamental limit* on the rate at which the transmission of reliable (error-free) messages can take place over a discrete, memoryless noisy channel.

Consider the case of a block code, in which an incoming message sequence is subdivided into sequential blocks k bits long and each k-bit block is mapped into a new n-bit block with $n > k$. The number of redundant bits added by the channel encoder to each transmitted block is $n - k$ bits. The ratio k/n is called the *code rate*. Using r to denote the code rate, we may thus write

$$r = \frac{k}{n} \tag{4.3}$$

where, of course, r is always less than unity. For a prescribed block length k, the code rate r (and therefore the system's coding efficiency) approaches zero as the encoded block length n approaches infinity.

The most unsatisfactory feature of the channel-coding theorem, however, is its nonconstructive nature. The theorem asserts the existence of good codes, but does not tell us how to find them. By *good codes*, we mean families of channel codes that are capable of providing reliable transmission of information (i.e., with an arbitrarily small probability of symbol error) over a noisy channel of interest at bit rates up to a maximum value less than the capacity of that channel. The *construction* of good codes is taken up in Sections 4.6 through 4.9.

4.4.3 Information Capacity Theorem

The source-coding theorem focuses on the efficient mapping of a source sequence. The channel-coding theorem focuses on the reliable transmission of a sequence over a noisy channel. Shannon's third theorem, the information capacity theorem, brings out the trade-off between channel bandwidth and signal-to-noise ratio at the channel output in the most insightful way.

Consider, then, the idealized setting of a discrete-time, memoryless, Gaussian channel, the output of which is described by the signal-to-noise ratio P/σ^2, where P is the average transmitted power and σ^2 is the variance of the zero-mean Gaussian-distributed channel noise. Let B denote the channel bandwidth, measured in hertz, and C denote the corresponding channel capacity, measured in bits per second. Then, according to the celebrated information capacity theorem, we may express the information capacity as

$$C = B\log_2\left(1 + \frac{P}{\sigma^2}\right) \text{ bits/s} \tag{4.4}$$

The information capacity theorem is one of the most remarkable results of information theory, for, in a single formula, it highlights most vividly the interplay among three key system parameters: channel bandwidth, average transmitted power, and channel noise variance $\sigma^2 = N_0 B$, where $N_0/2$ is the (two-sided) power spectral density of the additive channel noise. More specifically, the dependence of information capacity C on the channel bandwidth is *linear*, whereas its dependence on the signal-to-noise ratio P/σ^2 is *logarithmic*. Accordingly, it is easier to increase the information capacity of a wireless channel by expanding its bandwidth than increasing the average transmitted power, for a prescribed noise spectral density.

Equation (4.4) applies to a *single-input, single-output channel*. Extension of the information capacity theorem to *multiple-input, multiple-output channels* is discussed in Chapter 6.

4.4.4 Rate Distortion Theory

According to the source-coding theorem, for a discrete memoryless source, the average code-word length must be as large as the source entropy in order to represent the source perfectly. However, in many practical situations, there are constraints that force the source coding to be imperfect, thereby giving rise to unavoidable *distortion*. In situations of this kind, we speak of *source coding with a fidelity criterion*. The branch of information theory that deals with this criterion is referred to as *rate distortion theory*, which finds applications in two types of situations:

1. *Source coding*, wherein the permitted alphabet of the source code cannot represent the source output exactly, thereby forcing us to put up with *lossy data compression*.
2. *Information transmission*, which is required at a rate greater than the permissible channel capacity.

We may therefore view rate distortion theory as a natural extension of the source-coding and channel-coding theorems.

In the context of lossy data compression, we may think of a device that supplies a code with the least number of symbols for the representation of the source output, subject to an acceptable distortion. The data compressor thus retains the essential information content of the source output by blurring fine details in a deliberate, but controlled, manner. The data compression is *lossy*, in the sense that the source entropy is reduced. In an analog system, an example of lossy compression would be removing the frequency content of a voice signal above 3.1 kHz for a telephone call.

The entropy of an analog source is infinite by virtue of the fact that, in theory, its amplitude may occupy an infinitely large range. Hence, encoding of the source output at a *finite rate* will necessitate the use of a signal compression code. Stated another way, regardless of the sampling rate, it is impossible to digitally encode an analog signal with a finite number of bits without producing some distortion (i.e., loss of information). In addition to sampling, the digital representation of the analog signal involves the process of *quantization*, whereby the amplitude of each sample is rounded off to the nearest level selected from a finite set of levels.

Various types of quantizers have been proposed in the literature. However, insofar as speech is concerned—a matter that is of primary concern in wireless communications—*vector quantizers* are of special interest because of their improved signal-to-quantization noise ratio. A distinctive feature of this class of quantizers is that each block of consecutive samples of the speech signal is treated as a single entity, namely, a *vector*. The essential operation in a vector quantizer is the quantization of a random vector of samples by encoding it as a binary code word. The encoding is accomplished by comparing the vector with a *codebook* consisting of a set of stored reference vectors known as *code vectors* or *patterns*. Each pattern in the codebook is used to represent input vectors that are identified by the encoder as similar to the particular pattern, subject to the maximization of an appropriate fidelity criterion. The encoding process in a vector quantizer may thus be viewed as a *pattern-matching operation*. The code-excited linear predictive encoder, considered in the next section, is an example of a vector quantizer.

4.5   SPEECH CODING[2]

A recurrent theme in the design of digital wireless communication systems is the *efficient utilization of the allotted spectrum*. With this important objective in mind, digital wireless communication systems rely on the use of *speech coding* to remove almost all of the natural redundancy inherent in a speech signal, while maintaining a high-quality reproduction of the original signal on decoding. The most common approach to speech coding is to use a *linear predictive coding* (LPC) approach in one form or another. As the name implies, *linear prediction* is basic to this approach—hence the devotion of Subsection 4.5.1 to the essence of this operation. Subsections 4.5.2 and 4.5.3 discuss two extensions of linear prediction, namely, multipulse excited LPC and code-excited LPC.

4.5.1   Linear Prediction

One of the most celebrated problems in signal processing is that of *predicting* the present (or future) value of a discrete-time signal, given a set of past samples of the signal. In so doing, we have a powerful approach for building a *predictive model* for the underlying dynamics responsible for the generation of the signal. The smaller the prediction error in a statistical sense, the more reliable the model will be. The *prediction error* is defined as the difference between the actual future value of the signal and the predicted value produced by the model.

To be specific, consider a discrete-time signal represented by the set of samples $x(t), x(t - T_s),...,x(t - NT_s)$, where T_s denotes the sampling period. The sampling rate f_s is related to the sampling period as $f_s = 1/T_s$. In the *one-step* form of linear prediction, the requirement is to estimate the present value of the signal $x(t)$, given the N past samples $x(t - T_s), x(t - 2T_s),...,x(t - NT_s)$. Let $\hat{x}(t)$ denote the output of the predictive model. In linear prediction, $\hat{x}(t)$ is expressed as a linear combination of the N past samples via the formula

$$\hat{x}(t) = \sum_{n=1}^{N} a_n x(t - nT_s)$$

$$= \mathbf{a}^T \mathbf{x}$$

(4.5)

where the superscript T denotes matrix transposition and the N-by-1 signal vector $\mathbf{x}$ and parameter vector $\mathbf{a}$ are, respectively,

$$\mathbf{x} = [x(t - T_s), x(t - 2T_s), ..., x(t - NT_s)]^T$$

(4.6)

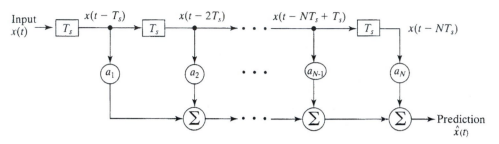

FIGURE 4.3   Structure of the FIR (i.e., tapped-delay-line) predictor.

and

$$\mathbf{a} = [a_1, a_2, ..., a_N]^T \tag{4.7}$$

The parameters $a_1, a_2, ..., a_N$ in effect define the N degrees of freedom available to us in designing the predictive model. Equation (4.5) readily suggests the *tapped-delay-line (TDL)* or *finite-duration impulse (FIR) filter*, shown in Fig. 4.3 as the structure for the predictive model.

The key question is: how do we determine the filter coefficients? To answer this question, we need a statistical criterion for optimizing the design of the filter. A criterion widely used in practice is the *mean-square-error (MSE) criterion*, defined by

$$\text{MSE} = \mathbf{E}[(x(t) - \hat{x}(t))^2] \tag{4.8}$$

where $\mathbf{E}$ denotes the statistical expectation operator and the difference $x(t) - \hat{x}(t)$ stands for the prediction error. It turns out that if $x(t)$ is the sample value of a stationary random process, then the optimum value of the parameter vector $\mathbf{a}$ is given by

$$\mathbf{a} = \mathbf{R}^{-1}\mathbf{r} \tag{4.9}$$

where the N-by-N matrix $\mathbf{R}$ is the correlation matrix of the tap inputs of the predictive model and the N-by-1 vector $\mathbf{r}$ has as its elements the autocorrelation of the input signal $x(t)$ for lags $T_s, 2T_s,..., NT_s$. The symbol $\mathbf{R}^{-1}$ in Eq. (4.9) stands for the *inverse* of the correlation matrix. (Note the similarity between a linear predictor exemplified by Eq. (4.9) and a smoothing filter defined by Eq. (3.89).)

If, however, the physical process responsible for the generation of the signal $x(t)$ is *nonstationary* (i.e., its statistics vary with time), then we require the use of an adaptive procedure whereby the model parameters are allowed to vary with time. A brief discussion of adaptive filtering algorithms is presented in Appendix E.

4.5.2 Multipulse Excited LPC

This form of speech coding exploits the *principle of analysis by synthesis*, which means that the encoder includes a replica of the decoder in its design. Specifically, the encoder consists of three main parts, as indicated in Fig. 4.4(a):

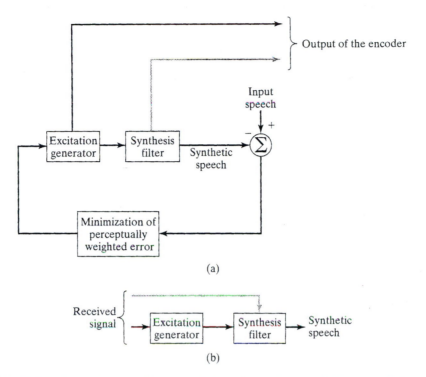

FIGURE 4.4   Multipulse excited linear predictive codec. (a) Encoder. (b) Decoder whose input (the received signal) consists of quantized filter parameters (signified by lighter arrows) and quantized excitation, as produced by the generator.

1. A *synthesis filter*, for the predictive modeling of speech. This may consist of an all-pole filter (i.e., a linear filter whose transfer function has poles only), which is designed to model the *short-term* spectral envelope of speech. The term "short-term" refers to the fact that the filter parameters are computed on the basis of predicting the present sample of the speech signal from 8 to 16 previous samples. The synthesis filter may also include a *long-term* predictor for modeling the fine structure of the speech spectrum. In such a case, the long-term predictor is connected in cascade with the short-term predictor. In any event, the function of the synthesis filter is to produce a synthetic version of the original speech that is configured to be of high quality.

2. An *excitation generator*, for producing the excitation applied to the synthesis filter. The excitation consists of a definite number of pulses every 5 to 15 ms. The amplitudes and positions of the individual pulses are adjustable.

3. *Error minimization*, for optimizing the perceptually weighted error between the original speech and the synthesized speech. The aim of this minimization is to optimize the amplitudes and positions of the pulses used in the excitation. Typically, a mean-square-error criterion (i.e., the mean-square value of the difference between an actual sample of the speech signal and its predicted value) is used for the minimization.

Thus, as shown in Fig. 4.4(a), the three parts of the encoder form a *closed-loop* optimization procedure, which permits the encoder to operate at a bit rate below 16 kb/s while maintaining high-quality speech on reconstruction.

The encoding procedure itself has two main steps:

1. The free parameters of the synthesis filter are computed with the use of the actual speech samples as input. This computation is performed outside the optimization loop over a period of 10 to 30 ms, during which the speech signal is treated as pseudostationary.

2. The optimum excitation for the synthesis filter is computed by minimizing the perceptually weighted error with the loop closed, as in Fig. 4.4(a).

Thus, the speech samples are divided into *frames* (10 to 30 ms long) for computing the filter parameters, and each frame is divided further into *subframes* (5 to 15 ms long) for optimizing the excitation. The quantized filter parameters and quantized excitation constitute the transmitted signal.

Note that by first permitting the filter parameters to vary from one frame to the next and then permitting the excitation to vary from one subframe to the next, the encoder is able to track the nonstationary behavior of speech, albeit on a batch-by-batch basis.

The decoder, located in the receiver, consists simply of two parts: the excitation generator and the synthesis filter, as shown in Fig. 4.4(b). These two parts are identical to the corresponding ones in the encoder. The function of the decoder is to use the received signal to produce a synthetic version of the original speech signal. This is achieved by passing the decoded excitation through the synthesis filter, whose parameters are set equal to those in the encoder.

To reduce the computational complexity of the *codec* (a contraction of coder/decoder), the intervals between the individual pulses in the excitation are constrained to assume a common value. The resulting analysis-by-synthesis codec is said to have a *regular-pulse excitation*.

4.5.3 Code-Excited LPC

Figure 4.5 shows the block diagram of the *code-excited LPC*, commonly referred to as CELP, which provides another method for speech coding. The distinguishing feature of CELP is the use of a predetermined *codebook* of stochastic (zero-mean white Gaussian) vectors as the source of excitation for the synthesis filter. The synthesis filter itself consists of two all-pole filters connected in cascade. One filter performs short-term prediction and the other performs long-term prediction.

As with the multipulse excited LPC, the free parameters of the synthesis filter are computed first, using the actual speech samples as input. Next, the choice of a particular vector (code) stored in the excitation codebook and the gain factor G in Fig. 4.5 is optimized by minimizing the average power of the perceptually weighted error between the original speech and the synthesized speech (i.e., the output of the synthesis filter). The address of the random vector selected from the codebook and the corresponding quantized gain factor, together with the quantized filter parameters, constitute the transmitted signal.

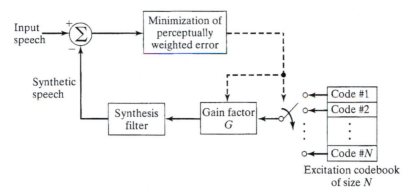

FIGURE 4.5   Encoder of the code-excited linear predictive codec (CELP). The transmitted signal consists of the address of the code selected from the codebook, the quantized gain factor G, and quantized filter parameters.

An identical copy of the codebook is made available to the decoder, and likewise for the synthesis filter. Hence, given the received signal, the decoder is able to parameterize its own synthesis filter and determine the appropriate excitation for the synthesis filter, thereby producing a synthetic version of the original speech signal.

CELP is capable of producing good-quality speech at bit rates below 8 kb/s. However, its computational complexity is intensive, because of the exhaustive search it makes of the excitation codebook. In particular, the weighted synthesized speech in the encoder has to be computed for all the entries in the codebook and then compared with the weighted original speech. Nevertheless, real-time implementation of CELP codecs has been made possible by virtue of advances in digital signal processing and VLSI technology.

4.6   ERROR-CONTROL CODING[3]

The next function to be addressed in the basic TDMA link of Fig. 4.1 is that of channel coding. There are two broadly defined categories of error-control coding techniques to be considered for this function:

1. *Forward error-correction (FEC) codes*, which rely on the controlled use of redundancy in the transmitted code word for both the *detection and correction* of errors incurred during the course of transmission over a noisy channel. Irrespective of whether the decoding of the received (noisy) code word is successful, no further processing is performed at the receiver. Accordingly, channel-coding techniques suitable for FEC require only a one-way link between the transmitter and the receiver.

2. *Automatic-repeat request (ARQ) schemes*, which use redundancy merely for the purpose of *error detection*. Upon the detection of an error in a transmitted code word, the receiver requests a repeat transmission of the word in question, thereby necessitating the use of a *return path* (i.e., feedback channel).

Unfortunately, the need to retransmit a code word introduces *latency* into the operation of the system, making the use of ARQ unsuitable for speech communications by wireless, which requires information transmission in real time. As for data, the ARQ technique is closely related to link-layer functions to be described in Chapter 7 and will therefore not be discussed here further.

Historically, FEC codes have been classified into *block codes* and *convolutional codes*. The distinguishing feature of this particular classification is the presence or absence of *memory* in the encoders for the two codes.

As mentioned in Section 4.1, to generate an (n,k) block code, the channel encoder accepts information in successive k-bit *blocks* and, to each block, adds $n - k$ redundant bits that are algebraically related to the k message bits, thereby producing an overall encoded block of n bits, where $n > k$. The channel encoder produces bits at the rate $R_0 = (n/k)R_s$, where R_s is the bit rate of the information source and $r = k/n$ is the *code rate*. The bit rate R_0, coming out of the encoder, is called the *channel data rate*. The code rate is a dimensionless ratio, whereas the data rate produced by the source and the channel data rate are both measured in bits per second.

In a convolutional code, the encoding operation may be viewed as the *discrete-time convolution* of the input sequence with the impulse response of the encoder. The duration of the impulse response equals the *memory* of the encoder. Accordingly, the encoder for a convolutional code operates on the incoming message sequence, using a "sliding window" equal in duration to its own memory. This, in turn, means that in a convolutional code, as opposed to what happens in a block code, the channel encoder accepts message bits as a continuous sequence and generates a continuous sequence of encoded bits at a higher rate.

Both linear block codes and convolutional codes find applications in wireless communication systems. In the next section, we discuss convolutional codes for which there exist powerful decoding schemes that are, in theory, capable of achieving optimal performance. In Section 4.12, we discuss an equally powerful class of linear block codes known as turbo codes.

4.6.1 Cyclic Redundancy Check Codes

An an example of a linear block code, we consider *cyclic codes*, which have the property that any cyclic shift of a code word in the code is also a code word. Cyclic codes are extremely well suited for *error detection*. We make this statement for two reasons: first, cyclic codes can be designed to detect many combinations of likely errors; second, the implementation of both encoding and error-detecting circuits is very simple with cyclic codes. It is for these reasons that many of the error-detecting codes used in practice are of the cyclic-code variety. A cyclic code used for error detection is referred to as a *cyclic redundancy check (CRC) code*.

We define an *error burst* of length B in an n-bit received word as a contiguous sequence of B bits in which the first and last bits or any number of intermediate bits are received in error. Binary (n,k) CRC codes are capable of detecting the following error patterns:

TABLE 4.1   CRC Codes.

Code	Generator Polynomial, $g(X)$	$n-k$
CRC-12 code	$1 + D + D^2 + D^3 + D^{11} + D^{12}$	12
CRC-16 code (USA)	$1 + D^2 + D^{15} + D^{16}$	16
CRC-ITU code	$1 + D^5 + D^{12} + D^{16}$	16

1. All error bursts of length $n-k$ or less.
2. A fraction of error bursts of length equal to $n-k+1$; the fraction equals $1 - 2^{-(n-k-1)}$.
3. A fraction of error bursts of length greater than $n-k+1$; the fraction equals $1 - 2^{-(n-k-1)}$.
4. All combinations of $d_{min} - 1$ (or fewer) errors, where d_{min} is the minimum distance that defines the error-correcting capability of the code.
5. All error patterns with an odd number of errors if the generator polynomial $g(D)$ for the code has an even number of nonzero coefficients, where D denotes a unit-delay operator; the generator polynomial of a code is responsible for generating all the code words in the code in response to the incoming information bits.

The issues of generator polynomial and minimum distance are discussed in greater detail later in the chapter.

Table 4.1 presents the generator polynomials of three CRC codes that have become international standards. All three codes contain $1 + D$ as a prime factor. The CRC-12 code is used for 6-bit characters, and the other two codes are used for 8-bit characters. CRC codes provide a powerful method of error detection for use in automatic-repeat request (ARQ) strategies.

4.7   CONVOLUTIONAL CODES

As already mentioned, the encoder in block coding accepts a k-bit message block and generates an n-bit code word. Thus, code words are produced on a block-by-block basis. Clearly, a provision must be made in the encoder to *buffer* an entire message block before generating the associated code word. There are applications, however, in which the message bits come in *serially* rather than in large blocks, in which case the use of a buffer may be undesirable. In such situations, the use of *convolutional coding* may be the preferred method. A convolutional coder generates redundant bits by using *modulo-2 convolutions*—hence the name of the coder.

The encoder of a binary convolutional code with rate $1/n$, measured in bits per symbol, may be viewed as a *finite-state machine (FSM)* that consists of an M-stage shift register with prescribed connections to n modulo-2 adders and a multiplexer that serializes the outputs of the adders. An L-bit message sequence produces a coded output sequence of length $n(L + M)$ bits. The code rate of the convolutional code is therefore given by

$$r = \frac{L}{n(L + M)} \text{ bits/symbol} \tag{4.10}$$

Typically, $L \gg M$. Hence, the code rate reduces to

$$r \approx \frac{1}{n} \text{ bits/symbol} \qquad (4.11)$$

The *constraint length* of a convolutional code, expressed in terms of message bits, is defined as the number of shifts over which a single message bit can influence the encoder output. In an encoder with an *M*-stage shift register, the *memory* of the encoder equals *M* message bits, and $K = M + 1$ shifts are required for a message bit to enter the shift register and finally come out. Hence, the constraint length of the encoder is *K*.

Figure 4.6 shows a convolutional encoder with $n = 2$ and $K = 3$. Hence, the code rate of this encoder is 1/2. The encoder operates on the incoming message sequence, one bit at a time. Note that the convolutional code generated by the encoder is *nonsystematic*, in that the message bits are transmitted in altered form.

Each path connecting the output to the input of a convolutional encoder may be characterized in terms of its *impulse response*, defined as the response of that path to a symbol 1 applied to its input, with each flip-flop in the encoder set initially in the zero state. Equivalently, we may characterize each path in terms of a *generator polynomial*, defined as the *unit-delay transform* of the impulse response. To be specific, let the *generator sequence* $(g_0^{(i)}, g_1^{(i)}, g_2^{(i)}, ..., g_M^{(i)})$ denote the impulse response of the *i*th path, where the coefficients $(g_0^{(i)}, g_1^{(i)}, g_2^{(i)}, ..., g_M^{(i)})$ equal 0 or 1, depending on the connections in the encoder. Correspondingly, the *generator polynomial* of the *i*th path is defined by

$$g^{(i)}(D) = g_0^{(i)} + g_1^{(i)}D + g_2^{(i)}D^2 + ... + g_M^{(i)}D^M \qquad i = 1,2,...,n \qquad (4.12)$$

where *D* denotes the unit-delay variable. The complete convolutional encoder is described by the set of generator polynomials $\{g^{(1)}(D), g^{(2)}(D), ..., g^{(n)}(D)\}$.

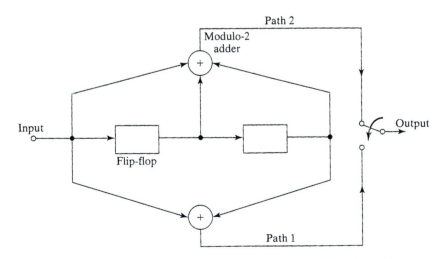

FIGURE 4.6   Convolutional encoder with constraint length 3 and rate 1/2.

EXAMPLE 4.1 Convolutional Code of Constraint Length $K = 3$

Consider the convolutional encoder of Fig. 4.6, which has two paths, numbered 1 and 2 for convenience of reference. The impulse response of path 1 (the lower path) is (1,0,1). Hence, the corresponding generator polynomial is given by

$$g^{(1)}(D) = 1 + D^2$$

The impulse response of path 2 (the upper path) is (1,1,1). Hence, the corresponding generator polynomial is given by

$$g^{(2)}(D) = 1 + D + D^2$$

For, say, the message sequence (10011), we have the polynomial representation

$$m(D) = 1 + D^3 + D^4$$

As with Fourier transformation, convolution in the time domain is transformed into multiplication in the D-domain, which is, of course, performed in a modulo-2 fashion.

The output polynomial of path 1 is given by

$$
\begin{aligned}
c^{(1)}(D) &= g^{(1)}(D)m(D) \\
&= (1 + D^2)(1 + D^3 + D^4) \\
&= 1 + D^2 + D^3 + D^4 + D^5 + D^6
\end{aligned}
$$

From this result, we immediately deduce that the output sequence of path 1 is (1011111). Similarly, the output polynomial of path 2 is given by

$$
\begin{aligned}
c^{(2)}(D) &= g^{(2)}(D)m(D) \\
&= (1 + D + D^2)(1 + D^3 + D^4) \\
&= 1 + D + D^2 + D^3 + D^6
\end{aligned}
$$

The output sequence of path 2 is therefore (1111001). Finally, multiplexing the two output sequences, we get the encoded sequence

$$\mathbf{c} = (11, 10, 11, 11, 01, 01, 11)$$

Note that the message sequence of length $L = 5$ bits produces an encoded sequence of length $n(L + K - 1) = 14$ bits. Note also that, for the shift register to be restored to its zero initial state, a terminating sequence of $K - 1 = 2$ zeros is appended to the last input bit of the message sequence. The terminating sequence of $K - 1$ zeros is called the *tail of the message*, or the *flush bits*. ∎

Problem 4.1   In Example 4.1, we determined the encoded sequence $\mathbf{c}$ by operating in the transformed-D domain. Repeat the evaluation of $\mathbf{c}$ by operating exclusively in the time domain. ∎

To simplify the presentation, we find it convenient to use an *octal* representation to describe the generator polynomial of a convolutional code. Specifically, *octal* refers to the value assumed by the generator polynomial $g^{(i)}(D)$ of Eq. (4.12) when the unit-delay variable D is set equal to the number 2 and the result is then converted to base 8. For example, the generator polynomial of the convolutional encoder of Fig. 4.6 is referred to simply as (5,7). For path 1, putting $D = 2$ in $g^{(1)}(D)$ yields $1 + 2^2 = 5$, and putting $D = 2$ in $g^{(2)}(D)$ for path 2 yields $1 + 2 + 2^2 = 7$—hence reference to the convolutional code as (5,7).

4.7.1 Trellis and State Diagrams of Convolutional Codes

To develop an insight into the operation of a convolutional encoder, we can use a graphic representation known as a *trellis*. This diagram is so called because a trellis is a treelike structure with reemerging branches. Figure 4.7 shows the trellis diagram for the convolutional encoder of Fig. 4.6. The convention used in Fig. 4.7 to distinguish between input symbols 0 and 1 is as follows. A code branch produced by an input 0 is drawn as a *solid* line, whereas a code branch produced by an input 1 is drawn as a *dashed* line. Each input (message) sequence corresponds to a specific path through the trellis. For example, we readily see from the figure that the message sequence (10011) produces the encoded output sequence (11, 10, 11, 11, 01), which agrees with the result of Example 4.1.

A trellis is insightful in that it brings out explicitly the fact that the associated convolutional encoder is a *finite-state machine*. We define the *state* of a convolutional encoder of rate $1/n$ in terms of the $(K - 1)$ message bits stored in the encoder's shift register. At time j, the portion of the message sequence containing the most recent K bits is written as $m_{j-K+1}, \cdots, m_{j-1}, m_j$, where m_j is the *current* bit. The $(K - 1)$-bit state of the encoder at time j is therefore written simply as $m_{j-1}, \cdots, m_{j-K+2}, m_{j-K+1}$. In the case of the simple convolutional encoder of Fig. 4.6, we have $(K - 1) = 2$. Hence, the state of this encoder can assume any one of four possible values, as described in Table 4.2(a). The trellis contains $L + K$ levels, where L is the length of the incoming message sequence and K is the constraint length of the code. The *levels* of the trellis are labeled $j = 0,1,...,$ $L + K - 1$ in Fig. 4.7 for $K = 3$. Level j is also referred to as the *depth j*; both terms are used interchangeably. The first $K - 1$ levels correspond to the encoder's departure from

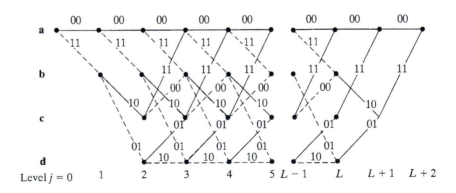

FIGURE 4.7   Trellis for the convolutional encoder of Figure 4.6.

an initial state, and the last $K-1$ levels correspond to its return to the final state. Clearly, not all the states can be reached in these two portions of the trellis. However, in the central portion, for which the level j lies in the range $K-1 \leq j \leq L$, all the states of the encoder are reachable. Note that the central portion of the trellis exhibits a fixed periodic structure.

Problem 4.2   The *state diagram* provides another graphical representation of a convolutional encoder. This second representation can be derived from the trellis diagram by focusing on a portion of the trellis from level j to level $j+1$, chosen in such a way that all possible states of the encoder are fully displayed. In the case of the encoder of Fig. 4.6, that requirement is attained for $j \geq 2$. The state diagram is obtained by coalescing the left and right nodes of the portion of the trellis diagram so chosen. The following convention is used in the construction of the state diagram: A transition from one state to another in response to input 0 is represented by a *solid* branch, whereas a transition in response to input 1 is represented by a *dashed* branch.

(1) Starting with the trellis diagram of Fig. 4.7, justify the construction of the state diagram shown in Fig. 4.8.

(2) Derive Table 4.2 for the encoder, with part (i) of the table showing the actual states of the encoder and part (ii) showing the pertinent state transitions.

(3) Starting at state *a*, corresponding to the *all-zero state*, walk through the state diagram of Fig. 4.8 in response to the message sequence (10011). Hence, verify that this sequence produces the encoded output sequence (11, 10, 11, 11, 01) in agreement with Example 4.1.   ■

TABLE 4.2   The Convolutional Encoder of Figure 4.6.

(i) State table

State	Binary Description
a	00
b	10
c	01
d	11

(ii) State-transition table

Input binary symbol	Old state	New state	Emitted symbol
0	*a*	*a*	00
	b	*c*	10
	c	*a*	11
	d	*c*	01
1	*a*	*b*	11
	b	*d*	01
	c	*b*	00
	d	*d*	10

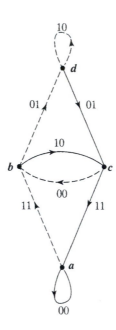

FIGURE 4.8   State diagram of the convolutional encoder of Figure 4.6.

4.7.2 Free Distance of a Convolutional Code

The performance of a convolutional code depends not only on the decoding algorithm used, but also on the error-correction properties of the code. In this context, the most important single measure of a convolutional code's ability to combat channel noise is the *free distance*, denoted by d_{free}. To pave the way for the definition of this important parameter, we need to introduce a couple of related definitions:

- *The Hamming weight* of a linear code is the number of nonzero elements in the code vector of the code.
- *The Hamming distance* between a pair of code vectors is the number of locations where their respective elements are different.

On this basis, the free distance of a convolutional code is formally defined as the *minimum (i.e., smallest) Hamming distance* between any two code vectors in the code. The important point to note here is that a convolutional code with free distance d_{free} can correct t errors only if d_{free} is greater than $2t$. Stated another way, a convolutional decoding algorithm will fail if the number of closely spaced errors in the received sequence exceeds $d_{free}/2$.

The free distance d_{free} can be determined from the state diagram of the convolutional encoder. Given the introductory nature of this book, the reader is referred elsewhere[4] for this determination. Suffice it to say, however, that the free distance of a convolutional code depends on the constraint length K of the code, as indicated in Table 4.3. In this table, the free distances of two types of convolutional codes—systematic

TABLE 4.3   Maximum Free Distances Attainable with Systematic and Nonsystematic Convolutional Codes of Rate 1/2.

Constraint Length K	Systematic	Nonsystematic
2	3	3
3	4	5
4	4	6
5	5	7
6	6	8
7	6	10
8	7[†]	10
9	‡	12

[†] The free distance given here is the best that can be achieved by a systematic nonrecursive code with rate 1/2, $K = 8$, and generator polynomials 400, 671 (Claude Berrou, private communication).
[‡] The free distance for a systematic convolutional code for $K = 9$ is not known to the authors.

and nonsystematic—are listed for varying K. In a *systematic convolutional code*, the incoming message bits are transmitted in unaltered form; this constraint is removed from the generation of a *nonsystematic convolutional code*. The convolutional code generated by the encoder of Fig. 4.6 is of the nonsystematic kind. An example of a systematic convolutional code is presented in Section 4.12.

According to Table 4.3, the nonsystematic convolutional code of constraint length $K = 3$, generated by the encoder of Fig. 4.6, has a free distance $d_{free} = 5$. Hence, this convolutional code is capable of correcting up to two errors in the sequence it receives over a binary symmetric channel, as will be demonstrated later in Example 4.2.

4.8   MAXIMUM-LIKELIHOOD DECODING OF CONVOLUTIONAL CODES

Now that we understand the operation of a convolutional encoder, the next issue to be considered is the decoding of a convolutional code. To that end, let **m** denote a *message vector* and **c** denote the corresponding *code vector* applied by the encoder to the input of a discrete memoryless channel. Let **r** denote the *received vector*, which may differ from the transmitted code vector due to channel noise. Given the received vector **r**, the decoder is required to make an *estimate* **m̂** of the message vector. Since there is a one-to-one correspondence between the message vector **m** and the code vector **c**, the decoder may equivalently produce an estimate **ĉ** of the code vector. We may then put **m̂** = **m** if and only if **ĉ** = **c**; otherwise, a *decoding error* is committed in the receiver. The *decoding rule* for choosing the estimate **ĉ**, given the received vector **r**, is said to be optimum when the *probability of decoding error* is minimized. For equiprobable messages, we may state that the probability of decoding error is minimized if, in accordance with the maximum likelihood principle, the estimate **ĉ** is chosen to maximize the *log-likelihood function* for the receiver. Let $p(\mathbf{r}|\mathbf{c})$ denote the conditional probability of receiving **r**, given that **c** was sent. Then, by definition, the log-likelihood function equals $\log p(\mathbf{r}|\mathbf{c})$. The *maximum-likelihood decoder* or decision rule may now be stated as follows:

Choose the estimate $\hat{\mathbf{c}}$ for which the (4.13)

log-likelihood function $\log p(\mathbf{r}|\mathbf{c})$ is maximum

To illustrate, consider the special case of a *memoryless binary symmetric channel* in which both the transmitted code vector $\mathbf{c}$ and the received vector $\mathbf{r}$ represent binary sequences, say, of length N. The channel is also characterized by equality of the conditional probabilities p_{01} and p_{10}, where p_{01} refers to the receiver making a decision in favor of symbol 0 when symbol 1 was sent, and p_{10} refers to the situation in which the reverse is true. Naturally, the two vectors $\mathbf{c}$ and $\mathbf{r}$ may differ from each other in some locations because of errors due to channel noise. Let $\mathbf{r}$ be the received code vector, and $\hat{\mathbf{c}}$ be a candidate code vector. Then, for each $\hat{\mathbf{c}}$, we compute the conditional probability

$$p(\mathbf{r}|\hat{\mathbf{c}}) = \prod_{i=1}^{N} p(r_i|\hat{c}_i) \tag{4.14}$$

which follows from the memoryless property. Correspondingly, the log-likelihood function for the convolutional decoder is

$$\log p(\mathbf{r}|\hat{\mathbf{c}}) = \sum_{i=1}^{N} \log p(r_i|\hat{c}_i) \tag{4.15}$$

Let the transition probability be defined as

$$p(r_i|\hat{c}_i) = \begin{cases} p & \text{if } r_i \neq \hat{c}_i \\ 1-p & \text{if } r_i = \hat{c}_i \end{cases} \tag{4.16}$$

Suppose also that the received vector $\mathbf{r}$ differs from the candidate code vector $\hat{\mathbf{c}}$ in exactly d positions. The number d defines the *Hamming distance* between vectors r and $\hat{\mathbf{c}}$. Then we may rewrite the log-likelihood function in Eq. (4.15) as

$$\log p(\mathbf{r}|\hat{\mathbf{c}}) = d\log p + (N-d)\log(1-p)$$
$$= d\log\left(\frac{p}{1-p}\right) + N\log(1-p) \tag{4.17}$$

In general, the probability of an error occurring is low enough for us to assume that $p < 1/2$. We also recognize that $N\log(1-p)$ is a constant for all $\hat{\mathbf{c}}$. Accordingly, we may restate the maximum-likelihood decoding rule for the binary symmetric channel as follows:

Choose the estimate $\hat{\mathbf{c}}$ that miminizes the Hamming distance d (4.18)

between the candidate code vector $\hat{\mathbf{c}}$ and the received vector $\mathbf{r}$.

That is, for a binary symmetric channel, the maximum-likelihood decoder reduces to a *minimum-distance decoder*. In such a decoder, the received vector **r** is compared with each possible candidate code vector **ĉ**, and the particular one closest to **r** is chosen as an estimate of the transmitted code vector. The term "closest" is used in the sense of minimum number of differing binary symbols between the code vectors **ĉ** and **r**—that is, the Hamming distance between them.

4.9 THE VITERBI ALGORITHM[5]

The equivalence between maximum-likelihood decoding and minimum-distance decoding for a binary symmetric channel implies that we may use a trellis to decode the sequence received over a binary symmetric channel in response to the transmission of a convolutionally encoded sequence characterized by that trellis. For a given convolutional code, the decoding is performed by choosing the particular path in the trellis whose coded sequence differs from the received sequence in the fewest number of places. The search procedure through the trellis of a convolutional code for maximum-likelihood decoding of the received sequence is a dynamic programming problem. The standard approach to solving this problem in coding applications is known as the *Viterbi algorithm*.

The Viterbi algorithm operates by computing a *metric* (i.e., a measure) for discrepancy for every possible path in the trellis. The metric for a particular path is defined as the Hamming distance between the coded sequence represented by that path and the received sequence. Thus, for each node (state) in the trellis of the encoder, the algorithm compares the two paths entering that particular node. The path with the lower metric is retained, and the other path is discarded. This computation is repeated for every level j of the trellis in the range $M \leq j \leq L$, where $M = K - 1$ is the encoder's memory and L is the length of the incoming message sequence. The paths that are retained by the algorithm are called *survivor paths* or *active paths*. For a convolutional code of constraint length K, no more than 2^{K-1} survivor paths and their metrics will be stored. This list of 2^{K-1} paths is always guaranteed to contain the maximum-likelihood choice—that is, a binary sequence closest to the transmitted sequence in Euclidean distance.

A summary of the Viterbi algorithm is presented in Table 4.4; its application is best illustrated by way of an example that follows the table.

TABLE 4.4   Summary of the Viterbi Algorithm.

Initialization

Label the states of the trellis from top to bottom as shown in Fig. T.1. Label the leftmost column of the trellis as time $j = 0$. Initialize the *cumulative path metric* to state $s = 0, 1, ..., S - 1$, at time step 0 to

$$J_0(s) = \left\{ \begin{array}{ll} 0 & \text{if } s = 0 \\ \infty & \text{if } s \neq 0 \end{array} \right\}$$

If the encoder produces L bits per transition, then we index the received vector **r** in groups of L bits; that is, $\mathbf{r} = (r_{1,1}, ..., r_{1,L}, r_{2,1}, ..., r_{2,L}, r_{3,1}, ...)$. We also label the L bits that are output from the encoder on a transition

TABLE 4.4   Summary of the Viterbi Algorithm.

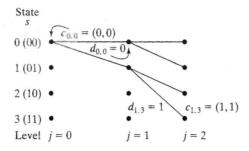

FIGURE T.1   Labeling of trellis states and transitions.

from state q to s as $(c_{q,s,1},c_{q,s,2},...,c_{q,s,L})$ and let $d_{q,s}$ be the corresponding bits at the input to the encoder for this transition, as shown in figure. Note that $d_{q,s}$ and $c_{q,s,1}$ are predetermined by the definition of the code. Define the survivor path to state s at time 0 to be the empty set, $\varnothing$.

Computation step $j + 1$

Let $j = 0, 1, 2,...$, and suppose that at the previous step j we have done only two things:

- *Identify all survivor paths.* The survivor path to state s has the smallest cumulative path metric to state s; that is,

$$J_{j+1}(s) = \min_q\{J_j(q) + H_{j+1}(q, s)\}$$

The smallest cumulative path metric is determined from the cumulative path metrics at the previous step and the branch metric for the current step. The branch metric at step $j + 1$ from state q to state s is given by

$$H_{j+1}(q, s) = \text{Hamming distance}\{(r_{j+1,1}, ..., r_{j+1,L}), (c_{q,s,1}, ..., q,s,L)\}$$

which is the Hamming distance between the received vector in symbol period $j + 1$ and the symbols on the trellis branch from state q to s. If there is no trellis branch from state q to s, then $H_{j+1}(q, s) = \infty$.

- *For each state, the survivor path and its metric are stored.* If $q_0(s)$ is the optimal solution to the minimization step, then the surviving path is given by the ordered set

$$\hat{m}_{j+1}(s) = \left(\hat{m}_j(q_0(s)), d_{q_0(s), s}\right)$$

and its metric is $J_{j+1}(s)$.

Final Step

Continue the computation until the algorithm completes its forward search through the trellis and therefore reaches the termination node (i.e., an all-zero state), at which time it makes a decision on the maximum-likelihood path. Then, as with a block decoder, the sequence of symbols associated with that path is released to the destination as the decoded version of the received sequence. In this sense, it is therefore more correct to refer to the Viterbi algorithm as a *maximum-likelihood sequence estimator*.

**EXAMPLE 4.2 Decoding of a Received Binary Sequence with
                  Two Errors**

Suppose that the encoder of Fig. 4.6 generates an all-zero sequence that is sent over a binary symmetric channel, and that the received sequence is (0100010000 ...). There are two errors in the received sequence due to noise in the channel: one in the second bit and the other in the sixth bit. We wish to show that this double-error pattern is correctable through the application of the Viterbi decoding algorithm.

In Fig. 4.9, we show the results of applying the algorithm for level (i.e., time step) $j = 1,2,3,4,5$. We see that for $j = 2$ there are (for the first time) four paths, one for each of the four states of the encoder. The figure also includes the metric of each path for each level in the computation.

In the left side of Fig. 4.9, for $j = 3$ we show the paths entering each of the states, together with their individual metrics. In the right side of the figure, we show the four survivors that result from application of the algorithm for level $j = 3,4,5$.

Examining the four survivors in Fig. 4.9 for $j = 5$, we see that the all-zero path has the smallest metric and will remain the path of smallest metric from this point forward. This clearly shows that the all-zero sequence is the maximum likelihood choice of the Viterbi decoding algorithm, which agrees exactly with the transmitted sequence.   ∎

Problem 4.3   Suppose that in Example 4.2 the received sequence is (1100010000 ...), which contains three errors compared to the transmitted all-zero sequence. This time, show that the Viterbi algorithm fails to decode the transmitted sequence correctly. That is, a triple-error pattern is uncorrectable by the Viterbi algorithm when applied to a convolutional code of constraint length $K = 3$. (The exception to this rule is a triple-error pattern spread over a time span longer than one constraint length, in which case it is likely to be correctable.)   ∎

4.9.1   Modifications of the Viterbi Algorithm

In reality, the Viterbi algorithm is a maximum likelihood sequence estimator, which means that the algorithm waits until the entire sequence at the channel output has been received before making a decision. However, when the received sequence is very long (near infinite), the storage requirement of the Viterbi algorithm becomes too high, and some compromises must be made. The approach usually taken is to "truncate" the path memory of the decoder as described here. A *decoding window* of acceptable length l is specified, and the Viterbi algorithm operates on a corresponding frame of the received sequence, always stopping after l steps. A decision is then made on the "best" path and the symbol associated with the first branch on that path is released to the user. Next, the decoding window is moved forward one time interval, and a decision on the next code frame is made, and so on. The decoding decisions made in this way are no longer truly maximum likelihood, but they can be made almost as good provided that the decoding window is long enough. Experience and analysis have shown that satisfactory results are obtained if the decoding window length l is on the order of 5 times the constraint length K of the convolutional code or more.

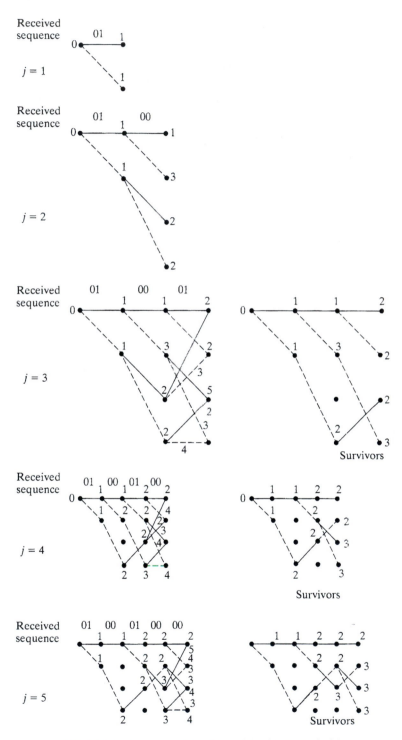

FIGURE 4.9 Steps in the Viterbi algorithm for Example 4.2.

4.10 INTERLEAVING[6]

Previous sections of this chapter have shown us how a digital wireless communication system can be separated by function into source-coding and channel-coding applications on the transmitting side and the corresponding inverse functions on the receiving side. We have also learned how analog signals can be captured in a digital format. The motivation behind these techniques is to minimize the amount of information that has to be transmitted over the wireless channel. Such minimization has at least two potential benefits in the two primary resources—transmit power and channel bandwidth—available to wireless communications:

1. *Reducing the amount of data that must be transmitted, which usually means that less power has to be transmitted.* Power consumption is always a serious concern for mobile terminals, which are typically battery operated.

2. *Reducing the spectral (or radio frequency) resources that are required for satisfactory performance.* This reduction enables us to increase the number of users who can share the same, but limited, channel bandwidth.

Moreover, insofar as channel coding is concerned, forward error-correction (FEC) coding provides a powerful technique for transmitting information-bearing data reliably from a source to a sink across the wireless channel.

However, to obtain the maximum benefit from FEC coding in many wireless channels, we require an additional technique known as *interleaving*. The need for this new technique is justified on the grounds that, in light of the material presented in Chapter 2, we know that wireless channels have *memory* due to multipath fading—the arrival of signals at the receiver via multiple propagation paths of different lengths. Of particular concern is *fast* fading, which arises out of reflections from objects in the local vicinity of the transmitter, the receiver, or both. The term *fast* refers to the speed of fluctuations in the received signal due to these reflections, relative to the speeds of other propagation phenomena. Compared with transmit data rates, even fast fading can be relatively slow. That is, fast fading can be approximately constant over a number of transmission symbols, depending upon the data transmission speed and the terminal velocity. Consequently, fast fading may be viewed as a time-correlated form of channel impairment, the presence of which results in statistical dependence among contiguous (sets of) symbol transmissions. That is, instead of being isolated events, transmission errors due to fast fading tend to occur in *bursts*.

Now, most FEC channel codes are designed to deal with a limited number of bit errors, assumed to be randomly distributed and statistically independent from one bit to the next. To be specific, in Section 4.9 on convolutional decoding, we indicated that the Viterbi algorithm, as powerful as it is, will fail if there are $d_{free}/2$ closely spaced bit errors in the received signal, where d_{free} is the free distance of the convolutional code. Accordingly, in the design of a *reliable* wireless communication system, we are confronted with *two conflicting phenomena: a wireless channel that produces bursts of correlated bit errors and a convolutional decoder that cannot handle error bursts.* Interleaving is an indispensable technique for resolving this conflict. First and foremost, however, it is important to note that, for interleaving, we do *not* need an exact statistical characterization of the wireless channel. Rather, we require only a knowledge of the *coherence time* for fast fading. Using the Clarke model for fast fading, considered in Example 2.11 of Chapter 2, we found that the *coherence time* for fast fading is approximately

$$T_{\text{coherence}} \approx \frac{0.3}{2f_D} \tag{4.19}$$

where f_D is the maximum Doppler shift. The Doppler shift depends upon the relative velocity between the transmitter and the receiver. Consequently, we would expect a "bad" signal period to have approximately the length of the Doppler shift. Equivalently, an error burst would typically have a duration $T_{\text{coherence}}$.

EXAMPLE 4.3 Interleaver Design Considerations

A mobile radio transmits data at 19.2 kbits/s in the 400-MHz band and must operate at vehicle speeds up to 100 km/hour. The radio design includes an off-the-shelf FEC chip with a rate-1/2, constraint-length-7 convolutional code. What is the expected duration of an error burst at top speed? How many closely spaced errors can this codec correct?

The maximum Doppler shift is given by

$$f_D = \frac{v}{c}(f_c) = \frac{100 \text{ km/hr}}{3 \times 10^8 \text{ m/s}}(400 \text{ MHz}) = 37 \text{ Hz}$$

The expected length of an error burst is therefore

$$R_b T_{\text{coherence}} = 19.2 \text{ kbits/s} \times \frac{0.3}{2 \times 37} = 78 \text{ bits}$$

On average, we would expect only half of these bits to be in error. (Why?) From Table 4.3, the free distance of a nonsystematic constraint-length $K = 7$ code is 10, so it can be expected to handle a maximum of only five closely spaced errors. ∎

To reconcile the two conflicting phenomena illustrated in Example 4.3, we do two things:

- use an *interleaver* (i.e., a device that performs interleaving), which randomizes the order of encoded bits *after* the channel encoder in the transmitter, and
- use a *deinterleaver* (i.e., a device that performs deinterleaving), which undoes the randomization *before* the data reach the channel decoder in the receiver.

Interleaving has the net effect of breaking up any error bursts that occur during the course of data transmission over the wireless channel and spreading them over the duration of operation of the interleaver. In so doing, the likelihood of a correctable received sequence is significantly improved. The positions of the interleaver in the transmitter and the deinterleaver in the receiver are shown in Fig. 4.1.

Three types of interleaving are commonly used in practice, as discussed next.

4.10.1 Block Interleaving

In basic terms, a *classical block interleaver* acts as a memory buffer, as shown in Fig. 4.10. Data are written into this $N \times L$ rectangular array from the channel encoder in column fashion. Once the array is filled, it is read out in row fashion and its contents

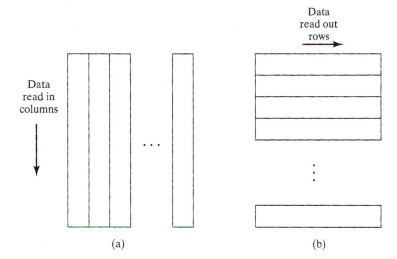

FIGURE 4.10   Block interleaver structure.
(a) Data "read in."
(b) Data "read out."

are sent to the transmitter. At the receiver, the inverse operation is performed: The contents of the array in the receiver are written rowwise with data, and once the array is filled, it is read out columnwise into the Viterbi decoder. Note that the (N,L) interleaver and deinterleaver described herein are both *periodic* with fundamental period $T = NL$.

Suppose the correlation time, or error-burst-length time, corresponds to L received bits. Then, at the receiver, we expect that the effect of an error burst would corrupt the equivalent of one row of the deinterleaver block. However, since the deinterleaver block is read out columnwise, all of these "bad" bits would be separated by $N - 1$ "good"' bits when the burst is read into the Viterbi decoder. If N is greater than the constraint length of the convolutional code being employed, then the Viterbi decoder will correct all of the errors in the error burst.

In practice, due to the frequency of error bursts and the presence of other errors caused by channel noise, the interleaver should ideally be made as large as possible. However, *an interleaver introduces delay* into the transmission of the message signal, in that we must fill the $N \times L$ array before it can be transmitted. This is an issue of particular concern in real-time applications such as voice, because it limits the usable block size of the interleaver and necessitates a compromise solution.

EXAMPLE 4.4 Interleaving Example

Figure 4.11(a) depicts an original sequence of encoded words, with each word consisting of five symbols. Figure 4.11(b) depicts the interleaved version of the encoded sequence, with the symbols shown in reordered positions. An error burst occupying four symbols, caused by channel impairment, is shown alongside Fig. 4.11(b). Note that the manner in which the encoded symbols are reordered by the interleaver is the same from one word to the next.

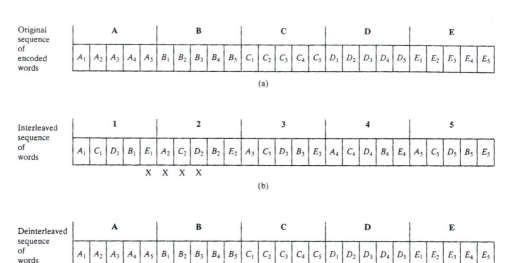

FIGURE 4.11   Interleaving example. (a) Original sequence. (b) Interleaved sequence. (c) Deinterleaved sequence.

On deinterleaving in the receiver, the shuffling of symbols is undone, yielding a sequence that resembles the original sequence of encoded symbols, as shown in Fig. 4.11(c). This figure also includes the new positions of the transmission errors. The important thing to note here is that the error burst is dispersed as a result of deinterleaving.

This example teaches us that (1) the burst of transmission errors is acted upon only by the deinterleaver and (2) insofar as the encoded symbols are concerned, the deinterleaver cancels the shuffling action of the interleaver.   ∎

Problem 4.4   Consider a mobile terminal moving at 30 km/hr and transmitting at 1.9 GHz. What is the expected coherence time of fading in the channel? The modem employed has a constraint-length-7 convolutional code. What is the minimum block size of the interleaver that you would recommend? If the data rate is 9.6 kbps, what is the delay introduced by this interleaver?

Ans. Coherence time is approximately 9.6 milliseconds. The $N \times L$ interleaver should have $N = 12$ and L corresponding to 9.6 milliseconds. The interleaver size does not depend on the data rate (Why?) but the delay would be 115.2 milliseconds.   ∎

4.10.2 Convolutional Interleaving

The block diagram of a *convolutional interleaver/deinterleaver* is shown in Fig. 4.12. Defining the period

$$T = LN,$$

we refer to the interleaver as an $(L \times N)$ convolutional interleaver, which has properties similar to those of the $(L \times N)$ block interleaver.

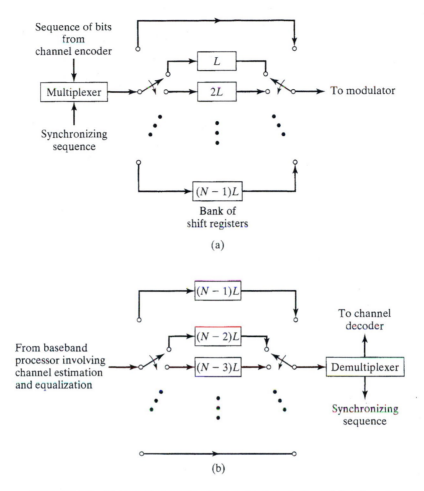

FIGURE 4.12 (a) Convolutional interleaver. (b) Convolutional deinterleaver.

The sequence of encoded bits to be interleaved in the transmitter is arranged in blocks of L bits. For each block, the encoded bits are sequentially shifted into and out of a bank of N registers by means of two synchronized input and output *commutators*. The interleaver, depicted in Fig. 4.12(a), is structured as follows:

1. The zeroth shift register provides no storage; that is, the incoming encoded symbol is transmitted immediately.

2. Each successive shift register provides a storage capacity of L symbols more than the preceding shift register.

3. Each shift register is visited regularly on a periodic basis.

With each new encoded symbol, the commutators switch to a new shift register. The new symbol is shifted into the register, and the oldest symbol stored in that register is shifted out. After finishing with the $(N-1)$th shift register (i.e., the last register), the commutators return to the zeroth shift register. Thus the switching–shifting procedure is repeated periodically on a regular basis.

The deinterleaver in the receiver also uses N shift registers and a pair of input/output commutators that are synchronized with those in the interleaver. Note, however, that the shift registers are stacked in the *reverse* order of those in the interleaver, as shown in Fig. 4.12(b). The net result is that the deinterleaver performs the inverse operation of that in the transmitter.

An advantage of convolutional over block interleaving is that in convolutional interleaving, the total *end-to-end delay* is $L(N-1)$ symbols and the memory requirement is $L(N-1)/2$ in both the interleaver and deinterleaver, which are one-half of the corresponding values in a block interleaver/deinterleaver for a similar level of interleaving.

The description of the convolutional interleaver/deinterleaver in Fig. 4.12 is presented in terms of shift registers. The actual implementation of the system can also be accomplished with *random access memory* (RAM) units in place of shift registers. This alternative implementation simply requires that access to the memory units be appropriately controlled.

4.10.3 Random Interleaving

In a *random interleaver*, a block of N input bits is written into the interleaver in the order in which the bits are received, but they are read out in a random manner. Typically, the permutation of the input bits is defined by a *uniform distribution*. Let $\pi(i)$ denote the permuted location of the ith input bit, where $i = 1,2\dots,N$. The set of integers denoted by $\{\pi(i)\}_{i=1}^{N}$, defining the order in which the stored input bits are read out of the interleaver, is generated according to the following two-step algorithm:

1. Choose an integer i_1 from the uniformly distributed set $\mathcal{A} = \{1,2,\dots,N\}$, with the probability of choosing i_1 being $P(i_1) = 1/N$. The chosen integer i_1 is set to $\pi(1)$.
2. For $k > 1$, choose an integer i_k from the uniformly distributed set $\mathcal{A}_k = \{i \in \mathcal{A}, i \neq i_1, i_2,\dots, i_{k-1}\}$, with the probability of choosing i_k being $P(i_k) = 1/(N-k+1)$. The chosen integer i_k is set to $\pi(k)$. Note that the size of the set $\mathcal{A}_k$ is progressively reduced for $k > 1$. When $k = N$, we are left with a single integer, i_N, that is set to $\pi(N)$.

To be of practical use in communications, random interleavers are configured to be *pseudorandom*, meaning that, within a block of N input bits, the permutation is random as just described, but the permutation order is exactly the same from one block to the next. Accordingly, pseudorandom interleavers are designed *off-line*. Pseudorandom interleavers are of interest in the construction of turbo codes,[7] which are discussed in Section 4.12.

4.11 NOISE PERFORMANCE OF CONVOLUTIONAL CODES

In Fig. 4.13, the simulated performance of several convolutional codes is compared with uncoded performance in an AWGN channel. The codes all have a rate of 1/2 and constraint lengths (K) of 3, 5, 7, and 9. The corresponding generator polynomials of the

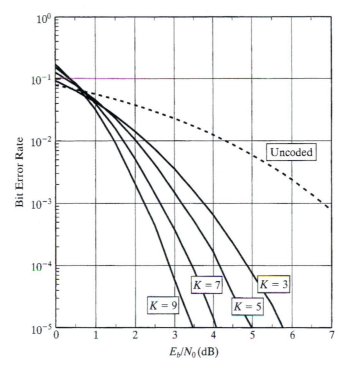

FIGURE 4.13   Comparison of uncoded and coded performance in AWGN channel.

encoders are given in Table 4.5. (The term "octal" used in the table was explained in Section 4.6, and the (5,7) convolution code of constraint length $K = 3$ was discussed in Example 4.1.) The horizontal axis in the figure represents the ratio of energy per information bit, E_b, to one-sided noise spectral density N_0. For the uncoded case, an information bit is the same as a channel bit. For a rate-1/2 code, there are two channel bits for each information bit. Binary PSK modulation with coherent detection is assumed in both the coded and uncoded cases.

In Fig. 4.14, the corresponding results are shown for a Rayleigh-fading channel. The coding also includes the equivalent of infinite interleaving; that is, there is no correlation of the fading process from one channel bit to the next. Here again, binary PSK modulation with coherent detection is assumed in both the coded and uncoded cases. Coherent detection is usually difficult to perform in fading channels, so the curves presented in the figure should be viewed as a limit on noise performance. Note also that the codes used in the simulations are all nonsystemic.

TABLE 4.5   Generators for Convolutional Codes Used in Figs. 4.13 and 4.14.

Constraint Length K	Generator in Octal
3	5,7
5	23, 35
7	133, 171
9	561, 753

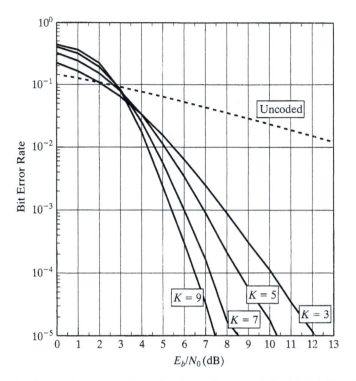

FIGURE 4.14 Comparison of uncoded and coded performance in Rayleigh-fading channel (with infinite interleaving).

Examining Figs. 4.13 and 4.14, we may make the following observations:

1. For low values of E_b/N_0, the uncoded performance is better than the coded performance, irrespective of whether the channel is an ANGN channel or a fading channel. This statement presumes the use of a maximum-likelihood decoder (e.g., Viterbi decoder). The coded performance improves on the uncoded performance only when the E_b/N_0 is large enough for the pairwise error probability of the code to assume exponentially decreasing values. The *pairwise error probability* of a code is defined as the probability that the received code vector $\mathbf{r}$ is closer (in the sense of Euclidean distance) to a candidate code vector $\hat{\mathbf{c}}$ in the code than the true code vector for $\hat{\mathbf{c}} \neq \mathbf{c}$.

2. For a prescribed E_b/N_0, the noise performance (measured in terms of the bit error rate) improves with increasing constraint length K for both AWGN and fading channels, which is intuitively satisfying.

3. For a prescribed constraint length K, the E_b/N_0 must be increased for the fading channel to exhibit a noise performance comparable to that attainable with the corresponding AWGN channel.

4. For constraint length $K = 9$ in the Rayleigh-fading channel, we can realize a bit error rate of 2×10^{-4} by using an $E_b/N_0 = 6$, which is manageable, thanks to the use of forward error-correction coding.

4.12 TURBO CODES[8]

As mentioned in Note 3, convolutional codes, discussed in Section 4.7, were first described by Elias in 1955. In this section, we discuss a new class of block codes known as *turbo codes*, which were discovered by Berrou *et al.* in 1993. Turbo codes are powerful codes by virtue of combining two important ideas, one built into the design of the transmitter and the other into the design of the receiver:

1. *Pair of encoders*, separated by an interleaver
2. *Iterative detection*, involving the use of feedback around a pair of decoders separated by a deinterleaver and an interleaver

The combination of these two ideas in a novel way has made it possible for turbo codes to provide significant improvements in the quality of data transmission over a noisy channel, yet the computational cost is modest compared with traditional encoding–decoding procedures.

4.12.1 Turbo Encoding

Turbo coding, illustrated in Fig. 4.15, uses a parallel FEC encoding scheme. With this scheme, the information is *systematically* encoded by two separate encoders. Often, the two encoders are identical, and to ensure that they do not produce the same output, the information bits are reordered through the use of a pseudorandom interleaver before the second encoding stage. This *pseudorandom interleaver* is often referred to as the *turbo interleaver* to distinguish it from *channel interleavers*, discussed in Section 4.10. The information bits and the parity bits generated by the two encoders are then transmitted over the channel. (In a typical turbo encoding scheme, the parity bits are punctured in a repeating pattern to increase the code rate, but this is not a requirement; *puncturing* refers to the omission of certain parity-check bits in the code so as to increase the data rate.)

Although any valid code is permissible, the component codes that are recommended for turbo encoding are short-constraint-length, systematic, recursive convolutional codes. With a systematic code, the original information bits are part of the

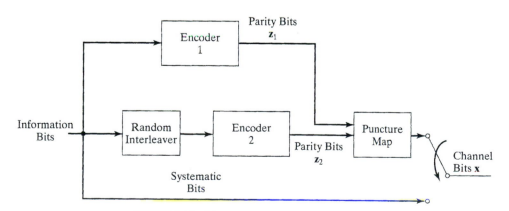

FIGURE 4.15   Illustration of turbo encoding strategy.

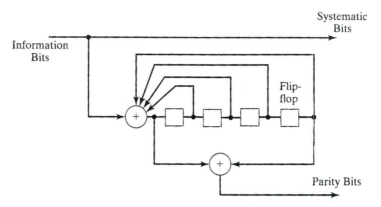

FIGURE 4.16   A systematic recursive convolutional code structure; delay T is equal to the symbol duration.

transmitted sequence; this is what is assumed in Fig. 4.15. *A recursive convolutional code means that the code has a feedback structure*, as opposed to the feedforward structure that characterizes the convolutional codes discussed in Section 4.7. The original turbo component code described by Berrou et al. is illustrated in Fig. 4.16. This is a simple 16-state code. The feedback nature of these component codes means that a single bit error corresponds to an infinite sequence of channel errors. To see this, consider an information sequence consisting of the binary digit 1, followed by an infinite number of zeros, as the input to the encoder shown in Fig. 4.16. If the encoder starts in the all-zero state, then it is straightforward to show that it never returns to that state with that kind of input.

Although the component codes are convolutional, turbo codes are inherently block codes, with the block size being determined by the size of the turbo interleaver. The block nature of the code also raises a practical problem of how to terminate the trellises of both encoders correctly. Often, the trellis of the second encoder is left unterminated.

4.12.2 Turbo Decoding

In addition to turbo codes having a novel parallel encoding structure, the decoding strategy applied to turbo codes is highly innovative. The scheme draws its name from the analogy of the structure of the decoding algorithm to the turbo engine principle. A block diagram of the decoder structure is shown in Fig. 4.17.

To illustrate the processing of the turbo decoder, let $(\mathbf{x}, \mathbf{z}_1, \mathbf{z}_2)$ be the vector of outputs from the turbo encoder of Fig. 4.15, where, for a particular turbo block of data, we have

- $\mathbf{x}$ as the vector of information (systematic) bits,
- $\mathbf{z}_1$ as the vector of parity bits from the first encoder, and
- $\mathbf{z}_2$ as the vector of parity bits from the second encoder.

Corresponding to these inputs applied to the channel, let $(\mathbf{u}, \zeta_1, \zeta_2)$ be the vector of noisy outputs from the *demodulator after matched filtering*.

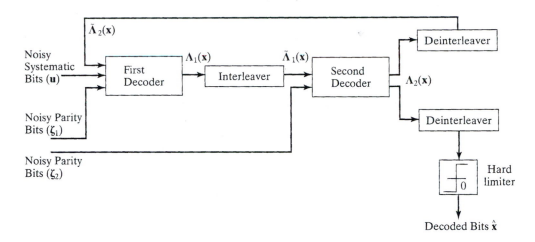

FIGURE 4.17   Block diagram of Turbo decoder.

In Fig. 4.17, the inputs to the first decoding stage are the channel samples $(\mathbf{u}, \boldsymbol{\zeta}_1)$ corresponding, respectively, to the systematic (information) bits $(\mathbf{x})$, the channel samples corresponding to the parity bits of the first encoder $(\mathbf{z}_1)$, and the *extrinsic information* about the systematic bits that were determined from previous decoder stages $(\tilde{\Lambda}_2(\mathbf{x}))$. We will define extrinsic information later, but suffice it to say that, on the first decoding iteration, the extrinsic information is zero. (If puncturing was applied at the transmitter, then the parity bits must be stuffed with zeros in the appropriate positions.)

Crucial to the performance of turbo decoding is the use of a *soft-input, soft-output (SISO) decoding algorithm*. That is, not only does the algorithm accept a soft input—a real value whose magnitude indicates the reliability of the input—but it also produces a soft output, generally the probability that a particular bit is a 0 or a 1. In the next subsection, we will describe a *maximum a posteriori probability decoding algorithm* (hereafter referred to as the *MAP algorithm*) that meets this requirement.

The first decoding stage uses the MAP algorithm to produce a refined (soft) estimate, $\text{Prob}(x(j)\,|\,\mathbf{u}, \boldsymbol{\zeta}_1, \tilde{\Lambda}_2(\mathbf{x}))$, of the systematic bits. However, for the implementation, it is more convenient to express this soft estimate as the equivalent log-likelihood ratio

$$\Lambda_1(x(j)) = \log\left(\frac{\text{Prob}(x(j)=1\,|\,\mathbf{u}, \boldsymbol{\zeta}_1, \tilde{\Lambda}_2(\mathbf{x}))}{\text{Prob}(x(j)=0\,|\,\mathbf{u}, \boldsymbol{\zeta}_1, \tilde{\Lambda}_2(\mathbf{x}))}\right) \qquad (4.20)$$

In practice, calculation of this log-likelihood ratio is much simpler than Eq. (4.20) would suggest. Before entering the second stage of processing on the first iteration, the vector of refined estimates, Λ_1, is reordered to compensate for the turbo interleaving at the transmitter, whereafter it is combined with the second set of parity samples from the channel as the input to the second decoding stage.

The second decoding stage uses the MAP algorithm and the second set of parity bits to produce a further refined estimate, $\text{Prob}(x(j)\,|\,\tilde{\Lambda}_1((\mathbf{x}), \boldsymbol{\zeta}_2))$, of the systematic bits. This second refined estimate is also expressed as a log-likelihood ratio, namely, $\Lambda_2(x(j))$. The estimate so produced can be hard-detected, if so desired, to provide bit estimates at this point. Alternatively, the output of the second stage can be used to provide extrinsic information to the first stage. In either case, the estimates must be reordered to compensate for the turbo interleaving.

As can be seen from Fig. 4.17, the first decoding stage, the interleaver, the second decoding stage, and the deinterleaver constitute a single-loop *feedback system*, thereby making it possible to iterate the decoding process in the receiver as many times as is deemed necessary for a satisfactory performance. Indeed, this iterative process constitutes the *turbo coding principle* at the receiver. Note, however, that during each set of iterations, the noisy input vector **u** is unaltered.

The decoding scheme of Fig. 4.17 relies on the implicit assumption that the bit probabilities remain independent from one iteration to another. To increase the independence of the inputs from one processing stage to the next, the turbo algorithm makes use of the concepts of intrinsic and extrinsic information. *Intrinsic information refers to the information inherent in a sample prior to a decoding operation. On the other hand, extrinsic information refers to the incremental information obtained through decoding.* To maintain as much independence as is practically possible from one iteration to the next, *only extrinsic information is fed from one stage to the next.*

An important property of the MAP algorithm is that it includes both intrinsic and extrinsic information in a product relationship (including a scaling factor). However, from a computational perspective, it is more convenient to express this relationship in terms of log-likelihood ratios, because then the product becomes a sum and the scaling term assumes a nonsignificant role. In particular, the extrinsic information at the output of the second stage is

$$\tilde{\Lambda}_2(\mathbf{x}) = \Lambda_2(\mathbf{x}) - \tilde{\Lambda}_1(\mathbf{x}) \tag{4.21}$$

That is, the extrinsic information (expressed in logarithmic form) is the difference between the input and output log-likelihood ratios for the systematic bits. (On the first pass, $\tilde{\Lambda}_1(\mathbf{x}) = \Lambda_1(\mathbf{x})$.) Similarly, at the output of the first stage, the extrinsic information supplied to the second stage by the first stage is given by

$$\tilde{\Lambda}_1(\mathbf{x}) = \Lambda_1(\mathbf{x}) - \tilde{\Lambda}_2(\mathbf{x}) \tag{4.22}$$

The basic MAP decoding algorithm is not changed in the turbo decoder. Details of the actual computation of Λ_1 and Λ_2 are omitted in Fig. 4.17 to simplify the exposition. The differences occur only in the processing, which is performed on the inputs to, and outputs from, the MAP algorithm.

On the last iteration of the decoding process, a hard decision is applied to the output of the second decoder (after deinterleaving) to produce an estimate of the *j*th information bits:

$$\hat{x}(j) = \text{sign}(\Lambda_2(x(j))) \tag{4.23}$$

4.12.3 Noise Performance

The impressive aspect of turbo codes is their noise performance. In Fig. 4.18, an example of that performance is presented. The results correspond to the encoding strategy described in Section 4.12.1, for an interleaver size of 64 kilobits. The performance is

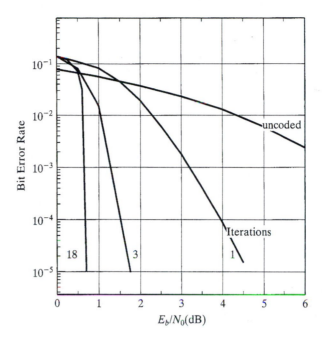

FIGURE 4.18 Performance of turbo codes as a function of number of iterations. (Reproduced from Berrou *et al.*, 1993, with permission of the IEEE.)

quantified as a function of the number of iterations. It is clear from the figure that the performance with a small number of iterations of the decoding process is not particularly impressive, but as the number of iterations is increased, the performance continually improves. With 18 iterations, the code achieves a bit error rate (BER) of 10^{-5} at an E_b/N_0 of only 0.7 dB. This is only about 0.7 dB away from the Shannon limit for this code-particular rate.

The performance of turbo codes at low E_b/N_0 is a function of the interleaver size. With smaller interleavers, the performance tends to degrade. However, even with an interleaver size of 300 bits, a BER of 10^{-4} can be obtained at an E_b/N_0 of 2.3 dB, which is indeed remarkable.

4.12.4 Maximum a Posteriori Probability Decoding

The *maximum a posteriori probability (MAP) decoder* considered herein makes certain assumptions about the transmission system. In particular, we assume $\mathbf{x} = \mathbf{x}_{[1,J]}$ is the information (data) to be transmitted and $\hat{\mathbf{x}}$ is the vector of corresponding data estimates. The transmitted code word is represented by $\mathbf{c} = \mathbf{c}_{[1,N]}$.

The N transmitted bits will be grouped together into T symbols, $\mathbf{v}_{[1,T]}$. This grouping will arise naturally from the structure of the code. For example, with a rate-1/2 code, T is $N/2$. The noisy received bits will be denoted by $\mathbf{r} = \mathbf{r}_{[1,N]}$ and the corresponding received symbols by $\mathbf{w}_{[1,T]}$. The detection (decoding) algorithm provides an estimate $\hat{\mathbf{x}}$ of the data, based on the sequence $\mathbf{r}$ (or $\mathbf{w}$) and prior knowledge of the

FEC code. The channel is assumed to be an AWGN channel. The channel encoder is assumed to have a trellis structure similar to that illustrated in Fig. 4.7 for four states. It is assumed that *the trellis starts and ends in the zero state* and that the appropriate null bits are added to the information stream to ensure proper termination of the trellis. Since the trellis is terminated, we are effectively assuming only block codes, even if a convolutional code is used.

The number of trellis transitions T, corresponding to the N code bits, is a function of the encoding strategy. Its relationship to the index of the channel bits or the information bits depends on the code rate and how the trellis is truncated. At time j, each channel symbol $v(j)$ corresponds to a state transition from $s(j-1)$ to $s(j)$. For example, with a rate-2/3 convolutional code, each channel symbol will have three bits and each pair of information bits will correspond to a state transition until the decoder reaches the termination stage. In the termination stage, information bits that are zeros are appended to the input, and the corresponding channel symbols are generated until the trellis returns to the all-zero state.

For the turbo decoder application, the objective of the MAP algorithm is to determine the conditional probability Prob $(x(j)|\mathbf{r})$ where $\mathbf{r} = (\mathbf{u}, \boldsymbol{\zeta}_1, \tilde{\mathbf{\Lambda}}_2)$ in one decoding instance and $\mathbf{r} = (\tilde{\mathbf{\Lambda}}_1, \boldsymbol{\zeta}_2)$ in the second decoding instance.

The *MAP algorithm includes a backward and forward recursion*, unlike the Viterbi algorithm, which contains only a forward recursion. The fundamental assumption of the MAP algorithm is that the channel encoding can be represented as a trellis, as illustrated in Fig. 4.7,where the next state depends only on the current state and the new input bit. The states of the trellis are indexed by the integer $s = 0,1,...,S-1$. The state of the trellis at time j is denoted by $s(j)$, and the output symbol due to the transition from $s(j-1)$ to $s(j)$ is denoted by $v(j)$. A state sequence from time j to i' is denoted by $\mathbf{s}_{[j,\,i']}$, and the corresponding output sequence is $\mathbf{v}_{[j,\,i']}$.

Although the ultimate goal of the MAP algorithm is to estimate the probability that a given symbol or information bit was transmitted, it is simpler to first derive the *a posteriori* probabilities of the states and transitions of the trellis, based on the observation $\mathbf{r}$. From these results, most of the probabilities of interest can be obtained by performing summations over selected subsets of states or transitions. To proceed, we denote the S-vector of state probabilities at time j based on the set of observations by

$$\boldsymbol{\lambda}(j) = \text{Prob}(\mathbf{s}(j)|\mathbf{r}) \qquad \boldsymbol{\lambda}(j) \in R^S \qquad (4.24)$$

where the jth element is

$$\lambda_{s(j)} = \text{Prob}(s(j) = s|\mathbf{r}) \qquad (4.25)$$

Then, for a rate $(1/n)$ convolutional code without feedback, assuming that the transmitted information bit is the least significant bit (LSB) of the state, the probability that a 1 was the information bit is given by

$$\text{Prob}(x(j) = 1|\mathbf{r}) = \sum_{s:\text{LSB}(s)\,=\,1} \lambda_s(j) \qquad (4.26)$$

where LSB(s) is the least significant bit of the state s. For a recursive code, the summation in Eq. (4.26) is over the set of transitions that correspond to a 1 at the input. (Transition probabilities will be explained shortly.)

Next, we define the *forward estimator* of state probabilities as the S-vector

$$\boldsymbol{\alpha}(j) = \text{Prob}[\mathbf{s}(j)|\mathbf{r}_{[1,j]}] \qquad \boldsymbol{\alpha}(j) \in R^M \qquad (4.27)$$

Similarly, we define the *backward estimator* of state probabilities as

$$\boldsymbol{\beta}(j) = \text{Prob}[\mathbf{s}(j)|\mathbf{v}_{[j,1]}] \qquad \beta(j) \in R^M \qquad (4.28)$$

Equations (4.27) and (4.28) are the estimates of the state probabilities at time j based on past and future observations, respectively. The important result related to these quantities is their relationship to $\lambda(j)$. First, however, let us define the vector product $\mathbf{c} = \mathbf{a} \cdot \mathbf{b}$ as $c(s) = a(s)b(s)$ for all s and define the L^1 norm for probability vectors as

$$\|\boldsymbol{a}\| = \sum_s a(s) \qquad (4.29)$$

Then we may write

$$\boldsymbol{\lambda}(j) = \frac{\boldsymbol{\alpha}(j) \cdot \boldsymbol{\beta}(j)}{\|\boldsymbol{\alpha}(j) \cdot \boldsymbol{\beta}(j)\|_1} \qquad (4.30)$$

That is, the state probabilities at time j are completely defined by the forward and backward estimators at that time.

A proof of the *separation theorem* embodied in Eq. (4.30) is presented in Appendix F. One aspect of this theorem is intuitively obvious: By definition, for a *Markov process*, the state distribution at time j, given the past, is independent of the state distribution at time j, given the future. However, the theorem says more: It says that there is a simple way of combining the forward and backward estimates to obtain a complete estimate; specifically, that estimate is just the normalized product of the state distributions based on the past and the future.

Thus, the objective is to find simple algorithms for obtaining the forward and backward estimates, which would be a solution to the MAP detection problem. If we represent the state transition probability at time j by

$$\gamma_{s',s}(j) = \text{Prob}(s(j) = s, w(j)|s(j-1) = s') \qquad (4.31)$$

and denote the matrix of these probabilities as

$$\boldsymbol{\Gamma}(j) = [\gamma_{s',s}(j)] \qquad \boldsymbol{\Gamma}(j) \in R^{S \times S}, \qquad (4.32)$$

then we have the following recursion equations for calculating the forward and backward state estimates:

$$\boldsymbol{\alpha}^T(j) = \frac{\boldsymbol{\alpha}(j-1)^T \boldsymbol{\Gamma}(j)}{\|\boldsymbol{\alpha}(j-1)^T \boldsymbol{\Gamma}(j)\|_1} \qquad (4.33)$$

and

$$\boldsymbol{\beta}(j) = \frac{\boldsymbol{\Gamma}(j+1)\boldsymbol{\beta}(j+1)}{\|\boldsymbol{\Gamma}(j+1)\boldsymbol{\beta}(j+1)\|_1} \tag{4.34}$$

Together, these two equations define the algorithm for MAP decoding. In particular, the steps are as follows:

1. $\boldsymbol{\alpha}(0)$ and $\boldsymbol{\beta}(T)$ are initialized according to the trellis structure. For a trellis beginning and ending in the all-zero state, we have $\boldsymbol{\alpha}_0(0) = 1$ and $\boldsymbol{\alpha}_s(0) = 0$ for all $s \neq 0$, and similarly for $\boldsymbol{\beta}(T)$.

2. When $r(j)$ is received, the decoder computes $\boldsymbol{\Gamma}(j)$ and then $\boldsymbol{\alpha}(j)$, using Eq. (4.33). The computed values of $\boldsymbol{\alpha}(j)$ are stored for all j and s. Note that $\mathbf{r}(j)$ is a vector of length L, defined by

$$\mathbf{r}(j) = [r_{j(1)}, r_{j(2)}, \dots r_{j(L)}]^T$$

where L is itself defined by the number of bits produced per transition by the encoder.

3. After the complete sequence $\mathbf{r}_{[1,T]}$ has been received, the decoder recursively computes $\boldsymbol{\beta}(j)$, using Eq. (4.34). Then, when the $\boldsymbol{\beta}(j)$ have been computed, they are multiplied by the appropriate $\boldsymbol{\alpha}(j)$ to obtain $\boldsymbol{\lambda}(j)$, using Eq. (4.30).

The MAP algorithm, also referred to as the *BCJR algorithm* in recognition of its originators,[9] involves a double recursion—one in the forward direction and the other in the reverse. Consequently, it has at least twice the complexity of the Viterbi algorithm in its most general form. However, it produces soft outputs, a feature that is the key to its usefulness in the turbo decoding algorithm.

4.13 COMPARISON OF CHANNEL-CODING STRATEGIES FOR WIRELESS COMMUNICATIONS

Channel-coding strategies, exemplified by convolutional codes and turbo codes, have established themselves as an essential design tool for reliable communications, each in its own way. Convolutional codes have a long history, dating back to the classic paper by Elias in the 1950s (see Note 3). In contrast, turbo codes are of recent origin, having been discovered in the early 1990s as a result of pragmatic experimentation by Berrou and Glavieux (see Note 8).

With classical convolutional codes covered in Section 4.9 and turbo codes in Section 4.12, it is apropos here that we discuss the relative merits of these two coding strategies for the correction of transmission errors. However, before proceeding with this discussion, we note that classical block codes such as Hamming codes, Bose–Chandhuri–Hocquenghem (BCH) codes, and Reed–Solomon codes, powerful as they are, have not been emphasized in the presentation largely because of their limited application to modern wireless channels. There are two reasons for not covering these classical block codes in this chapter:

- Block codes can be designed to handle error bursts that may occur on a wireless channel, but their ability to handle large numbers of distributed errors is rather limited.

- Block codes are typically algebraic in their formulation, using hard decisions that tend to destroy information; consequently, they do not obtain the performance advantage that is available through the use of soft decisions.

Nevertheless, there are some wireless channels to which block codes are applied. In high-SNR channels, for example, where the errors are few, block codes are often used to correct errors that do occur. In other wireless communication systems, block codes may be used as part of a concatenated coding scheme. In a two-level form of concatenation used in such systems, the block code is used as the outer code, and the inner code is typically a convolutional code. With such a configuration, the (inner) convolutional decoder in the receiver corrects the majority of the errors produced by the wireless channel. Then the (outer) block decoder corrects the few errors that remain at the output of the convolutional decoder.

To proceed with the task at hand, in what follows we present a comparative evaluation of classical convolutional codes and turbo codes in terms of encoding and decoding considerations, as well as other matters that pertain to signal transmission over wireless channels.

4.13.1 Encoding

A convolutional encoder can assume one of two forms:

- *Nonrecursive nonsystematic*, in which form the convolutional encoder distinguishes itself by, first, the use of feedforward paths only and, second, the information bits losing their distinct identity as a result of the convolution process.
- *Recursive systematic*, in which form, first, the convolutional encoder uses feedforward as well as feedback paths and, second, the k-tuple of information bits appears as a subset of the n-tuple of output bits.

On the one hand, for historical reasons, nonrecursive nonsystematic schemes have been advocated for classical convolutional encoders. On the other hand, turbo codes use recursive systematic convolutional (RSC) encoders, as discussed in Section 4.12. Comparing turbo codes with convolutional codes, we find some strengths, weaknesses, and similarities between the two codes:

- By virtue of the fact that the turbo code uses a pseudorandom interleaver to separate its own two convolutional encoders, the turbo code is closer to the random code originally advocated by Shannon in his formulation of communication theory than it is to the classical convolutional code. However, unlike Shannon's random codes, turbo codes are decodable, hence their practical importance.
- Experimentation shows that turbo codes work better than classical convolutional codes when the code rates are high or the signal-to-noise ratios are low.
- Both types of codes require the use of flush bits to return them to the state 0 at the end of the incoming information bits. With the parallel encoding structure of turbo codes, it is not straightforward to flush the second encoder, so flushing is often not done.
- Unlike convolutional codes, turbo codes have an error floor. That is, the BER drops very quickly at the beginning, but eventually levels off and decreases at a

much slower rate. The leveling-off point is often in the BER range of 10^{-5} to 10^{-8}, depending on factors such as how flushing is done, how long the block size is, and which turbo interleaver is being used. Achieving very low error rates (less than 10^{-10}) is difficult with a turbo code; but then, this kind of error performance is rarely required for wireless communications, except perhaps in computer communications, which demand a high degree of reliability over a wireless channel.

- Both types of codes perform better with soft decisions. Turbo codes rely on soft inputs to work. Convolutional codes can work with either soft-decision or hard-decision inputs, but the penalty for using the latter in an AWGN channel is approximately 2 dBs.

4.13.2 Decoding

The traditional approach to decoding a convolutional code is to use the Viterbi algorithm, which operates as a maximum-likelihood state estimator. The complexity of the Viterbi algorithm is directly proportional to the number of states, which in turn depends exponentially on v, denoting the number of memory units in the shift register of the code. Since the minimum free distance (i.e., the error-correction capability) of the convolutional code increases with v, there is a trade-off between performance and decoder complexity. Current wireless systems often use $v = 6$ or 8 (constraint length $K = 7$ or 9) convolutional codes, corresponding to 64-state and 256-state decoders, respectively.

By contrast, turbo decoding relies on the exchange of extrinsic information between two soft-input, soft-output (SISO) decoding stages on an iterative basis. The decoder may use a maximum a posterior probability (MAP) algorithm or its approximation. The advantage of turbo codes is that they can achieve large coding gains with very simple component codes typically, 8-state or 16-state codes. The MAP decoder complexity is proportional to the number of states and is approximately twice the complexity of the Viterbi algorithm for the same number of states; the multiplying factor of two is due to the fact that the MAP algorithm, or its approximation, operates in the forward as well as backward direction. The overall complexity of the turbo decoder is equal to the computational complexity of one complete decoding iteration, multiplied by the number of iterations. In practice, between four and eight iterations usually are employed, although optimum performance often would require twice that number of iterations or more.

With the decoding process of a turbo code being iterative in nature, it would be highly desirable to have at our disposal a tool for describing the convergence analysis of the code and its design. The *extrinsic information transfer (EXIT) chart* provides such a tool in graphical form. The development of the EXIT chart assumes that the size of the turbo interleaver is infinitely large. In particular, the chart describes the exchange of extrinsic information between the first stage and second stage in the turbo decoder, and vice versa. For each constituent decoding stage, the EXIT chart is defined as *the function that maps the prior information to the extrinsic information applied to the decoder in question*, with the information capacity of its communication channel treated as a parameter.[10]

4.13.3 AWGN Channel

For an additive white Gaussian noise (AWGN) channel, which, for example, is closely approximated by a line-of-sight satellite communication link, turbo codes outperform convolutional codes by a significant margin in their ability to achieve near-optimum performance. Optimality is defined in terms of theoretical limits imposed by Shannon's information capacity theorem. The theoretical limit for a block of information bits that is infinitely long is measured in terms of E_b/N_0 required for a prescribed code rate and minimum bit error rate (BER). When information blocks of a finite size are considered, the performance is measured in terms of the increase in E_b/N_0 (measured in dBs with respect to the theoretical limit) required for prescribed values of the code rate, block size, and frame error rate (FER). The FER is the number of frames that have at least one bit error, divided by the total number of frames used in the calculation.

For example, for a code rate of 2/3 and minimum bit error rate of 10^{-4}, the theoretical limit (assuming an information block of infinite size) is about −0.8 dB. For the same code rate, but a block size of 500 information bits, using a turbo code, we need to increase E_b/N_0 by about 1.3 dB (i.e., a net value for E_b/N_0, equal to $1.3 - 0.8 = 0.5$ dB) in order to maintain an FER of 10^{-4}. This increase in E_b/N_0 is indeed modest in comparison with that for convolutional codes.

4.13.4 Fading Wireless Channels

Wireless channels are prone to error bursts, while both convolutional and turbo decoders work best with independent error events. Hence, both types of codes require the use of interleavers in the transmitter and corresponding deinterleavers in the receiver, as illustrated in Fig. 4.1, to combat error bursts that occur on wireless channels. In the absence of problems such as loss of synchronization, nonstationary characteristics of the channel, or other disruptions in the communications link, larger interleavers imply better performance for both decoding techniques.

The performance curves of turbo codes, which plot BER against E_b/N_0 in dBs, have a brick-wall shape. The corresponding performance curves of convolutional codes exhibit a slow roll-off characteristic, which matches the behavior of fading wireless channels reasonably well; whether this feature provides an advantage or not depends on the ratio of the data rate and block length to the fading rate. Moreover, for short block lengths, which are most robust for communication over fading wireless channels, the improvement offered by turbo codes over convolutional codes is usually small.

Both types of codes are present in third-generation (3G) wireless standards.

4.13.5 Latency

The issue of latency refers to the delay incurred by a channel decoder in processing the received signal in order to recover the original sequence of information bits. Latency is particularly serious in the transmission of voice signals over a wireless channel, as close-to-real-time communication is a necessary requirement, but perhaps less so in the transmission of data.

Latency is proportional to the interleaver size, since the decoder usually must receive the complete interleaved block of bits before it can start decoding. For turbo codes that include two interleavers, namely, a turbo interleaver and a channel interleaver, the delay is proportional to the larger of the two. Latency may also be related to decoder speed, but for both types of codes, practical decoders ordinarily operate at the channel rate or faster, and this is usually not a distinguishing feature.

Smaller block sizes mean smaller latency. Convolutional codes usually achieve their maximum gain at a smaller block size. Turbo codes, in contrast, require large block sizes to achieve their maximum gain, which is usually greater than that attained by convolutional codes.

4.13.6 Joint Equalization and Decoding[11]

An elegant feature of the turbo coding principle is that it provides an iterative basis for the joint implementation of channel equalization and channel decoding in the receiver in a way that makes a significant difference to the overall receiver performance. This issue is discussed in a theme example in Section 4.18.

1. Turbo codes typically have a very small free distance, but, upon decoding, they have far fewer error events at this free distance than convolutional codes have. It is this latter characteristic that gives rise to the brick-wall nature of turbo code performance. However, asymptotically, the performance of turbo codes is still characterized by their minimum free distance. Since this distance is often small, it causes the BER performance of the turbo decoder to level off at high SNR and then appear as an error floor compared with the initial brick-wall performance.

2. The convolutional encoders used in the turbo encoder are recursive, which are therefore non-return-to-zero encoders in that they return to their initial state only with probability 2^{-v}, where v is the number of memory units in the shift register of the code. Furthermore, the turbo code may be endowed with a cyclic trellis, which means that the temporal representation of the states assumed by the encoder for time steps $j = 1$ to $j = k$, where k, the number of information bits, repeats itself in a cyclic manner. Consequently, the turbo code becomes a block code, and, most important, the non-return-to-zero sequence produced by the encoder affects all of the redundant bits introduced by the encoder in such a way that the probability of the turbo decoder failing to recover the original non-return-to-zero sequence is negligibly small.

4.14 RF MODULATION REVISITED

With source (speech) encoding and channel encoding in place, the next operation that is performed in the transmitter is modulation to prepare the signal for transmission over the wireless channel. Here, we have a wide range of strategies to consider, as discussed in Chapter 3. However, with *spectral efficiency* of narrowband wireless communications being an issue of primary concern, it is necessary that adjacent-channel interference (i.e., the spillage of the spectrum of a modulated signal into adjacent channels) be minimized.

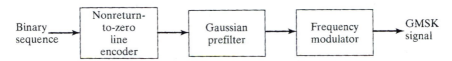

FIGURE 4.19   Block diagram of Gaussian minimum-shift keying (GMSK) signal generator.

Partial-response modulation provides one approach to accomplishing this practical requirement. The basic idea of partial-response modulation is to ensure that the *phase response of the modulated signal is spread over several symbol periods*. Stated another way, the phase response in any signaling interval is "partial"—hence the term "partial-response modulation." Note, however, that the use of partial-response modulation has the effect of a deliberate introduction of *controlled* intersymbol interference (ISI), which typically requires the use of an equalizer at the receiver for its removal. *Gaussian minimum-shift keying* (GMSK) is an example of partial-response modulation. (See Section 3.7 for its detailed description.) A block diagram of the GMSK signal generator is shown in Fig. 4.19.

4.15 BASEBAND PROCESSING FOR CHANNEL ESTIMATION AND EQUALIZATION[12]

The generated RF modulated signal $s(t)$ is transmitted over the narrowband wireless channel whose midband is centered on the carrier frequency f_c. Thereupon, the transmitted RF signal is convolved with the impulse response of the channel, $h(t)$. (The use of convolution to describe signal transmission over the channel is justified, since the wireless channel is essentially linear, as discussed in Chapter 2.) The signal resulting from this convolution operation plus additive white Gaussian noise (AWGN) $w(t)$ at the channel output constitutes the received RF signal

$$x(t) = s(t) \otimes h(t) + w(t) \tag{4.35}$$

where, again, the symbol $\otimes$ denotes convolution. The signal $x(t)$ is ready for processing by the receiver.

With the ever-increasing availability of digital signal-processing devices powered by continuing improvements in computer hardware (in the form of silicon chips) and software, the trend nowadays is to convert the received RF signal $x(t)$ into *baseband* form. Simply put, not only is the use of digital signal processing cost effective, but it also provides flexibility unmatched by analog devices. The RF-to-baseband conversion is accomplished with the *quadrature demodulator*, depicted in Fig. 4.20, in accordance with the discussion of the complex representation of band-pass signals and systems presented in Section 3.4. The demodulator yields a *complex baseband signal* $\tilde{x}(t)$ which is equivalent to the RF modulated signal $x(t)$ in that there is no loss of information. The baseband signal $\tilde{x}(t)$ is complex in that it consists of two orthogonal components, as shown by the formula

$$\tilde{x}(t) = x_I(t) + x_Q(t) \tag{4.36}$$

where $x_I(t)$ is the *in-phase component* and $x_Q(t)$ is the *quadrature component*. Note that both $x_I(t)$ and $x_Q(t)$ are low-pass real signals whose common bandwidth is one-half that of the RF-modulated signal $x(t)$.

Recall from the discussion presented in Section 3.4 that RF-to-baseband conversion has an effect on the narrowband wireless channel centered on the carrier frequency f_c, which is similar to the channel output $x(t)$. Specifically, the real impulse response of the channel, $h(t)$, is transformed into the complex equivalent baseband form

$$\tilde{h}(t) = h_I(t) + jh_Q(t) \tag{4.37}$$

where $h_I(t)$ is the in-phase component and $h_Q(t)$ is the quadrature component. In a manner similar to the baseband representation of the modulated signal $x(t)$, both $h_I(t)$ and $h_Q(t)$ are real low-pass impulse responses occupying the same frequency band as the complex baseband signal $\tilde{x}(t)$.

The analog baseband quadrature signals $x_I(t)$ and $x_Q(t)$ are converted into digital form, whereafter the baseband signal processing is carried out in the receiver. (The analog-to-digital converter is not shown in Fig. 4.20, as it is considered part of the RF-to-baseband converter.)

With the digitized in-phase and quadrature components of the received signal at hand, the baseband signal processing can begin in the receiver. The objective of this processing is to compute a set of *transition metrics* that, in turn, are employed in a Viterbi equalizer designed to remove the effects of any controlled ISI (e.g., due to the partial-response modulation) and the channel-induced ISI, thereby facilitating data recovery. To that end, the baseband processor solves two intermediate problems, with the solution to the first leading to the solution of the second. The first problem is that of estimating the unknown impulse response of the channel. The second problem involves generating all the possible signals that would emanate from a channel having

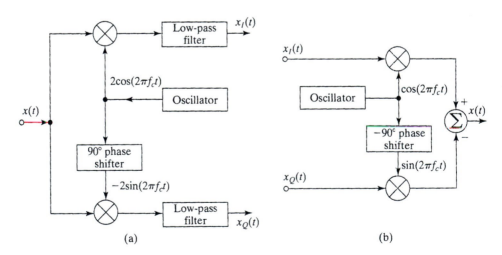

FIGURE 4.20 (a) Scheme for deriving the in-phase and quadrature components of RF-modulated signal $x(t)$. (b) Scheme for reconstructing the RF-modulated signal from its in-phase and quadrature components.

such an estimate for its impulse response. The related issues of channel estimation and estimated signal generation are addressed in what follows in that order.

4.15.1 Channel Estimation

To proceed with the estimation of the channel impulse response, the baseband signal $\tilde{x}(t)$ is first *demultiplexed* into two parts, as illustrated in Fig. 4.21. The part denoted by $\tilde{x}_{channel}(t)$ relates to the sounding signal employed for channel estimation. The other part, denoted by $\tilde{x}_{data}(t)$, contains information on the original source signal $m(t)$.

Before reaching the baseband processor, the channel-probing signal is introduced into the "packetizer" in the transmitter and is then modulated on the RF carrier, transmitted over the narrowband channel, and, finally, converted into complex baseband form; typically, the probing signal has an impulselike autocorrelation function, for reasons to be outlined in the discourse that follows. To simplify the discussion, we ignore the effect of channel noise. Under this assumption, the signal $\tilde{x}_{channel}(t)$ may be expressed as the convolution of two complex time functions:

$$\tilde{x}_{channel}(t) = \tilde{c}(t) \otimes \tilde{h}(t) \qquad w(t) = 0 \qquad (4.38)$$

Here, $\tilde{c}(t)$ and $\tilde{h}(t)$ are, respectively, the baseband versions of the probing signal $c(t)$ and channel impulse response $h(t)$. The signal $\tilde{x}_{channel}(t)$ is applied to a *matched filter*, whose impulse response, $\tilde{q}(t)$, is equal to the complex conjugate of a time-reversed version of $\tilde{c}(t)$. Appropriately delayed to satisfy causality (see Appendix D), the impulse response is

$$\tilde{q}(t) = \tilde{c}^*(T_c - t) \qquad (4.39)$$

where T_c is the duration of the sounding signal $c(t)$ and the asterisk denotes complex conjugation. The matched-filter output is therefore

$$\tilde{z}(t) = \tilde{q}(t) \otimes \tilde{x}_{channel}(t)$$
$$= \tilde{c}^*(T_c - t) \otimes \tilde{c}(t) \otimes \tilde{h}(t) \qquad (4.40)$$

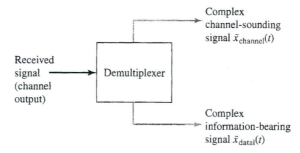

FIGURE 4.21   Block diagram of demultiplexer at the receiver input, following the quadrature demodulator.

The following is an important property of matched filters (see Appendix D):

> *The output of a matched filter, in response to an input signal to which the filter is matched, is equal to the autocorrelation function of the input signal.*

In the context of our present discussion, this statement means that

$$\tilde{c}^*(T_c - t) \otimes \tilde{c}(t) = r_{\tilde{c}}(T_c - t) \tag{4.41}$$

where $r_{\tilde{c}}(\tau)$ is the *autocorrelation function* of the complex envelope $\tilde{c}(t)$ of the probing signal for lag $\tau = T_c - t$. Hence, Eq. (4.40) reduces to

$$\tilde{z}(t) = r_{\tilde{c}}(T_c - t) \otimes \tilde{h}(t) \tag{4.42}$$

In general, the autocorrelation function of a complex signal has a complex value. However, by having the original sounding signal $c(t)$ exhibit *even symmetry* about its midpoint $t = T_c/2$, the autocorrelation function $r_{\tilde{c}}(\tau)$ assumes a *real* value, which simplifies the design of the baseband processor considerably. To emphasize this point, we set

$$r_{\tilde{c}}(\tau) = \rho(\tau) \qquad \text{for all } \tau \tag{4.43}$$

where $\rho(\tau)$ is a real autocorrelation function that is an even function of τ, that is,

$$\rho(-\tau) = \rho(\tau) \tag{4.44}$$

Accordingly, we may rewrite Eq. (4.42) in terms of this new function as

$$\tilde{z}(t) = \rho(t - T_c) \otimes \tilde{h}(t) \tag{4.45}$$

In words, the output $\tilde{z}(t)$ of the matched filter is equal to the real-valued, delayed autocorrelation function $\rho(t - T_c)$, convolved with the complex baseband impulse response $\tilde{h}(t)$ of the channel.

Suppose now that the autocorrelation function $\rho(\tau)$ not only is real, but also is in the form of a *delta function* $\delta(\tau)$. Then Eq. (4.45) simplifies further to the ideal result

$$\tilde{z}(t) = \delta(t - T_c) \otimes \tilde{h}(t)$$
$$= \tilde{h}(t - T_c) \tag{4.46}$$

where we have used the fact that the convolution of a time function with a delta function leaves that time function unchanged, except for a possible time shift.

The result of Eq. (4.46) is idealized in that it is derived under two special conditions:

1. The channel noise $w(t)$ is zero.
2. The probing signal $c(t)$ is long enough for the autocorrelation function of its complex envelope $\tilde{c}(t)$ to approach a delta function.

In practice, the operation of a wireless communication system deviates from this ideal setting—hence the need to speak of an "estimate" of the complex impulse response $\tilde{h}(t)$ at the output of the matched filter. Hereafter, we refer to this estimate as $\tilde{h}_{est}(t)$. Ignoring the effect of channel noise, we now note from Eqs. (4.45) and (4.46) that

$$\tilde{h}_{est}(t) = \rho(t) \otimes \tilde{h}(t + T_c) \tag{4.47}$$

where $\tilde{h}(t)$ is the actual value of the complex baseband impulse response of the channel. The effect of ignoring $w(t)$ is justified only when the signal-to-noise ratio is high.

4.15.2 Viterbi Equalization

As mentioned previously, a primary objective of the baseband processor is to undo the convolution performed on the transmitted signal by the channel, which is indeed a task well suited for the Viterbi algorithm functioning as an equalizer. This is yet another novel application of the algorithm, building on what we said earlier in Section 4.9 on convolutional decoding: the Viterbi algorithm is a maximum-likelihood sequence estimator.

Consider, then, a channel with *memory l*, requiring the use of a *Viterbi equalizer* with a window of length l. (For the application at hand, a value of 4 is considered typical for l.) Correspondingly, the equalizer has 2^l possible states, with each state consisting of l symbols. As with convolutional decoding, we need a metric for the design of the equalizer, which, in turn, requires the generation of two kinds of waveforms:

1. *Estimated received waveforms.* This set of waveforms is generated by cycling a local modulator and channel model through all the possible l-bit sequences for every bit period. The combination of local modulator and channel model based on the channel estimate $\tilde{h}_{est}(t)$ is termed the *estimated waveform generator*, with its output denoted by the complex waveform $\tilde{\xi}_{est}(t)$.

2. *Compensated received waveform.* From Eq. (4.45), we note that, except for a delay, the complex impulse response estimate $\tilde{h}_{est}(t)$ equals the actual complex impulse response $\tilde{h}(t)$, convolved with the real autocorrelation function $\rho(t)$. Since the estimated waveform generator embodies the channel model, it follows that the actual received signal $\tilde{x}_{data}(t)$ should also be convolved with $\rho(t)$. Then the compensated version of $\tilde{x}_{data}(t)$, namely,

$$\tilde{\xi}_{data}(t) = \rho(t) \otimes \tilde{h}(t) \tag{4.48}$$

would be on par with the estimated received waveform $\tilde{\xi}_{est}(t)$.
Note that $\tilde{\xi}_{data}(t)$ and $\tilde{\xi}_{est}(t)$ are both continuous-time signals.

In order to generate fairly accurate digital representations of $\tilde{\xi}_{data}(t)$ and $\tilde{\xi}_{est}(t)$, it is necessary that they be sampled at a rate of n times the incoming bit rate, so as to prevent aliasing. That is, each bit of data (actual as well as estimated) is represented by n samples, where n is an integer equal to or greater than two.

The *squared Euclidean distance* between the ith sample of the kth bit in the actual received waveform $\tilde{\xi}_{data}(t)$ and the corresponding ith sample of the estimated received waveform $\tilde{\xi}_{est}(t)$ pertaining to a possible state v of the equalizer may be expressed as the sum of two squared terms, one due to the in-phase components of these two waveforms and the other due to their quadrature components:

$$\mu_{k,v}^2(i) = (\tilde{\xi}_{data,I}(k, i) - \tilde{\xi}_{est,I}(v, i))^2 + (\tilde{\xi}_{data,Q}(k, i) - \tilde{\xi}_{estQ}(v, i))^2$$

With η samples per bit, the sample index i ranges from 0 to $\eta - 1$. Hence, we may define the transition metric for bit k of the actual signal and possible state v of the equalizer as

$$\mu_{k,v} = \sum_{i=0}^{\eta-1} \mu_{k,v}^2(i) \tag{4.49}$$

where $v = 0,1,...,2^{l-1}$

Putting the ideas discussed here together, we may now formulate the block diagram of Fig. 4.22 for the baseband processor for channel estimation and equalization, leading to data recovery. The processor consists of three subsystems: the *estimated waveform generator*, *transition metric computer*, and *Viterbi equalizer*.

Building on the idea of the Viterbi algorithm as a maximum-likelihood sequence estimator, we may now describe the way in which the Viterbi equalizer performs its computations. The basic difference between the Viterbi equalizer and the Viterbi decoder (discussed in Section 4.9) lies in what is used for the transition metric. Specifically, we use $\mu_{k,v}$ (defined in Eq. (4.49)) for equalization and the Hamming distance for hard-decision-based convolutional decoding. Accordingly, the steps involved in the Viterbi equalization are as follows:

1. Compute the transition metric $\mu_{k,v}$ for bit k of the actual received signal and state v of the equalizer, where $v = 0,1,...,2^{l-1}$ and l stands for the window length of the equalizer.

2. Compute the accumulated transition metric for every possible path in the trellis representing the equalizer. The metric for a particular path is defined as the squared Euclidean distance between the estimated received waveform represented by that path and the actual received waveform. For each node in the trellis

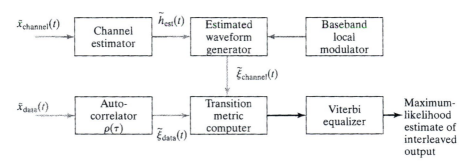

FIGURE 4.22   Block diagram of baseband processor for channel estimation and equalization. The lighter arrows indicate complex signals.

(i.e., each state of the equalizer), the Viterbi algorithm compares the two paths entering that node. The path with the lower metric is retained, and the other path is discarded.

3. Repeat the computation for every bit of the received signal.

4. The survivor, or active, path discovered by the algorithm defines the l-bit sequence applied to the local modulator in Fig. 4.22 for which the estimated received waveform $\tilde{\xi}_{est}(t)$ is the closest to the actual received waveform $\tilde{\xi}_{data}(t)$ in Euclidean distance. With this sequence at hand, the tasks of channel estimation and equalization are completed.

One last comment is in order: The window length l assigned to the equalizer depends not only on the memory of the channel, but also on whether the modulator used in the transmitter has memory of its own or not (i.e., partial-response modulation). Let l_{mem} denote the memory of the modulator and $l_{channel}$ denote the memory of the channel. Then we may express the window length of the equalizer as

$$l = l_{mem} + l_{channel} \tag{4.50}$$

For example, $l_{mem} \approx 2$ for GMSK. With $l_{channel} = 4$, for example, we thus have $l \approx 6$.

4.16 TIME-DIVISION MULTIPLE ACCESS

The discussion thus far has focused on specific functional blocks (i.e., speech coding, channel shaping, coding, and modulation) that are basic to the design of a digital communication system. With this material at hand, we are now ready to discuss TDMA, a widely used form of multiple access for wireless communications.

The purpose of a *time-division multiple-access (TDMA) system* is to permit a number of users, say, N, to access a wireless communication channel of bandwidth B on a time-shared basis. The immediately apparent features that distinguish TDMA from FDMA are twofold:

1. Each user has access to the full bandwidth B of the channel, whereas in FDMA each user is assigned a fraction of the channel bandwidth, namely, B/N.

2. Each user accesses the channel for only a fraction of the time that it is in use and on a periodic and orderly basis, with the transmission rate being N times the user's required rate. By contrast, in FDMA, each user accesses the channel on a continuous-time basis.

Both of these features have significant implications for the operation of a TDMA wireless communication system. Access to the full bandwidth of the wireless channel means that we may now be dealing with wideband data transmission, which makes the TDMA system vulnerable to frequency-selective fading. In contrast, FDMA deals with narrowband transmission, which means that the fading channels are typically frequency flat. To combat the frequency-selective fading problem requires the use of sophisticated signal-processing techniques. Access to the channel on a time-shared basis has implications of its own. In particular, the transmission of information-bearing data over the channel takes place in the form of *bursts*, which, in turn, further complicates the requirement of synchronizing the receiver to the transmitter.

In the context of implementation, unfortunately, there is no TDMA structure applicable to all TDMA wireless systems in operation. Nevertheless, they do share a common feature: Each *frame* of the TDMA structure contains N time *slots* of equal duration. It is in the detailed structure of each slot and in the way in which the transmitting and receiving slots are assigned in time that TDMA systems differ from one another.

Typically, the bits constituting each slot of a TDMA frame are divided into two functional groups:

- *Traffic data bits*, which represent digitized speech or other forms of information-bearing data.
- *Overhead bits*, whose function is to assist the receiver in performing some auxiliary functions that are essential for satisfactory TDMA operation.

The auxiliary functions include synchronization and channel estimation. Specifically, the *synchronization* bits in a slot enable the receiver to recover sinusoidal carrier and bit-timing information, which are needed for coherent demodulation. The *framing* bits are used to estimate the unknown impulse response of the channel, which is needed for estimating the transmitted signal. As already mentioned, in TDMA systems, the transmission of information-bearing signals pertaining to any user is not continuous in time. Rather, it is *discontinuous*, requiring the use of a *buffer-and-burst* strategy. The burst form of data transmission over the channel results in an increase in the synchronization overhead, as each receiver is required to piece the transmitted signal (e.g., speech) together as it is received over a succession of frames.

Up to now, the discussion of the TDMA framing structure has been of a generic nature. Theme Example 1, presented in Section 4.17, describes the frame structure of a specific system.

4.16.1 Advantages of TDMA over FDMA

The following are some of the advantages TDMA has over FDMA:

1. With TDMA, the use of a diplexer can be avoided at the mobile terminal. A diplexer is a complicated and expensive arrangement of filters that allows the mobile terminal to transmit and receive data at the same time without jamming its own information-bearing signal. In TDMA, the terminal need not transmit and receive at the same time; hence, a diplexer is not needed.

2. With TDMA, only one RF carrier at a time is present in the channel. If the channel includes a nonlinearity, then the effects of the nonlinearity are much reduced on a single carrier than if multiple carriers were present, as in FDMA. Examples of such nonlinearities are the power amplifiers employed in base stations or those in satellite transponders.

3. With voice, a significant portion of the call consists of quiet time, when neither party is speaking. With a TDMA strategy, special processing techniques can be employed to fill the quiet times with data or other voice calls to improve the channel's efficiency.

4. With FDMA, the base station must have a channel unit (transmitter/receiver pair) for each active session. With TDMA, the same channel unit is shared between multiple sessions; thus, the base station hardware can be significantly simplified.

To achieve some of these advantages requires the use of complicated signal-processing techniques, which, in turn, necessitates a reliance on digital signal-processing technology in the form of silicon chips for cost-effectiveness. Moreover, this same enabling technology has made it possible to implement other functional needs of TDMA systems efficiently:

- sophisticated timing and fast acquisition operations for synchronizing the receiver to the transmitter, and
- source coding and channel coding techniques for the efficient and reliable transmission of information over the channel.

Putting all these operational advantages and practical realities together has made TDMA preferable to FDMA. However, a major disadvantage of TDMA is that its deployment requires an increased rate of data transmission across the wireless channel, which, in turn, may result in increased intersymbol interference (ISI), making channel equalization in the receiver a necessary practical requirement in TDMA systems.

4.16.2   TDMA Overlaid on FDMA

From the discussion presented thus far, it may appear that TDMA is implemented in a rigorous, pure form. In reality, however, TDMA is implemented in an overlaid fashion on FDMA, for practical reasons. To appreciate this point, it is important that we first recognize that the usable radio spectrum extends from tens of hertz to tens of gigahertz, which represents over nine orders of magnitude. By international agreement, this spectrum is shared by allotting certain portions of it to certain applications. For example, in North America, the band from 118 to 130 MHz is dedicated to aeronautical safety communications, and the bands from 824–849 to 869–894 MHz are dedicated to public telephony. In a very high level sense, this is a form of FDMA: *sharing the spectrum on the basis of frequency.*

Hence, every wireless communication system has an FDMA baseline, and multiple-access schemes such as TDMA are overlaid on this baseline. One of the issues is the *granularity of the underlying FDMA structure.* In this sense, TDMA comes in three basic forms:

1. *Wideband TDMA.* In this form of TDMA, there is only one or a small number of frequency channels, typically several megahertz wide. Wideband TDMA has been used in satellite communication systems in which the TDMA service occupies the full bandwidth of the satellite transponder.

2. *Medium-band TDMA.* In this form of TDMA, there is a significant number of frequency (FDMA) channels, but the bandwidth of each channel is still large enough (100 to 500 kHz) that frequency-selective fading can be expected. The GSM system discussed as a theme example in Section 4.17 is an example of medium-band TDMA, with sufficient FDMA channels being available to assign different-frequency channels to different cells and to perform the necessary task of interference management.

3. *Narrowband TDMA.* This last form of TDMA is a simple step up from a pure FDMA system. The number of users time-sharing a single channel is small, and the number of frequency channels is typically large. The bandwidth of a channel in narrowband TDMA is relatively small (usually less than 50 kHz), and, as a consequence, we can usually assume the multipath phenomenon to be flat fading. The North American IS-54 digital telephone system is an example of a narrowband TDMA system. (See Note 2 of Chapter 3.)

The appropriate choice of granularity for the underlying FDMA systems depends upon several factors:

- In a cellular system, the granularity has to be sufficient to allow different frequency assignments in a neighboring cell and perform flexible interference management.
- System complexity increases with the channel bandwidth and data-transmission rate, with the increase in complexity occurring in both the synchronization and processing aspects of the system. Lower bandwidths tend to imply lower cost solutions and lower power requirements.
- Propagation conditions may favor higher bandwidth systems, but only if appropriate measures are implemented to use this advantage. Frequency-selective fading that occurs in medium and wideband TDMA systems can provide a diversity advantage, but only if the receiver includes an effective equalizer.

One final comment is in order: TDMA is the not the only choice of multiple access for overlaying on an FDMA baseline. In Chapter 5, we will present code-division multiple access (CDMA), which can be considered a wideband system, but, in reality, is still overlaid on an FDMA baseline. Moreover, wireless communication is not limited to a single overlay. For example, from the discussion to be presented in Section 4.17, we will see that the GSM system is not simply TDMA overlaid on FDMA; rather, it also includes a third multiple-access strategy known as frequency hopping (FH); that is, GSM is, in reality, an FDMA/TDMA/FH system. Frequency hopping is also discussed in Chapter 5.

4.17 THEME EXAMPLE 1: GSM[13]

The *Global System for Mobile (GSM) communications* is a digital wireless communication system that is used all over the world. Figure 4.23 displays the basic TDMA frame structure of GSM. The structure is composed of eight 577-μs slots, which makes the total frame duration equal to 4.616 ms. The 1-bit *flag* adjacent to each data burst of 57 bits is used to identify whether the data bits are digitized speech or some other information-bearing signal. The 3 *tail* bits, all logical zeros, are used in convolutional decoding of the channel-encoded data bits. The 26-bit training sequence in the middle of the time slot is used for channel equalization. Finally, the *guard time*, occupying 8.25 bits, is included at the end of each slot to prevent data bursts received at the base station from mobile users from overlapping with each other; this is achieved by transmitting no signal at all during the guard time.

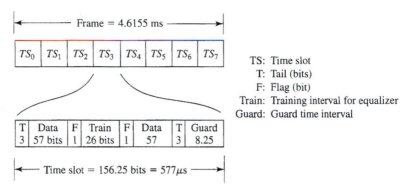

FIGURE 4.23   Frame structure of GSM communications.

The *frame efficiency* of a TDMA system is defined as the number of bits representing information-bearing signals (e.g., digitized speech), expressed as a percentage of the total number of bits (including the overhead) that are transmitted in a frame. With each slot consisting of 156.25 bits, of which 40.25 bits are overhead (ignoring the 2 flag bits), the frame efficiency of GSM is

$$\left(1 - \frac{40.25}{156.25}\right) \times 100 = 74.24\%$$

It is important to note, however, that (as remarked in Section 4.16.2) GSM is *not* a pure TDMA system. Rather, it combines TDMA with frequency hopping. Accordingly, a physical channel is *partitioned in both time and frequency.* The channel is partitioned in time because, with eight slots in a TDMA frame, each carrier frequency supports eight physical channels mapped onto the eight slots. A time slot assigned to a particular physical channel is naturally used in every TDMA frame for as long as that channel is engaged by a mobile user. Consequently, partitioning of the channel in frequency arises because the carrier assigned to such a slot changes its frequency from one frame to the next in accordance with a frequency-hopping algorithm.

In Section 4.10, we introduced the idea of interleaving as a way of combatting the Rayleigh fading problem. Frequency hopping combined with interleaving enables a TDMA system to combat the fading problem even more effectively. In the context of a TDMA system, the *principle of frequency hopping* embodies the following two considerations:

1. The carrier used to modulate a TDMA frame changes its frequency from one frame to the next.
2. If a particular TDMA frame happens to be in a deep fade, then it is highly unlikely that the next TDMA frame will also be in a deep fade, provided that the change in carrier frequency applied by the frequency-hopping algorithm from the particular frame in question to the next one is sufficiently large.

For uplink transmission, in Europe, GSM uses the frequency band 890 to 915 MHz, and for downlink transmission, it uses the frequency band 935 to 960 MHz. In either case, the maximum frequency change from one frame to the next is 25 MHz. Expressed as a percentage of the mean carrier frequency, the *maximum frequency hopping* for the downlink is approximately

$$\frac{25}{900} \times 100 = 2.8\%$$

With this percentage of maximum frequency hopping, it turns out that the time spent by a rapidly moving mobile user in a deep fade is reduced to about 4.6 ms, which is essentially the frame duration. In the case of slowly moving mobile users (e.g., pedestrians), the frequency-hopping algorithm built into the design of GSM produces substantial gains against fades.

GSM employs a moderately complicated, 13-kilobits/s regular pulse-excited speech codec (coder/decoder) with a long-term predictor. To provide error protection for the speech-encoded bits, concatenated convolutional codes and multilayer interleaving are employed. An overall speech delay of 57.5 ms occurs in the system.

Turning next to the type of digital modulation used in GSM, we find that the method of choice is Gaussian minimum-shift keying (GMSK), which was discussed in Sections 3.7 and 4.14. For GSM, the time–bandwidth product WT of GMSK is standardized at 0.3, which provides the best compromise between increased bandwidth occupancy and resistance to cochannel interference. Ninety-nine percent of the radio frequency (RF) power of GMSK signals so specified is confined to a bandwidth of 250 kHz, which means that, for all practical purposes, the sidelobes of the GMSK signal are insignificant, for all practical purposes, outside this frequency band.

The available spectrum is divided into 200-kHz-wide subchannels, each of which is assigned to a GSM system transmitting data at 271 kb/s. Figure 4.24 depicts the power spectrum of a channel in relation to its two adjacent channels; this plot is the

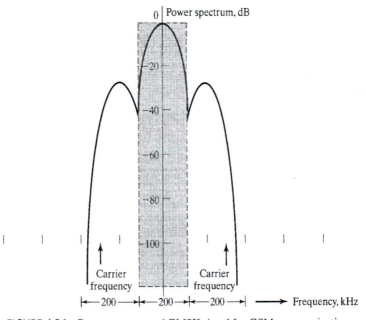

FIGURE 4.24   Power spectrum of GMSK signal for GSM communications.

passband version of the baseband power spectrum of Fig. 3.21 corresponding to $WT_b = 0.3$. From Fig. 4.24, we may make the following important observation: The RF power spectrum of the subchannel shown shaded is down by an amount close to 40 dB at the carrier frequencies of both adjacent subchannels, which means that the effect of cochannel interference in GSM is small.

4.18 THEME EXAMPLE 2: JOINT EQUALIZATION AND DECODING[14]

The material presented in Section 4.12 has taught us an important principle in digital communication theory, hereafter referred to as the *turbo coding principle*. The principle may be stated as follows:

> *The performance of the receiver of a digital communication system, embodying the plot of bit error rate (BER) versus transmitted signal energy per bit-to-noise spectral density ratio, E_b/N_0, may be significantly improved by using*
>
> *(i) a concatenated encoding strategy at the transmitter and*
>
> *(ii) an iterative receiver, with all of its components operating in soft-input, soft-output (i.e., analog) form.*

The iterative receiver is the hallmark of the turbo coding principle.

In Fig. 4.15, the concatenated encoding strategy is implemented in *parallel* form, so called because encoder 1 and encoder 2 operate in parallel on their respective inputs. Moreover, the two encoder inputs are essentially statistically independent by virtue of the turbo interleaver that separates them.

Alternatively, we can implement the concatenated encoding strategy in *serial* form, as illustrated in Fig. 4.25(a). Although, at first sight, this structure looks familiar

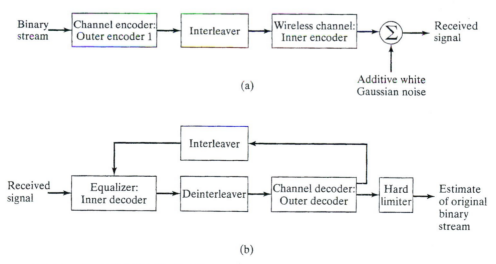

(a)

(b)

FIGURE 4.25   Joint equalization-and-decoding problem.
(a) Turbo encoder of the serial form, with the channel viewed as the inner encoder.
(b) Iterative Turbo decoder, highlighting the application of feedback around the two decoding stages.

in the context of a fast-fading wireless communication system, the viewpoint embodied in its description as a *two-stage encoder* can be justified along the following lines:

- The channel encoder, introduced into the transmitter chain to improve the reliability of communication, is viewed as the *outer encoder*.

- The wireless channel, essential for the communication process, is viewed as the *inner encoder*.

- The *channel interleaver*, introduced to disperse the burst of errors produced by the possible presence of the fast-fading phenomenon in the wireless channel, separates the two encoders in accordance with the transmit part of the turbo-coding principle.

Correspondingly, the *two-stage decoder* is configured as an *iterative receiver*, as shown in Fig. 4.25(b), in accordance with the receiver part of the turbo-coding principle. Herein lies the basis of a novel receiver structure made up of the following constituents:

- A soft-input, soft-output *channel equalizer*, designed to mitigate the effect of intersymbol interference (ISI) produced by the transmission of the encoded-interleaved signal across the channel; the equalizer acts as the *inner decoder*.

- A soft-input, soft-output *channel decoder*, designed to improve the estimates of encoded data symbols; the channel decoder acts as the *outer decoder*.

- A *channel deinterleaver*, designed to undo the permutation that is present in soft outputs produced by the equalizer, so as to facilitate proper channel decoding.

- An *interleaver*, designed to repermute the soft outputs produced by in the channel decoder, so that the feedback signal applied to the equalizer assumes a form consistent with the received signal.

Now, if we were to open the feedback loop in Fig. 4.25(b) by removing the interleaver in the *feedback path*, we would be left with a conventional receiver defined by the *forward path* made up of the channel-equalizer, channel-deinterleaver, channel-decoder chain. The practical advantage of the iterative receiver is that it performs *joint equalization and decoding*, thereby offering the potential for improving the performance of the receiver by virtue of the feedback around the two stages of processing: channel equalization and channel decoding.

In particular, the reduction in bit error rate through joint equalization and decoding performed iteratively can be explained by observing that each component in the receiver, namely, the equalizer and the decoder, helps to bootstrap the performance of the other. The *bootstrap* action manifests itself as follows:

- The equalizer uses *frequency diversity* in the channel to improve the decoder performance through *ISI reduction*.

- The decoder uses *time diversity* in the code to improve the equalizer performance through *improved estimates of uncoded data symbols*.

The net result of this bootstrap action is that, in the course of three to five iterations, a significant reduction in the bit error rate is accomplished, as is illustrated in a simple, yet insightful, computer experiment described next.

4.18.1 Computer Experiment

Consider the serial concatenated encoder of Fig. 4.25(a), with the following specifications:

1. Channel encoder (outer encoder): convolutional encoder
   Code rate = 1/2
   Constraint length, $K = 3$
   Generator polynomials:
   $g^{(1)}(D) = 1 + D^2$
   $g^{(2)}(D) = 1 + D + D^2$

2. Interleaver:
   Type: pseudorandom interleaver
   Block size: 1000 bits

3. Wireless channel: Tapped-delay-line model with the following tap weights (see Fig. 4.26, where T denotes the symbol duration):

$$w_0 = 0.93$$
$$w_1 = -0.17$$
$$w_2 = 0.35$$

Euclidean norm of the tap-weight vector $\mathbf{w}$:

$$\|\mathbf{w}\| = (w_0^2 + w_1^2 + w_2^2)^{1/2}$$
$$= ((0.93)^2 + (-0.17)^2 + (0.35)^2)^{1/2}$$
$$\approx 1$$

4. Modulation (not shown in Fig. 4.25(a)): Binary phase-shift keying (BPSK). With this simple method of modulation, the baseband model of the system assumes a real-valued form throughout the system.

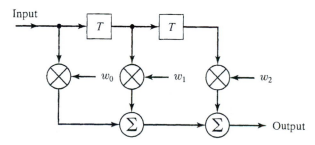

FIGURE 4.26   Tapped-delay-line model of wireless channel with three tap-weights; the blocks labeled T act like unit-delay operations.

The iterative two-stage receiver of Fig. 4.25(b) was implemented as follows:

1. Equalizer (inner decoder).
   - The channel impulse response, assumed to be known.
   - The decoding trellis, formed on the basis of the channel impulse response (i.e., tap weights of the tapped-delay-line model) and BPSK.
2. Deinterleaver, designed to deinterleave the soft outputs produced by the equalizer.
3. Channel decoder (outer decoder).
   - The decoding trellis, formed on the basis of the convolutional encoder's generator polynomials $g^{(1)}(D)$ and $g^{(2)}(D)$
   - Construction of the decoding trellis, discussed in Section 4.7
4. Interleaver, designed to interleave the soft outputs produced by the channel decoder.
5. Decoding algorithm for both the equalizer and channel decoder: The logarithmic form of the maximum a posterior probability (MAP) algorithm, discussed in Section 4.12.

Using computer simulations of the encoder/decoder system of Fig. 4.25, we plot the receiver performance, in terms of BER versus E_b/N_0, in Fig. 4.27, on the basis of which we may make the following observations:

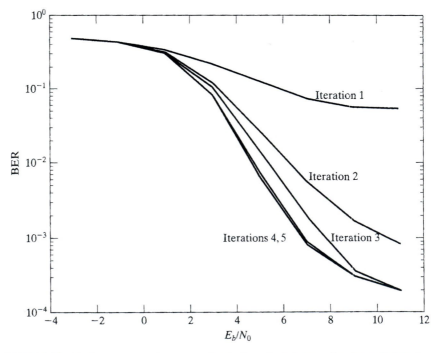

FIGURE 4.27   Performance receiver curves for the iterative joint equalization-and-decoding experiment.

1. Iterative detection, performed in accordance with the turbo coding principle, provides a significant improvement in receiver performance measured with respect to the first iteration; in effect, iteration 1 represents what is achievable with a noniterative (i.e., conventional) receiver.
2. The receiver converges in about five iterations.
3. Little change in receiver performance occurs in going from iteration 4 to iteration 5.

Problem 4.5   The baseband model used in the computer experiment on joint equalization and decoding is real valued, which is justified for BPSK modulation. To improve spectral efficiency, QPSK modulation is commonly used. Discuss the modifications that would have to be made to the baseband model in order for it to handle QPSK modulation.   ∎

The computer experiment just presented assumes that the receiver has perfect knowledge of the *channel state information* (CSI). In practice, we have to deal with a wireless channel that is typically nonstationary, in which case the equalizer structure has to be expanded to include a CSI estimator. (See Problem 4.24.)

4.19 THEME EXAMPLE 3: RANDOM-ACCESS TECHNIQUES

There are many instances in multiaccess communications in which a user terminal is required to send a packet of information to the base station at a random instant in time. Such instances occur, for example, when the terminal wishes to log onto the system or when the user wishes to make a telephone call. The system must provide a means by which these random requests can be serviced. This could be done in a number of ways:

1. The system could permanently assign one channel to each user.
2. The system could *poll* each user at regular intervals to see if he or she had anything to transmit.
3. The system could provide a *random-access* channel that the users could access at any time.

Since a typical user has a low duty cycle, the first approach is wasteful of spectrum. The second approach could result in long delays if there is a large number of users, and if the users are mobile, the polling process can become complicated. In this section, we will consider the third approach of assigning a random-access channel.

4.19.1 Pure Aloha[15]

Consider the following model of the random-access channel: Let us assume that there is a large population of user terminals that operate independently of each other and that each terminal has no knowledge of when the other terminals will transmit. Each terminal transmits random packets of length P, and the average transmission rate is λ. That is, there are, on average, λ packets transmitted per second by the entire user population.

This situation is commonly modeled as a *Poisson process*. From Appendix C, with Poisson arrivals, the probability that there are k arrivals in the period $[0,t]$ is given by

$$\text{Prob}(X(t) = k) = \frac{(\lambda t)^k}{k!} e^{-\lambda t} \tag{4.51}$$

Poisson processes have a *memoryless property*; that is, the probability that there are k arrivals in the period $[0,t]$ is the same as the probability that there are k arrivals in the period $[s, t+s]$ for some arbitrary s.

   If two packets collide (i.e., they overlap in time), it is assumed that the information in both packets is lost. It is of interest to determine the throughput S of this channel; *throughput* is the number of packets that can be transmitted per slot, on average. If there was only one user terminal, then, clearly, the maximum throughput would be unity. In the case of a large number of terminals, however, we must consider the probability that two or more packets will collide. In Fig. 4.28, we show several types of collisions. From the figure, it should be clear that, for a packet transmitted at time t_0, any packet transmitted in the interval $[t_0 - T, t_0 + T]$, where T is the packet duration, will cause a collision.

   With a Poisson model for packet arrival times, if a packet is being transmitted, then the probability that no additional packets arrive during the period $[t_0 - T, t_0 + T]$ is given by Eq. (4.51) with $k = 0$, or

$$\text{Prob}(\text{No additional packets in time } 2T) = e^{-2\lambda T} \tag{4.52}$$

Consequently, the throughput of an Aloha system is given by the product of the packet arrival rate and the probability that a packet is successfully received; that is,

$$S = \lambda e^{-2\lambda T} \tag{4.53}$$

If we normalize this equation to packets received per packet time T, then the *normalized throughput* is given by

$$\begin{aligned} S_0 &= \lambda T e^{-2\lambda T} \\ &= G e^{-2G} \end{aligned} \tag{4.54}$$

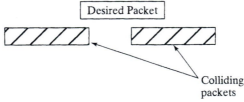

FIGURE 4.28   Illustration of packet collision zone.

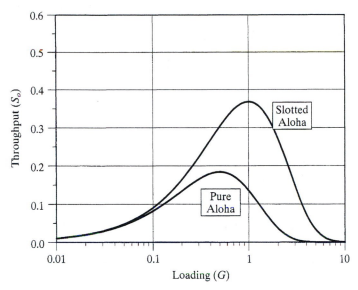

FIGURE 4.29   Normalized throughput for unslotted and slotted Aloha.

where $G = \lambda T$ is the *normalized loading per packet period*—that is, the average number of packets per slot time T. This throughput is plotted as a function of the offered load in Fig. 4.29. The peak throughput occurs at $G = 1/2$, yielding $S_0 = (2e)^{-1} \approx 0.184$ packets per packet time. That is, with an Aloha random-access channel, the maximum throughput is less than 19% of the full channel capacity. In practice, the throughput is maintained at a much smaller value so as to ensure stability of the approach.

4.19.2 Slotted Aloha

The performance of an Aloha system can be improved if a framing structure is provided. This framing structure includes fixed slot times, and user terminals are required to synchronize their transmissions with the slot times. Often, the timing of the Aloha frame is based on the timing of a forward-link broadcast channel. This form of random access with framing is referred to as *slotted Aloha*.

With slotted Aloha, a collision occurs only if the two user terminals transmit during the same T second slot. In a manner analogous to the development for the unslotted Aloha case, the normalized throughput of the slotted Aloha is given by

$$S_0 = Ge^{-G} \tag{4.55}$$

The normalized throughput of unslotted Aloha is also plotted in Fig. 4.29. The peak throughput with slotted Aloha, $S_{max} = 1/e \approx 0.36$ packet per slot, is double that of unslotted Aloha.

4.19.3 Carrier-Sense Multiple Access[16]

The Aloha protocol was first applied in satellite networks in which the user terminals were dispersed over a wide geographic area. The transmissions were received by the satellite and then re-broadcast over the whole area. There are two consequences of this approach. The first is that user terminals can hear the broadcast by the satellite and thus determine immediately whether a collision occurred and there is no need to transmit an acknowledgment. Second, these systems use geostationary satellites that have an altitude of approximately 36,000 kilometers, resulting in a significant transmission delay. Consequently, the user terminal can "hear" a collision, but it is too late to avoid it.

In terrestrial networks, it is often the case that each user terminal can hear the transmissions of all the other user terminals. This process of listening to the channel is referred to as *sensing*. Such a situation led to the development of another random-access protocol known as *carrier-sense multiple access* (CSMA). In its simplest form, this protocol has the following three steps:

1. If the channel is sensed as *idle*, the user terminal transmits the packet.
2. If the channel is sensed as *busy*, the transmission is scheduled for a later time according to a specified random distribution.
3. At the new point in time, the user terminal senses the channel and repeats the algorithm.

If transmission were instantaneous, then collisions would occur in the CSMA protocol only if two terminals transmitted at exactly the same time; this should be a rare occurrence. Although the transmission delay is smaller in terrestrial networks than it is in satellite networks, it is not negligible. Let τ be the maximum transmission delay between any pair of user terminals. Then collisions can occur between packets that have the timing shown in Fig. 4.30.

To analyze the throughput of CSMA, we use the following model: As in the Aloha case, we assume that packet arrivals have a Poisson distribution with average rate λ and packet duration T. Analogously to transmission in the Aloha case, a packet is transmitted successfully if it is the only packet to be transmitted in time τ. Consequently, the probability that a packet is transmitted successfully is given by

$$\text{Prob(No additional packets in time } \tau) = e^{-\lambda \tau} \qquad (4.56)$$

Because of the sensing strategy, the average throughput calculation is more complex than in the Aloha case. Since a packet arriving at a terminal does not mean that it will be transmitted immediately, we have to calculate the average transmission rate; this involves calculating the average busy time per packet and the average idle time per packet. The sum of the two gives the average time between packet transmissions.

The average busy time of a channel is the packet duration, plus the propagation delay, plus the relative delay of the last colliding packet, as illustrated in Fig. 4.30. The relative delay of the last colliding packet is a random variable denoted by Y. To determine the distribution of this random variable, we note that, due to the memoryless property of the Poisson process,

$$\text{Prob (No additional packets in the interval } (t, t+s)) = e^{-\lambda s} \qquad (4.57)$$

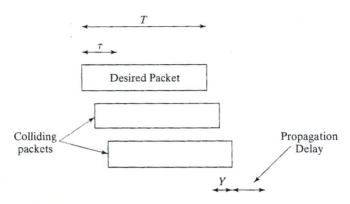

FIGURE 4.30   Illustration of collision conditions for CSMA.

So the probability that the last packet is transmitted at time y or before is equivalent to the probability that no packets are transmitted in the interval $(y, \tau]$, or

$$\text{Prob}(Y \le y) = e^{-\lambda(\tau - y)} \qquad \text{for } 0 < y < \tau \qquad (4.58)$$

By determining the probability density function of random variable Y from Eq. (4.58), we may compute the average relative delay of the last colliding packet (see Problem 4.33):

$$\mathbf{E}[Y] = \tau - \frac{1 - e^{-\lambda\tau}}{\lambda} \qquad (4.59)$$

Combining all of the components, we find that the average busy time of a channel per transmission is

$$T_{\text{busy}} = T + \tau + \mathbf{E}[Y]$$

$$= T + 2\tau - \frac{1 - e^{-\lambda\tau}}{\lambda} \qquad (4.60)$$

For a Poisson distribution, the average idle time is given by

$$T_{\text{idle}} = \frac{1}{\lambda} \qquad (4.61)$$

The throughput of the CSMA channel is therefore

$$S = \frac{\text{Prob(successful transmission)}}{T_{\text{busy}} + T_{\text{idle}}}$$

$$= \frac{e^{-\lambda\tau}}{T + 2\tau - \dfrac{1 - e^{-\lambda\tau}}{\lambda} + \dfrac{1}{\lambda}} \qquad (4.62)$$

$$= \frac{\lambda e^{-\lambda\tau}}{\lambda(T + 2\tau) + e^{-\lambda\tau}}$$

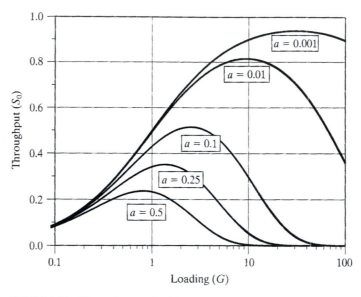

FIGURE 4.31 Throughput of the CSMA channel for varying parameter a.

If we normalize the transmission delay by setting $a = \tau/T$ and $G = \lambda T$, then the normalized throughput, in packets per packet time, is given by

$$
\begin{aligned}
S_0 &= \frac{Ge^{-aG}}{G(1 + 2a) + e^{-aG}} \\
&= \frac{G}{1 + (1 + 2a)Ge^{aG}}
\end{aligned}
\tag{4.63}
$$

We plot this normalized throughput for various delay parameters a in Fig. 4.31 as a function of the load offered. Intuitively, the smaller a is, the larger is the expected throughput of CSMA, which is evident in the figure.

4.19.4 Other Considerations with Random-Access Protocols

The analysis to this point applies to a datagram type of service: packets are sent and forgotten. In many systems, if there is a collision, the packet is retransmitted. If the number of retransmissions is small, then the previous results hold. But if the collision rate is high, retransmissions can add appreciably to the total traffic presented to the network at any particular time.

With random-access protocols, collisions cause delays in delivering the message. There is a trade-off between throughput and delay. With high loading, the throughput approaches zero, and the average delay per packet approaches infinity. The delay depends upon the retransmission strategy and the propagation delay. Consequently, it is difficult to compare the delays of different random-access strategies. Retransmission cannot occur as soon as a collision is detected, as that will most certainly result in another collision. Instead, there has to be a random back-off such that the user terminal waits a random period of time before retransmitting.

Note also that the results presented herein have assumed that there are no packet errors due to noise.

4.20 SUMMARY AND DISCUSSION

In this chapter, we discussed digital wireless communications built around time-division multiple access (TDMA). Both the functional blocks that constitute the transmitter of a TDMA wireless communication system and the corresponding ones in the receiver involve features that follow from some powerful theorems in Shannon's information theory. Among these features are the following:

- *Source coding*, which is used to remove redundant bits inherent in an information-bearing signal in accordance with the source-coding theorem and rate distortion theory, thereby improving the spectral efficiency of the system.
- *Channel coding*, which involves the purposeful addition of redundant bits to the transmitted signal in a controlled manner in accordance with the channel-coding theorem, thereby providing protection against channel noise.
- *Interleaving*, which involves the pseudorandomization of the bits in a TDMA frame so as to combat the fading problem.

From the preceding features, it is apparent that the deployment of a TDMA wireless communication system offers a practical means of improving receiver performance at the expense of a significant increase in system complexity. However, system complexity is not an issue of concern, because electronics are inexpensive, thanks to the ever-increasing computing power and cost-effectiveness of silicon chips and computer software.

The other major issue discussed in the chapter is that of channel estimation and equalization. Since TDMA channels are typically wider than FDMA channels, they are more likely to be frequency selective, hence posing a more difficult channel-estimation problem. The need for channel estimation arises because the impulse response of the channel is unknown. This matter is taken care of by transmitting a known sequence over the channel. The issue of equalization relates to the need for undoing the convolution performed on the transmitted signal by the channel. The Viterbi equalizer, based on the Viterbi algorithm acting as a maximum-likelihood sequence estimator, provides a powerful method for solving this second problem.

GSM was discussed in the chapter as the first of three theme examples. In GSM, TDMA is combined with frequency hopping so as to combat the problem of deep fades in a more effective manner than would be possible otherwise. The combined use of time slots in TDMA frames and changes in the carrier frequency from one frame to the next in accordance with a frequency-hopping algorithm results in partitioning of the physical channel in both time and frequency. For the transmission of TDMA frames over the channel, GSM uses Gaussian minimum-shift keying (GMSK), which is a spectrally efficient form of continuous-phase frequency-shift keying. A combination of regular pulse-excited speech codecs (with long-term prediction), concatenated convolutional codecs, and multilayer interleavers/deinterleavers is used for the processing of individual speech channels.

The second theme example addressed a novel application of the turbo coding principle to solve the joint equalization-and-decoding problem in an iterative manner. The procedure described therein is different from the traditional approach, in that the channel equalizer and channel decoder operate inside a closed feedback loop, helping to bootstrap the performance of each other. The net result is a receiver that achieves, for a prescribed E_b/N_0, a bit error rate significantly lower than that attainable by traditional means (typically, in three to five iterations).

The third theme example addressed random-access techniques in which each user terminal is required to send a packet of information to the base station at a random instant in time. In this third example, we discussed the following configurations with increasing levels of performance:

- Pure Aloha, involving a large population of user terminals that operate independently of each other and that have no knowledge of when the other terminals will transmit.
- Slotted Aloha, using fixed slot times and requiring the user terminals to synchronize their transmissions with the slot times.
- Carrier-sense multiple access, following three steps:
  1. Transmit the packet if the channel is idle
  2. Schedule the transmission for a later time if the channel is busy
  3. Sense the channel at a new point in time, and repeat the algorithm.

According to the approach taken to describe the channel coding process in this chapter, channel encoding is performed separately from modulation; likewise for demodulation and decoding in the receiver. Moreover, the provision for error correction is made by transmitting additional redundant bits (i.e., parity-check bits) in the code, which has the effect of lowering the spectral efficiency in bits per second per hertz. That is, bandwidth utilization is traded for increased power efficiency. To attain a more efficient utilization of the two communication resources, namely, channel bandwidth and transmit power, the processes of channel coding and modulation would have to be treated as a single entity rather than two separate ones. This treatment is precisely what is done in trellis-coded modulation.[17] When this method of modulation is applied to wireless communications, the usual procedure is to to insert an interleaver between the encoder and signal-space mapper so as to overcome the multipath fading problem.

The interleaving process may be viewed as a form of time diversity introduced into the transmitted signal. (A detailed treatment of diversity, with emphasis on spatial diversity, is presented in Chapter 6.) In a variant of coded modulation known as the bit-interleaved coded modulation (BICM), the interleaving process is applied at the level of encoded bits rather the encoded symbols.[18] The motivation behind BICM is to produce a coded system with a high degree of time diversity, whereby the bit error rate of the receiver is governed by some product of d_{free} terms, where d_{free} refers to the free binary Hamming distance of the code. It turns out that for the same E_b/N_0, BICM outperforms the baseline approach for coded modulation by producing a smaller BER.

NOTES AND REFERENCES

[1] The celebrated classic paper of Shannon (1948) laid down the foundations of information theory, and with it, a principled approach to the design of digital communication systems. For a detailed treatment of Shannon's theory, see Cover and Thomas (1991). An introductory treatment of the subject is presented in Chapter 9 of Haykin (2001).

[2] Chapter 3 of Steele and Hanzo (1999) presents a detailed discussion of speech-coding techniques, with an emphasis on their relevance to wireless communications.

[3] For detailed treatments of traditional error-control coding techniques, see Clark and Cain (1981), Lin and Costello (1983) and Michelson and Levesque (1985). These techniques are also discussed in Chapter 4 of Steele and Hanzo (1999), which emphasizes their relevance to wireless communications. Convolutional codes were first described by Elias (1955).

[4] The free distance of convolutional codes is discussed in Viterbi and Omura (1979), Benedetto and Biglieri (1999), and Proakis (1995).

[5] The classic paper on the Viterbi algorithm is due to Viterbi (1967). For a tutorial paper on this algorithm, see Forney (1973).

[6] Interleaving, of both the block and convolutional types, is discussed in some detail in Clark and Cain (1981) and in lesser detail in Sklar (2001).

[7] Discussions of pseudorandom interleaving in the context of turbo coding are presented in Vucetic and Yuan (2000) and Heegard and Wicker (1999). We may also mention the so-called S-constraint interleaver, devised by Divsalar and Pollara (1995). This new interleaver is based on the generation of N uniformly distributed integers, subject to an S-constraint defined by the minimum interleaving distance. Specifically, in the construction of a turbo code, S is chosen to correspond to the maximum input error pattern length to be broken by the interleaver.

[8] Turbo codes were invented by Berrou et al. (1993); see also Berrou and Glavieux (1996, 1998). Heegard and Wicker (1999), Vucetic and Yuan (2000), and Hanzo et al. (2002) discuss turbo codes as well. For a detailed treatment of iterative decoding, see Chugg et al. (2001). Hanzo et al. discusses the performance of turbo codes over fading channels.

The discovery of turbo processing has emerged as a revolution, not only affecting the development of good codes for reliable communications and channel equalization as discussed in this chapter, but also affecting other design aspects of digital communication systems, as summarized here:

- Turbo-like codes for source coding as well as joint source-channel coding
- Iterative timing recovery and phase estimation for synchronizing the receiver to the transmitter
- Turbo-like MIMO (multiple-input, multiple output) wireless communications

Turbo-BLAST, an example of turbo-MIMO wireless communications, is discussed in Chapter 6.

[9] The BCJR algorithm is named in honor of its four originators: Bahl, Cocke, Jelinek, and Raviv, who coauthored the first forward–backward method for implementing maximum a posterior probability decoding in 1974.

[10] For the original exposition of the EXIT chart, see ten Brink (1999).

[11] For matters relating to issues 1 and 2 discussed under Subsection 4.13.6 on joint equalization and decoding, see Benedetto and Montrosi (1996) and Berrou (2003), respectively.

[12] The material presented in this section follows Chapter 6 of Steele and Hanzo (1999).

[13] The abbreviation GSM originally stood for the French name *Groupe de travail pour les Services Mobiles.* In recognition of the widespread use of GSM communications all over the globe, it was later renamed *Global System of Mobile* communications.

[14] The issue of an iterative receiver for dealing with joint equalization and decoding is discussed in Chugg *et al.* (2001), pp. 105–110.

[15] The term "Aloha" is Hawaiian, and it literally means "love," but it is commonly used as a greeting or farewell. The original ALOHA system was developed at the University of Hawaii by Abramson (1970) as a protocol for a terrestrial radio system that broadcast packets for computer communications. It was later adapted to satellite communications. The collection of papers edited by Abramson (1993) has numerous contributions describing the ALOHA protocol variants and their performance.

[16] In carrier-sense multiple-access, if every terminal cannot hear every other one, then there can be a degradation in performance. In Tobagi and Kleinrock (1975), it is shown that if users can be divided into two groups *A* and *B* such that members within a group can hear each other, but cannot hear all the members in the other group, then the performance of CSMA rapidly degrades such that it is worse than slotted Aloha. There is further degradation with three or more groups.

[17] Trellis-coded modulation was discovered by Ungerboeck; see his seminal paper published in 1982; see also the two-part tutorial paper (Ungerboeck, 1987).

[18] Bit-interleaved coded modulation (BICM), using a rate-2/3, 8-PSK modulator, was first described in Zehavi (1992). For a detailed information-theoretic treatment of BICM, see Caire *et al.* (1998). These two papers discuss two different ways of implementing BICM, depending on how the bit-interleaving process is itself implemented:

- The encoded bits are interleaved separately.
- A common bit interleaver is used.

The latter implementation makes it possible to take a more general approach to the treatment of BICM.

ADDITIONAL PROBLEMS

Convolutional coding

Note: For Problem 4.6 through 4.10. the same message sequence, 10111..., is used so that we may compare the outputs of different encoders for the same input.

Problem 4.6   Consider the convolutional encoder of Fig. 4.32 with rate $r = 1/2$ and constraint length $K = 2$. Find the encoder output produced by the message sequence 10111....

Problem 4.7   Figure 4.33 shows a convolutional encoder with rate $r = 1/2$ and constraint length $K = 4$. Determine the encoder output produced by the message sequence 10111....

Problem 4.8   Consider the convolutional encoder of Fig. 4.34 with rate $r = 2/3$ and constraint length $K = 2$. Determine the code sequence produced by the message sequence 10111....

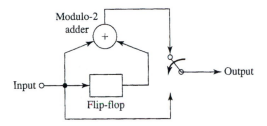

FIGURE 4.32 Diagram for problem 4.6.

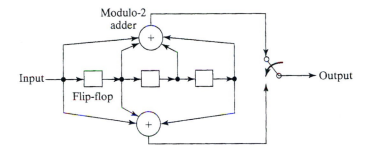

FIGURE 4.33 Diagram for problem 4.7.

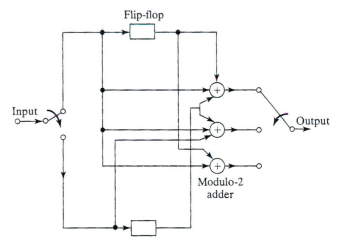

FIGURE 4.34 Diagram for problem 4.8.

Problem 4.9 Construct the trellis diagram for the encoder of Fig. 4.33, assuming a message sequence of length 5. Trace the path through the trellis corresponding to the message sequence 10111.... Compare the resulting encoder output with that found in Problem 4.7.

Problem 4.10   Construct the state diagram for the encoder of Fig. 4.33. Starting with the all-zero state, trace the path that corresponds to the message sequence 10111..., and compare the resulting code sequence with that found in Problem 4.7.

Problem 4.11   The code rate of the convolutional codes discussed in Section 4.7 is $1/n$, where n is the number of modulo-2 adders used in the encoder. A new convolutional code with code rate k/n is required, where k is an integer. How would you generate such a convolutional code? Justify your answer.

Problem 4.12   A convolutional code has the constraint length $K = 5$.

   (a) Assuming that the code is of a nonsystematic nature and using the Viterbi algorithm, how many errors can be corrected by the code for a "bursty" wireless channel?
   (b) Repeat the problem for a systematic code.

Problem 4.13   Consider the nonsystematic convolutional codes listed in Table 4.3 for constraint length $K = 6$ and 7.

   (a) Construct the generator polynomials for these two codes.
   (b) Find the free distance for each of the codes.
   (c) Identify which, if any, of the two codes can deal with a fading channel that produces a burst of five transmission errors.

Interleaving

Problem 4.14   The expected length of an error burst produced by a fast-fading channel is inversely proportional to the speed of a mobile unit and directly proportional to the bit rate transmitted by the unit. Justify the validity of this statement.

Problem 4.15

   (a) Continuing with the classical block interleaver/deinterleaver discussed in Subsection 4.10.1, show that the end-to-end delay produced by the interleaving/deinterleaving action is $2LN$ bits and that the memory requirement is LN.
   (b) Show that the corresponding results for the convolutional interleaver discussed in Subsection 4.10.2 are LN and $LN/2$, respectively.

Problem 4.16   The classical block interleaver can be operated in four different permuted ways, depending on the order in which the incoming symbols are written into the rows of the matrix of memory elements and the order in which they are read out along the columns:

   (a) Left-to-right write in/top-to-bottom read out (LR/RB)
   (b) Left-to-right write in/bottom-to-top read out (LR/BT)
   (c) Right-to-left write in/top-to-bottom read out (RL/TB)
   (d) Right-to-left write in/bottom-to-top read out (RL/BT)

Consider an interleaver in which $N = 4$ rows and $L = 4$ columns.

(a) Given the input sequence {0,1,2,3,4,...,14,15,16,...}, construct the output sequence for each of the preceding four permutations.

(b) With $N = L$, the LR/TB and RL/BT permutations are self-inverse; that is,

$$\mathbf{I}_{LR/TB}^{-1} = \mathbf{I}_{LR/TB} \quad \text{and} \quad \mathbf{I}_{RL/TB}^{-1} = \mathbf{I}_{RL/TB}$$

where $\mathbf{I}$ is the square matrix describing the interleaver and $\mathbf{I}^{-1}$ is the inverse matrix describing the deinterleaver. Demonstrate the self-inverse property for the example specified in part (a).

Problem 4.17 The classical block (N,L) interleaver has an important property: *A burst of less than L contiguous transmission errors results in isolated errors at the deinterleaver output that are separated from each other by at least N symbols.* Demonstrate this property for each of the following block interleavers:

(a) $N = 4, L = 4$
(b) $N = 4, L = 5$
(c) $N = 4, L = 6$

Problem 4.18 Compare the advantages and disadvantages of block interleavers with those of convolutional interleavers in the context of four issues:

(a) End-to-end delay
(b) Memory requirement
(c) Synchronization of commutators in the receiver with those in the transmitter
(d) Complexity of implementation

Baseband processing for channel estimation and equalization

Problem 4.19 Estimation of the impulse response of a wireless communication channel requires the transmission of a sounding sequence, preferably with the following desirable features:

1. The autocorrelation function of the sequence closely approximates a delta function.
2. The sequence can be locally generated in the receiver in synchrony with the transmitter.

Describe a sounding sequence that satisfies these two requirements, and justify the sequence.

Problem 4.20 Equation (4.50) states that the window length of the Viterbi equalizer equals the sum of two memories, one due to the shaping filter used in the partial-response modulation and the other due to the wireless channel. Justify the validity of that equation.

Turbo decoding

Problem 4.21 Construct a table comparing the features that distinguish the Viterbi algorithm and the BCJR algorithm, discussed in Sections 4.9 and 4.12, respectively.

Problem 4.22   Suppose that a turbo encoder of the parallel form is used as the channel encoder (i.e., outer encoder) in the serial two-stage encoder of Fig. 4.25(a).

(a) Develop the structure of the new iterative receiver.
(b) In what ways does this receiver differ from that of Fig. 4.25(b)?
(c) What can be improved by the new receiver? What about limitations?

Problem 4.23   The iterative joint equalization-and-decoding receiver of Fig. 4.25 is an example of a closed-loop feedback system. In such a system, the receiver may become *unstable* (i.e., diverge). In qualitative terms, explain how such a phenomenon can arise in practice.

Problem 4.24   In the computer experiment presented in Section 4.18, it was assumed that the receiver "knows" about the state of the channel. In TDMA wireless communication systems, this requirement is usually satisfied by including a training sequence in each packet transmitted. For example, in GSM, the training sequence, which is "known" to the receiver, occupies 20 percent of each packet.

   With this background, the requirement is to postulate a procedure for estimating the channel impulse response.

(a) Adapting Eq. (4.9) to the problem at hand, formulate a procedure for performing this estimation.
(b) Given the joint equalization-and-decoding strategy described in Fig. 4.25, discuss how the use of bootstrapping (made possible by the strategy) can be used to improve the estimate of the channel impulse response.

Problem 4.25   In this problem, we revisit the turbo principle applied to the joint equalization-and-decoding problem discussed as Theme Example 2 in Section 4.18. Specifically, we look at a more difficult multipath channel represented by the tapped-delay-line model of Fig. 4.26 with the following tap weights:

$$w_0 = 0.5$$
$$w_1 = \sqrt{0.5}$$
$$w_2 = 0.5$$

One way to tackle this problem is to consider the following system specifications:

1. Channel code: rate-1/2, 16-state convolutional code with the following two generator polynomials:

$$g^{(1)}(D) = 1 + D^3 + D^4$$
$$g^{(2)}(D) = 1 + D^2 + D^4$$

2. Block interleaver of size 57×30
3. QPSK modulation with Gray coding
4. Packet size: 57 symbols

Contrast the issues involved in the equalization-and-decoding problem embodied in this problem versus those discussed in the computer experiment in Section 4.18.

Time-division multiple access

Problem 4.26   The composition of a TDMA frame permits the use of a single-carrier frequency for both forward and reverse transmissions of information-bearing signals. Describe how such a two-way communication can be accomplished.

Problem 4.27   A TDMA frame uses a preamble of n_P bits, N time slots, and n_{T_1} trail bits. Each time slot contains n_{T_2} trail bits, n_G guard bits, n_S synchronization and channel estimation bits, and n_I information-bearing bits. Derive formulas for (a) the framing efficiency η of the system and (b) the size of a TDMA frame.

Problem 4.28   As mentioned in Section 3.8, FDMA relies on the use of highly selective bandpass filters for its operation. A bandpass filter is said to be highly selective if its *quality factor* or, simply *O-factor*, is high compared with unity; the Q-factor is defined as the ratio of the midband frequency of the filter to its bandwidth.

     In this context, TDMA enjoys an advantage over FDMA in that it relaxes this requirement. Discuss how TDMA attains its advantage.

Random-access techniques

Problem 4.29   Suppose a system with a user population of 100 terminals plans to use short packets for making reservations on a longer demand-assigned TDMA channel. There are two options for making the reservations: a polling method or slotted Aloha. Assume that the packet length is equal to T in either case. Which method would be the most efficient and under what conditions?

Problem 4.30

   (a) Prove that the peak throughput of a pure Aloha system is $1/2e$, where e is the base of the natural algorithm.
   (b) Suppose a system has two different packet lengths: T and $2T$. Find the throughput of a slotted Aloha system in this case.

Problem 4.31   Suppose that, in a slotted Aloha system, there is a 10% packet error rate due to noise, in addition to the error rate associated with those packet errors due to collisions.

   (a) Discuss how the system throughput will be affected.
   (b) Repeat the discussion for a carrier-sensitive multiple-access (CSMA) system.

Problem 4.32   Assume that, in a local area network using carrier-sensitive multiple access (CSMA), the average number of packets offered per second is 1500, with a packet size of 50 microseconds. If the maximum diameter of a local area network is 200 meters, determine the expected throughput of the system in packets per second.

Problem 4.33   Following the discussion on page 247, show that, in a CSMA system, the average relative delay of the last colliding packet is

$$E[Y] = \tau - \frac{1 - e^{-\lambda\tau}}{\lambda}$$

Spread Spectrum and Code-Division Multiple Access

5.1 INTRODUCTION

In many systems, due to a limited spectrum, multiple users must share the same band of frequencies. Some examples are cellular telephones, radio dispatch offices for taxi companies, and air-traffic control communications. There are many ways of sharing the spectrum efficiently, each with its own advantages and disadvantages. Among the techniques described in previous chapters are frequency-division multiple access (FDMA), in which each user is given a small portion of the total available spectrum, and time-division multiple access (TDMA), in which each user is allowed full use of the available spectrum, but only during certain periods. In some systems, such as GSM communications, a combination FDMA/TDMA system is used, as discussed in Chapter 4.

In this chapter, we focus on a third approach to sharing the radio spectrum: code-division multiple access (CDMA). CDMA refers to a multiple-access technique in which the individual terminals use spread-spectrum techniques and occupy all of the spectrum whenever they transmit. *It is the interference attenuation property of spread spectrum that allows multiple users to occupy the same spectrum at the same time.* The CDMA approach is used in some current cellular telephone systems.

Spread-spectrum systems encompass modulation techniques in which the signal of interest, with an *information bandwidth* R_b, is spread to occupy a much larger *transmission bandwidth* R_c. Spread-spectrum systems, in general, are *digital* communication systems; in a loose sense, the comparable analog technique is frequency modulation (FM), although it is not usually considered a form of spread spectrum.

To compare a spread-spectrum system with an FDMA system, consider a service for which the available bandwidth is $W = R_c$. An FDMA would divide this bandwidth into N channels of width $R_b = R_c/N$, and each user would be allotted a channel of bandwidth R_b. Ordinarily, R_b would closely match the minimum bandwidth required by the user. With spread-spectrum techniques, the spectrum is not divided. Rather, each user is permitted to occupy all or any part of the spectrum when transmitting, and more than one user is allowed to transmit simultaneously.

Spread-spectrum techniques were originally developed for military applications,[1] but commercial interest in such techniques has increased recently, due mainly to their promise of greater tolerance for interference. The impetus for this interest has been twofold: the growth of cellular telephony systems; and regulatory rulings that have allowed the unlicensed use of some frequency bands for low-power transmitters, as long as the signal is spread to reduce interference.

There are two basic spread-spectrum techniques: *direct sequencing* (DS) and *frequency hopping* (FH). Also, a variety of hybrid techniques use different combinations of these two basic techniques. With direct-sequence spreading, the original signal is *multiplied by a known signal* of much larger bandwidth. With frequency-hopped spreading, the *center frequency of the transmitted signal is varied* in a pseudorandom pattern.

Spread-spectrum techniques have a number of advantages:

- increased tolerance to interference
- low probability of detection or interception
- increased tolerance to multipath
- increased ranging capability.

Direct-sequence modulators process a narrowband signal to spread it over a much wider bandwidth. With this approach, each user terminal is assigned a *unique spreading signature* that makes each user's communications approximately orthogonal to those of other users. This is similar to the way in which distinct carrier frequencies and time slots make users' transmissions approximately orthogonal in FDMA and TDMA, respectively. Spreading the signal de-sensitizes the original narrowband signal to some potential channel degradations and to interference; this property becomes advantageous as the demand for spectrum reuse increases. The transmitted energy remains the same, but due to the much larger bandwidth, the signal spectrum is often below the noise floor of receivers. *The signal looks like noise to any receiver that does not know the signal's structure.* This makes the signal difficult to detect, even if one is looking for it; a characteristic that has distinct advantages for military applications. Since multipath can be viewed as a form of interference, increased tolerance to interference also means increased tolerance to multipath. In fact, as we will see with some receiver structures, multipath energy may be used to advantage to improve performance. The fourth advantage, increased ranging capability, is due to the fact that timing error Δt is inversely proportional to the signal bandwidth,

$$\Delta t \propto \frac{1}{R_c} \qquad (5.1)$$

and that timing error corresponds directly to range error. This property permits some spread-spectrum techniques to measure distance or terminal location, through a method known as triangulation.

Frequency-hopped systems take a different approach: Their modulators process the narrowband signal and change the carrier frequency every few symbols. A pseudorandom hopping pattern that is known by the receiver is used by the transmitter. To an outside observer, the signal appears to be transmitting on randomly selected frequencies, although the *hop time* on each frequency is usually constant.

Multiple FH transmitters share the same frequency band by using different hopping patterns. If the transmitters are synchronized, then the hopping patterns can be selected so that there are few or no *collisions* (both transmitters communicating on the same frequency and at the same time). If they are not synchronized, then collisions will occur, but, for a sufficiently wide hopping bandwidth, so infrequently that the resulting errors can be corrected by a forward error-correcting code.

Technology plays an important role in these systems. FH systems that transmit multiple symbols during each hop period are referred to as *slow frequency hoppers*. Typically, early FH systems were slow frequency hoppers, because they were constrained by the rate at which frequency synthesizers could be switched. With fast frequency hopping, the same symbol is generally transmitted on multiple hops. Due to phase continuity limitations on the frequency synthesizer switching, FH systems typically use a noncoherent form of modulation, such as FSK, rather than BPSK or QPSK.

In Section 5.2, the basics of direct-sequence (DS) modulation are presented. This is followed by a description of several types of spreading codes in Section 5.3. In Section 5.4, we investigate the advantages of DS for wireless applications, particularly as related to multiple-access interference and multipath. In Sections 5.5 through 5.8, we review the application of synchronization, channel estimation, power control, and forward error-correction coding in the CDMA context. In Section 5.9, we introduce the notion of multiuser detection, and in Section 5.10, various aspects of operating CDMA in a cellular environment are considered. In Section 5.11, FH systems are discussed. We end the chapter with five theme examples related to spread spectrum, thereby emphasizing the wide applicability of this communications technique.

5.2 DIRECT-SEQUENCE MODULATION[2]

In the discussion that follows, it is assumed that the underlying modulation is BPSK or QPSK. Spread-spectrum techniques can be used with other modulation formats, but most practical applications are limited to BPSK and QPSK.

5.2.1 The Spreading Equation

In Chapter 3, we saw that, in complex envelope form, one symbol of a BPSK signal may be represented as

$$\tilde{s}(t) = b\sqrt{E_b}g(t) \qquad\qquad 0 \leq t \leq T \qquad\qquad (5.2)$$

where b is the data symbol, $g(t)$ is the *symbol-shaping function*, and T is the symbol duration. For BPSK modulation, the data symbol b is either $+1$ or -1. If the symbol-shaping function is assumed to be rectangular, then

$$g(t) = \begin{cases} \sqrt{\dfrac{1}{T}}, & 0 \leq t \leq T \\ 0 & \text{otherwise} \end{cases} \qquad\qquad (5.3)$$

Chapter 3 examined the more general case in which the symbol-shaping function is chosen to have a root-raised cosine spectral shape to minimize both spectral bandwidth and intersymbol interference after detection. The model can also be extended to a complex envelope representation of QPSK modulation by allowing b to take the complex values $b \in \{\pm 1 \pm j\}$.

For the rectangular pulse shape, the transmit spectrum is given by

$$S_g(f) = T\mathrm{sinc}^2(fT) \tag{5.4}$$

and the effective bandwidth of the signal is determined by the symbol period T. The corresponding spectrum is shown in the top right of Fig. 5.1. This is the conventional BPSK spectrum with rectangular pulse shaping.

For a direct-sequence signal, the symbol-shaping function is selected to be a sequence of Q rectangular pulses

$$g(t) = \sum_{q=1}^{Q} c(q)g_c(t - qT_c) \tag{5.5}$$

where $\{c(q)\}$ is the *spreading sequence* composed of the individual *chips* $c(q)$ that are $+1$ or -1 and T_c is the chip duration. The constant Q, denoting the upper limit of summation in Eq. (5.5), is the *spreading factor* and usually satisfies the relation $QT_c = T$. The chip shape, $g_c(t)$, is also often assumed to be rectangular; that is,

$$g_c(t) = \begin{cases} \sqrt{\dfrac{1}{T}}, & 0 \le t \le T_c \\ 0 & \text{otherwise} \end{cases} \tag{5.6}$$

where the scaling is chosen to provide unit energy in the pulse $g(t)$.

To sum up, with DS modulation, instead of the rectangular pulse shape of Eq.(5.3), the pulse shape is a sequence of shorter rectangular pulses called chips. The chips are a pseudorandom sequence of $+1$'s and -1's known at the receiver and often having a repetition period equal to the symbol period. In Fig. 5.1, the application of a

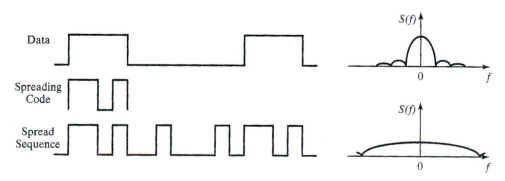

FIGURE 5.1 Spreading by a factor of four in the time and frequency domains.

spreading sequence of length 4 to a data sequence is illustrated, together with the approximate spectrum.

Problem 5.1   What is the equation for the spectrum of the spreading sequence given by Eq. (5.5)? (*Hint: Use the convolution theorem.*)
Ans. $S_g(f) = (T/Q^2)\,\mathrm{sinc}^2(fT_c)$ ∎

5.2.2 Matched-Filter Receiver

In Chapter 3, we saw that if the received signal was

$$\tilde{x}(t) = \tilde{s}(t) + \tilde{w}(t) \tag{5.7}$$

where $\tilde{w}(t)$ is the complex envelope representation of white Gaussian noise, then the optimum receiver corresponds to the processing

$$y = \int_0^T \tilde{x}(t)\tilde{s}^*(t)\,dt \tag{5.8}$$

When the data portion of the signal is not known, but there is linear modulation such as that described by Eq. (5.2), we can equivalently write the optimum processing as

$$y = \int_0^T \tilde{x}(t)g^*(t)\,dt$$

$$= b\sqrt{E_b}\int_0^T |g(t)|^2\,dt + \int_0^T \tilde{w}(t)g^*(t)\,dt \tag{5.9}$$

$$= b\sqrt{E_b} + \eta$$

where b is the data symbol of BPSK or QPSK and η is a noise sample. This result holds for an isolated symbol, regardless of the symbol-shaping function $g(t)$. Consequently, if we choose $g(t)$ to be a spreading sequence, as defined by Eq. (5.5), or a rectangular pulse, as defined by Eq. (5.3), the optimum receiver principles still hold.

The outstanding issue with direct-sequence spreading is that of characterizing the noise sample η. From Eq. (5.9), we readily see that the mean of the noise sample η is zero. The variance of η is given by

$$\sigma_\eta^2 = E\left[\left|\left(\int_0^T \tilde{w}(t)g^*(t)\,dt\right)\right|^2\right] = \int_0^T\int_0^T E\left[\tilde{w}(t)\tilde{w}^*(s)\right]g(t)g^*(s)\,dt\,ds$$

$$= \int_0^T\int_0^T N_0\delta(t-s)g(t)g^*(s)\,dt\,ds = N_0\int_0^T |g(t)|^2\,dt = N_0 \tag{5.10}$$

where the spectral density of the complex baseband noise $\tilde{w}(t)$ is N_0. From Eq. (5.10), the noise power does not depend on the symbol shape, which is the essence of matched filtering. (See Appendix D). The conclusion is that, *in AWGN, the transmission performance of direct-sequence spread spectrum (DS-SS) is identical to that of nonspread systems.* The corresponding signal-to-noise ratio with optimum detection is

$$\text{SNR} = \frac{(\mathbf{E}[y])^2}{\sigma_\eta^2} = \frac{E_b}{N_0} \tag{5.11}$$

In Fig. 5.2, we illustrate an implementation of a baseband direct-sequence modulator and corresponding demodulator. For the modulator, the raw data sequence in rectangular format is multiplied by the spreading sequence and transmitted. At the demodulator, the received signal-plus-noise is multiplied by the same spreading sequence, and the data are detected with an *integrate-and-dump filter* to maximize the signal-to-noise ratio, as suggested by Eq. (5.9).

Problem 5.2   Filtering with an integrate-and-dump filter is equivalent to convolving with a rectangular pulse of length T. Show, by using Parseval's theorem, that the noise bandwidth of an integrate-and-dump circuit is $1/T$. (Parseval's theorem is discussed in Appendix A.)   ■

5.2.3 Performance with Interference

As the previous development indicates, the SNR at the output of a DS receiver is identical to that obtained with a nonspread BPSK or QPSK receiver. Consequently, for coherent detection, the uncoded BER performance of a DS-SS system is given by

$$P_e = 0.5 \text{ erfc}\left(\sqrt{E_b/N_0}\right) \tag{5.12}$$

when the underlying modulation is either BPSK or QPSK modulation. Recall that, although the bit error rates for BPSK and QPSK are the same, the symbol error rates differ, as discussed in Chapter 3.

One advantage of DS modulation is the reduced receiver sensitivity to interference. This advantage is due to the fact that the de-spreading circuit acts as a spreading circuit for any signal to which it is not matched. If the receiver signal includes an in-band tone (CW) interferer, then the de-spreading circuit in the receiver spreads this interference over the whole band, and the subsequent integrator rejects all but a narrow portion of the interference.

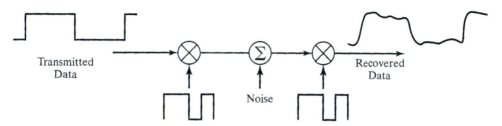

FIGURE 5.2   A simple CDMA modulator and optimum demodulator, with representative waveforms.

To illustrate this behavior, suppose the received signal includes a jamming tone, given in complex envelope representation by

$$\tilde{\xi}(t) = \sqrt{\frac{A_\xi}{T}} e^{-j2\pi f_\xi t} \tag{5.13}$$

where f_ξ is the frequency of the jammer and $\sqrt{A_\xi/T}$ is its amplitude. The received signal, including interferer, is given by

$$\tilde{x}(t) = \tilde{s}(t) + \tilde{\xi}(t) + \tilde{w}(t) \tag{5.14}$$

Since the matched receiver of Eq. (5.9) is a linear receiver, we may consider the effects of the jamming tone independently of the desired signal. That is, we may consider the jamming term

$$y_\xi = \int_0^T \tilde{\xi}(t) g^*(t) dt \tag{5.15}$$

independently of the signal and noise terms. After multiplication by the de-spreading sequence as shown in Fig. 5.2, the jamming tone becomes

$$
\begin{aligned}
\tilde{\xi}(t) g^*(t) &= \sqrt{\frac{A_\xi}{T}} e^{-j2\pi f_\xi t} g^*(t) \\
&= \sqrt{\frac{A_\xi}{T}} e^{-j2\pi f_\xi t} \sum_{q=1}^{Q} c^*(q) g_c(t - qT_c) \\
&= \sqrt{\frac{A_\xi}{T}} \sum_{q=1}^{Q} c^*(q) g_c(t - qT_c) e^{-j2\pi f_\xi t} \qquad 0 \le t \le T
\end{aligned}
\tag{5.16}
$$

The corresponding spectrum is given by

$$
\begin{aligned}
S_{\xi g}(f) &= \frac{A_\xi}{T} \times Q \times \frac{T}{Q^2} \text{sinc}^2((f + f_\xi) T_c) \\
&= \frac{A_\xi}{Q} \text{sinc}^2((f + f_\xi) T_c)
\end{aligned}
\tag{5.17}
$$

That is, the de-spreading circuit performs a spreading function on the jammer. The jamming energy is spread over a bandwidth proportional to $1/T_c$ and centered at f_ξ. To estimate the contribution of the jammer to the noise after demodulation, we make two assumptions:

1. The frequency offset, f_ξ, is small, so the noise spectrum at baseband, after spreading is approximately $S_\xi(f) = A_\xi/Q$.
2. The noise bandwidth of the integrate-and-dump circuit is $1/T$ (See Problem 5.2.)

Let $S_p(f)$ represent the spectrum of the integrate-and-dump detection filter. After the integrate-and-dump circuit, the equivalent noise contribution of the jamming is given by the frequency-domain equation

$$\sigma_{\xi}^2 = \int_{-\infty}^{\infty} S_{\xi g}(f) S_p(f) df$$

$$\approx \frac{A_{\xi}}{Q} \times \frac{1}{T} \tag{5.18}$$

$$= \frac{A_{\xi}}{P_G T}$$

where $P_G = T/T_c = R_c/R_b = Q$ is defined as the *processing gain*. In a nonspread system, the interference power after the integrate-and-dump circuit is

$$\sigma_{\xi}^2 = \int_{-\infty}^{\infty} A_x \mathrm{sinc}^2((f + f_{\xi})t) S_p(f) df \tag{5.19}$$

$$\approx \frac{A_{\xi}}{T}$$

In conclusion, *in a DS-SS system, the interference is reduced by the processing gain P_G relative to its effect in a nonspread system.* This increased interference tolerance is an important motivation for using DS-SS communications.

Problem 5.3   Fill in the missing details in the development of Eq. (5.19).   ∎

5.3   SPREADING CODES

Code division is a multiple-access strategy similar to time division and frequency division, and the previous section has shown the advantage of DS modulation in terms of improved tolerance to interference. The objective of every multiple-access strategy is to allow multiple users to access the radio resource in a manner that maximizes the use of the resource and minimizes the interference among users. These users must be, in a sense, approximately orthogonal. *With TDMA, the users are time orthogonal; with FDMA, the users are approximately frequency orthogonal*, as is conceptually illustrated in Fig. 5.3a. *With CDMA, the users are approximately code orthogonal*, as defined in the discussion that follows.

To achieve this orthogonality, a different symbol-shaping function

$$g_k(t) = \sum_{q=1}^{Q} c_k(q) g_c(t - qT_c) \qquad 0 \le t \le T \tag{5.20}$$

is assigned to each user k. The sequence $\{c_k(q)\}$ is referred to as the *spreading code* or *signature sequence* for user k. The approximate orthogonality of $g_j(t)$ and $g_k(t)$ for different time offsets τ can be represented as the requirement

$$R_{jk}(\tau) = \int_{-\infty}^{\infty} g_j(t + \tau) g_k^*(t) dt \approx 0 \qquad \text{for } j \ne k \tag{5.21}$$

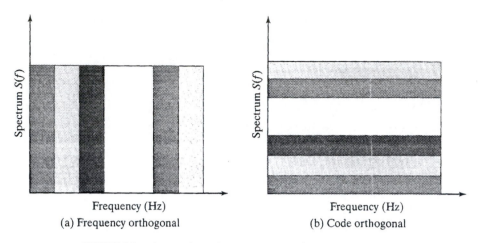

Frequency (Hz)   Frequency (Hz)

(a) Frequency orthogonal   (b) Code orthogonal

FIGURE 5.3   Comparison of spectrum use of CDMA and FDMA.

This relationship assumes that the receiver is using a matched filter. Equation (5.21) says that *the cross-correlation of two different symbol shapes is approximately zero for all time offsets* τ. An additional self-orthogonality requirement that minimizes a number of practical channel and receiver effects is

$$R_{kk}(\tau) \approx 0 \qquad \text{for } \tau > 0 \tag{5.22}$$

That is, for all time offsets greater than zero, each spreading code is orthogonal to itself. *The orthogonality conditions of Eqs. (5.21) and (5.22) mean that different user messages can be separated at the receiver,* even though they occupy the same frequency channel and the same period.

The degree to which Eqs. (5.21) and (5.22) are satisfied depends on the choice of the spreading sequences $\{c_k(q)\}$. In this section, we define codes $\{c_k(q)\}$ such that the orthogonality relationships of Eqs. (5.21) and (5.22) are satisfied as perfectly as possible (i.e., $R_{jk}(\tau)$ and $R_{kk}(\tau)$ are both as close to zero as possible).

EXAMPLE 5.1 Orthogonality of Messages

Suppose bits are transmitted simultaneously over the same channel by two users such that the received signal is

$$\tilde{r}(t) = b_1 g_1(t) + b_2 g_2(t) + \tilde{w}(t) \tag{5.23}$$

where the subscript denotes either user 1 or user 2. If the signal $\tilde{r}(t)$ is processed by user 1's receiver, we obtain

$$\begin{aligned} y &= \int_0^T \tilde{r}(t) g_1^*(t) dt \\ &= \int_0^T (b_1 g_1(t) + b_2 g_2(t) + \tilde{w}(t)) g_1^*(t) dt \end{aligned} \tag{5.24}$$

Assuming that the sequences $g_1(t)$ and $g_2(t)$ are approximately orthogonal, then, in the light of Eq.(5.21), we have

$$y = b_1\sqrt{E_b}\int_0^T |g_1(t)|^2 dt + b_2\sqrt{E_b}\int_0^T g_1^*(t)g_2(t)dt + \int_0^T w(t)g_1^*(t)dt$$

$$\approx b_1\sqrt{E_b} + \eta$$

The processing described by Eq. (5.24) provides the basis for a *conventional receiver* for CDMA. In particular, we see that it provides the same performance as a single-user receiver, provided that the orthogonality condition of Eq. (5.21) is satisfied.   ∎

Problem 5.4   Suppose that, in Example 5.1, the two users were asynchronous. That is, user 2 had a delay τ relative to user 1. Does the result still hold?
Ans. Yes   ∎

5.3.1 Walsh–Hadamard sequences

Equation (5.21) is a requirement to design approximately orthogonal spreading sequences. To pursue this issue of orthogonality further, consider first the case in which the users are time synchronous. This condition implies that only the special case of $\tau = 0$ has to be considered in the orthogonality relation of Eq. (5.21). It turns out that for this special case, the orthogonality relations can be satisfied exactly, and the resulting codes are known as the *Walsh–Hadamard sequences*.

To construct a Walsh–Hadamard sequence, we begin with sequences of length 2. We may readily construct two orthogonal sequences of length 2, given by the rows of the matrix

$$\mathbf{H}_1 = \begin{bmatrix} 1 & 1 \\ 1 & -1 \end{bmatrix} \tag{5.25}$$

There are various other ways that two orthogonal vectors could be expressed, but, in reality, they are equivalent to Eq. (5.25). The motivation for the Walsh–Hadamard codes comes from noting that we can construct four orthogonal sequences of length 4 from the sequences of length 2 as follows: Let

$$\mathbf{H}_2 = \begin{bmatrix} \mathbf{H}_1 & \mathbf{H}_1 \\ \mathbf{H}_1 & -\mathbf{H}_1 \end{bmatrix} = \begin{bmatrix} 1 & 1 & 1 & 1 \\ 1 & -1 & 1 & -1 \\ 1 & 1 & -1 & -1 \\ 1 & -1 & -1 & 1 \end{bmatrix} \tag{5.26}$$

That is, the matrix of orthogonal sequences of length 4 is constructed from four submatrices built on $\mathbf{H}_1$ and its negative. Each row of Eq. (5.26) is orthogonal to the other three rows. In general, we may construct 2^n orthogonal sequences of length 2^n

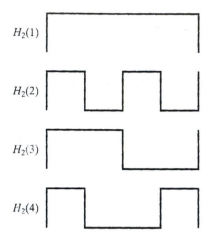

FIGURE 5.4   Walsh–Hadamard codes of length 4.

from sequences of length 2^{n-1} by the operation

$$\mathbf{H}_n = \begin{bmatrix} \mathbf{H}_{n-1} & \mathbf{H}_{n-1} \\ \mathbf{H}_{n-1} & -\mathbf{H}_{n-1} \end{bmatrix} \tag{5.27}$$

Equations (5.25) and (5.27) define an algorithm for constructing 2^n orthogonal Walsh–Hadamard sequences of length 2^n. Figure 5.4 shows the Walsh–Hadamard codes of length 4.

From inspection of the figure, it is clear that some of the Walsh–Hadamard codes will have poor autocorrelations and cross-correlations at time offsets other than zero. For example, if $\mathbf{H}_n(j)$ represents the jth row of $\mathbf{H}_n$, Eq. (5.22) will not be satisfied with $\mathbf{H}_n(1)$ for $\tau > 0$.

EXAMPLE 5.2 Cross-Correlation Between Walsh–Hadamard Codes

The cross-correlation function of a pair of length–128 (i.e., $\mathbf{H}_7$) sequences that are four times oversampled is shown in Fig. 5.5. (That is, each chip is represented by four samples. The samples are all +1 if the chip is +1, and −1 if the chip is −1.) The cross-correlation at zero time offset, $R_{jk}(0)$ is zero, but there are two large peaks at small time offsets that approach the peak autocorrelation value $R_{kk}(0) = 1$. This is not desirable for many applications. ∎

Problem 5.5   (a) Write a program to compute and plot the cross-correlation between two different Walsh–Hadamard sequences of length 2^6 with four times oversampling. Repeat for sequences of length 2^8. (b) What can be hypothesized about the cross-correlation as a function of sequence length? ∎

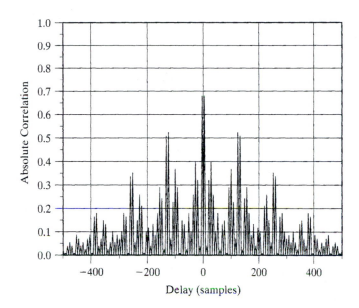

FIGURE 5.5 Normalized absolute cross-correlation of $H_7(63)$ and $H_7(64)$ with four times oversampling.

5.3.2 Orthogonal Variable Spreading Factors

In some applications, we may want to combine messages having different data rates in an orthogonal manner. In particular, a message of rate r_1 is to be spread by a factor s_1 to produce an overall chip rate of ρ, while a second message of rate r_2 is to be spread by a factor s_2 to produce the same overall chip rate ρ. If the spreading factors s_1 and s_2 are powers of 2, then this can be done with the Walsh–Hadamard sequences. The result so obtained is referred to as *orthogonal variable spreading factors* (OVSF) and is illustrated in Fig. 5.6.

In the figure, user 1 has a data rate r_1 and is spread by code 1. User 2 is transmitting at half this rate and using a different code. The periodic extension of the spreading code of user 1 is orthogonal to the spreading code of user 2. For OVSF, the orthogonality requirement for this case can be stated mathematically as

$$\int_0^{T_1} g_1(t)g_2^*(t + lT_1)\ dt = 0 \qquad \text{for } l = 0, \dots, T/T_1 \tag{5.28}$$

That is, code 1, of duration T_1, is orthogonal to all subsequences of code 2, of the same length, and offset by a multiple of T_1, the length of code 1.

Suppose we need OVSF codes of length n_1 and n_2, with $n_1 < n_2$. Then the algorithm for constructing OVSF codes is described as follows:

1. Construct $\mathbf{H}_{n_1}$ by the usual Walsh–Hadamard algorithm.
2. Choose one row of the matrix $\mathbf{H}_{n_1}$ as the code of length 2^{n_1}. Let $\mathbf{H}'_{n_1}$ represent the Hadamard matrix with the selected row removed.

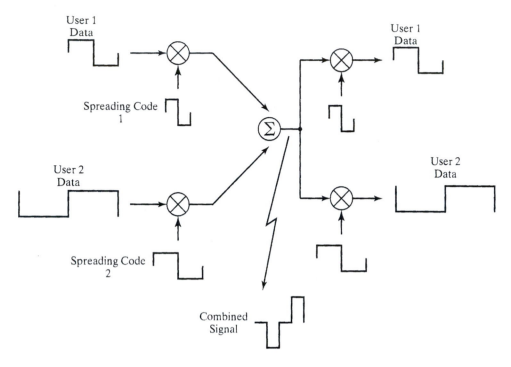

FIGURE 5.6   Illustration of orthogonal variable spreading factors.

3. Continue the Walsh–Hadamard algorithm with $\mathbf{H}'_{n_1}$; that is,

$$\mathbf{H}'_{n_1+1} = \begin{bmatrix} \mathbf{H}'_{n_1} & \mathbf{H}'_{n_1} \\ \mathbf{H}'_{n_1} & -\mathbf{H}'_{n_1} \end{bmatrix} \tag{5.29}$$

Then continue until the desired $\mathbf{H}'_{n_2}$ is constructed.

4. Choose any row of $\mathbf{H}'_{n_2}$ as the spreading code of length 2^{n_2}.

5. If a third code is needed, continue in a similar manner.

Problem 5.6   Show that two codes constructed by the algorithm for OVSF codes have the desired orthogonality properties. Do OVSF codes apply to asynchronous systems? Justify your answer.   ∎

5.3.3   Maximal-Length Sequences[3]

As the previous section illustrated, the *Walsh–Hadamard codes have good orthogonality properties when they align in time*. Their orthogonality properties can be poor when they do not align in time. Ideally, we would like sequences that are orthogonal for all time shifts, but it is sequences which are only approximately orthogonal that are practically achievable.

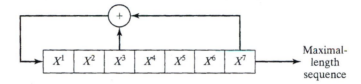

FIGURE 5.7   Shift register generator of maximal-length sequence.

Theoretically, a randomly chosen sequence will have good autocorrelation properties, on average. For communications applications, we need sequences that have properties similar to those of random sequences, but that can be generated simply at both the transmitter and the receiver. One class of sequences that satisfies this condition is the class of *maximal-length sequences* or *m-sequences*. These sequences can be generated by using a binary shift register with feedback, as shown in Fig. 5.7, where $\oplus$ denotes modulo-2 addition. If the memory of the shift register is m, then there are only 2^m possible states for the register. The input to the shift register is the modulo-2 sum of the output register and various internal registers. If the feedback taps are properly selected, then the shift register will cycle through all states except for the zero state before repeating any state; equivalently, it will generate a sequence of length $2^m - 1$ before repeating itself. (The shift register produces a sequence of 0's and 1's, and this sequence is mapped onto a sequence of −1's and 1's when it is used as a spreading code.)

For a given memory m, not all sets of feedback taps will generate a maximal-length sequence, although there is usually more than one set. Those sets of taps which will work can be derived theoretically from a topic in abstract algebra known as *Galois field theory*. There is a correspondence between maximal-length shift register design and *irreducible (non-factorable) polynomials* in Galois fields with binary coefficients. For example, the polynomial $x^2 + 1$ can be factored as $(x + 1)^2$ by means of binary arithmetic, while the polynomial $x^2 + x + 1$ cannot be factored with the use of only binary coefficients. Thus, the latter can be used to construct a maximal-length sequence, but the former cannot.

Maximal-length sequences have five important properties:

1. *Length property*: Each maximal length sequence is of length $2^m - 1$.
2. *Balance property*: Each maximal length sequence has 2^{m-1} ones and $2^{m-1} - 1$ zeros.
3. *Shift property*: The modulo-2 sum of an m-sequence and any circular-shifted version of itself produces another circular-shifted version of itself.
4. *Subsequence property*: Each maximal sequence contains a subsequence of 1, 2, 3,..., $m - 1$ zeros and ones.
5. *Autocorrelation property*: Define the circular autocorrelation function of a sequence $\{c(n)\}$ by

$$R_{jj}(k) = \frac{1}{Q} \sum_{q=0}^{Q-1} c(q)c((q+k)\bmod(Q)) \tag{5.30}$$

where $Q = 2^m - 1$. If the sequence is a maximal-length sequence, then the normalized autocorrelation is

$$R_{jj}(k) = \begin{cases} 1 & k = 0 \\ \dfrac{-1}{Q} & k \neq 0 \end{cases} \tag{5.31}$$

The autocorrelation property implies that, in comparing any nontrivial circular shift of the sequence with the original sequence, bit by bit, the number of disagreements always exceeds the number of agreements by 1. The larger Q is, the better the autocorrelation properties of the maximal-length sequences are, in the sense that the maximal length sequence comes ever so much closer to a random sequence. The initial state of the shift register of Fig. 5.7 can be any binary sequence of length m, except for the all-zeros sequence. This implies that *each maximal-length sequence provides 2^m-1 approximately orthogonal sequences obtained by different (circular) time shifts of the original sequence.*

As mentioned previously, the feedback taps are related to irreducible polynomials. In Table 5.1, we provide some binary irreducible polynomials up to order 12. Note that, for many orders, there is more than one irreducible polynomial. The cross-correlation properties of maximal-length sequences of the same length, but generated by different polynomials, are not necessarily as good as Eq. (5.31).

TABLE 5.1   Table of some irreducible polynomials with binary coefficients.

Order	Irreducible Polynomials
2	$x^2 + x + 1$
3	$x^3 + x + 1, x^3 + x^2 + 1$
4	$x^4 + x + 1$
5	$x^5 + x^2 + 1, x^5 + x^4 + x^3 + x^2 + 1$
6	$x^6 + x + 1, x^6 + x^5 + x^2 + x + 1$
7	$x^7 + x^3 + 1, x^7 + x^5 + x^4 + x^3 + x^2 + x + 1$
8	$x^8 + x^4 + x^3 + x^2 + 1, x^8 + x^6 + x^5 + x^3 + 1$
9	$x^9 + x^4 + 1, x^9 + x^6 + x^4 + x^3 + 1$
10	$x^{10} + x^3 + 1, x^{10} + x^7 + x^3 + x^2 + 1$
11	$x^{11} + x^2 + 1, x^{11} + x^{10} + x^9 + x^7 + x^6 + x^4 + x^3 + x^2 + x + 1$
12	$x^{12} + x^6 + x^4 + x + 1, x^{12} + x^{11} + x^9 + x^8 + x^7 + x^5 + x^2 + x + 1$

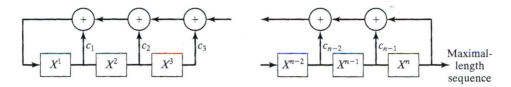

FIGURE 5.8   Maximal-length shift register corresponding to Eq. (5.32).

Given the irreducible polynomial

$$x^m + c_{m-1}x^{m-1} + c_{m-2}x^{m-2} + \ldots + c_1 x + 1 = 0 \tag{5.32}$$

which is of order m, we may construct the shift register for the corresponding m-sequence as shown in Fig. 5.8.

The autocorrelation properties of m-sequences extend to the case in which the shift is a fraction of a chip. In Fig. 5.9, the autocorrelation function is plotted for a maximal-length sequence that is four times oversampled. When the relative shift (delay) is four or more samples (one chip), the autocorrelation is $1/Q$, as described previously. For a shift of less than a chip, there is linearly increasing autocorrelation that peaks when the relative shift is zero.

Problem 5.7   Prove the autocorrelation property of Eq. (5.31) for m-sequences. (*Hint:* Use the preceding properties 1 through 5 as needed. ∎

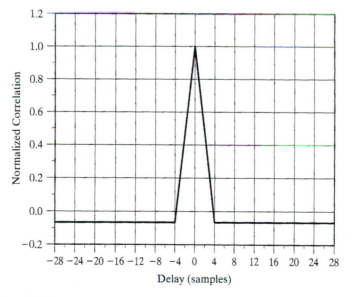

FIGURE 5.9   Circular autocorrelation function of maximal-length sequence of length $2^4 - 1$ with four times oversampling.

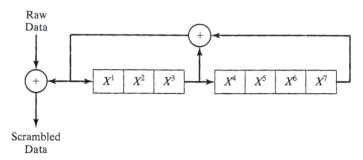

FIGURE 5.10 Scrambler implementation for $f(x) = x^7 + x^3 + 1$.

5.3.4 Scramblers

Although the discussion in this section has thus far focused on the use of maximal-length sequences for the spreading of a CDMA signal, these sequences are also used as data scramblers. The purpose of a *scrambler* is to prevent the transmission of long strings of zeros and ones that may appear in raw data. Long strings of zeros or ones can cause difficulties for tracking circuits in the receiver and can cause peaks in the transmit spectrum that may result in excessive interference into other services. (Note that scramblers should not be confused with interleavers, discussed in Chapter 4. Interleavers merely change the order in which the bits of data are streamed. By contrast, scramblers actually change the data. Both operations are completely reversible at the receiver.)

To implement a scrambler, the output of the maximal-length shift register is modulo-2-added with the data as shown in Fig. 5.10 to scramble the data. At the receiver, the same operation is repeated to undo the scrambling of the transmitter. The key to implementing the scrambler and de-scrambler is that they must be initialized to the same state at the same point in the data sequence. Scramblers can also be used as a very simple form of encryption if the initial state is known only to the transmitter and receiver.

Problem 5.8 To show that scramblers based on *m*-sequences are not very good encryption devices, determine the minimum number of consecutive bits that would need to be known to reconstruct the initial state. (Assume that the generating polynomial is known.) *Ans. m* ∎

5.3.5 Gold Codes[4]

In practical applications, we would like to assign transmitters different codes that have good cross-correlation properties. From Section 5.3.3, two methods for doing this come immediately to mind:

1. Maximal-length sequences have good correlation properties, and we could assign different shifts of the same sequence to different users. There are $2^m - 1$ such shifts/sequences. However, if the transmitters are uncoordinated, then they will not know each other's timing and two transmitters could reuse the same shifted sequence.

2. Alternatively, we could assign users different maximal-length sequences—that is, different generator polynomials. The difficulty with this approach is that these sequences often do not have good cross-correlation properties. There is also a limited number of such generators for each sequence length.

An alternative approach proposed by Gold is to sum two maximal-length sequences of the same length, but using different generators, as shown in Fig. 5.11. In that figure, two *m*-sequences of length $2^7 - 1$ are synchronized; otherwise, they operate independently. The two generator polynomials used in Fig. 5.11 are taken from Table 5.1. *Gold codes are created by summing the output of the two m-sequence generators to produce a new 0–1 sequence.* It is this latter sequence that is the new spreading code. Note that, although the two generators are synchronous and have the same repetition length, they can be started in different states. For each starting state of the first generator, there are $2^m - 1$ potential starting states of the second generator.

Gold was able to show that, for particular choices of generator polynomials, these sequences have good cross-correlation properties. Recall Eqs. (5.21) and (5.22), in which orthogonality conditions of two spreading codes were defined. If these equations are normalized such that $R_{kk}(0) = 1$, then the cross-correlation of the Gold codes is approximately bounded by

$$|R_{jk}(\tau)| \le \frac{1}{Q} + \frac{2}{\sqrt{Q}} \qquad j \ne k \tag{5.33}$$

where $Q = 2^m - 1$. (The approximation error incurred in Eq. (5.33) is very small for Q greater than 15.) Furthermore, the property described by Eq.(5.33) applies for all $2^m - 1$ cyclic shifts of either code. Thus, *for select pairs of maximal-length sequences, we can generate $2^m - 1$ distinct Gold codes* (i.e., codes that are not cyclic rotations of each other) that have good cross-correlation and autocorrelation properties.

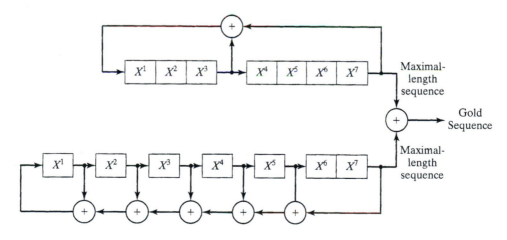

FIGURE 5.11   Generation of a Gold sequence.

Recall that the autocorrelation of m-sequences was equal to $1/Q$. From Eq. (5.33), the autocorrelation of Gold codes is asymptotically proportional to $2/\sqrt{Q}$. There is clearly a performance trade-off between the cross-correlation properties and the number of sequences.

EXAMPLE 5.3 Autocorrelation and Cross-Correlation of a Gold Code

In Fig. 5.12, we show the circular autocorrelation of a Gold code. The autocorrelation is generated from the two polynomials of order 7 in Table 5.1. In Fig. 5.13, the cross-correlation of this first Gold code with a second Gold code, generated from the same pair of polynomials, but with a different starting state, is shown. In both instances, it is clear that the correlation values are low and tightly controlled, although not quite as good as the original maximal-length sequences. ■

Problem 5.9   Write a computer program to do the following:
(a) Generate a sequence from the two fifth-order polynomials in Table 5.1. Is the sequence a Gold code? Repeat for the pair of ninth-order polynomials.
(b) Compute the average magnitude of the circular cross-correlation $\mathbf{E}[|R_{jk}(\tau)|]$ and autocorrelation $\mathbf{E}[|R_{jj}(\tau)|]$ for $\tau > 0$. ■

5.3.6 Random Sequences

In this section, we consider the case in which the sequences are chosen at random. For example, let $\{x_m\}$ be a random sequence of +1's and −1's of length Q. By definition, for

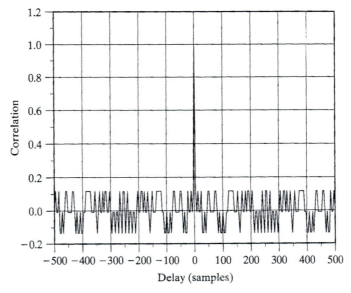

FIGURE 5.12   Normalized circular autocorrelation of a Gold code of length $2^7 - 1$.

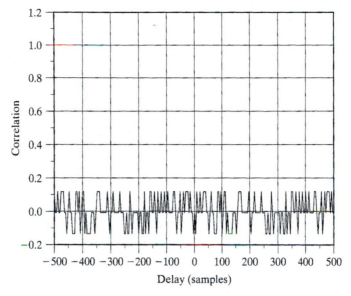

FIGURE 5.13 Normalized circular cross-correlation of two Gold codes of length $2^7 - 1$.

$m \neq n$, x_n and x_m are independent and have zero mean. For random sequences, we define the cross-correlation of the two sequences as

$$R_{xy}(k) = \mathbf{E}\left[\frac{1}{Q}\sum_{q=1}^{Q} x_q y_{q+k}\right] \tag{5.34}$$

(Compare Eq. (5.34) with Problem 5.10 for deterministic sequences.) The autocorrelation function of the random sequence $\{x_m\}$ is given by

$$R_{xx}(k) = \mathbf{E}\left[\frac{1}{Q}\sum_{q=1}^{Q} x_q y_{q+k}\right]$$

$$= \begin{cases} \dfrac{1}{Q}\displaystyle\sum_{q=1}^{Q} \mathbf{E}\left[x_q\right]\mathbf{E}\left[x_{q+k}\right] = 0 \times 0 = 0 & k \neq 0 \\[3mm] \dfrac{1}{Q}\displaystyle\sum_{q=1}^{Q} \mathbf{E}\left[x_q^2\right] = 1 & k = 0 \end{cases} \tag{5.35}$$

If we compute the cross-correlation properties of two random sequences $\{x_m\}$ and $\{y_m\}$, we similarly get

$$R_{xy}(k) = 0 \qquad \text{for all } k \tag{5.36}$$

So, as alluded to earlier, random sequences have ideal autocorrelation and cross-correlation properties. Thus, random sequences would be ideal as spreading codes. However, if a truly random sequence is used as a spreading code at the transmitter, then the code, in addition to the message, would have to be communicated to the receiver in order to do the de-spreading. Note that finite lengths of very long pseudorandom codes, such as m-sequences, do have characteristics similar to those of random sequences of an equivalent length.

The autocorrelation of two random sequences, given by Eq. (5.35), can be interpreted as the expected interference amplitude between the two users. Also of interest is the expected interference power. In what follows, let R_{xy} be the cross-correlation of two random sequences $\{x_m\}$ and $\{y_m\}$ of length Q, where we have dropped the dependence on the time shift τ, since a time-shifted random sequence is simply another random sequence. The average cross-correlation energy of two random sequences is given by $E\left[|R_{xy}|^2\right]$, over all sequences of length Q. If we assume that the autocorrelation is normalized such that $R_{xx} = 1$, then

$$
\begin{aligned}
E\left[|R_{xy}|^2\right] &= E\left[\left(\frac{1}{Q}\sum_{q=1}^{Q} x_q y_q\right)^2\right] \\
&= E\left[\frac{1}{Q^2}\sum_{q=1}^{Q}\sum_{m=1}^{Q} x_q y_q x_m y_m\right] \\
&= \frac{1}{Q^2}\sum_{m=1}^{Q} x_m^2 y_m^2 \\
&= \frac{1}{Q^2}\sum_{m=1}^{Q} 1 \times 1 \\
&= \frac{1}{Q}
\end{aligned}
\tag{5.37}
$$

where the third line follows from the second by the autocorrelation properties of random sequences. Consequently, we see that *the interference power between two random sequences is inversely proportional to the length of the sequence.* This property is another manifestation of the processing gain, and it is a property that will be used several times in the pages that follow.

Problem 5.10 Show that, for a spreading sequence composed of rectangular chips, the cross-correlation function of Eq. (5.21) is equivalent to

$$
R_{xy}(k) = \lim_{Q \to \infty}\left(\frac{E_b}{Q}\sum_{k=1}^{Q} x_m y_{m+k}\right)
$$

when the delay $\tau = kT_c$. Also, show that this limit is equivalent to

$$R_{xy}(k) = \frac{E_b}{Q} \sum_{q=0}^{Q-1} x_q y_{(q+k) \bmod Q}$$

for a periodic sequence of length Q.   ∎

5.4 THE ADVANTAGES OF CDMA FOR WIRELESS

CDMA is a multiple-access strategy for wireless communications based on DS-SS. In the previous sections, we made the following observations:

- The performance of a single-user DS-SS signal is identical to that of the corresponding BPSK and QPSK signal in AWGN. The spreading provided by the code simply changes the spectrum of the signal used to transmit the information.
- If the spreading codes are perfectly orthogonal, it is possible to achieve the single-user performance in the multiuser case.
- It is difficult to design perfectly orthogonal codes for all cases.

In this section, we will consider the effect on multiple-access performance when the spreading codes are not perfectly orthogonal.

5.4.1 Multiple-Access Interference

Suppose that there are K CDMA transmitters sharing the same channel, and let the received complex baseband signal be represented by

$$\tilde{x}(t) = \sum_{k=1}^{K} \alpha_k \tilde{s}_k(t) + \tilde{w}(t) \qquad 0 \le t \le T \qquad (5.38)$$

where the $\{\alpha_k\}$ represent the propagation losses of the K transmission paths. Then the individual transmitted signals reflect their respective data and spreading waveforms, given by

$$\tilde{s}_k(t) = b_k \sqrt{E_b} g_k(t) \qquad (5.39)$$

where the b_k and g_k represent different data sequences and spreading codes, respectively. Equation (5.38) describes a *synchronous CDMA system*; in particular, it assumes that the timing, frequency, and phasing of all the users are identical. This condition is often impractical, if not impossible, to achieve with independent transmitters. Nevertheless, it is useful for illustrating concepts. Much of what follows also applies to asynchronous systems, such as the two-user system shown in Fig. 5.14.

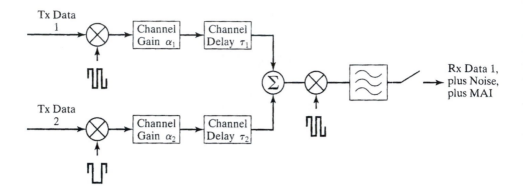

FIGURE 5.14 Illustration of two-user CDMA system with a conventional receiver.

Suppose the desired signal corresponds to $k = 1$ in Eq. (5.38). Then the conventional single-user receiver applies the correlator matched to signature sequence $g_1(t)$ to the received signal and obtains

$$
\begin{aligned}
y &= \int_0^T \tilde{x}(t) g_1^*(t) \, dt \\
&= \alpha_1 b_1 \sqrt{E_b} + \eta_1 + \sum_{k=2}^{K} \alpha_k b_k \sqrt{E_b} \int_0^T g_k(s) g_1^*(s) \, dt \\
&= \alpha_1 b_1 \sqrt{E_b} + \eta_1 + \sqrt{E_b} \sum_{k=2}^{K} \alpha_k b_k R_{1k} \qquad 0 \le t \le T
\end{aligned}
\tag{5.40}
$$

where the first two terms of the last line are equivalent to the single-user case (see Eq. (5.9))—that is, a signal term plus a noise term. The last term in the expression is a multiple-access interference term. The average amplitude at the correlator output is given by

$$
\begin{aligned}
\mathbf{E}[y] &= \alpha_1 b_1 \sqrt{E_b} + \mathbf{E}[\eta_1] + \mathbf{E}\left[\sqrt{E_b} \sum_{k=0}^{K} \alpha_k b_k R_{1k} \right] \\
&= \alpha_1 b_1 \sqrt{E_b} + 0 + 0 \\
&= \alpha_1 b_1 \sqrt{E_b}
\end{aligned}
\tag{5.41}
$$

where we have assumed that the data $\{b_k\}$ are independent and of zero mean. The variance of the output y is

$$\sigma_Y^2 = \mathbf{E}\left[Y - \mathbf{E}[y]^2 \right]$$

$$= \mathbf{E}\left[|\eta_1|^2 + 2\mathrm{Re}\left[\eta_1^* \left(\sqrt{E_b} \sum_{k=2}^{K} \alpha_k b_k R_{1k} \right) \right] + \left| \sqrt{E_b} \sum_{k=2}^{K} \alpha_k b_k R_{1k} \right|^2 \right]$$

$$= N_0 + 0 + E_b \sum_{k=2}^{K} \sum_{l=2}^{K} \alpha_k \alpha_l^* \mathbf{E}[b_k b_l^*] \mathbf{E}[R_{1k} R_{1l}^*] \tag{5.42}$$

$$= N_0 + E_b \sum_{k=2}^{K} \sum_{l=2}^{K} \alpha_k \alpha_l^* \delta(k-1) \mathbf{E}[R_{1k} R_{1l}^*]$$

$$= N_0 + E_b \sum_{k=2}^{K} |\alpha_k|^2 \mathbf{E}|R_{1k}|^2$$

where the second line follows from squaring the difference between Eqs. (5.40) and (5.41). The third line uses $E\left[|\eta_1|^2\right] = N_0$ and $E\left[b_k b_l^*\right] = \delta(k-l)$. The parameters $\{R_{jk}\}$ are the cross-correlations between the different spreading codes. If the spreading codes have properties similar to those of random codes, then, from Eq.(5.37), we have

$$E\left[|R_{jk}|^2\right] \approx \frac{1}{Q} \tag{5.43}$$

where there are Q chips per period T. The contribution of the multiple-access term to the noise variance is the second term on the right of Eq. (5.42), given approximately by

$$\sigma_{\mathrm{MAI}}^2 \approx E_b \sum_{k=2}^{K} |\alpha_k|^2 \frac{1}{Q} \tag{5.44}$$

At this point, we assume that the individual transmitters are *power controlled* (see Section 5.5.3), which means that the transmitted power for each user terminal is adjusted so that the received power is constant; that is,

$$\alpha_k = 1 \qquad \text{for all } k \tag{5.45}$$

Under this assumption, the first- and second-order statistics of the multiple-access interference contribution to the noise in Eq. (5.40) can be summarized as

$$\mu_{\mathrm{MAI}} = \mathbf{E}[y - \alpha_1 b_1 \sqrt{E_b}] = 0$$

and

$$\sigma_{\mathrm{MAI}}^2 = \frac{K-1}{Q} E_b \tag{5.46}$$

where the first equation follows from Eq. (5.41) and the second follows from Eqs. (5.44) and (5.45). From the central limit theorem (see Appendix C), as the number of terms, K, increases, the summation term of Eq. (5.40) will approach a Gaussian random variable. Consequently, for large K, the multiple-access interference behaves like an additive Gaussian white-noise term.

The signal-to-interference-plus-noise ratio SINR seen by the modem includes both the receiver noise and the multiple-access interference and, with $\alpha_k = 1$ and $b = \pm 1$, is given by

$$
\begin{aligned}
\text{SINR} &= \frac{(E[y])^2}{\sigma_Y^2} \\[2ex]
&= \frac{E_b}{\left(N_0 + \dfrac{K-1}{Q}E_b\right)} \\[2ex]
&= \frac{E_b}{N_0}\left(\frac{1}{1 + \dfrac{K-1}{Q}\dfrac{E_b}{N_0}}\right) \\[2ex]
&= \frac{E_b}{N_0}D_g
\end{aligned}
\tag{5.47}
$$

The parameter D_g is the degradation in single-user performance due to the multiple-access interference. This degradation is defined by

$$
D_g = \left(1 + \frac{K-1}{Q}\frac{E_b}{N_0}\right)^{-1}
\tag{5.48}
$$

The degradation depends upon the operating E_b/N_0; the number of users, K; and the spreading factor Q. The operating E_b/N_0 is the single-user SINR in the absence of any multiple-access interference. More users imply greater degradation, as we might expect. But a larger spreading factor (processing gain) reduces the effect of the interference and decreases the degradation. In Fig. 5.15, we show the performance degradation in bit error rate performance as a function of the operating E_b/N_0 for various ratios of users to spreading factors. The degradation depends on the ratio of the number of users to the spreading factor.

The Gaussian approximation (implied by the central limit theorem) for the multiple-access interference is truly valid only for a large number of users. However, as its name indicates, convergence is very rapid in the central portion of the distribution and less rapid in the tails of the distribution. Thus, we would expect that, for a small number of users, the approximation would be valid at high error rates, but that the accuracy would decrease at the lower error rates. This is indeed the case. In practice, CDMA systems usually employ forward error-correction (FEC) coding (discussed in Chapter 4); consequently, it is the high uncoded error rates ($>10^{-3}$) that are of most

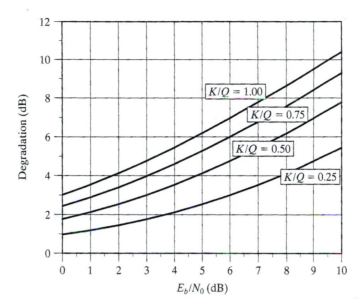

FIGURE 5.15   Performance degradation due to multiple-access interference (large Q).

interest. That is, at the operating points of interest, the Gaussian approximation is quite reasonable, even with a small number of users.

Problem 5.11   The multiple-access interference is a weighted sum of $K-1$ independent, identically distributed random variables. What is the mean and variance of the sum of K independent, identically distributed binary (± 1) random variables?

Ans. $\mu = 0$; $\sigma^2 = K$, *assuming that* ± 1 *are equally likely.*   ■

5.4.2 Multipath Channels

As described in Chapter 2, multipath is the reception of multiple, possibly interfering, copies of the same signal. Consequently, the tolerance of spread-spectrum techniques to interference extends also to a tolerance of multipath. With some receiver designs, multipath can even be used advantageously.

Recall the WSSUS multipath model described in Section 2.7.4. Taken as a snapshot in time, the complex envelope of the channel impulse response is given by

$$\tilde{h}(t) = \sum_{l=1}^{L} \alpha_l \delta(t - \tau_l) \tag{5.49}$$

That is, the channel consists of L multipath rays of amplitude $\{\alpha_l\}$ and relative delays $\{\tau_l\}$. The theoretical spectrum of this channel impulse response extends from $-\infty$ to ∞, with a coherence bandwidth inversely related to the delay spread, as discussed in Chapter 2. Let $H(f)$ denote the Fourier transform of Eq. (5.49). Since the channel is linear, only that portion of the amplitude response $H(f)$ affecting the signal is relevant. If the majority of the signal energy falls inside the band $|f| < R_c/2$, then, we define the windowed channel spectrum by

$$H_R(f) = \begin{cases} H(f) & |f| < R_c/2 \\ 0 & \text{otherwise} \end{cases}$$

(5.50)

Since $H_R(f)$ is band limited, by the Nyquist sampling theorem it follows that its time-domain equivalent can be represented by the sample sequence $\{h_n\}$, where

$$h_w(t) = \sum_{n=-\infty}^{\infty} h_n \operatorname{sinc}(R_c(t - n/R_c))$$

(5.51)

With $R_c = 1/T_c$, the chip rate, by the convolutional sampling theorem (see Appendix A), the received signal may be represented as

$$\begin{aligned} \tilde{x}_{mp}(t) &= \tilde{h}_w(t) \otimes \tilde{s}(t) + \tilde{w}(t) \\ &= \sum_{l=-\infty}^{\infty} h_l s(t - lT_c) + \tilde{w}(t) \\ &= \sum_{l=0}^{L'} h_l s(t - lT_c) + \tilde{w}(t) \end{aligned}$$

(5.52)

where it is assumed that there are only a small number(L') of significant samples in the channel impulse response. Consequently, we have converted from the arbitrary channel delays $\{\tau_l\}$ of Eq. (5.49) to uniformly spaced time delays $\{lT_c\}$ in Eq. (5.52).

EXAMPLE 5.4 Three-Ray Multipath

A three-ray multipath model is shown in Fig. 5.16. In the figure, each of the three paths has an independent (assumed distinct) delay $\{\tau_l\}$ and an independent complex gain $\{g_l\}$. The gains can be time varying, in general. The delays are not assumed to be multiples of the chip interval, but, by the preceding argument, if the chip rate is fast enough, the representation of Eq. (5.52) holds, although the number of taps may be larger than three.   ■

5.4.3 RAKE Receiver

A receiver structure that is often used for CDMA systems operating in multipath environments is the RAKE receiver, depicted in Fig. 5.17. This receiver has $(L + 1)$ fingers and gets the name "RAKE" from its resemblance to the common garden rake. Each finger of the receiver attempts to demodulate one path of the composite multipath signal.

With a RAKE receiver, the signal is sampled at the chip rate and then passed through a delay line with T_c-spaced elements. For the moment, we assume that the

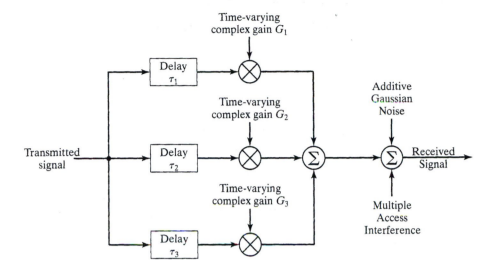

FIGURE 5.16   Multipath channel model.

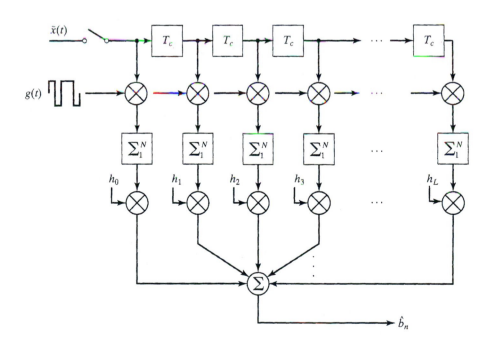

FIGURE 5.17   RAKE receiver for CDMA over multipath channels.

channel impulse response $\{h_l\}$ is time invariant, but, in practice, it can be time varying. *The output of each delay element is processed by a single-user receiver;* that is, for the ith finger (corresponding to delay iT_c), we have

$$
\begin{aligned}
y_i &= \int_{-iT_c}^{T-iT_c} \tilde{x}(t + iT_c)g^*(t)dt \\[2mm]
&= \int_{-iT_c}^{T-iT_c} \sum_{l=0}^{L} h_l \tilde{s}(t - lT_c + iT_c)g^*(t)dt + \int_{-iT_c}^{T-iT_c} \tilde{w}(t + iT_c)g^*(t)dt \\[2mm]
&= \sum_{l=0}^{L} h_l \int_{-iT_c}^{T-iT_c} \tilde{s}(t - lT_c + iT_c)g^*(t)dt + \eta_i \\[2mm]
&\approx h_i b \sqrt{E_b} + b\sqrt{E_b} \sum_{l \neq i} h_l R_{gg}(l - i) + \eta_i
\end{aligned}
\tag{5.53}
$$

where we have used the fact that $\tilde{s}(t) = b\sqrt{E_b}\,g(t)$ for one symbol period T and $\int \tilde{s}(t + \tau)g^*(t)dt \approx 0$ unless $\tau = 0$—that is, unless $lT_c = iT_c$ in the third line. This calculation assumes that the multipath delay spread is much less than the symbol period.

Equation (5.53) shows that the ith finger of the RAKE receiver provides an estimate of the data received over the path delayed by iT_c. This estimate is weighted by the channel tap h_i and includes Gaussian noise, plus a secondary noise due to the multipath interference from the other paths. The multipath interference is similar in form to the multiple-access interference described in Section 5.4.1. If the delay spread of the channel, LT_c, is much less than the symbol period T, then

$$
\int_0^T g(t - (l - i)T_c)g^*(t)dt \approx R_{gg}(l - i)
\tag{5.54}
$$

where $R_{gg}(t)$ is the autocorrelation of the spreading sequence. This relationship is illustrated pictorially in Fig. 5.18, where the desired signal provides a strong correlator output while the *self-interference* provides a weaker correlator output.

As with Eq. (5.42), it can be shown that the contribution of the self-interference to the total noise *for the first finger* is equal to

$$
\sigma_Y^2 = N_0 + E_b \sum_{l=1}^{L} |h_l|^2 |R_{gg}(lT_c)|^2
\tag{5.55}
$$

If the spreading code has cross-correlation properties similar to those of random codes, then we would expect the approximate result

$$
\sum_{l \neq j} |h_l|^2 |R_{gg}(lT_c)|^2 \approx \frac{1}{Q} \sum_{l \neq j} |h_l|^2
\tag{5.56}
$$

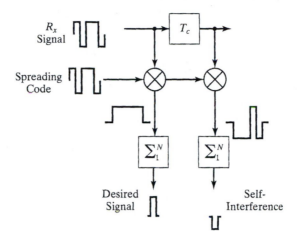

FIGURE 5.18   Illustration of multipath cross-correlation noise.

In many circumstances, the spreading factor Q is large enough to make this multipath interference negligible.

Returning to the RAKE receiver, we find that the combined estimate of the data is obtained by taking a weighted combination of the outputs of each finger of the RAKE receiver, as shown in Fig. 5.17. That is, the final estimate of the data is given by a weighted sum of the output of each finger of the RAKE receiver:

$$
\begin{aligned}
y &= \sum_{i=1}^{L} h_i^* y_i \\
&\approx \sqrt{E_b} \sum_{l=0}^{L} |h_l|^2 b + \sum_{l=0}^{L} h_l^* \eta_l
\end{aligned}
\tag{5.57}
$$

Here, we have assumed that the self-interference term is negligible. The processing described by Eq. (5.57) is known as *maximal-ratio combining* (MRC). The resulting output signal-to-noise ratio is given by

$$
\text{SNR} = \frac{(E[y]^2)}{\sigma_Y^2} = \frac{E_b \left(\sum_{l=0}^{L} |h_l|^2 \right)^2}{N_0 \sum_{l=0}^{L} |h_l|^2}
\tag{5.58}
$$

$$
= \frac{E_b \sum_{l=0}^{L} |h_l|^2}{N_0}
$$

The σ_Y in Eq.(5.58) differs from that in Eq.(5.55) because it assumes that the multipath interference is negligible, but it includes the weighted sum of receiver noise from each of the RAKE fingers. From Eq. (5.58), it is clear that the RAKE receiver makes use of all of the signal energy provided by the channel. If the channel is time invariant, then the BER performance is given by the standard expression for AWGN, with E_b/N_0 replaced by Eq. (5.58). The number of channel taps that need to be considered depends on the application of interest. In some cases, it may be wise to include only the strongest taps in the combining process, as the contributions from other taps may be dominated by the processing noise. Chapter 6 discusses maximal-ratio combining in greater detail.

Problem 5.12 Suppose a wireless channel is identified as a two-ray multipath channel with the second ray having half the power of the first; that is,

$$h(t) = \delta(t) + \delta(t - T_c)/\sqrt{2}$$

For a spreading factor $Q = 4$, compare the degradation observed with this multipath channel with that of a two-user AWGN channel with the same spreading factor. How do they compare when $Q = 128$? ∎

5.4.4 Fading Channels[5]

The assumption of time-variant channel taps is often unrealistic. In most RAKE applications, the channel is fading and the individual channel taps are modeled as independent Rayleigh-fading random variables. *Transmission systems are usually designed such that the channel is approximately constant, at least over the duration of a symbol interval.* Under this assumption, the fading performance of single-user spread spectrum BPSK or QPSK is the same as in the nonspread case.

As we have seen in Section 3.9, for the transmission of coherent BPSK over a single-ray Rayleigh-fading transmission path, the bit error rate performance is given by

$$P_e \approx \frac{1}{4E_b/N_0} \tag{5.59}$$

Now, consider a multipath channel as described in Section 5.4.2, where there are L taps in the channel. Suppose that each tap is independently time varying, but has equal average power. This is a situation known as *Lth-order diversity*. Under these circumstances, we do not expect all channels to be poor simultaneously, and consequently, there is a significant improvement over the performance in the single-ray case. In this case, BPSK performance (for $E_b/N_0 > 10$ dB) is given by (see Chapter 6)

$$P_e \approx \binom{2L-1}{L}\left(\frac{1}{4E_b/N_0}\right)^L \tag{5.60}$$

where the second factor on the right-hand side denotes a combinatorial function. That is, *with second-order diversity, there is an approximate squaring of the bit error rate;* with third-order diversity, there is a cubing of the bit error rate; and so on. In Eq. (5.60), E_b/N_0 represents the energy received per diversity path.

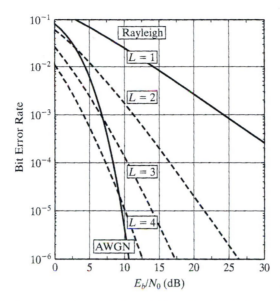

FIGURE 5.19   Comparison of performance in AWGN with that of Rayleigh fading with different diversity orders.

Figure 5.19 indicates that diversity affords a remarkable improvement in performance over a Rayleigh fading channel. Part of this improvement is because more energy is transmitted per bit. With $L = 2$, twice as much energy is transmitted; with $L = 4$, four times as much energy is transmitted. Even without this factor, there is a significant improvement in reliability with diversity. This is why we can achieve better performance than AWGN with higher order diversity. In practice, the diverse paths will typically not have the same power, and the channel's performance will fall somewhere in between that of single- and that of Lth-order diversity.

5.4.5 Summary of the Benefits of DS-SS

Thus far in this chapter, we have observed that DS-SS has a number of features that can be important in a multiple-access system:

1. The spectral density of the transmitted signal is reduced by a factor equal to the processing gain.
2. We can reduce the effect of interference by a factor, known as the processing gain, that is the ratio of the spread bandwidth to the original bandwidth.
3. Under ideal conditions, there is no difference in bit error rate performance between spread and nonspread forms of BPSK or QPSK.
4. In multipath channels, if multipath is treated as interference, then its effect can be reduced by the processing gain.
5. Through proper receiver design, we can use multipath to advantage to improve receiver performance by capturing the energy in paths having different transmission delays.

6. In fading channels, by the use of a RAKE receiver, a spread-spectrum receiver can obtain an important advantage in diversity.

7. The choice of spreading codes is critical to reducing multiple-access interference (spreading-code cross-correlations) and multipath self-interference (spreading-code autocorrelation).

From the preceding observations, it is clear that spread-spectrum techniques are an important tool to have in the communication engineer's toolbox. The one disadvantage of these techniques is their large bandwidth requirement.

Problem 5.13 Suppose a system receives a data packet via two independent radio paths. The probability of packet error on either path is p. What is the probability that the same packet will be received in error over both paths? Relate your answer to the diversity results of Eq. (5.60).
Ans. Probability of error $= p^2$ ∎

5.5 CODE SYNCHRONIZATION[6]

In previous sections of this chapter, it has been assumed that the phase of the signal, as well as the signal timing, is known. Recovering the phase to perform coherent detection is a problem that is common to most digital modulation techniques. The problem of determining signal timing—in particular, sequence time—is more closely associated with spread-spectrum techniques.

To simplify the illustration of code synchronization, we will assume that the spreading code is transmitted repeatedly, without any data modulation. That is, we will assume that $b = 1$ in Eq. (5.2) for all time. A circuit for recovering the timing of the spreading code is presented in Fig. 5.20. The passband received signal is first down-converted to complex baseband, using fixed in-phase and quadrature oscillators. The in-phase (real) and quadrature (imaginary) branches are then processed in an identical manner. The complex baseband received signal prior to the chip filters may be represented as

$$
\begin{aligned}
\tilde{x}(t) &= \sqrt{E_b}\, g(t) e^{j(2\pi\Delta f t + \phi)}, & 0 \le t < T \\
&= \sqrt{E_b}\, g(t-T) e^{j(2\pi\Delta f t + \phi)}, & T \le t < 2T
\end{aligned}
\tag{5.61}
$$

where Δf is the residual frequency error after down-conversion, ϕ is the residual phase error, and $g(t)$ is the spreading sequence given by Eq.(5.5). Equation (5.61) assumes that the spreading code is repeated every T seconds and there is no data modulation. To simplify matters, we will ignore the effects of noise. If we sample the signal $\tilde{x}(t)$ at the chip rate, then the output of the sampling device, prior to the correlation filter, at time $t = kT_c$ is given by

$$
\begin{aligned}
\tilde{x}(kT_c) &= \sqrt{E_b}\, g(kT_c \bmod T) e^{j(2\pi\Delta f k T_c + \phi)} \\
&= \sqrt{E_b}\, c(k \bmod Q) e^{j(2\pi\Delta f k T_c + \phi)}
\end{aligned}
\tag{5.62}
$$

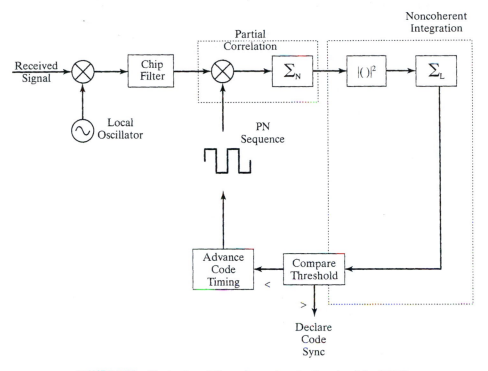

FIGURE 5.20   Illustration of the code synchronization circuit for BPSK.

where the modulo-Q in the expression represents the fact that the code sequence repeats every Q chips. Due to the frequency error, we perform a *partial correlation* over a length of the sequence in which the phase is almost constant. In what follows, it is assumed that the frequency error is small enough that the phase is approximately constant at ϕ_1 over the summation time from $[k_1, k_1 + \Delta k]$. The first in-phase and quadrature summations in Fig. 5.20 perform the partial-correlation operation

$$
C_x(k_1) = \sum_{k=k_1}^{k_1 + \Delta k} x(kT_c)c^*((m_0 + (k - k_1)) \bmod Q)
$$

$$
= E_b \sum_{k=k_1}^{k_1 + \Delta k} c(k \bmod Q)c^*((m_0 + (k - k_1)) \bmod Q)e^{j(2\pi \Delta f k T_c + \phi)} \qquad (5.63)
$$

$$
\approx E_b e^{j\phi'} \sum_{k=k_1}^{k_1 + \Delta k} c(k \bmod Q)c^*((m_0 + (k - k_1)) \bmod Q)
$$

where m_0 is the unknown code offset between the received signal and the local code generator. The value of $C_x(k_1)$ depends on whether the two sequences do or do not align. Ideally,

$$C_x(k_i) \approx \begin{cases} E_b \Delta k e^{j\phi'} & \text{if the sequences align} \\ 0 & \text{otherwise} \end{cases} \tag{5.64}$$

where ϕ' is the sum of the residual phase error ϕ and the phase rotation over the correlation time due to its residual frequency error. Because of the requirement of little phase rotation, it is likely necessary that $\Delta k < N$, so the partial correlation C_x may not be approximately zero if the sequences do not align. To increase confidence in the synchronization, we form a decision variable based on the noncoherent sum of $C_x(k_i)$ over a number of partial correlations; that is, the decision variable is

$$D = \sum_{l=1}^{L_1} |C_x(k_l)|^2 \tag{5.65}$$

and we compare it with a threshold. If D exceeds the threshold, the de-spreading and spreading sequences are assumed to align and code synchronization is declared. If not, then the timing is advanced to a different offset (m_0) and the process is repeated until code synchronization is obtained—that is, until the decision variable of Eq. (5.65) exceeds the threshold.

Problem 5.14   Under what conditions would Eq. (5.64) be exactly true?   ∎

5.6   CHANNEL ESTIMATION

We have characterized the effect of *a flat-fading channel* on the complex baseband signal by the time-varying complex gain (attenuation) $\alpha(t)$. This channel gain affects both the amplitude and phase of the received signal. If the receiver is moving relative to the transmitter, there will be, in addition, a frequency shift of the received signal that we model as the complex baseband signal $e^{j2\pi\Delta ft}$. A frequency shift could also be caused by frequency differences between transmitter and receiver oscillators.

In *a frequency-selective channel*, this effect is compounded by a factor L when we consider the complex representation of the channel impulse response (see Section 5.4.2):

$$\tilde{h}(t) = \sum_{l=1}^{L} \alpha_l(t)\delta(t - \tau_1)$$
$$\approx \sum_{l=1}^{L'} h_l(t)\delta(t - lT_c) \tag{5.66}$$

By definition, a coherent receiver requires accurate phase and frequency estimation. A non-coherent receiver does not require phase estimation, but it does require approximate frequency estimation. If an amplitude modulation strategy is used, then amplitude estimation is also required. In the RAKE receiver, we have seen that, to perform maximal-ratio combining requires knowledge of the channel coefficients $\{h_l\}$, regardless of the modulation strategy. Consequently, a significant number of parameters often must be estimated.

The mobile channel effects on phase, frequency, and amplitude are slowly time-varying. *It is critical to receiver performance to have an accurate channel estimation and tracking strategy to track these variations.* This channel estimation problem is an extension of the channel estimation problem of Section 3.8. In Chapter 3, there was only one channel to estimate; with the model of Eq.(5.66), there are now L' channels to estimate. We will consider an example of how a RAKE receiver can be modified to provide estimates of the channel taps $\{h_l\}$.

In Fig. 5.21, we show a modified RAKE receiver that provides estimates of the channel taps $h_l(t)$. In this circuit, the incoming data stream proceeds along the same processing paths as in the normal RAKE receiver. If the channel taps were known, the output would be the maximal-ratio combined sum from each of the fingers. The additional circuitry shown in the figure works on the data stream after it is combined. First, it estimates the data; if the data are BPSK, estimation involves making a hard decision—that is, ±1. The estimated data are then used to remove the modulation from a stored copy of the signal in each of the RAKE fingers; the stored copy is indicated by the delay element T in the diagram. For a particular finger, the product of the

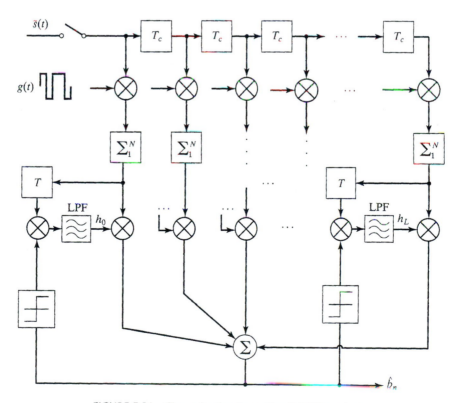

FIGURE 5.21   Channel estimation with a RAKE receiver.

hard decision and the stored sample provides a sample of the impulse response at that particular delay. It is assumed that the channel taps vary slowly in time relative to the data rate, so the channel-tap estimates are low-pass filtered (LPF) to improve their accuracy before being used as channel weighting.

The preceding description provides the general idea of the channel-tracking algorithm and how it should work. There are additional issues, however, that a practical algorithm would have to address:

- To start the algorithm, we must have reliable data estimates. This requirement usually involves transmitting a training sequence to adapt the channel estimates initially.

- The bandwidth of the low-pass filter must be chosen appropriately—neither so narrow that it cannot track the time variations of the channel nor so wide that it lets too much noise into the processing.

- If the data stream is channel encoded to operate at a lower SNR, then the reliability of the uncoded data at the output of the RAKE receiver is liable to be poor at all times. To wait until the data stream is decoded before making the data estimate would incur too much delay, as well as large storage requirements. Alternative approaches, such as transmitting an unmodulated training sequence on an intermittent basis or transmitting it continuously in an orthogonal channel, have been used in a number of practical systems. (See Sections 5.12 and 5.15.)

Problem 5.15   Suppose a spread-spectrum service is operating at 2 GHz and the maximum user terminal velocity is expected to be less than 80 km/hour. What should the bandwidth of the filter in the channel estimation circuit of the corresponding RAKE receiver be? *Ans. 148 Hz.*   ∎

5.7   POWER CONTROL: THE NEAR–FAR PROBLEM

As illustrated in Chapter 2, the received power level in a terrestrial wireless application can vary by as much as 70 dB or more, depending on the distance between the transmitter and the receiver and depending on the transmission path. In conventional FDMA systems, transmitters that reuse the same channel are separated such that their mutual interference is sufficiently attenuated. With CDMA, the intent is to reuse the same frequency channel within the same area. To investigate the effects that this might have on performance, recall Eq. (5.38), where the complex baseband received signal is

$$\tilde{x}_c(t) = \sum_{k=1}^{K} \alpha_k \tilde{s}_k(t) + \tilde{w}(t) \tag{5.67}$$

and where each user is affected by a different channel gain α_k. If we follow the same development of Section 5.4.1, the SINR of the first user is given by

$$\text{SINR} = \frac{(E[y])^2}{\sigma_Y^2}$$

$$= \frac{\alpha_1^2 E_b}{N_0 + \frac{1}{Q} E_b \sum_{k=2}^{K} \alpha_k^2}$$

$$= \frac{\alpha_1^2 E_b}{N_0} \frac{1}{1 + \frac{K-1}{Q} \frac{\alpha_1^2 E_b}{N_0} \left(\frac{1}{(K-1)} \sum_{k=2}^{K} \left(\frac{\alpha_k}{\alpha_1} \right)^2 \right)}$$

$$= \alpha_1^2 \frac{E_b}{N_0} D_g'$$

(5.68)

where

$$D_g' = \frac{1}{1 + \frac{K-1}{Q} \frac{\alpha_1^2 E_b}{N_0} \left(\frac{1}{K-1} \sum_{k=2}^{K} \left(\frac{\alpha_k}{\alpha_1} \right)^2 \right)}$$

(5.69)

In Eq. (5.68), $\alpha_1^2 E_b/N_0$ is the SNR in the absence of multiple-access interference. The factor D_g is the degradation due to multiple-access interference. Comparing the definition of D_g' in Eq.(5.69) with the definition of D_g in Eq.(5.48), we see that

- The received energy for the desired signal is now $\alpha_1^2 E_b$ instead of E_b. This is simply a result of not setting $\alpha_1 = 1$ in Eq. (5.48).
- The other, more important, difference is that the denominator of D_g' includes the factor

$$\frac{1}{(K-1)} \sum_{k=2}^{K} \left(\frac{\alpha_k}{\alpha_1} \right)^2$$

(5.70)

The term in Eq. (5.70) is a multiplier applied to the multiple-access interference. If all transmitted signals are received with equal power, then this term is unity. However, if any of the signal strengths is significantly greater than the desired signal (i.e., $\alpha_k >> \alpha_1$), then there will be a significant increase in the multiple-access interference. This phenomenon is known as the *near–far problem*. Previously, the near–far problem was mentioned in Chapter 3 in reference to adjacent-channel interference in FDMA systems, but it clearly occurs in CDMA as well. That is, signals coming from transmitters significantly closer to the receiver can cause excessive interference with the desired signal. To illustrate the problem, we plot the degradation in performance with CDMA for the case of $K = 2$ users in Fig. 5.22. The degradation of the first user is plotted against relative power of the second user, namely, $20\log_{10}(\alpha_2/\alpha_1)$, for several E_b/N_0 ratios. The figure illustrates that the losses can be quite significant, with power differences as low as 15 dB.

One solution to the near–far problem is *power control*, in which each user's transmit power is adjusted individually such that the received power, $\alpha_k^2 E_b$, is constant and thus independent of k. As one would suspect, power control can improve interference management significantly.

There are two implementation issues related to power control that need attention:

1. The first issue is *latency*: a delay between the time the power is estimated and the time the correction is applied. To implement power control, the power must be measured at the receiver and then relayed to the transmitter to adjust the transmit power. This process incurs the round-trip delay associated with the distance between the transmitter and the receiver. If power levels vary significantly faster than the round-trip delay, power control may not be effective. In some cases, power control is not suitable for compensating fast Rayleigh fading due to terminal motion, although it can compensate for the slower variation of median signal variations as the terminal moves through the service area.

2. The second issue is *accuracy*. To minimize processing delay, simple processing must be performed on the raw signal at the receiver, which means minimal averaging of a potentially very noisy signal. Consequently, the power estimates are somewhat inaccurate.

Techniques for performing power control are discussed in Chapter 7.

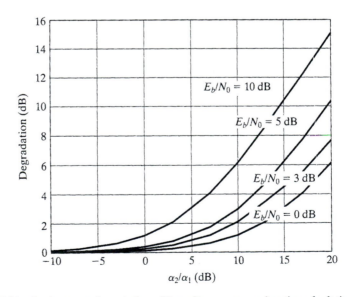

FIGURE 5.22 Performance degradation of first of two users as a function of relative power and processing gain.

Problem 5.16   Suppose that a wireless service is operating at 800 MHz, with a maximum terminal velocity of 100 km/hour. What is the maximum transmitter–receiver separation that would permit reasonable power control compensation of Rayleigh fading, assuming that measurement and processing times are negligible? What would be negligible in this case?
Ans. The maximum separation is 20 km, assuming that the tracking delay is less than 1 percent of a period of the maximum Doppler frequency. The measurement and processing delay must be significantly less than 135 μs; satisfying this requirement is perhaps unrealistic.   ∎

5.8   FEC CODING AND CDMA

Spread-spectrum and forward error correction encoding both expand the number of bits to be transmitted or, equivalently, the transmit spectrum. DS-SS increases the bandwidth by a large factor, but adds no redundancy. Forward error correction (FEC) coding generally increases the bandwidth by a factor of two to three, but includes redundancy. As we have seen already, both of these techniques can improve a signal's performance under various circumstances. The question is: can they be combined effectively, and if so, how should that be done?

In most circumstances, there is a maximum bandwidth available which limits the amount of spreading that can be applied. Previously, the maximum bandwidth was designated R_c and the maximum spreading rate was $Q = R_c/R_b$. We now decompose this maximum spreading rate into two parts:

$$Q = Q_s \times \frac{1}{r} \tag{5.71}$$

In Eq. (5.71), Q_s is the factor due to DS spreading, and $r < 1$ is the FEC code rate. This split of the spreading factor between two different mechanisms can be made transparent to our previous calculations if we replace E_b by E_s, where E_b is the energy per information bit and E_s is the energy per channel bit. The two variables are related by $E_s = rE_b$; that is, the energy per channel bit is simply r times the energy per information bit.

Under these conditions, namely, the degradation of single-user performance with multiple-access interference (equal-power interferers), if we note that each channel bit (i.e., each FEC-encoded bit) is spread by the factor Q_s, we can write Eq. (5.48) as

$$D_g = \left(1 + \left(\frac{K-1}{Q_s}\right)\frac{E_s}{N_0}\right)^{-1}$$

$$= \left(1 + \left(\frac{K-1}{Q/r}\right)\frac{E_b}{rN_0}\right)^{-1} \tag{5.72}$$

$$= \left(1 + \left(\frac{K-1}{Q}\right)\frac{E_b}{N_0}\right)^{-1}$$

The conclusion drawn from Eq. (5.72) is that *the general expression for the degradation due to multiple-access interference when coding is present is unchanged* from the uncoded expression. *However, since the operating E_b/N_0 is typically much lower in coded situations, there is significant improvement in overall performance.*

EXAMPLE 5.5 Improved Multiple-Access Performance with FEC Coding

Suppose a system has an information rate $R_b = 4800$ bps and $Q = R_c/R_b = 32$. The system is error protected by a rate-1/2 convolutional code. Compare the degradation D_g with and without FEC coding at a BER of 10^{-5} when there are seven interfering users.

Without FEC encoding, the single-user $E_b/N_0 = E_s/N_0$ required to obtain a BER of 10^{-5} is 9.6 dB with BPSK modulation. The spreading factor in this case is $Q_s = Q$. The resulting degradation factor is

$$D_g = \left(1 + \frac{8-1}{32}10^{9.6/10}\right)^{-1} = 0.33 \sim -4.8 \text{ dB}$$

With rate-1/2 constraint-length-7 convolutional FEC encoding, the single-user E_b/N_0 required to obtain a BER of 10^{-5} is 4.5 dB with BPSK modulation. With FEC encoding, the spreading factor Q_s would be reduced to 16, but the total spreading Q remains at 32. The resulting degradation factor is

$$D_g = \left(1 + \frac{8-1}{32}10^{4.5/10}\right)^{-1} = 0.62 \sim -2.1 \text{ dB}$$

Thus, the use of FEC coding improves performance over the case without coding by 2.7 dB.

In Fig. 5.23, we plot the degradation versus the loading factor K/Q for two cases: with and without FEC coding. A rate-1/2 FEC code is used for the coded cases, with the characteristics shown in Table 5.2.

The conclusion to be drawn from Fig. 5.23 is that, for either BER, FEC provides a large improvement in performance. That is, the FEC code does not detract from the processing gain. ∎

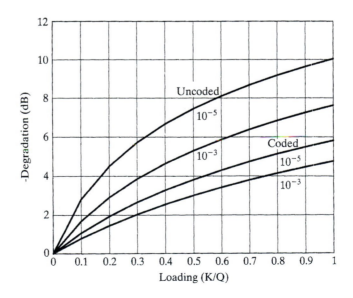

FIGURE 5.23   Degradation due to multiple-access interference with and without FEC for BER of 10^{-3} and 10^{-5}. The FEC code is a rate-1/2 memory-6 convolutional code.

TABLE 5.2   Bit error rates for uncoded and coded examples.

BER	Required uncoded E_b/N_0(dB)	Required coded E_b/N_0(dB)
10^{-3}	6.8	3.0
10^{-5}	9.6	4.5

Forward error correction coding is usually part of a power-versus-bandwidth trade-off in the system design. With CDMA, since coding can be included with no bandwidth penalty and can still provide a power gain, there is an added bonus. The benefit is even greater in fading channels, for which *FEC coding often provides a form of time diversity.*

Problem 5.17   Do you expect FEC codes to have a greater or lesser benefit in Rayleigh-fading channels? Discuss your answer.   ■

5.9   MULTIUSER DETECTION[7]

Previous sections have indicated that multiple-access interference can be a serious detriment to the performance of a CDMA system, particularly if the objective is to maximize spectral efficiency. In this section, methods of ameliorating the effects of multiple-access interference are introduced. We consider a system with K independent terminals, as shown in Fig. 5.24.

In the figure, each terminal, with its own modulating waveform (spreading sequence), transmits over a channel with arbitrary relative gain $\{\alpha_k\}$ and delay $\{\tau_k\}$. The signals are summed at the receiver, together with additive noise. The received signal is then simultaneously de-spread by the filters matched to each of the spreading waveforms to produce the matched-filter output $\{y_k(i)\}$ for each user at each instant of time, i. The matched-filter outputs then undergo further processing in the form of *multiuser detection*, to reduce the multiple-access interference.

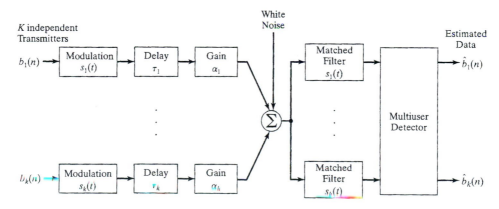

FIGURE 5.24   Multiuser system model.

In the case of synchronous CDMA, it can be shown that an equivalent baseband discrete-time model for the system over one symbol period is

$$\mathbf{y} = \mathbf{A}\mathbf{R}\mathbf{b} + \eta \qquad 0 \leq t \leq T \tag{5.73}$$

where $\mathbf{b} = [b_1, b_2, ..., b_K]^T$ is the vector of bits corresponding to the K terminals. The matrix

$$\mathbf{A} = \begin{bmatrix} \alpha_1 & 0 & ... & 0 \\ 0 & \alpha_2 & ... & 0 \\ \vdots & \vdots & & \vdots \\ 0 & 0 & ... & \alpha_K \end{bmatrix} \tag{5.74}$$

is the diagonal matrix of channel gains, assuming either a constant channel or a channel that is approximately constant over one symbol interval. The matrix $\mathbf{R}$ is the cross-correlation matrix of the symbol-shaping waveforms of the different users. That is, the ijth element of $\mathbf{R}$ is given by

$$R_{ij} = \int_0^T g_i(t)g_j^*(t)dt \tag{5.75}$$

and η is a vector of correlated Gaussian noise samples with elements

$$\eta_i = \int_0^T g_i^*(t)\tilde{w}(t)dt \tag{5.76}$$

The elements y_i of the vector $\mathbf{y}$ represent the matched filter output of the ith user's data after de-spreading and integration.

Equation (5.73) is an example of a *multiple-input, multiple-output (MIMO) system*, on which more will be said in Chapter 6. *The object of multiuser detection is to find ways of processing* $\mathbf{y}$ *to recover the transmitted data* $\mathbf{b}$ *accurately and efficiently.* If the spreading sequences were perfectly orthogonal, then $\mathbf{R}$ would be the identity matrix, and the approach discussed in this chapter would work ideally. In practice, we have seen that, with m-sequences and Gold codes, the spreading sequences are *not* perfectly orthogonal, and if there is a large number of users, then the cumulative interference can be significant.

The objective of a multiuser detection algorithm is to perform the equivalent of interference cancellation. In general, it is assumed that the spreading codes of all users are known, but the data streams are not. The objective is to jointly detect the data stream of interest, removing, as much as possible, the effect of one user's data on the others. To meet this objective, we identify three methods:

Method 1. The *conventional receiver* simply estimates each user's bits on the basis of the sign of the individual elements of $\mathbf{y}$; that is,

$$\hat{\mathbf{b}} = \text{sign}\{\mathbf{y}\} \tag{5.77}$$

Equation (5.77) effectively assumes that the multiple-access interference is negligible—that is, that $\mathbf{R}$ is approximately a diagonal matrix. A conventional receiver is the approach that we have used in previous sections.

Method 2. The *optimum (maximum-likelihood) detector* for the data vector $\mathbf{b}$ is given by

$$\hat{\mathbf{b}} = \arg \min_b |\mathbf{y} - \mathbf{A}\mathbf{R}\mathbf{b}|^2 \qquad (5.78)$$

where the notation $\arg \min_x\{...\}$ refers to the value of x that minimizes the quantity within the braces. The method based on Eq. (5.78) searches over all possible $\mathbf{b}$ vectors to determine the one that minimizes the square error between the matched-filter outputs $\mathbf{y}$ and the predicted value. This is a computationally intensive approach, since we must evaluate the square error for all possible vectors $\mathbf{b}$.

EXAMPLE 5.6 Optimum Multiuser Detection

Suppose there are $K = 2$ users and the measured quantities are

$$\mathbf{y} = \begin{bmatrix} -0.08 \\ -0.47 \end{bmatrix}, \qquad \mathbf{R} = \begin{bmatrix} 1 & 0.33 \\ 0.33 & 1 \end{bmatrix}, \qquad \mathbf{A} = \begin{bmatrix} 1 & 0 \\ 0 & 1 \end{bmatrix}, \qquad (5.79)$$

If we evaluate the square error of Eq. (5.78) for the four vectors $[1\ 1]^T$, $[1\ -1]^T$, $[-1\ 1]^T$, and $[-1\ -1]^T$, we find that $\mathbf{b} = [1\ -1]^T$ produces the smallest error. This is clearly a different answer than would be produced by the conventional receiver. ∎

Method 3. The third approach, which lies between the preceding two approaches, is the *de-correlating detector*. With this approach, the receiver attempts to invert the channel and estimates the data on the basis of

$$\hat{\mathbf{b}} = \text{sign}\{\mathbf{R}^{-1}\mathbf{y}\} \qquad (5.80)$$

There are advantages and drawbacks to these three approaches and to their extensions to more practical situations. At present, multiuser detection is an active research area.[8]

Problem 5.18   For Example 5.6, does the de-correlating receiver produce the same estimate as the conventional receiver? as the optimum receiver? or does it produce something different? ∎

5.10 CDMA IN A CELLULAR ENVIRONMENT

One of the advantages of CDMA is that *interference caused by another CDMA signal appears to be approximately equivalent to additive Gaussian noise.* This property is used advantageously in cellular systems. In Section 2.8.5, we described how to define reuse distances for cellular FDMA systems. With CDMA systems, the same-frequency channel can be reused in the adjacent cell, as long as multiple-access interference (MAI) is kept below a given level. This *MAI is directly proportional to the channel loading.*

We will analyze multiple access in a cellular environment for the uplink from a mobile terminal to a base station located at the center of the cell, as shown in Fig. 5.25, where two types of interference affect performance. Let us define *intracellular interference* as the interference caused by users within the same cell. If there is perfect power control such that all users are received at equal power at the base station, then the intracellular interference is given by Eq. (5.46), which we express here as

$$I_{\text{intracell}} = \frac{K-1}{Q}E_b \approx \frac{K}{Q}E_b \qquad (5.81)$$

where E_b is the common received power level.

Intercellular, or other-cell, interference is the second type of interference, and it is due to reuse of the same CDMA channel in the surrounding cells. Most of this interference will come from the first and second tiers of surrounding cells. The interference from more distant cells suffers more attenuation, due to the greater propagation loss. The signals causing intercellular interference are received at different power levels, because they are power controlled relative to other base stations. As a consequence, other-cell interference depends on the propagation losses from a mobile terminal to two different base stations. In general, the relative power from radios in other cells will be attenuated relative to the power from the intracell radios due to their greater distance. Let f be the *relative other-cell interference factor*, and define it as

$$f = \frac{I_{\text{other-cell}}}{I_{\text{intracell}}} \qquad (5.82)$$

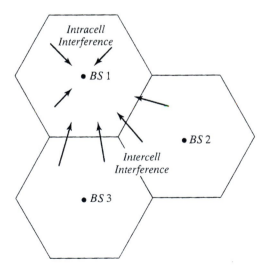

FIGURE 5.25 Intracell and intercell interference.

where it is assumed that the traffic loading in all cells is the same. The size of the factor f depends upon a number of quantities:[9]

1. *The propagation-loss exponent.* The larger the propagation loss exponent is, the more quickly adjacent-cell interference will be attenuated. Built-up environments will tend to decrease other-cell interference more quickly than suburban and rural environments will; this is advantageous, as the built-up environments tend to have the highest density of users.

2. *The variations in signal strength due to shadowing.* Shadowing between the mobile terminal and the desired base station will be compensated for by power control. Shadowing of the same mobile terminal signal relative to other base stations will not be compensated and will cause a significant and variable effect on other-cell interference. The larger the variation due to shadowing, the greater the potential is to cause problems.

3. *The handoff technique between cells.* Since, with CDMA, the same frequencies are reused in every cell, it is possible to send and receive the same call from multiple base stations. This property becomes particularly important at the boundary between cells, where there is handover of communication responsibilities between base stations as a mobile terminal moves from one cell to the next. A *hard handover* refers to the case where communications are terminated with one base station immediately upon acquisition of the second base station. A hard handover is typically what is done in a TDMA system. However, a hard handover may result in a ping-pong effect, with the mobile terminal switching back and forth between the two base stations as the strength of the signal varies at the boundary between the two cells. A *soft handover* can occur in CDMA, since the same frequencies are used in each cell; in this case, the mobile terminal maintains communications with both base stations until one of them becomes significantly stronger than the other. A soft handover prevents the ping-pong behavior, and the dual base-station capability is a form of diversity that can increase capacity in a heavily loaded system and also increase coverage in a lightly loaded system. In fact, the diversity effect permits a significant reduction in the margin allowed for shadowing; the resulting decrease in power means that considerably less other-cell interference appears. (Other techniques for avoiding the ping-pong effect are discussed in Chapter 7.)

The other-cell interference factor ranges from 0.5 to 20, depending upon the foregoing contributing factors.

Combining Eqs. (5.81) and (5.82), we may write the total interference density as

$$
\begin{aligned}
I_0 &= I_{\text{intracell}} + I_{\text{other-cell}} \\
&\approx (1+f)\frac{K}{Q}E_b
\end{aligned}
\tag{5.83}
$$

Ultimately, the impact of this interference on a cellular CDMA is felt at the individual receiver. The signal-to-interference-plus-noise ratio that is seen at the individual receiver is given by

$$\text{SINR} = \frac{E_b}{N_0 + I_0}$$

$$= \frac{E_b}{I_0\left(1 + \dfrac{N_0}{I_0}\right)} \tag{5.84}$$

Cellular CDMA systems are often *interference limited*; that is, the operating conditions are such that $I_0 > N_0$, typically 6 to 10 dB higher. The ratio I_0/N_0 depends upon the cell size. With large cells and battery-operated terminals, most of the transmit power is used to achieve the desired range. Thus, large cells tend to be noise limited. Smaller cells, in contrast, tend to be interference limited, and the interference floor seen at the receiver is typically greater than the noise floor of the receiver. Substituting Eq. (5.83) into Eq. (5.84), we obtain

$$\text{SINR} = \frac{1}{(1+f)\dfrac{K}{Q}\left(1 + \dfrac{N_0}{I_0}\right)} \tag{5.85}$$

Equation (5.85) shows the three factors that affect the SINR seen at the receiver, and these, in turn, limit spectral efficiency. The three factors are the channel loading K/Q, the other-cell interference, and the operating I_0/N_0. These three factors are related to system design. The channel loading is clearly a design parameter that needs to be maximized in a commercial system; the other-cell interference depends on the environment, but also on the handover technique; and, as mentioned, the third factor, I_0/N_0, is related to cell size, another design parameter.

In Fig. 5.26, we plot the SINR from Eq. (5.85) as a function of cell loading for different f and I_0/N_0 ratios. The figure illustrates a number of interesting points:

- For a constant SINR, moving from a noise-limited system ($I_0/N_0 = 0$ dB) to an interference-limited system ($I_0/N_0 = 10$ dB) increases the permissible channel loading (capacity). The loading must be significantly decreased to support a noise-limited system at the same SINR. Thus, large cells must have lighter loading than small cells.

- For a constant SINR, increasing other-cell interference significantly reduces the permissible channel loading. Although it is not evident from the figure, when there is significant shadowing, soft handoffs are the key to reducing margins and keeping other-cell interference low.

- Reducing the SINR required by the receiver can significantly improve the permissible channel loading. Sections 5.4 and 5.8 have described methods of reducing the required SINR, such as RAKE receivers and FEC coding.

The results of Fig. 5.26 do not include effects such as voice activation and cell sectorization. *Voice activation* refers to the fact, that in a telephone conversation, each speaker

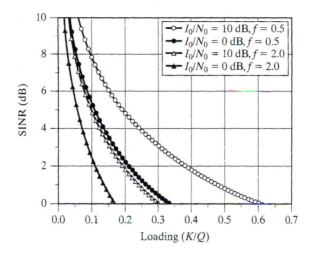

FIGURE 5.26   Maximum channel loading as a function of the other-cell interference factor f and the ratio of the interference noise density to that of the receiver.

is speaking only 40% of the time, on average. If a terminal transmits only when the user speaks, then it will contribute to the interference just 40% of the time. Consequently, we would expect that a system consisting mainly of telephone cells and using voice activation would see a 40% reduction in interference, on average, and could tolerate proportionately higher cell loadings.

Cell sectorization refers to the fact that, to improve system capacity, base stations are sometimes installed with directional antennas as opposed to omnidirectional antennas. Typically, directional antennas over a sector of 120°, radiating and receiving signals only in that sector. This directionality reduces the interference generated and received by the base station and could increase the cell loading by a factor close to three.

Problem 5.19   Define the *cellular spectral efficiency* v, in bits/s/Hz/cell; this is the total number of bits/s/Hz transmitted by all users in a cell. For a QPSK base modulation, assume that the spectral efficiency of a single CDMA user is $2/Q$ bits/s/Hz, where Q is the length of the spreading code. Suppose the receiver requires a specified SINR. Using Eq. (5.85), develop an expression for v that depends on the received I_0/N_0, SINR, and f. Why does the result not depend explicitly on Q? How does it depend implicitly on Q?

Ans. $v = \dfrac{2}{(1+f)\left(1 + \dfrac{N_0}{I_0}\right)\text{SINR}}$

■

5.11 FREQUENCY-HOPPED SPREAD SPECTRUM[10]

In Section 5.1, we mentioned that direct sequence is not the only form of spread spectrum; frequency hopping (FH) is another method of spreading a relatively low-data-rate digital signal over a large bandwidth. In some respects, a frequency-hopped spread spectrum (FH-SS) is similar to the narrowband modulation techniques discussed in Chapter 3, the difference being that the carrier frequency f_c of the transmitted signal is changed at regular intervals.

The advantages of FH-SS are as follows:

- *High tolerance of narrowband interference.* Since the FH signal varies its center frequency, a fixed-tone or narrowband interference signal centered at a frequency f_ξ will cause degradation only when it aligns with the carrier frequency $f_\xi \approx f_c$.
- *Relatively straightforward interference avoidance.* If a certain band of the radio spectrum is known to contain interference, the hopping frequencies can be selected to avoid this band.
- *Current technology permits frequency-hopped bandwidths on the order of several gigahertz.* This spread bandwidth is at least an order of magnitude greater than what can be achieved with DS systems.

When the relative power in the interference is greater than the processing gain of the DS signal, it will cause the DS-SS system to fail. With FH, a single tone will cause degradation only at one hop frequency, regardless of its strength relative to the desired signal. The interference tolerance of FH thus may be greater than that of DS.

FH-SS has the following disadvantages:

- *Noncoherent detection.* Frequency synthesizer implementations often limit the continuity of the signal across hops. As a result, modulation techniques such as M-ary FSK with non-coherent detection are employed. This transmission strategy suffers a performance penalty compared with coherent detection strategies.
- *Higher probability of detection.* Since a single hop period is similar to a narrowband carrier, it is not hidden by the noise and is simpler to detect than a DS signal. However, if the hopping frequency is relatively fast, it is still difficult to intercept (i.e., it is difficult to detect the transmitted data without a prior knowledge of the hopping pattern).

The higher probability of detection depends upon the spread bandwidth of the frequency-hopped signal. Although the individual hops are not spread in frequency and thus can have significant spectral peaks, an intercept receiver that is searching for such a peak must have an input bandwidth proportional to the spread bandwidth. A wide input bandwidth means considerably more received noise that may drown out the peaks of the FH signal.

Early FH systems had hop rates of several Hz, but current systems are capable of hop rates of 100 kHz and higher. These high hop rates make it difficult for a rogue receiver to intercept the transmission without prior knowledge of the hopping pattern.

5.11.1 Complex baseband representation of FH-SS

As with the DS-SS situation, there are two components to the frequency-hopped complex baseband signal: a modulation portion and a frequency-spreading portion. For a single symbol interval, the complex baseband signal can be represented as

$$\tilde{s}(t) = m(t)\exp(j2\pi\zeta_k t + \theta_k) \qquad 0 \leq t \leq T \qquad (5.86)$$

where $m(t)$ is the data modulation, ζ_k is the frequency of the current hop, and θ_k is the phase. Due to the limitations of frequency synthesizer technology, it cannot be assumed that the phase is continuous between hops. For linear modulation, we have

$$m(t) = b\sqrt{E_s}g(t) \qquad 0 \leq t \leq T \qquad (5.87)$$

where b is the data symbol and $g(t)$ is the pulse shape. For frequency-hopped systems, it is more likely that M-ary FSK will be used, in which case we have

$$m(t) = \sqrt{E_s}\exp(j2\pi f_i t) \qquad 0 \leq t \leq T \qquad (5.88)$$

where f_i is the frequency representing the current symbol where E_s is the energy per symbol. (See Section 3.7.2.)

The data symbols have a period T. The *hop period*, T_{hop}, is the period when the transmit frequency is constant—that is, $\zeta = \zeta_k$, where $\{\zeta_k\}$ is the set of permissible hop frequencies.

The hop period can be larger than the symbol time (i.e., $T_{hop} > T$), in which case there are multiple data symbols transmitted on each hop. This situation is referred to as *slow-frequency hopping*. Alternatively, the hop period can be shorter than the symbol period ($T_{hop} \leq T$), in which case the same symbol is transmitted on multiple hops and the phenomenon is referred to as *fast-frequency hopping*. For the same data rate, fast frequency hopping places greater requirements on the frequency synthesizer switching times, but it provides a diversity advantage to the transmission strategy.

The term *chip* is also used in frequency-hopped systems, but the meaning may appear different from the DS meaning of the word, depending on whether we have a slow or a fast-frequency-hopped system. In an FH/MFSK system, "chip" refers to the tone of shortest duration. If R_c represents the chip rate, R_s the symbol rate, and R_h the hop rate, then

$$R_c = \max\{R_s, R_h\} \qquad (5.89)$$

That is, for slow-frequency-hopped systems, the chip rate is equal to the symbol rate, but for fast-frequency-hopped systems, the chip rate is equal to the hopping rate.

A simplistic view of an FH transmitter is illustrated in Fig. 5.27, which depicts the data being modulated with an M-ary FSK strategy. The resulting signal is then upconverted to the hop frequency. Usually, the frequency synthesizer uses the combination of the pseudonoise (PN) code generator output and the MFSK modulator output to

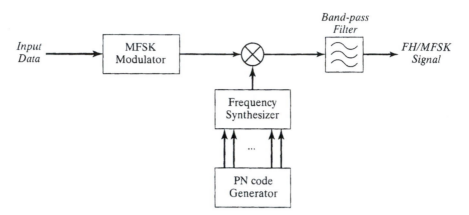

FIGURE 5.27   Block diagram of frequency-hopped transmitter.

produce a tone at the desired transmission frequency; this eliminates the need for a subsequent upconversion. Mathematically, the tone (signal) may be expressed as

$$
\begin{aligned}
s(t) &= \sqrt{E_s}\,\mathrm{Re}\{m(t)\exp(j(2\pi(\zeta_k+f_c)t+\phi_k))\} \\
&= \sqrt{E_s}\,\mathrm{Re}\{\exp(j(2\pi(f_i+\zeta_k+f_c)t+\phi_k))\} \qquad 0\le t\le T
\end{aligned}
\tag{5.90}
$$

where f_c is the center frequency of the band.

The hop sequence is chosen in a pseudorandom pattern. Often, the number of hop frequencies is selected to be a power of two, 2^m. This hop sequence is generated by a maximal-length (ML) shift register. Recall that in the DS case, the ML sequence bits were used one at a time as the chipping sequence. With frequency hopping, the bits are used m at a time, and the m bits are used as an integer index for selecting the hop frequency.

If the ML shift register has a memory m, then using m bits at a time is equivalent to using the state of the ML shift register for selecting the hop frequency. At the receiver, once the frequency of a single hop is determined, the state of the shift register is known, and then the hopping pattern can be predicted. If two transmitters wish to share the same spectrum, they can use the same hopping pattern, but they must ensure that there is a time offset in the initial state. In practice, more sophisticated hopping patterns are employed that use a much longer ML sequence, with only an m-bit segment of the state being used to select the hopping frequency.

5.11.2 Slow-Frequency Hopping

With M-ary FSK modulation, the M tones must be selected such that they are an integer number of symbol rates apart. This spacing is required to maintain the orthogonality between the tones and to permit reliable noncoherent detection. Such a spacing implies that the minimum bandwidth of an MFSK signal is approximately MR_s. With

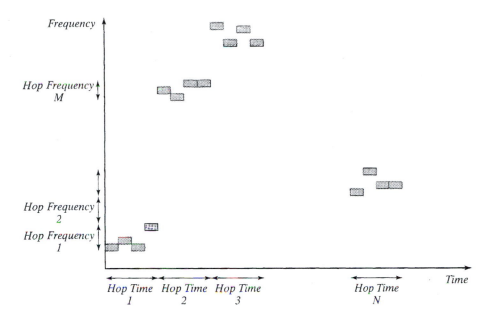

FIGURE 5.28   Frequency–time plot of output of FH/MFSK system.

FH/MFSK systems, the hop frequencies are often chosen such that they are separated by the bandwidth of the MFSK signal; that is,

$$\min\{\zeta_k - \zeta_j\} \approx MR_s \qquad k \neq j \tag{5.91}$$

Figure 5.28 depicts the output of a frequency-hopped transmitter on a frequency-versus-time plot. The time axis shows that four frequencies (symbols) are transmitted sequentially during each hop period. These frequencies are part of an M-ary FSK group and thus are confined to the same narrowband bandwidth during the same hop interval. In the next hop interval, the next four symbols are transmitted, and so on.

EXAMPLE 5.7 Slow-Frequency-Hopping System

A military system is to be designed to transmit 64 kbps, using 16-FSK. A hop rate of 1 kHz is used over an available spectrum of 2 GHz, with a negligible synthesizer switching time between hops. How many data symbols are transmitted per hop? How many nonoverlapping hop frequencies are there? What would be the expected symbol and bit error rates in the presence of a high-power jamming tone that could occur anywhere in this 2-GHz bandwidth?

With 16-FSK, there are 4 bits per symbol; thus, a bit rate of 64 kHz implies a symbol rate of 16 kHz. A hop rate of 1 kHz with negligible switching time implies a hop time of 1 millisecond, or 16 symbols per hop.

From Section 3.7.2, the minimum bandwidth of an M-ary FSK signal is approximately $B_m \approx MR_s$, where R_s is the symbol rate. Consequently, $B_m \approx 256$ kHz, and the number of non-overlapping hops is

$$n_{FH} = \frac{B_{ss}}{B_m} = \frac{2 \text{ GHz}}{256 \text{ kHz}} \approx 7812 \tag{5.92}$$

A tone interferer with an arbitrary frequency in this band will corrupt one out of 7812 hops, on average. Thus, the expected symbol error, if the noise levels are negligible, is $1/7812 = 1.3 \times 10^{-4}$. On average, one-half of the bits of each corrupted symbol are in error, so the expected bit error rate is 6.5×10^{-5}. ∎

5.11.3 Fast-Frequency Hopping

With fast-frequency hopping, the same symbol is transmitted on multiple hops. The detector for individual hops is similar to that used for slow-frequency hopping, but for combining the outputs from multiple hops, two algorithms may be considered:

1. *Majority logic detector.* With this detector, a hard decision on the symbol is made every hop interval, and the outputs of the different hops are combined by choosing the symbol that appears the most often.
2. *Noncoherent combining.* With this detector, a soft measure related to the likelihood is used for combining the outputs. Often, this soft measure is related to the signal energy in a given frequency bin.

5.11.4 Processing gain

In military applications, an intentional interferer is referred to as a *jammer*. If a jammer has a power J available for jamming purposes and spreads this interference power over the whole spread bandwidth, then the resulting interference density is $J_0 = J/W_{ss}$. If this interference is the dominant source of degradation, then the received signal-to-noise ratio is

$$\begin{aligned} \text{SNR} &= \frac{E_s}{J_0} \\ &= \frac{C/R_s}{J/W_{ss}} \\ &= \frac{C}{J}\left(\frac{W_{ss}}{R_s}\right) \end{aligned} \tag{5.93}$$

where E_s is the energy per symbol and C is the carrier power for the desired signal. For this system, the processing gain is given by

$$P_G = \frac{W_{ss}}{R_s} \tag{5.94}$$

As intuitively expected, the processing gain is the ratio of the spread bandwidth to the symbol rate.

5.12 THEME EXAMPLE 1: IS-95[11]

Cellular wireless communications have evolved from analog techniques to the more flexible digital techniques that are currently employed. Future developments are aimed at further enhancing these digital techniques to integrate voice, messaging, and high-speed data. These cellular systems have been classified into three generations:

1. *Analog systems*, which are usually referred to as *first-generation systems*.
2. *Initial digital systems*, such as GSM, IS-54, and IS-95, which are primarily for voice and are called *second-generation systems*.
3. *Systems that integrate voice and data*, which provide higher data rates than second-generation systems and are called *third-generation systems*.

In North America, the uplink frequency band for cellular communications is 869–894 MHz. The downlink frequency for all channels is 45 MHz below the uplink frequency. These frequencies are used for first- and second-generation systems. The uplink frequency band for personal communications services (PCS's) is 1930–1990 MHz, and the downlink band is 80 MHz below the uplink frequency. These bands are intended for third-generation systems, but some second-generation systems have migrated to them.

If a service provider is using a CDMA wireless link, then it is likely based on a standard that is known as IS-95. The basic mode for this standard is the transmission of data at a rate of 9.6 kbps. The majority of traffic is voice signals, but data form an increasing portion of the IS-95 traffic. The data are direct-sequence spread to a chip rate of 1.2288 megachips per second (Mcps), and the allocated channel bandwidth is 1.25 MHz. The second-generation IS-95 standard is the foundation for an important third-generation standard known as CDMA2000.

5.12.1 Channel Protocol

The IS-95 standard defines a variety of communications channels for both downlink and uplink transmission. All of the different channel types use the CDMA access scheme. Some of the more important downlink and uplink channels defined in the IS-95 standard are described in Table 5.3.

The need for the traffic channels is clear, but the purpose of the remaining channels may not be so obvious. The purpose of the remaining channels becomes more evident when we consider the procedure for making a call to or from a mobile terminal. The typical *procedure for making a call* from a mobile terminal consists of the following steps:

1. The mobile terminal searches for the Pilot channel and synchronizes to it.
2. The mobile terminal then immediately locks onto the Sync channel, since it is synchronized with the Pilot channel; from the Sync channel, the terminal gets system information, such as the spreading code of the access channel and the paging channel.
3. Once the mobile terminal has the system information, it sends a request to set up a telephone call (i.e., the number of the party to be called) over the Access channel.

TABLE 5.3   Main communication channels for IS-95 standard.

Channel Type	Function
Pilot channel	This is a downlink reference channel for synchronization and tracking purposes. It is typically the strongest channel, with 10 to 20% of the total power, and it provides the capability for soft handoff and coherent detection. This channel does not carry any data.
Sync channel	This downlink channel provides the mobile station with system information such as short PN offset, system time, long code state, paging channel data rate, system ID, and other related information.
Paging channel	This downlink channel is used for sending short messages, such as public messages, details of registration procedures, pages, and various other short messages for individual mobile terminals.
Downlink traffic channel	The main downlink traffic-bearing channel, this channel provides a dedicated link between the base and the mobile terminal. Supplemental channels of this type are added dynamically to meet the required data rate.
Access channel	The system includes many of these uplink random-access channels on which mobile terminals communicate short messages, such as information on registration, call originations, and responses to pages, to the base station. The access channel uses a prearranged long code offset.
Uplink traffic channel	Similar to the downlink traffic channel, this uplink channel is intended to transport dedicated data.

4. The mobile terminal then listens to the Paging channel to hear the response and the possible traffic channel assignment.

5. Once the downlink and uplink traffic channels are assigned, the mobile terminal starts transmitting on the uplink channel, and the base station tunes a receiver to that channel. The opposite occurs in the downlink direction.

The procedure just described illustrates the use of the six different types of channel, some shared and others dedicated. The same six channels can be used for receiving a call. The typical *procedure for receiving a call* at a mobile terminal consists of the following steps:

1. The base station transmits a short message on the paging channel, indicating that a call is arriving; the mobile terminal monitors this channel at regular intervals.

2. If the mobile terminal is monitoring the Paging channel and receives the page, it responds on the Access channel.

3. From this point on, the procedure becomes the same as that involved in making a call from the mobile terminal.

The foregoing procedure illustrates some features that are common to most wireless telephone systems. In particular, there are a number of *signaling channels* that are dedicated to setting up the telephone link, but that do not carry any user traffic. In data networks, the signaling channels often are not present or are part of the traffic channels. The reason for their use in telephone networks is to speed the response time of the network. Telephone calls have real-time requirements arising from the limited amount of time people are willing to wait to make a connection.

5.12.2 Pilot Channel

One key feature of the IS-95 CDMA transmission strategy is the Pilot channel, which permits very fast synchronization and reliable channel tracking. To make efficient use of the channel resources, the terminal must be able to synchronize quickly, which can be difficult in CDMA with long spreading codes. In Section 5.5.1, we showed how this code synchronization might be done in the absence of data modulation. When data are present, it becomes a much harder problem. The presence of this unmodulated Pilot channel simplifies the problem.

The Pilot channel also plays an important role in channel tracking. In Section 5.4, we illustrated how a RAKE receiver requires proper weights for the different fingers in order to perform maximal ratio combining. In Section 5.6, a method is described for estimating these weights. The drawback of the method is that it requires estimates of the channel data. If the estimated data are reliable, then the scheme works well. In the IS-95 standard, the data are FEC encoded to allow operation at low signal-to-noise ratios. Consequently, in the receiver, the data estimates prior to decoding can be unreliable. If the receiver uses the Pilot channel to track the channel variations, there is no need to estimate the data, since the Pilot channel is unmodulated. Since the user terminal receives a number of different downlink channels, all of which are synchronized, switching from one channel to the other is straightforward, with no need to resynchronize. Consequently, the synchronization portion of the receiver tracks only the pilot channel.

To minimize interference, many of the downlink channels are transmitted at very low power, and FEC coding is used to obtain acceptable performance. At these low powers, unreliable receiver synchronization could actually limit the channel's performance. However, the Pilot channel is transmitted at much higher power— approximately 20% of the total transmit power for all channels from a particular base station. Consequently, synchronization occurs quickly and reliably. Since the same Pilot channel is shared by all user terminals, the average increase in power per terminal required to achieve this performance is very small.

Because the user terminals are mobile and IS-95 is a cellular system, the terminals will occasionally traverse cell boundaries, whereupon they must hand over operations to a new base station. From a performance viewpoint, the handover should be transparent to the user; that is, users should be unaware that they have crossed a cell boundary. An added benefit of pilot channels is that, since they are transmitted by each base station, a terminal can monitor the pilot channels of all the surrounding base stations and select the one that is strongest. In IS-95 user terminals, such a procedure is often implemented by having a *tracking receiver* that both monitors the strongest pilot channel and provides the synchronization information necessary for demodulating the traffic and other channels. A second *search receiver* is continuously testing the pilot channels from adjacent cells and measuring their signal strength to determine whether and when a handover should occur. We will discuss handover algorithms in greater detail in Chapter 7.

5.12.3 Downlink CDMA Channels

With the motivation for having different types of channel given in Section 5.12.1, Fig. 5.29 illustrates how the downlink channels are generated, prior to combining them in the base station transmitter.

The four downlink channels are shown in the figure. In the downlink direction, all channels that are emitted by the same base station can be synchronized. Consequently, the system uses Walsh–Hadamard spreading codes of length 64 in that direction to reduce the downlink intracell multiple-access interference to zero.

The four downlink channels are configured as follows:

- The Pilot channel is an unmodulated channel that is spread by the Walsh–Hadamard code #0, as shown at the top of the figure.
- The Sync channel contains low-rate data that are convolutionally encoded and interleaved before being spread by the Walsh–Hadamard code #32.
- The Paging channel contains low-rate data that are also coded and interleaved. Prior to being spread, the paging data are randomized with a scrambler that is specific to the terminal for which the page is intended. The data are then spread by Walsh–Hadamard code #1.

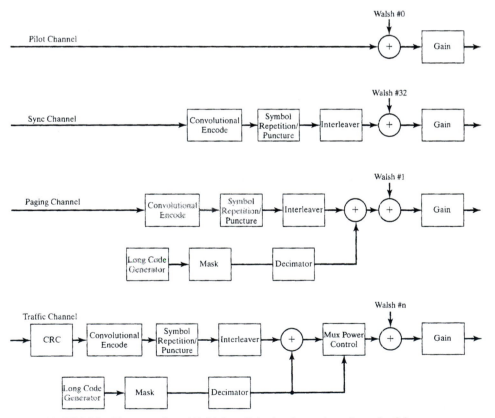

FIGURE 5.29   Bit processing of IS-95 downlink, showing various channels of the system.

- The Traffic channel is FEC encoded and interleaved and then scrambled with a sequence that is specific to the intended receiver. Also multiplexed in with the traffic channel are control bits used as feedback for adjusting the power of the mobile transmitter. The traffic channel is then spread with an assigned Walsh–Hadamard sequence that is orthogonal to all the previously mentioned Walsh–Hadamard codes.

This structure of Fig. 5.29 illustrates the application of many of the ideas that are presented earlier in the chapter. First, since all channels are generated at the same source, they can be synchronized. This allows us to use the Walsh–Hadamard codes for spreading and combining without generating any multiple-access interference. The standard defines Walsh–Hadamard functions that are of length 64; when these functions are combined with rate-1/2 FEC encoding, the overall chip rate is 1.2288 MHz. The Walsh–Hadamard functions can be derived with the use of the algorithm described in Section 5.3.1.

The base data rate of the system is 9.6 kbps; all downlink data are encoded with a rate-1/2 constraint length-9 convolutional code. This produces a base channel bit rate of 19.2 kbps. The two generators for this convolutional code are

$$g_1(x) = x^8 + x^7 + x^6 + x^5 + x^3 + x + 1$$

$$g_2(x) = x^8 + x^6 + x^5 + x^4 + 1$$

(5.95)

As the block diagram of Fig. 5.29 indicates, all data-bearing channels use FEC encoding and interleaving. This combination improves the robustness of the communication and reduces its sensitivity to multiple-access interference and fading, as described earlier.

To illustrate the processing that occurs in a CDMA transmission scheme, we will consider one of the downlink channels, referred to as a *traffic channel*. The corresponding transmitter blocks for this channel are illustrated in Fig. 5.30.

The first processing block shown in the figure is a *convolutional encode* block. The downlink channel uses a rate-1/2 constraint length-9 convolutional code to protect against transmission errors. This encoding process has the effect of doubling the rate to 19.2 kbps.

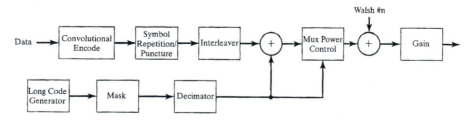

FIGURE 5.30   Processing blocks of downlink traffic channel of IS-95.

Following the encoding block is a *symbol repetition/puncture* block. The purpose of this second block is to either repeat or puncture (remove) some of the bits produced by the convolutional encoder. The motivation here is to produce a constant symbol rate for the subsequent processing stages, regardless of what the input rate is. For the baseline channel rate of 9.6 kbps, the second block has no effect. However, the systems can also support data rates of 1200, 2400, 4800, and 14,400 bps; with these different rates, there is an effect. If the data rate is 4800 bps, then each bit would be repeated twice. If the higher rate of 14.4 kbps is used, then the second block removes (punctures) one out of every three channel bits.

The output of the symbol repetition block is fed to the *interleaver*, which has a time span of 20 ms and contains 384 channel bits. It is a row–column interleaver, as described earlier, with 64 rows and six columns. The array is filled in column order, but the rows are read out in a permuted order. The 20-ms interleaver size matches the size of the voice frame generated by the vocoder in this system.

The output of the interleaver is then scrambled. The IS-95 standard defines a very long maximal-length sequence of length $2^{42} - 1 \approx 4 \times 10^{12}$ for this purpose. The generator polynomial for this sequence is

$$
\begin{aligned}
g(x) = \; & 1 + x^7 + x^9 + x^{11} + x^{15} + x^{16} + x^{17} + x^{20} + x^{21} + x^{23} + \\
& x^{24} + x^{25} + x^{26} + x^{32} + x^{35} + x^{36} + x^{37} + x^{39} + x^{40} + x^{41} + x^{42}
\end{aligned}
\tag{5.96}
$$

This scrambler is referred to as the *long code*; it is synchronized and used systemwide by all base stations and all mobile terminals. The base station applies modulo-2 operations, constituting a different *mask* to the state of this scrambler for each mobile terminal to generate the scrambler bits. The mask is determined by the mobile terminal's ID number. This can be shown to be equivalent to using a different initial state of the scrambler for each user terminal. Consequently, all communications that are intended for a specific terminal are scrambled with a sequence known only at that terminal. The scrambler does not spread the data. The long code generator is clocked at the chip rate and thus runs 64 times faster than the channel bit rate. For this reason, the output of the long code mask is decimated before being used as a scrambling sequence.

5.12.4 Power Control

The transmission format for the downlink traffic channel includes bits that are used for *closed-loop power control*. These bits control the transmit power of the user terminal. To transfer the power control information from the base station to the mobile terminal, power control bits are multiplexed with the traffic bits at a low rate. This is done after scrambling and FEC encoding, to minimize the delay in the power control loop. Power control bits are included by replacing traffic bits at a very low data rate. In IS-95 the power control bits are added at a rate of 800 Hz. Since the traffic bits are FEC encoded and the discarded traffic bits are rare, we end up with a minimal effect on the traffic channel performance.

At the user terminal, a power control bit that is a binary 1 means "decrease the power," while a 0 means "increase the power." Since these bits are not protected by

FEC encoding they can have a relatively high error rate; consequently, they are averaged over time to determine the direction in which the power should be adjusted.

5.12.5 Cellular Considerations

We have described how the different traffic channels and pilot channel can be combined to eliminate multiple-access interference in the downlink direction. The IS-95 system operates in a *cellular environment*, so the question remains, How does a mobile terminal distinguish the signals produced by one base station from those produced by another, since they are based on the same scrambling and spreading codes?

The answer to this question may be explained by Fig. 5.31, in which four downlink channels are shown being combined. Since all the channels are synchronous and use the orthogonal Walsh–Hadamard functions for spreading, they can be added directly as shown in the figure, generating no multiple-access interference. The output of the summer is no longer bipolar, as it is for the sum of a number of spread signals. The real output of the summer is then I- and Q-modulated, with the same signal transmitted on both the in-phase (I) and quadrature (Q) channels.

After the combined channels are I-and Q-modulated, the resulting sequence is further scrambled with a short *m*-sequence, referred to as the *short code*. The IS-95 standard defines two distinct short codes of length 2^{15}, one each for the in-phase and quadrature channels. The generators are, respectively, as follows:

$$g_I(x) = 1 + x^2 + x^6 + x^7 + x^8 + x^{10} + x^{15}$$
$$g_Q(x) = 1 + x^3 + x^4 + x^5 + x^9 + x^{10} + x^{11} + x^{12} + x^{15}$$

(5.97)

To obtain a sequence of length 2^{15}, an extra zero is added whenever 14 zeros occur at the output of either generator of Eq. (5.97).

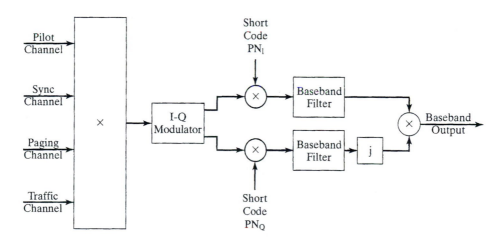

FIGURE 5.31   Downlink channel combining and short code scrambling for IS-95 standard.

The purpose of the short codes is to differentiate the signals emitted by different base stations. The period of the short code is 26.66 milliseconds. All of the base stations use the same short code, but with different timing offsets relative to a common reference. The receiver attempts to synchronize to this short code, and the strongest correlation identifies the nearest base station. There are 512 different offsets, corresponding to approximately 52 microseconds each; this offset is longer than any expected multipath delay, so channel effects are unlikely to cause synchronization to the wrong base station.

5.12.6 Uplink

The previous subsections have concentrated on the IS-95 downlink, but an equal number of important considerations apply to the uplink. In the uplink direction, the transmitters are not synchronized, and consequently, there is more multiple-access interference. The block diagram of the modulator functions for the uplink traffic channel is shown in Fig. 5.32. Most components are qualitatively similar to the downlink, but there are two important differences:

1. A rate-1/3, constraint length-9 convolutional code is employed. While this code has the same asymptotic coding gain as the rate-1/2 code used in the downlink, it has slightly greater gain over a large SNR range. The rate-1/3 code implies that the basic channel bit rate is 28.8 kbps for the uplink channel.

2. The system does not use Walsh–Hadamard sequences for orthogonal spreading. This particular use would not be practical on the uplink, since user terminals are not synchronized. The uplink uses a form of orthogonal modulation. In particular, every six bits at the interleaver output are used to select one of 64 Walsh–Hadamard orthogonal sequences for transmission. In this case, the orthogonality is employed to assist detection over fading channels, rather than to minimize multiple-access interference. The result is a 1.2288-MHz chip sequence similar to that utilized in the downlink.

The modulation and scrambling of the chip sequence are very similar to their use in the downlink, except for the use of offset-QPSK modulation. (See Section 3.3.4.) The near-constant envelope properties of offset QPSK make transmission more efficient for a power-limited mobile terminal.

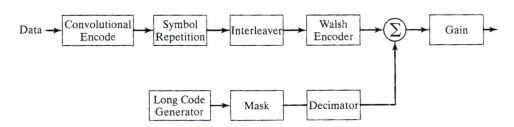

FIGURE 5.32   Block diagram of uplink traffic channel processing for IS-95.

EXAMPLE 5.8 Capacity in an IS-95 Cell

Suppose an IS-95 system is interference limited, with a ratio of total intracell-plus-intercell interference to receiver noise of 6 dB. What is the average number of users allowed per cell if the other-cell interference factor is 0.55 and the required SINR at the receiver is 8 dB?

From Eq. (5.85),

$$\text{SINR} = \frac{1}{(1+f)\left(1+\frac{N_0}{I_0}\right)\frac{K}{Q}} \tag{5.98}$$

Solving Eq. (5.98) for the number of users per cell, K, we obtain

$$K = \frac{Q}{(1+f)\left(1+\frac{N_0}{I_0}\right)\text{SINR}} \tag{5.99}$$

With IS-95, the total spreading factor $Q = 128$ including the Walsh codes and FEC coding; so, substituting this value of Q and the other given quantities into Eq. (5.99), we obtain

$$K = \frac{128}{(1+0.55)(1+0.25)6.3} \tag{5.100}$$
$$= 10.5 \text{ users per cell}$$

This result assumes continuously transmitting users. If voice activation is taken into consideration, the value of K increases by a factor of approximately 2.5 to 26 users per cell. ∎

Problem 5.20   What is the repetition period of the long code?
Ans. Approximately 41.4 days. ∎

5.13 THEME EXAMPLE 2: GPSS[12]

The *Global Positioning Satellite System* (GPSS) is an example of a system that uses DS-SS to perform ranging functions and position location. Using a handheld GPSS receiver, a person can determine his or her position to a root-mean-square (rms) accuracy of 25 meters or less. By performing differential measurements, such a receiver can also provide accurate estimates of velocity (including direction of motion). Although GPSS is primarily a navigation system rather than a communications system, it is often used in conjunction with communications terminals. In this capacity, it provides a good example of how CDMA can be used effectively.

The GPSS consists of 24 satellites operating in a low-earth orbit, each transmitting a direct-sequence signal with a center frequency of 1575 MHz. There are two modes of operation, depending on the accuracy desired and the capabilities of the device. With the *Standard Positioning Service* (SPS), the chip rate is 1.023 Megachips/s. With the *Precise Positioning Service (PPS)*, the chip rate is 10.23 Megachips/s. To obtain a position fix in three dimensions requires a minimum of four satellites to be visible.

All satellites broadcast data regarding their position and the time at a rate of 50 bps. For the SPS service, a coarse acquisition (C/A) code of 1023 chips, with a period of 1 millisecond, is used for spreading this data. Each satellite uses a different Gold code for its C/A code, but all codes are derived from the following two generators:

$$g_1(x) = 1 + x^3 + x^{10}$$

$$g_2(x) = 1 + x^2 + x^3 + x^6 + x^8 + x^9 + x^{10}$$

(5.101)

Each satellite is assigned a specific offset (initial state) of the first code relative to the second, and the offsets are known by the mobile terminals.

A GPSS receiver searches for the presence of GPSS satellite signals by correlating the received signal with the different Gold sequences. Since the period of the data (20 milliseconds) is relatively long compared with the repetition period of the code (1 millisecond), the correlator is typically run over several code periods to provide to improve the signal-to-noise ratio.

The receiver will obtain a correlation peak for all visible satellites. Since the satellite transmissions are synchronized, the time differences between the different correlation peaks are measures of the relative path differences to the different satellites. With four visible satellites, a mobile terminal can compute three differential delays between the satellites. This, together with satellite position information, allows a position to be estimated. If more than four satellites are visible, the calculations use the extra information to corroborate and improve the position estimate.

The system is designed for a minimum receiver sensitivity of −130 dBm. The system noise figure must therefore not only allow for receiver noise, but also include the multiple-access interference of up to 11 or more satellites sharing the same band.

Problem 5.21   Assume a two-dimensional flat Earth with two satellites in known, fixed positions. Let t_1 be the time taken for radio waves to travel from Satellite 1 to the terminal position. Let t_2 be the time to the other satellite, as shown in Fig. 5.33. Suppose the time difference $\Delta t_{12} = t_1 - t_2$ is known. Derive an equation describing where the terminal could be located.

Ans. Let $r_0 = (x,y,0)$, $r_1 = (x_1,y_1,z_0)$, and $r_2 = (x_2,y_2,z_0)$. Then $c(t_1 - t_2) = |r_1 - r_0| - |r_2 - r_0|$, where c is the speed of light. ∎

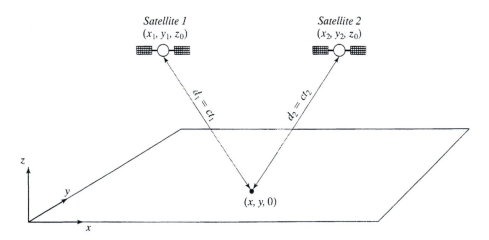

FIGURE 5.33   Illustration for Problem 5.21.

5.14 THEME EXAMPLE 3: BLUETOOTH

The Bluetooth wireless system, also known as IEEE standard 802.15.1, was developed as a wireless replacement for short-range cables linking portable consumer electronic products. Applications in which Bluetooth may be used include connecting cell phones to headsets and connecting a cell phone to a laptop computer. Bluetooth can also be adapted to eliminate the cables associated with keyboards, printers, games, etc. Its ultimate goal is to make many of these human–machine interfaces more portable and user friendly, and to avoid the inconvenience and expense of cabling.

Bluetooth is designed to operate in the band of frequencies from 2400 to 2483.5 MHz, one of the frequency bands designated for industrial, scientific, and medical (ISM) applications. Regulators have permitted unlicensed use of this band if the transmit power is kept low and the signal uses spread-spectrum techniques. To meet a low-power requirement for unlicensed operation, Bluetooth systems have a maximum transmit power of 2.5mW and are designed for a range of 10 meters or less. Consequently, Bluetooth has a much shorter range and different areas of application than other wireless standards such as the IEEE 802.11a Standard. (A special class of Bluetooth terminal transmits up to 100 mW and has a range of up to 100m.)

The frequency band for Bluetooth has a number of potential sources of interference. One is microwave ovens, the fundamental frequency of which falls almost directly in the middle of the band. To avoid interference and to minimize whatever interference is generated, unlicensed users of this band are required to use some form of spread-spectrum modulation.

To meet this spread-spectrum requirement, Bluetooth uses frequency-hopping spread-spectrum modulation with 1600 hops per second. It hops among 79 channels that are spaced 1 MHz apart. The frequency-hopping strategy means that, if strong interference with a bandwidth of 1MHz or less is present, it will affect only one, or at most two, channels. This would effectively degrade the throughput by less than 2.5%. Such an approach is particularly attractive if the interference is narrowband and potentially much stronger than the desired signal. The period of this pseudorandom hopping sequence is almost 24 hours.

The transmission rate is 1 Mbps with Gaussian filtered binary FSK with a nominal modulation index of 0.32. This technique is similar to that of GMSK modulation described in Chapter 3. With the aforementioned modulation index, there is approximately a 150-kHz separation between the mark and space frequencies (i.e., the frequencies used to represent the symbols 0 and 1). If we ignore the effects of filtering, the spread-spectrum hopping equation can be written in the complex envelope form as

$$\tilde{s}(t) = \exp\left(j2\pi h_0 \sum_{n=1}^{N} b_n g(t-nT)\right)\exp(j2\pi\zeta_k t) \qquad (5.102)$$

The first factor on the right-hand side of Eq. (5.102) is the GMSK data modulation, where $g(t)$ is the Gaussian pulse shape, as described in Section 3.7.5, and h_0 is the modulation index. The second factor is the hopping frequency ζ_k, which changes every dwell time $T_{hop} = KT$, where there are K symbols per hop. For the Bluetooth system, the hopping frequencies $\{\zeta_k\}$ are spaced by 1 MHz, and there are 79 potential values. The symbol period is one microsecond, and the hop time is 625 microseconds, corresponding to 625 symbol periods. The transmitted signal is then given by

$$s(t) = \text{Re}[\tilde{s}(t)\exp(j2\pi f_c t)] \tag{5.103}$$

where f_c is the center frequency of the transmit band—approximately 2.45 GHz.

The Bluetooth packet structure is shown in Fig. 5.34. It consists of 72 bits of access code, a 54-bit header, and 0 to 2754 bits of data payload. During one slot (hop) time, the transmission scheme can transmit slightly less than 625 bits if we allow for guard time. Longer multislot packets are handled by maintaining the same hop frequency throughout the packet; this simplifies the data detection. The transmitter (and receiver) "catches up" to the hopping pattern at the end of the multi-slot packet.

To protect against errors in transmission, forward error correction is applied to both the packet header and the payload. A rate-1/3 *repetition code* is applied to the header; that is, each bit is transmitted three times. An optional rate-2/3 *Hamming code*[13] can be applied to the payload. This (15,10) Hamming code is a block code that encodes 10 payload bits into 15 channel bits; it is capable of correcting single errors and detecting double errors in each 15-bit code block. Bluetooth also supports the option of *automatic-repeat request* (ARQ), whereby a retransmission can be requested by the receiving terminal if a packet is received incorrectly. The data throughput of Bluetooth is lower than 1 Mbps because of this overhead. Bluetooth can support an asymmetric link with up to 721 kbps (channel bits) in one direction and 57.6 kbps in the other, or a symmetric link with 432.6 kbps in both directions.

Up to eight Bluetooth-enabled devices can automatically configure themselves and form a *piconet* consisting of one master and seven slaves. The piconet is distinguished from other piconets in the vicinity by its hopping sequence.

The Bluetooth service can be used for data or voice transmission. Since Bluetooth is an RF link, it overcomes some of the limitations of competing technologies, such as infrared links, which require line-of-sight links and work only over very short distances. Bluetooth will work with devices such as printers and computers, which may be "hidden" from one another, yet are in very close proximity. The speed of Bluetooth is adequate for printing and file transfer applications, but the service is unable to support higher speed applications such as live video.

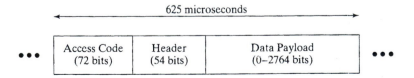

FIGURE 5.34   Bluetooth packet structure (adapted, with permission, from IEEE Standard 802.15.1).

Problem 5.22   Two Bluetooth devices, being operated in close proximity, are used to transmit a packet at the same time. What is the probability that their packets collide? When the packets collide, the ARQ mechanism causes both Bluetooth devices to retransmit almost immediately. What is the probability that there are collisions on both the first and second attempts?

Ans. Probability of collision on first attempt = 0.01266.
   Probability of collision on both first and second attempts = 1.6×10^{-4} ∎

5.15 THEME EXAMPLE 4: WCDMA[14]

Third-generation mobile wireless systems are often referred to as *universal mobile terrestrial telecommunication systems* (UMTS's). The term UMTS includes all aspects of the system, including services and applications, network planning and architecture, protocols, and the physical layer. The UMTS system intends to integrate all forms of mobile communications, including terrestrial, satellite, and indoor communications. Consequently, UMTS must support a number of different *air interfaces*—that is, physical layer implementations. One of the main air interfaces for this system is referred to as wideband CDMA (WCDMA), which is the topic of this section.

Objectives of third-generation systems include providing the user with higher data rates and seamlessly integrating data and voice services. To achieve these objectives, the data are spread over a much wider frequency range. In Table 5.4, we compare some physical layer properties of IS-95, a second-generation system, with those of the third-generation WCDMA system.

Second-generation systems were first implemented in the 800- and 900-MHz bands, but some migration to the PCS bands has occurred as the lower frequency bands have become congested. The third-generation systems will be implemented in the PCS bands from 1800 MHz to 2000 MHz, but the exact frequency ranges will depend upon the jurisdiction.

TABLE 5.4   Comparison of second- and third-generation standards.

	IS-95	WCDMA
Channel bandwidth	1.25 MHz	5 MHz
Chip rate	1.2288 MHz	3.84 MHz
Data rates	up to 9.6 kbps	up to 2 Mbps
Frame size	20ms	10 ms
Pulse shaping	48-tap FIR	22% root-raised cosine
Spreading factor	64	up to 512
Number of channels /per terminal	1	variable
Downlink/uplink sharing	FDD	FDD/TDD
Downlink modulation	QPSK/Pilot	QPSK/Pilot
Uplink modulation	OQPSK/Orthogonal	QPSK/Pilot
Downlink FEC	$r = 1/2, K = 9$ convol. code	$r = 1/2, 1/3$ convol. or turbo
Uplink FEC	$r = 1/3, K = 9$ convol. code	$r = 1/2, 1/3$ convol. or turbo

At a high level, the implementation of the physical layer is somewhat similar to its implementation in the second- and third-generation standards. In what follows, we will concentrate on the significant differences between these two systems, as indicated in Table 5.4.

5.15.1 Bandwidth and Chip Rate

WCDMA increases the chip rate by more than a factor of three, and the channel bandwidth by a factor of four, compared with IS-95. The faster chip rate implies that the WCDMA receiver can provide greater multipath resolution, and with a RAKE receiver, this implies greater frequency diversity. Consequently, performance is expected to be more robust in wireless channels.

5.15.2 Data Rates and Spreading Factor

The second-generation system has a standard data rate that is spread by a fixed spreading code of length 64; lower data rates are provided by repeating the bits. WCDMA provides a whole range of data rates through its ability to adjust the spreading factor. On the downlink, the spreading factor can range from 4 up to 256. Different users are combined on the downlink via orthogonal variable spreading factor (OVSF) codes, as described in Section 5.3. This combination provides multiples of the basic data rate while minimizing the intracell interference on the downlink, thereby permitting bit rates up to 384 kbps on circuit-switched connections and up to 2 Mbps on packet-switched networks.

On the uplink, the spreading factor can be as high as 512. The higher spreading rate permits a lower, but more robust, data rate. Individual user channels are not orthogonal on the uplink, because they are asynchronous. Each terminal is capable of transmitting multiple channels on the uplink; the channels are kept orthogonal using the OVSF codes.

5.15.3 Modulation and Synchronization

Recall that the IS-95 system uses a pilot channel in the downlink direction to provide synchronization, channel tracking, and handover functions. In the uplink direction, orthogonal modulation is employed, which permits the more robust noncoherent demodulation to be used. WCDMA employs coherent detection on both uplink and downlink; it is able to do this because a pilot channel is included in both directions. In the downlink direction, each base station transmits a pilot channel, as well as multiplexing pilot symbols with the data channels. In the uplink, each terminal multiplexes pilot symbols with its data.

5.15.4 Forward Error-Correction Codes

Both IS-95 and WCDMA use convolutional codes for performing forward error-correction (FEC) coding. There are however, three differences:

1. The IS-95 standard provides a small number of data rates that are implemented by repeating the data symbols or puncturing the code bits by means of a simple pattern.

2. The WCDMA standard allows for a wide variety of data rates by allowing variable puncture patterns based on 1/2-rate and 1/3-rate convolutional codes of constraint length 9.

3. The WCDMA standard includes the option for applying the recently developed and more powerful turbo codes for forward error correction. The eight-state rate-1/3 version of the turbo codes is employed in this system; turbo codes are discussed in Section 4.12.

The WCDMA physical layer is required to support variable bit rate transport channels, to offer bandwidth-on-demand services, and to be able to multiplex several services to one connection. With these multiple data streams, forward error correction is independently applied to each data stream, not to the aggregate. The reasoning behind this differentiation is that the various data streams may require different qualities of service. For example, if a voice channel and a video channel are transmitted simultaneously, the voice channel can generally tolerate a higher error rate than the video channel; consequently, it would use a less powerful (higher rate) FEC code.

WCDMA allows for two layers of interleaving, the first over 10 milliseconds and the second, when delay constraints allow, over 20, 40, or 80 milliseconds. These two interleaving layers are present on both the uplink and the downlink.

5.15.5 Channel Types

Second-generation standards have a number of different types of channels—for example, paging and traffic channels—but there is a direct correspondence between the physical channel and its logical function. A second-generation mobile terminal is capable of transmitting only one channel at a time. The same is usually true on the receiving side if the pilot channel is excluded.

WCDMA uses the concept of *transport channels*. Typically, different transport channels either have different logical functions or service different applications. Several of these transport channels may be multiplexed on a single transmission. The different transport channels may have different data rates, use different coding, etc., and any or all may change on a frame-by-frame basis. To manage this association of transport channels and their properties with a physical channel, each frame has a transport format combination indicator (TFCI), which is transmitted by the physical control channel to inform the receiver about which transport channels are active for the current frame.

There are two types of transport channels: dedicated channels and common channels. A common channel—used for functions such as broadcasting, paging, and random access—is shared among all users or a group of users in a cell. A dedicated channel is for only one user.

5.15.6 Uplink

On the uplink, physical layer control information is carried by the dedicated physical control channel (DPCCH), with a fixed spreading factor of 256. The higher layer information, including user data, is carried on dedicated physical data channels (DPDCH's) possibly with varying spread factors.

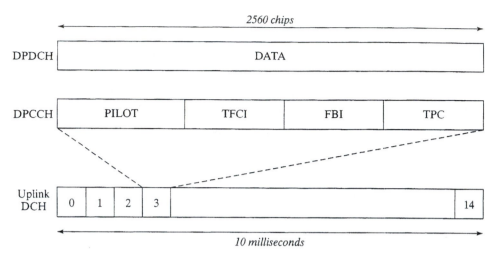

FIGURE 5.35 Structure of the uplink dedicated channels (adapted with permission of ETSI).

An uplink transmission consists of one DPCCH and one or more DPDCH channels, as illustrated in Fig. 5.35. The combination of one physical control channel and one or more physical data channels is called a coded composite transport channel (CCTRCh). The structure of the DPCCH divides each 10-ms frame into 15 slots is repeated every slot. In each slot, the DPCCH transmits

  (i) four pilot bits that are used for synchronization and tracking,

 (ii) TFCI (Transport Format Combination Indicator) bits, which define the format of the DPDCH channels,

(iii) TPC (Transmission Power Control) bits, which carry the power control function for the downlink, and

(iv) FBI (Feedback Information) bits, which are employed when closed-loop transmission diversity is used on the downlink.

The slot structure permits the fast response time required for some control functions. All DPDCH data channels are sent at the same power level, which means that demands for different quality of service have to be addressed by the coding and information rate.

5.15.7 Downlink

The downlink also has both dedicated and common components, but in this case, the two types of channel are multiplexed as shown in Fig. 5.36, with user information alternating with control information. Pilot symbols are included here, as well as on a separate pilot channel. These pilots bits have a specific pattern, differing in each slot, but repeating every frame, and can be used for frame synchronization.

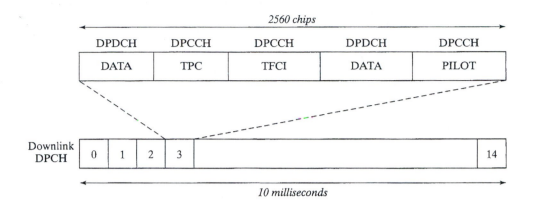

5.15.8 Multicode Transmission

The simultaneous transmission of two or more CDMA channels by the same terminal is referred to as *multicode* transmission. This form of transmission tends to increase the *peak-to-average ratio* (PAR) of the transmitted waveform, thereby affecting the efficiency power amplifier of the user terminal. Power amplifier efficiency is a serious issue for the mobile terminal, for which size and battery weight are critical. If only two channels are transmitted—the I channel and the Q channel—then, with the use of a special complex scrambling code, the peak-to-average ratio of the WCDMA signal is kept small, even if the two channels are transmitted at different power levels.

When more than two channels are transmitted, the PAR becomes even larger. The PAR is more of a concern in the uplink direction; in the downlink direction, the transmitter does not usually rely on battery power.

5.15.9 Cellular Considerations

Several features of WCDMA are expected to offer performance improvements over second-generation systems in a cellular environment:

1. *With downlink power control,* WCDMA can minimize the power radiated by the base station. Less radiated power means less interference power present in surrounding cells.

2. The synchronization strategy of WCDMA allows for *asynchronous base stations.* In IS-95, all base stations used the same scrambling code (Short Code) to distinguish among their transmissions, but with different timing offsets. The latter were fixed, requiring that the base stations be synchronized by using a GPSS receiver at each station. In WCDMA, a specially formatted synchronization channel has the information embedded in it that determines the timing offset of the local base station.

3. This third-generation system also has several options in its scrambling and spreading strategy that will simplify the use of techniques such as *transmit diversity* and *multiuser detection*.

Problem 5.23 One advantage of higher spread bandwidths is their ability to handle higher information rates. Are there any other advantages? Are there any disadvantages? ∎

5.16 THEME EXAMPLE 5: WI-FI

The abbreviation *Wi-Fi* stands for wireless fidelity and refers to wireless local area network technology for home, office, and transient users. For example, Wi-Fi base stations are being set up in locations such as airports, hotels, coffee shops, and other public areas to connect transient users. A Wi-Fi network can be used to connect computers to each other, to the Internet, and to wired Ethernet networks.

Wi-Fi networks comprise the radio technologies associated with IEEE Standards 802.11a, 802.11b, and 802.11g to provide secure, reliable, fast wireless connections. Equipment based on the IEEE 802.11a standard operates in the unlicensed 5-GHz radio band and can provide data rates up to 54 Mbps; this standard was discussed in Section 2.11. Equipment based on the IEEE 802.11b standard operates in the unlicensed 2.4-GHz radio bands and can provide data rates up to 11 Mbps. The more recent IEEE 802.11g standard provides up to 54 Mbps in the 2.4-GHz band. The objective of these standards is to furnish a service similar to the basic wired Ethernet networks available in many offices.

In this section, we will discuss the IEEE 802.11b component of Wi-Fi. This standard, which applies to operation in the unlicensed 2.4-GHz band, requires a minimum 10-dB processing gain by regulation. The minimum processing gain forces some spreading of the transmitted signal to reduce interference. There are, in fact, two spreading options provided in the standard for spread spectrum operation at 2.4 GHz: a direct-sequence approach and a frequency-hopped approach. We will describe the direct-sequence approach here.

Basic service with the 802.11b gives a data throughput rate of either 1 Mbps or 2 Mbps, depending upon whether BPSK or QPSK modulation is used. The packet structure is illustrated in Fig. 5.37.

The fields in this packet structure are defined as follows:

- *Sync.* This is part of the preamble and consists of 128 bits used for bit, frequency, and code synchronization. Sync is a field of all ones that has been scrambled.
- *SFD, or start-of-frame delimiter.* The rest of the preamble, this is a 16-bit field used for frame synchronization.

128 bits	16 bits	8 bits	8 bits	16 bits	8 bits	3 – 8191 bits
SYNC	SFD	SIGNAL	SERVICE	LENGTH	CRC	DATA PAYLOAD

FIGURE 5.37 Packet structure for IEEE 801.11b standard (adapted with permission, from IEEE).

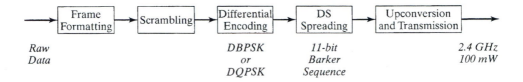

FIGURE 5.38   Data processing for IEEE 802.11b (DS).

- *Signal.* This 8-bit field defines the data rate; the integer value of the field represents the bit rate divided by 100 kbps.
- *Service.* This 8-bit field defines what transmission options are used. At present, most of these bits are reserved for future use, but the nonreserved ones indicate which of various options are implemented.
- *Length.* This 16-bit field describes the length of the data field in microseconds. For 1 Mbps, length maps directly to the number of bits. The maximum length is $2^{13} - 1$.
- *CRC.* This is the cyclic redundancy check, applied to all of the header information. The CRC is not applied to the preamble fields.

The preamble and header fields are transmitted at 1 Mbps with BPSK modulation. The modulation of the payload field, either BPSK or QPSK for the basic service, is determined from the data rate defined in the Signal field. The transceiver operates in a half-duplex mode, either transmitting or receiving, but not both simultaneously.

In North America and Europe, this service operates in the band from 2.4 to 2.4835 GHz, with seven channels. The channel center frequencies are given by the expression

$$f_c(\text{MHz}) = 2407 + 5i \qquad (5.104)$$

where the channel number i ranges from 1 to 7. The data processing for this service is depicted in Fig. 5.38. The data are first formatted into frames in a process that includes the calculation of the frame header and the addition of the preamble bits. The resulting bits are then scrambled. There is no FEC encoding of the data at the base rates of 1 and 2 Mbps. The scrambled bits are differentially encoded then spread with an 11-bit sequence, with a chipping rate of 11 MHz.

Wi-Fi service has several interesting variants, compared with what has been discussed previously in this chapter:

- *Barker sequence.* The spreading sequence is an 11-bit Barker sequence given by $\{+1,-1,+1,+1,-1,+1,+1,+1,-1,-1,-1\}$. Barker sequences[15] of various lengths that have good autocorrelation properties.
- *Self-synchronizing scrambler.* This service uses an unusual scrambling technique, quite unlike the approach described in Section 5.3.4. As illustrated in Fig. 5.39, with this technique preceding bits are used to scramble subsequent bits. The scrambler can be started in any arbitrary state except the all-ones state. This

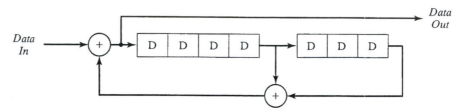

FIGURE 5.39   Data-scrambling technique for IEEE 802.11b (DS) (adapted, with permission, from IEEE Standard 802.11b).

scrambler has the advantage that it is self-synchronizing. That is, by feeding the received bits directly into the scrambler, the initial state is automatically recovered if there are no bit errors.

The standard allows transmission of up to 100 milliwatts of power in North America and Europe. Power levels up to 1 watt are permitted in the United States if power control is implemented. The transmit spectral mask for this service is shown in Fig. 5.40; the spectral mask defines the maximum amount of power that a service can transmit at a given frequency that is offset from the center frequency. The mask indicates that, at frequencies between 11 and 22 MHz from the carrier frequency, the transmitted power must be 30 dB below the power of the carrier frequency. This restriction places limits on the amount of interference that a signal can cause in the adjacent channel, Also shown in the figure is the $\text{sinc}^2 x$ power spectrum that one expects with rectangular chips. Clearly, rectangular chips will not meet the transmit mask requirements, and some filtering of the transmit signal is required.

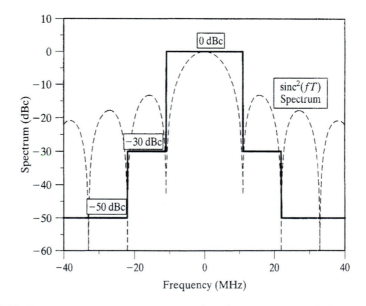

FIGURE 5.40   Spectral mask for IEEE 802.11b (DS) (adapted with permission of IEEE).

This theme example has shown some similarities and some differences relative to the direct sequencing techniques discussed in previous sections of the chapter:

- Barker codes were introduced as an alternative spreading sequence.
- A differential modulation and detection strategy has been proposed.

Problem 5.24   What is the major disadvantage of the self-synchronizing scrambler described in Section 5.16 when used with unreliable wireless channels?   ∎

Problem 5.25   Compute the autocorrelation properties of the four-times-oversampled Barker code described in Section 5.16. The sequence $\{+++++--++-+-+\}$ is a Barker code of length 13. What are the cross-correlation properties of the two sequences?   ∎

5.17 SUMMARY AND DISCUSSION

In this chapter, we have looked at spread-spectrum modulation and its application to code-division multiple-access strategies.

1. We have presented the two major forms of spread-spectrum modulation:
   - Direct-sequence spread-spectrum modulation, in which each symbol of the data sequence is modulated by a known spreading sequece.
   - Frequency-hopped spread-spectrum modulation, wherein the center frequency of the signal is changed (hopped) at fixed intervals.

   The spreading sequence and the hopping pattern are both controlled by a pseudorandom code sequence.

2. We have shown that, in the absence of other impairments, *spread-spectrum performance is identical to that of narrowband modulation techniques in AWGN*.

3. When other channel impairments, such as intentional interference, multipath, and multiple-access interference, are present, we observe the following:
   - The effect of these impairments is reduced by the processing gain of the DS-SS system relative to the narrowband case.
   - A similar processing gain benefit occurs in FH-SS. However, if the interference is narrowband, then FH systems can tolerate much higher levels.

4. Good spreading code design is essential to maximizing the processing gain of a DS-SS system. The key to good multiple-access performance is low cross-correlations between spreading codes. A number of spreading codes were considered:
   - Walsh–Hadamard codes provide ideal correlation performance when users are synchronous.
   - Maximal-length sequences have very good correlation properties, but are limited in number.
   - Gold codes are formed from pairs of maximal-length sequences, have good correlation properties, and are more numerous than maximal-length codes.

- Random codes also have close-to-ideal performance, but must be known at the receiver. They are useful for modeling the performance that may be obtained with very long pseudorandom codes.

5. With DS-SS, there are opportunities and challenges:
   - Forward error correction coding enhances the performance of a DS-SS system, always improving the processing gain.
   - Code synchronization to a pseudorandom spreading sequence can be a challenging problem.
   - When used as a multiple-access strategy, DS systems are subject to the near–far problem and, consequently, require the use of power control.
   - In a frequency-selective environment, a DS-SS employing *a RAKE receiver could use all of the energy* provided by the channel.

6. A frequency-hopped SS system has the following properties:
   - It generally requires a noncoherent modulation technique between hops; consequently, the performance is not as good as that of DS.
   - It can easily avoid known interference.

As is made plain from the conclusion of this chapter and the many theme examples, spread-spectrum modulation has a wide range of applications in wireless systems.

NOTES AND REFERENCES

[1] Spread-spectrum techniques began to make their appearance just before and during World War II. There were two applications in particular: encryption and ranging. Initial spread-spectrum encryption techniques were often analog in nature, with a noiselike signal multiplying a voice signal. The noise signal or the information characterizing it was often sent on a separate channel to allow decryption at the receiver. The second application used spread-spectrum techniques to enhance the ranging capabilities of radars and reduce the effect of chaff during bombing missions. After World War II, spread-spectrum investigations were motivated by the potential to provide highly jam-resistant communication systems. Scholtz (1982) provides some history on the origins of spread spectrum communications.

[2] Petersen, Ziemer, and Borth (1995) is a good general introduction to spread-spectrum systems. Dixon (1994) describes both direct-sequence and FH systems and has an interesting chapter on the trade-offs that are involved in using spread spectrum for commercial applications.

[3] Lee and Miller (1998) offer a thorough explanation of many aspects of the IS-95 system and, in the process, gives an in-depth discussion of CDMA design principles. Chapter 6 of the Lee and Miller book provides a detailed investigation of the properties of pseudorandom sequences.

[4] The original Gold codes were presented in Gold (1968). Chapter 3 of Petersen, Ziemer, and Borth (1995) explains these sequences in greater detail. Gold codes are also discussed in Holmes (1990).

[5] Before the advent of satellite systems, HF radio was one of the few methods of communicating long distance without wires, and it tended to be unreliable. Thus, there was a need for diversity to improve its performance. This was the motivation for investigations into diversity systems for

fading channels, such as those described in Chapter 7 of Proakis (1989) and Chapters 10 and 11 of Schwartz, Bennett, and Stein (1995). The recent development of personal wireless communications has rekindled this interest in diversity for similar reasons. Diversity is discussed in greater detail in Chapter 6 of the current text.

[6] Viterbi (1995) covers many of the topics of this chapter at a more advanced level, with some emphasis on the IS-95 standard. Of particular interest is Chapter 3, on synchronization. More advanced material on synchronization techniques for DS systems can be found in Holmes (1990) and Dixon (1994).

[7] One of the seminal papers on multiuser detection was written by Verdu (1986). The paper spawned a large research effort, and Verdu (1998) is devoted to explaining many of the basic aspects of multiuser detection.

[8] By combining multiuser detection and forward error correction in an iterative turbo-like fashion, some amazing capacity results can be achieved; see Moher (1998) for more information on this technique.

[9] In Viterbi (1995), the effect of propagation factors on other-cell interference and on various types of handoff between cells, as well as their effect, in turn, on performance is described in greater detail.

[10] The three-volume set by Simon, Omura, Scholtz, and Levitt (1985) covers all aspects of spread-spectrum communications, and frequency-hopped systems are covered in Volume 1. Much of this material is repeated in the single-volume work by Simon, Omura, Scholtz, and Levitt (1994).

[11] We have mentioned the books by Viterbi (1995) and Lee and Miller (1998) that cover various aspects of the IS-95 standard. Complete details of the IS-95 standard may be found in TIA (1999).

[12] Kaplan (1996) discusses numerous aspects of the GPSS, many of which are relevant to direct-sequence systems in general.

[13] The generator polynomial for the Hamming code is

$$g(D) = (D + 1) \times (D^4 + D + 1)$$

As described in Section 4.5, if the incoming 10 information bits are arranged as the coefficients of a ninth-order polynomial in D denoted by $m(D)$, then the 15 encoded bits to be transmitted are the binary coefficients of the product polynomial $g(D)\, m(D)$.

[14] The WCDMA standards can be found on the website *www.etsi.org*. For a summary of these standards, see Holma and Toskala (2000). Chapter 6 of the Holma and Toskala book provides a more detailed view of the physical layer of this system.

[15] The original Barker codes were presented in Barker (1953). These codes are discussed and compared with other short sequences with good autocorrelation properties in Chapter 14 of Spilker (1977). Chapter 2 of Stüber (2001) also compares Barker codes, Gold sequences, and several types of sequences not mentioned here.

ADDITIONAL PROBLEMS

Processing Gain

Problem 5.26   Suppose a jammer causes wideband interference in a direct-sequence signal. The interference is noiselike with a bandwidth $2R_c$ and a density J_0 for $|f| < R_c$. What is the processing gain?

Problem 5.27   Write a program to simulate an uncoded spread-spectrum system using a rectangular chip shape with a spreading factor of 31. Include an additive white-noise channel. Plot the performance of the system and compare it against the theoretical performance. Now assume that A/D sampling following the receiver chip-matched filter has only 1-bit resolution. Implement this feature and plot the performance. What is the degradation relative to ideal? If the spreading factor is increased to 127, what is the degradation?

Spreading Codes

Problem 5.28 Compute the autocorrelation of a sequence generated from a memory-7 shift register with several randomly selected sets of feedback taps. How do the autocorrelation results compare with m-sequence correlations.

Problem 5.29 The length of an m-sequence is $2^m - 1$, which is often a prime number. Powers of two frequently simplify computations, compared with prime numbers. Suppose a zero is added to the m-sequence to include the subsequence of m zeros; this lengthens the sequence to a power of 2 and also balances the number of zeros and ones. What are the autocorrelation properties of the resulting sequence? How does the new sequence compare with the original m-sequence and with a Gold sequence of the same length? Repeat the problem for $m = 7$ and $m = 9$.

Problem 5.30 Given the defining equations for the Walsh–Hadamard matrices H_1 and H_n, prove that the rows of H_{n+1} are orthogonal. (*Hint.* Use proof by induction.)

Problem 5.31 If, from a set of Walsh–Hadamard codes of length 128, two with a spreading factor of 8 are chosen, how many codes with a spreading factor of 16 remain?

Problem 5.32 Consider the set of Walsh–Hadamard codes of length 2^{n_3}. If one needs one code of length 2^{n_1} and three codes of length 2^{n_2}, where $n_1 < n_2$, how many codes of length 2^{n_3} are available?

Problem 5.33 Suppose a receiver may need any one of the 2^n Walsh–Hadamard codes of length 2^n at any given time, but does not need to store them all in memory. Is there a more direct way of computing a particular sequence of length 2^n without going through the n steps of the algorithm described in Section 5.3.1?

Problem 5.34 What is the expected autocorrelation of a random selected sequence of 1's and -1's of length N. What is the covariance on this autocorrelation at each time offset?

Problem 5.35 Since the period of a maximal-length shift register is $2^m - 1$, it must not include one of the 2^m possible states. Which state is not included? Why?

Problem 5.36 Write a program that generates two m-sequences of length $2^8 - 1$. Confirm that the sequences have the autocorrelation properties of m-sequences. Plot their cross-correlations when four times oversampled. What can be said about the sequences?

Problem 5.37 Compare the cross-correlation values of m-sequences and Gold codes for randomly chosen codes, and comment on the differences. What happens as $Q = 2^m - 1$ becomes large?

Problem 5.38 Suppose a system uses OVSF to combine users with different data rates with spreading factors Q_1 and Q_2, where $Q_1 < Q_2$. Compare the effect of multipath and multiple-access noise on the two users.

Multiple-Access Interference

Problem 5.39 Suppose that, in a consideration of multiple-access interference, users are power controlled, but are not phase coherent; that is, each user's bits b_k have a random phase rotation $e^{j\theta_k}$, relative to the other user bits. How will this affect the multiple-access noise? Assume that the phase rotation is constant over the period of the spreading code.

Problem 5.40 One suggested application of CDMA techniques is for overlaying a new service on top of an existing service. Suppose a point-to-point microwave link at 4 GHz currently exists. (This link may be referred to as the primary user.) This microwave link has a 1-MHz bandwidth and 40-dB margin to account for rain fades. A second service provider wants to reuse the same spectrum to distribute four 10-kHz audio channels with a total baseband bandwidth of 40 kHz (secondary user). Assume that the required transmit power of the second service is approximately $10 \log_{10}(40 \text{ kHz}/1 \text{ MHz})$ lower than the first. If the audio channels are spread over the full bandwidth, what would be the degradation of the primary service under clear-sky conditions? What interference would the primary service cause in the secondary service. That is, what would the signal-to-interference ratio (SIR) of the secondary user be? Should the primary user accept the application of the secondary user? Under what conditions, if any?

Multipath

Problem 5.41 Suppose a multipath channel can be characterized by the impulse response

$$\tilde{h}(t, \tau) = \alpha_0 \delta(\tau) + \alpha_1 \delta(\tau - T_c) + \alpha_2 \delta(\tau - 2T_c) + \alpha_3 \delta(\tau - 3T_c) \tag{5.105}$$

Develop an expression for E_b/M_0 in a DS system, where M_0 is the multipath noise density and E_b corresponds to the energy per bit of the shortest path. What is the corresponding expression for the second finger of a RAKE receiver?

Problem 5.42 Suppose a DS system has both multipath and multiple-access noise. Develop an expression for the total degradation for a non-RAKE receiver. Plot for the same cases as depicted in Fig. 5.15.

Problem 5.43 In many diversity scenarios, the received signals are not of equal strength, and the analysis of performance is more difficult. To simplify analysis, some authors propose an entropic measure of the channel by defining the diversity as

$$D = -\sum_{l=0}^{L} \frac{|h_l|^2_{avg}}{h_T} \log \frac{|h_l|^2_{avg}}{h_T} \tag{5.106}$$

where $|h_l|^2$ is the average power of the lth path and

$$h_T = \sum_{l=0}^{L} |h_l|^2_{avg} \tag{5.107}$$

Show that this definition of diversity reduces to the known case when the powers are equal. Discuss the merits of this approach for other cases by considering a couple of simple examples.

Problem 5.44   Suppose that a multipath channel has an exponential power-delay profile; that is, $|h_l|^2 = |h_0|^2 a^l$, where $0 < a < 1$. Develop an expression for the entropic diversity of this channel when L is very large. Develop a program that simulates the performance of coherently detected BPSK under these channel conditions. Compare the simulated BER performance with that predicted from the diversity estimate. What is the prediction error? Where is the model accurate?

Code Synchronization

Problem 5.45   Practical implementations will separate the real and imaginary processing in the code synchronization circuit. Redraw Fig. 5.20, showing the separate processing of the real and imaginary parts. How would the circuit differ if the modulation were QPSK and the code sequence were different in the I and Q channels?

Problem 5.46   Define the partial correlation of a sequence with itself to be

$$C_x(m_0) = \frac{1}{N_1} \sum_{m=1}^{N_1} c(m)c(m_0 + m) \tag{5.108}$$

Write a program that generates an m-sequence of length $2^7 - 1$, and compute the partial correlations of this sequence with itself. Let the partial-correlation lengths N_1 equal 8, 16, 32, and 64. Let the offsets m_0 be 10, 20, and 40. Compare the results with the full autocorrelation of the m-sequence.

Problem 5.47   Repeat the previous problem for a Gold sequence of length $2^7 - 1$. Make a statement comparing the partial-correlation performance of m-sequences and Gold codes.

Problem 5.48   The derivation of the code synchronization circuit made a number of simplifying assumptions. Identify those assumptions which are likely to affect performance in practice, and make qualitative statements about their effect on performance.

Forward Error - Correction FEC Coding

Problem 5.49   Describe how the presence of rate-1/4 FEC coding would affect the implementation and performance of a RAKE receiver.

Problem 5.50   A frequency-hopped system with 80 different hop frequencies is to be shared among three different users. The user transmissions are not coordinated so collisions between users can be assumed to occur at random. The system is a fast-frequency-hopped system with one binary symbol transmitted per hop. For a given user, determine the probability of collision and, from this, the probability of a bit error, assuming that all errors are caused by collisions. Suppose users apply a (15,11) Hamming code to their transmissions. This code can correct up to two errors in every 15-bit block. What is the probability of bit error with the Hamming code?

Cellular Systems

Problem 5.51 Suppose that, in a highly urban area, the average "other cell" interference factor $f = 0.77$. Compare the cellular spectral efficiency in this case with that in a rural area where the "other cell" interface factor $f = 1.6$. How would the difference affect the system design?

Problem 5.52 In the previous problem, suppose the capacity was enhanced by installing sectored antennas. How would this approach affect the calculations? Consider how it would affect both the intracell and intercell interference.

Problem 5.53 Explain why fast Rayleigh fading is not considered in intercell interference. What role does it play?

Problem 5.54 In Section 2.8.5, we saw that FDMA frequency reuse was limited by carrier-to-interference (C/I) requirements and propagation characteristics. In Section 5.10, the cell loading of a CDMA system depends on the intracell and intercell interference, the latter of which also depends upon propagation characteristics. For the FDMA system, develop an expression for the spectral efficiency $v = 2/N$, where N is the frequency reuse factor. Compare your result with the CDMA expression for v developed in Problem 5.19. Assume that the other-cell interference factor f varies from 1.6 to 0.5 with changes in the path loss exponent from 3 to 5. For which technique, FDMA or CDMA, is spectral efficiency more sensitive to propagation?

Theme Examples

Problem 5.55 What is the Walsh–Hadamard sequence $H_{64}(1)$? What does your answer imply about the modulation of the pilot channel at the baseband output of Fig. 5.39? How would this be useful for synchronizing the receiver?

Problem 5.56 In the IS-95 uplink, instead of using complete on/off voice activation, the transmitter user simply lowers the data rate and transmit power level, so that the receiver can maintain lock. Comment on how this approach will affect both multiple-access interference and the tracking performance requirements of the receiver.

Problem 5.57 Since all IS-95 base stations use the same short code, this simplifies the acquisition procedure for the mobile terminal. To ensure that there is no ambiguity among base stations, their short codes are offset in time. If 40 km is the maximum separation between base stations, what is the minimum time offset between the short codes of the two base stations that will prevent them from aligning in the coverage area. How many distinct timing offsets of this size are there?

Problem 5.58 The short PN code of IS-95 is of length $2^{15} - 1$ chips with a chip rate of 1.2288 MHz. Show a block diagram of a circuit that could be used to recover the short-sequence timing. Suppose the receiver oscillator has an accuracy of 10 ppm. How would this affect your design?

Problem 5.59 Compare the capacity of the IS-95 system of Example 5.6 with that of an FDMA system that has 128 channels, but also has a 1-in-19 cellular reuse pattern. What if the reuse pattern is 1-in-12? 1-in-7? Why is voice activation not a factor in capacity calculations for an FDMA system?

Problem 5.60 In IS-95 receivers, the pilot channel is typically used for channel estimation. With reference to Section 5.6, explain how this might be done. Discuss the advantages it might have over the method of Section 5.6.

Problem 5.61 Compare the advantages and disadvantages of IS-95 and GSM for cellular telephone applications.

Problem 5.62 Suppose a third satellite were present in Problem 5.21 and Δt_{13} and Δt_{23} were known, as well as the position of the satellite. What can now be said about the position of the terminal?

Problem 5.63 In a large open office that is 30 m by 60 m, a wireless access node is located in the center, as depicted in Fig. 5.41. A user terminal is located along one wall. What is the longest multipath delay relative to the direct path? Assume that only multipath with a single reflection is significant. If a path loss exponent of 3.1 applies to both the direct path and the multipath, what is the attenuation of this longest-delay multipath relative to that of the direct path?

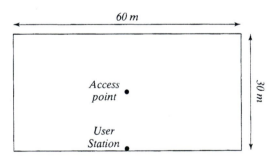

FIGURE 5.41 Diagram for Problem 5.63.

C H A P T E R 6

Diversity, Capacity and Space-Division Multiple Access

6.1 INTRODUCTION

Up to now, we have emphasized the multipath fading phenomenon as an inherent characteristic of the wireless channel. Given this physical reality, how do we make the communication process across the wireless channel into a *reliable* operation? The answer to this fundamental question lies in the use of *diversity*, which may be viewed as a form of redundancy. In particular, if several replicas of the information-bearing signal can be transmitted simultaneously over independently fading channels, then there is a good likelihood that at least one of the received signals will not be severely degraded by channel fading. There are several methods for making such a provision. In the context of the material covered in this book, we may identify three approaches to diversity:

1. Frequency diversity
2. Time (signal-repetition) diversity
3. Space diversity

In *frequency diversity*, the information-bearing signal is transmitted by means of several carriers that are spaced sufficiently apart from each other to provide independently fading versions of the signal. This may be accomplished by choosing a frequency spacing equal to or larger than the coherence bandwidth of the channel. The frequency-hopping form of spread-spectrum modulation, discussed in Chapter 5, is an example of frequency diversity.

In *time diversity*, the same information-bearing signal is transmitted in different time slots, with the interval between successive time slots being equal to or greater than the coherence time of the channel. If the interval is less than the coherence time of the channel, we can still get some diversity, but at the expense of performance. In any event, time diversity may be likened to the use of a repetition code for error-control coding. In a more general setting, we may view channel coding with interleaving, discussed in Chapter 4, as a form of time diversity.

In *space diversity*, multiple transmit or receive antennas, or both, are used, with the spacing between adjacent antennas being chosen so as to ensure the independence of possible fading events occurring in the channel. In practice, however, we find that antenna spacings which result in correlations as high as 0.7 may incur a performance loss of at most half a decibel, compared with the ideal case of independent channels. Of the three kinds of diversity, space diversity is the subject of interest in this chapter. Depending on which end of the wireless link is equipped with multiple antennas, we may identify three different forms of space diversity:

1. *Receive diversity*, which involves the use of a single transmit antenna and multiple receive antennas.
2. *Transmit diversity*, which involves the use of multiple transmit antennas and a single receive antenna.
3. *Diversity on both transmit and receive*, which combines the use of multiple antennas at both the transmitter and receiver. Clearly, this third form of space diversity includes transmit diversity and receive diversity as special cases.

In the literature, a wireless channel using multiple antennas at both ends is commonly referred to as a *multiple-input, multiple-output (MIMO) channel*. Technology built around MIMO channels resolves the fundamental issue of having to deal with two practical realities of wireless communications:

- a user terminal of limited battery power, and
- a channel of limited RF bandwidth.

Given fixed values of transmit power and channel bandwidth, this new technology offers a sophisticated approach to exchanging increased system complexity for *boosting* the channel capacity (i.e., the spectral efficiency of the channel, measured in bits per second per hertz) up to a value significantly higher than that attainable by any known method based on a single-input, single-output channel. More specifically, when the wireless communication environment is endowed with rich Rayleigh scattering, the MIMO channel capacity is roughly proportional to the number of transmit or receive antennas, whichever is smaller. That is to say, we have a spectacular increase in spectral efficiency, with the channel capacity being roughly doubled by doubling the number of antennas at both ends of the link.

Another approach to increasing the spectral efficiency of wireless communications is to use highly *directional antennas*, whereby user terminals are separated in space by virtue of their angular directions. This approach is the basis of *space-division multiple access* (SDMA), discussed in the latter part of the chapter.

The chapter is organized as follows: Section 6.2 discusses the notion of space diversity on receive, using four techniques for its implementation, namely, selection combining, maximal-ratio combining, equal-gain combining, and square-law combining. Section 6.3 describes a mathematical model of MIMO wireless communications. This discussion is followed by Section 6.4 on the channel capacity of MIMO systems, assuming that the receiver "knows" the state of the channel. Section 6.5 presents another viewpoint of the input–output relation of the MIMO channel by applying a transformation known as singular-value decomposition to the channel matrix; the resulting decomposition is insightful in the context of fading correlation. Section 6.6 discusses space–time block codes for the joint coding of multiple transmit antennas in MIMO wireless communications. Section 6.7 examines differential space-time block codes, used to simplify the receiver design; such an approach eliminates the need for the receiver to "know" the state of the channel.

Space-division multiple access and the related use of smart antennas are discussed in Section 6.8. The next three sections, 6.9 through 6.11, present three theme examples respectively dealing with (1) a type of coherent MIMO wireless communication system popularized as a BLAST architecture, (2) the practical merits of different antenna diversity techniques and the spectral efficiency of space–time block codes and BLAST systems, and (3) keyhole channels that arise when the channel matrix of a MIMO wireless link is rank deficient.

6.2   "SPACE DIVERSITY ON RECEIVE" TECHNIQUES

In "space diversity on receive," multiple receiving antennas are used, with the spacing between adjacent antennas chosen so that their respective outputs are essentially independent of each other. This requirement may be satisfied by spacing the adjacent receiving antennas by as much as 10 to 20 radio wavelengths or less apart from each other. Typically, an elemental spacing of several radio wavelengths is deemed adequate for space diversity on receive. The much larger spacing is needed for elevated base stations, for which the angle spread of the incoming radio waves is small; note that the spatial coherence distance is inversely proportional to the angle spread. Through the use of diversity on receive as described here, we create a corresponding set of fading channels that are essentially independent. The issue then becomes that of combining the outputs of these statistically independent fading channels in accordance with a criterion that will to provide improved receiver performance. In what follows, we describe four diversity-combining techniques: selection combining, maximal-ratio combining, equal-gain combining, and square-law combining; the first three involve the use of linear receivers, and the fourth utilizes a nonlinear receiver.

6.2.1   Selection Combining

The block diagram of Fig. 6.1 depicts a diversity-combining structure that consists of two functional blocks: N_r linear receivers and a logic circuit. This diversity system is said to be of a *selection-combining* kind, in that, given the N_r receiver outputs produced by a common transmitted signal, the logic circuit *selects* the particular receiver output with the *largest signal-to-noise ratio* as the received signal. In conceptual terms, selection combining is the simplest form of "space diversity on receive" techniques.

To describe the benefit of selection combining in statistical terms, we assume that the wireless communication channel is described by a *frequency-flat, slowly fading Rayleigh channel*. The implications of this assumption are threefold:

1. The frequency-flat assumption means that all the frequency components constituting the transmitted signal are characterized by the same random attenuation and phase shift.
2. The slow-fading assumption means that fading remains essentially unchanged during the transmission of each symbol.
3. The fading phenomenon is described by the Rayleigh distribution.

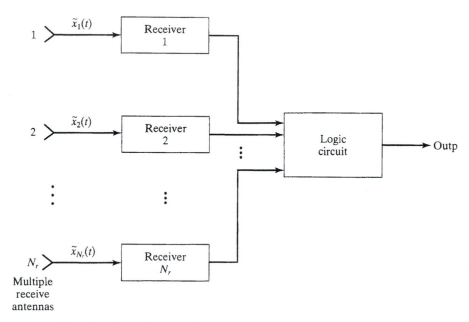

FIGURE 6.1 Block diagram of selection combiner, using N_r receive antennas.

Let $\tilde{s}(t)$ denote the complex envelope of the modulated signal transmitted during the symbol interval $0 \le t \le T$. Then, in light of the assumed channel, the complex envelope of the received signal of the kth diversity branch is defined by

$$\tilde{x}_k(t) = \alpha_k e^{j\theta_k}\tilde{s}(t) + \tilde{w}_k(t) \qquad \begin{matrix} 0 \le t \le T \\ k = 1, 2, \dots, N_r \end{matrix} \qquad (6.1)$$

where, for the kth diversity branch, the fading is represented by the multiplicative term $\alpha_k e^{j\theta_k}$ and the additive channel noise is denoted by $\tilde{w}_k(t)$. With the fading assumed to be slowly varying relative to the symbol duration T, we should be able to estimate and then remove the unknown phase shift θ_k at each diversity branch with sufficient accuracy, in which case Eq. (6.1) simplifies to

$$\tilde{x}_k(t) \approx \alpha_k \tilde{s}(t) + \tilde{w}_k(t) \qquad \begin{matrix} 0 \le t \le T \\ k = 1, 2, \dots, N_r \end{matrix} \qquad (6.2)$$

The signal component of $\tilde{x}_k(t)$ is $\alpha_k \tilde{s}(t)$ and the noise component is $\tilde{w}_k(t)$. The average signal-to-noise ratio at the output of the kth receiver is therefore

$$(\text{SNR})_k = \left(\frac{\mathbf{E}[|\alpha_k \tilde{s}(t)|^2]}{\mathbf{E}[|\tilde{w}_k(t)|^2]} \right) = \frac{\mathbf{E}|\tilde{s}(t)|^2}{\mathbf{E}|\tilde{w}_k(t)|^2}E[\alpha_k^2] \qquad k = 1, 2, \dots, N_r \qquad (6.3)$$

Ordinarily, the mean-square value of $\tilde{w}_k(t)$ is the same for all k. Accordingly, we have

$$(\text{SNR})_k = \frac{E}{N_0} \mathbf{E}[\alpha_k^2] \qquad k = 1, 2, ..., N_r \tag{6.4}$$

where E is the symbol energy and N_0 is the one-sided noise spectral density. For binary data, E equals the transmitted signal energy per bit, namely, E_b.

Let γ_k denote the *instantaneous* signal-to-noise ratio measured at the output of the kth receiver during the transmission of a given symbol. Then replacing the mean-square value $\mathbf{E}[|\alpha_k|^2]$ by the instantaneous value $|\alpha_k|^2$ in Eq. (6.4), we may write

$$\gamma_k = \frac{E}{N_0} \alpha_k^2 \qquad k = 1, 2, ..., N_r \tag{6.5}$$

Under the assumption that the random amplitude α_k is Rayleigh distributed, the squared amplitude α_k^2 will be *exponentially distributed* (i.e., chi-squared with two degrees of freedom; see Section 2.6.1). If we further assume that the average signal-to-noise ratio $(\text{SNR})_k$ over the short-term fading is the same, namely, Γ_{av}, for all the N_r diversity branches, then we may express the probability density functions of the random variables Γ_k pertaining to the individual branches as

$$f_{\Gamma_k}(\gamma_k) = \frac{1}{\gamma_{\text{av}}} \exp\left(-\frac{\gamma_k}{\gamma_{\text{av}}}\right) \qquad \begin{array}{l} \gamma_k \geq 0 \\ k = 1, 2, ..., N_r \end{array} \tag{6.6}$$

Problem 6.1   Following the material presented in Section 2.6.1, derive Eq. (6.6).   ∎

For some signal-to-noise ratio γ, the associated cumulative distributions of the individual branches are

$$\begin{aligned} \text{Prob}(\gamma_k \leq \gamma) &= \int_{-\infty}^{\gamma} f_{\Gamma_k}(\gamma_k) d\gamma_k \\ &= 1 - \exp\left(-\frac{\gamma}{\gamma_{\text{av}}}\right) \qquad \gamma \geq 0 \end{aligned} \tag{6.7}$$

Since, by design, the N_r diversity branches are essentially statistically independent, the probability that all the diversity branches have a signal-to-noise ratio less than the threshold γ is the product of the individual probabilities that $\gamma_k < \gamma$ for all k; that is,

$$\begin{aligned} \text{Prob}(\gamma_k < \gamma \text{ for } k = 1, 2, ..., N_r) &= \prod_{k=1}^{N_r} \text{Prob}(\gamma_k < \gamma) \\ &= \prod_{k=1}^{N_r} \left[1 - \exp\left(-\frac{\gamma}{\gamma_{\text{av}}}\right)\right] \\ &= \left[1 - \exp\left(-\frac{\gamma}{\gamma_{\text{av}}}\right)\right]^{N_r} \qquad \gamma \geq 0 \end{aligned} \tag{6.8}$$

which decreases in numerical value with increasing N_r.

The cumulative distribution function of Eq. (6.8) is the same as the cumulative distribution function of the random variable Γ_{sc} described by the value

$$\gamma_{sc} = \max\{\gamma_1, \gamma_2, \ldots, \gamma_{N_r}\} \tag{6.9}$$

which is less than the threshold γ if, and only if, the individual signal-to-noise ratios $\gamma_1, \gamma_2, \ldots, \gamma_{N_r}$ are all less than γ. Indeed, the cumulative distribution function of the selection combiner (i.e., all of the N_r diversity branches that have a signal-to-noise ratio less than γ) is given by

$$F_\Gamma(\gamma_{sc}) = \left[1 - \exp\left(-\frac{\gamma_{sc}}{\gamma_{av}}\right)\right]^{N_r} \qquad \gamma_{sc_i} \geq 0 \tag{6.10}$$

By definition, the probability density function $f_\Gamma(\gamma_{sc})$ is the derivative of the cumulative distribution function $F_\Gamma(\gamma_{sc})$ with respect to the argument γ_{sc}. Hence, differentiating Eq. (6.10) with respect to γ_{sc} yields

$$\begin{aligned} f_\Gamma(\gamma_{sc}) &= \frac{d}{d\gamma_{sc}} F_\Gamma(\gamma_{sc}) \\ &= \frac{N_r}{\gamma_{av}} \exp\left(-\frac{\gamma_{sc}}{\gamma_{av}}\right)\left[1 - \exp\left(-\frac{\gamma_{sc}}{\gamma_{av}}\right)\right]^{N_r-1} \qquad \gamma_{sc} \geq 0 \end{aligned} \tag{6.11}$$

For convenience of graphical presentation, we use the scaled probability density function

$$f_X(x) = \gamma_{av} f_{\Gamma_{sc}}(\gamma_{sc})$$

where the normalized variable x is defined by

$$x = \gamma_{sc}/\gamma_{av}$$

Figure 6.2 plots $f_X(x)$ versus x for a varying number of receive-diversity branches, N_r, under the assumption that the short-term signal-to-noise ratios for all the N_r branches share the common value γ_{av}. From the figure, we can make the following observations:

1. As the number of diversity branches, N_r, is increased, the probability density function $f_X(x)$ of the normalized random variable $X = \Gamma_{sc}/\gamma_{av}$ moves progressively to the right.

2. The probability density function $f_X(x)$ becomes more and more symmetrical, and Gaussian, as N_r is increased.

Stated another way, a frequency-flat, slowly fading Rayleigh channel is modified through the use of selection combining into a Gaussian channel, provided that the number N_r of diversity channels is sufficiently large. Realizing that a Gaussian channel is a "digital communication theorist's dream," we can now see the practical benefit of using selection combining.

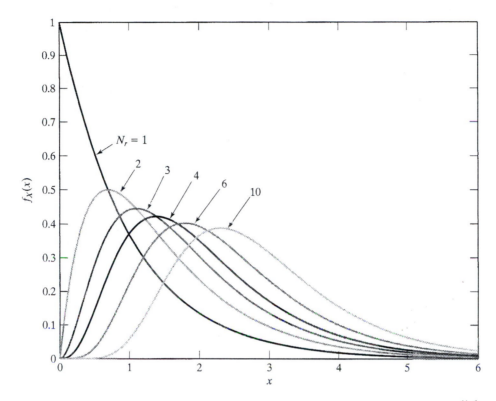

FIGURE 6.2 Normalized probability density function $f_X(x) = N_r \exp(-x)(1 - \exp(-x))^{N_r-1}$ for a varying number N_r of receive antennas.

According to the theory described herein, the selection-combining procedure requires that we monitor the receiver outputs in a continuous manner and, at each instant of time, select the receiver with the strongest signal (i.e., the largest instantaneous signal-to-noise ratio). From a practical perspective, such a selective procedure is rather cumbersome. We may overcome this practical difficulty by adopting a *scanning* version of the selection-combining procedure as follows:

- Start the procedure by selecting the receiver with the strongest output signal.
- Maintain the procedure by using the output of this particular receiver as the combiner's output, so long as its instantaneous signal-to-noise ratio does not drop below a prescribed threshold.
- As soon as the instantaneous signal-to-noise ratio of the combiner falls below the threshold, select a new receiver that offers the strongest output signal, and continue the procedure.

This technique has a performance similar to that of the nonscanning version of selective diversity.

EXAMPLE 6.1 Outage Probability of Selection Combiner

The *outage probability* of a diversity combiner is defined as *the percentage of time the instantaneous output signal-to-noise ratio of the combiner is below some prescribed level for a specified number of branches.* Using the cumulative distribution function of Eq. (6.10), Fig. 6.3 plots the outage curves for the selection combiner with N_r as the running parameter. The horizontal axis of the figure represents the instantaneous output signal-to-noise ratio of the combiner relative to 0 dB (i.e., the 50-percentile point for $N_r = 1$), and the vertical axis represents the outage probability, expressed as a percentage. From the figure, we observe that the fading depth introduced through the use of space diversity on receive diminishes rapidly with the increase in the number of diversity branches. ∎

6.2.2 Maximal-Ratio Combining[2]

The selection-combining technique just described is relatively straightforward to implement. However, from a performance point of view, it is not optimum, in that it ignores the information available from all the diversity branches except for the particular branch that produces the largest instantaneous power of its own demodulated signal.

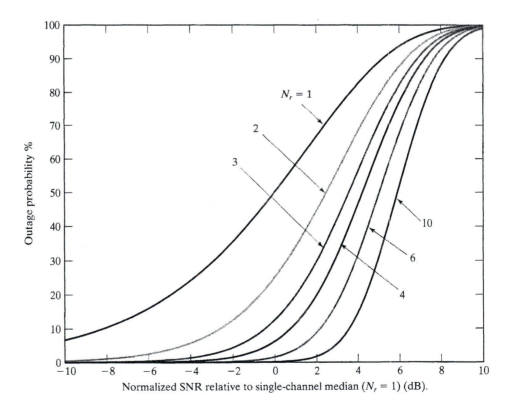

FIGURE 6.3 Outage probability for selector combining for a varying number N_r of receive antennas.

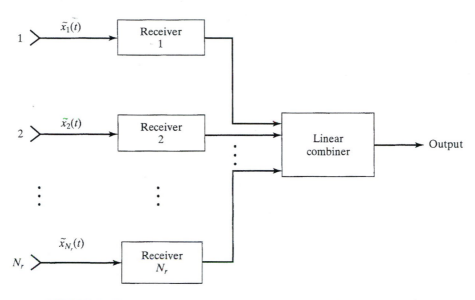

FIGURE 6.4   Block diagram of maximal-ratio combiner using N_r receive antennas.

This limitation of the selection combiner is mitigated by the *maximal-ratio combiner*, the composition of which is described by the block diagram of Fig. 6.4. The maximal-ratio combiner consists of N_r linear receivers, followed by a linear combiner. Using the complex envelope of the received signal at the kth diversity branch given in Eq. (6.1), we find that the corresponding complex envelope of the linear combiner output is defined by

$$\tilde{y}(t) = \sum_{k=1}^{N_r} a_k \tilde{x}_k(t)$$

$$= \sum_{k=1}^{N_r} a_k [\alpha_k e^{j\theta_k} \tilde{s}(t) + \tilde{w}_k(t)] \qquad (6.12)$$

$$= \tilde{s}(t) \sum_{k=1}^{N_r} a_k \alpha_k e^{j\theta_k} + \sum_{k=1}^{N_r} a_k \tilde{w}_k(t)$$

where the a_k are *complex weighting parameters* that characterize the linear combiner. These parameters are changed from instant to instant in accordance with variations in signals in the N_r diversity branches over the short-term fading process. The requirement is to design the linear combiner so as to maximize its output signal-to-noise ratio at each instant of time. From Eq. (6.12), we note the following two points:

1. The complex envelope of the output signal equals $\tilde{s}(t) \sum_{k=1}^{N_r} a_k \alpha_k e^{j\theta_k}$.

2. The complex envelope of the output noise equals $\displaystyle\sum_{k=1}^{N_r} a_k \tilde{w}_k(t)$.

Assuming that the $\tilde{w}_k(t)$ are mutually independent for $k = 1,2,...,N_r$, the output signal-to-noise ratio of the linear combiner is therefore

$$(\text{SNR})_c = \frac{\mathbf{E}\left[\left|\tilde{s}(t)\sum_{k=1}^{N_r} a_k \alpha_k e^{j\theta_k}\right|^2\right]}{\mathbf{E}\left[\left|\sum_{k=1}^{N_r} a_k \tilde{w}_k(t)\right|^2\right]}$$

$$= \frac{\mathbf{E}[|\tilde{s}(t)|^2]}{\mathbf{E}[|\tilde{w}_k(t)|^2]} \cdot \frac{\mathbf{E}\left[\left|\sum_{k=1}^{N_r} a_k \alpha_k e^{j\theta_k}\right|^2\right]}{\mathbf{E}\left[\sum_{k=1}^{N_r} |a_k|^2\right]} \tag{6.13}$$

$$= \left(\frac{E}{N_0}\right)\frac{\mathbf{E}\left[\left|\sum_{k=1}^{N_r} a_k \alpha_k e^{j\theta_k}\right|^2\right]}{\mathbf{E}\left[\sum_{k=1}^{N_r} |a_k|^2\right]}$$

where E/N_0 is the *symbol energy-to-noise spectral density ratio*.

Let γ_c denote the *instantaneous output signal-to-noise ratio* of the linear combiner. Then, using

$$\left|\sum_{k=1}^{N_r} a_k \alpha_k e^{j\theta_k}\right|^2 \quad \text{and} \quad \sum_{k=1}^{N_r} |a_k|^2$$

as the instantaneous values of the expectations in the numerator and denominator of Eq. (6.13), respectively, we may write

$$\gamma_c = \left(\frac{E}{N_0}\right)\frac{\left|\sum_{k=1}^{N_r} a_k \alpha_k e^{j\theta_k}\right|^2}{\sum_{k=1}^{N_r} |a_k|^2} \tag{6.14}$$

The requirement is to maximize γ_c with respect to the a_k. This maximization can be carried out by following the standard differentiation procedure, recognizing that the weighting parameters a_k are complex. However, we choose to follow a simpler procedure based on the Cauchy–Schwarz inequality, as described next.

Let a_k and b_k denote any two complex numbers for $k = 1,2,...,N_r$. According to the *Cauchy–Schwarz inequality* for complex numbers, we have

$$\left| \sum_{k=1}^{N_r} a_k b_k \right|^2 \leq \sum_{k=1}^{N_r} |a_k|^2 \sum_{k=1}^{N_r} |b_k|^2 \tag{6.15}$$

which holds with equality for $a_k = cb_k^*$, where c is some arbitrary complex constant and the asterisk denotes complex conjugation.

Thus, applying the Cauchy–Schwarz inequality to the instantaneous output signal-to-noise ratio of Eq. (6.14), with a_k left intact and b_k set equal to $\alpha_k e^{j\theta_k}$, we obtain

$$\gamma_c \leq \left(\frac{E}{N_0} \right) \frac{\displaystyle\sum_{k=1}^{N_r} |a_k|^2 \sum_{k=1}^{N_r} \left| \alpha_k e^{j\theta_k} \right|^2}{\displaystyle\sum_{k=1}^{N_r} |a_k|^2} \tag{6.16}$$

Cancelling common terms in Eq. (6.16) readily yields

$$\gamma_c \leq \left(\frac{E}{N_0} \right) \sum_{k=1}^{N_r} \alpha_k^2 \tag{6.17}$$

Equation (6.17) proves that, in general, γ_c cannot exceed $\sum_k \gamma_k$, where γ_k is as defined in Eq. (6.5). The equality in Eq. (6.17) holds for

$$\begin{aligned} a_k &= c(\alpha_k e^{j\theta_k})^* \\ &= c\alpha_k e^{-j\theta_k} \qquad k = 1, 2, ..., N_r \end{aligned} \tag{6.18}$$

where c is some arbitrary complex constant. Equation (6.18) defines the complex weighting parameters of the maximal-ratio combiner. On the basis of this equation, we may state that the optimal weighting factor a_k for the kth diversity branch has a magnitude proportional to the amplitude α_k of the signal and a phase that cancels the signal's phase θ_k to within some value that is identical for all the N_r diversity branches. The phase alignment just described has an important implication: It permits the *fully coherent addition* of the N_r receiver outputs by the linear combiner.

Equation (6.17) with the equality sign defines the instantaneous output signal-to-noise ratio of the maximal-ratio combiner, which is written as

$$\gamma_{\text{mrc}} = \left(\frac{E}{N_0} \right) \sum_{k=1}^{N_r} \alpha_k^2 \tag{6.19}$$

According to Eq. (6.5), however, $(E/N_0) \alpha_k^2$ is the instantaneous output signal-to-noise ratio of the kth diversity branch. Hence, the maximal-ratio combiner produces an instantaneous output signal-to-noise ratio that is the sum of the instantaneous signal-to-noise ratios of the individual branches; that is,

$$\gamma_{mrc} = \sum_{k=1}^{N_r} \gamma_k \tag{6.20}$$

The term "maximal-ratio combiner" has been coined to describe the combiner of Fig. 6.4 that produces the optimum result given in Eq. (6.20). Indeed, if follows from this result that the instantaneous output signal-to-noise ratio of the maximal-ratio combiner can be large even when the signal-to-noise ratios of the individual branches are small. The selection combiner of Section 6.2.1 is clearly inferior in performance to the maximal-ratio combiner, since the instantaneous signal-to-noise ratio produced by the selection combiner is simply the largest among the N_r terms of Eq. (6.20).

The maximal signal-to-noise ratio γ_{mrc} is the sample value of a random variable denoted by Γ_{mrc}. According to Eq. (6.19), γ_{mrc} is equal to the sum of N_r exponentially distributed random variables for a frequency-flat, slowly fading Rayleigh channel. From probability theory, the probability density function of such a sum is known to be *chi-square with $2N_r$ degrees of freedom* (see Appendix C); that is,

$$f_{\Gamma_{mrc}}(\gamma_{mrc}) = \frac{1}{(N_r-1)!} \frac{\gamma_{mrc}^{N_r-1}}{\gamma_{av}^{N_r}} \exp\left(-\frac{\gamma_{mrc}}{\gamma_{av}}\right) \tag{6.21}$$

Note that for $N_r = 1$, Eqs. (6.11) and (6.21) reduce to the same value, which is to be expected.

Figure 6.5 plots the scaled probability density function $f_X(x) = \gamma_{av} f_{\Gamma_{mrc}}(\gamma_{mrc})$ against the normalized variable $x = \gamma_{mrc}/\gamma_{av}$ for varying N_r. On basis of this figure, we may make observations similar to those for the selection combiner, except for the fact that, for any N_r, the scaled probability density function for the maximal-ratio combiner is radically different from that for the selection combiner.

EXAMPLE 6.2 Outage Probability for Maximal-Ratio Combiner

The cumulative distribution function for the maximal-ratio combiner is defined by

$$\text{Prob}(\gamma_{mrc} < x) = \int_0^x f_{\Gamma_{mrc}}(\gamma_{mrc}) d\gamma_{mrc}$$
$$= 1 - \int_x^\infty f_{\Gamma_{mrc}}(\gamma_{mrc}) d\gamma_{mrc} \tag{6.22}$$

where the probability density function $f_{\Gamma_{mrc}}(\gamma_{mrc})$ is itself defined by Eq. (6.21). Using Eq. (6.22), Fig. 6.6 plots the outage probability for the maximal-ratio combiner with N_r as a running parameter. Comparing this figure with Fig. 6.3 for selection combining, we see that the outage-probability curves for these two diversity techniques are superficially similar. The *diversity gain*, defined as the savings in E/N_0 at a given bit error rate, provides a measure of the effectiveness of a diversity technique on an outage-probability basis. ∎

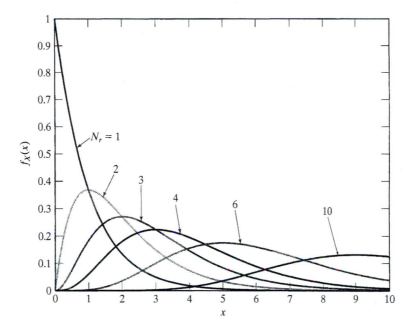

FIGURE 6.5 Normalized probability density function $f_X(x) = \dfrac{1}{(N_r-1)!} x^{N_r-1} \exp(-x)$ for a varying number N_r of receive antennas.

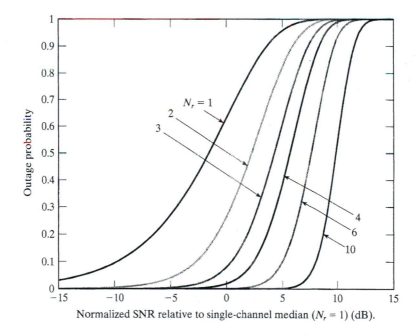

Normalized SNR relative to single-channel median ($N_r = 1$) (dB).

FIGURE 6.6 Outage probability of maximal-ratio combiner for a varying number N_r of receive antennas.

As a point of comparison of the outage performances of selection combining and maximal-ratio combining, consider a diversity-on-receive system with $N_r = 6$ and normalized output signal-to-noise ratio = 5dB. Examination of Figs. 6.3 and 6.6 for these settings reveals the following outage probabilities:

Selection combiner: 50%
Maximal-ratio combiner: 10%

These numbers clearly illustrate the highly superior outage performance of the maximal-ratio combiner over the selection combiner.

EXAMPLE 6.3 Bit Error Rate of Coherent Binary FSK

In this example, we determine the average probability of symbol error for the case of coherent binary frequency-shift keying (BFSK) over a frequency-flat, slowly fading Rayleigh channel. The use of maximal-ratio combining at the receiver is assumed. This simple case is amenable to an analytic formulation. (Note that the requirement for a coherent phase reference for coherent analytic evaluation detection makes the use of selection combining somewhat meaningless—hence the interest in only maximal-ratio combining in the context of coherent BFSK.)

Adapting the formula for the probability of symbol error for BFSK over an additive white Gaussian-noise channel for the problem at hand, we may write

$$\text{Prob}(\text{error}|\gamma_{\text{mrc}}) = \frac{1}{2}\text{erfc}\left(\sqrt{\frac{1}{2}\gamma_{\text{mrc}}}\right) \qquad (6.23)$$

which is obtained by substituting γ_{mrc} for the signal energy-to-noise spectral density ratio E/N_0 in the formula for coherent BFSK in Table 3.4.

We next recognize that the instantaneous output signal-to-noise ratio γ_{mrc} is in fact a random variable. To determine the average probability of symbol error, we must average the conditional probability of error of Eq. (6.23) with respect to γ_{mrc}, or

$$P_e = \mathbf{E}[\text{Prob}(\text{errorr}|\gamma_{\text{mrc}})] \qquad (6.24)$$

where the expectation also is with respect to γ_{mrc}. This expectation is found by multiplying the conditional probability $\text{Prob}(\text{error}|\gamma_{\text{mrc}})$ by the probability density function of γ_{mrc} and then integrating the product with respect to γ_{mrc}. That is, we write

$$P_e = \int_0^\infty \text{Prob}(\text{error}|\gamma_{\text{mrc}})f_\Gamma(\gamma_{\text{mrc}})d\gamma_{\text{mrc}} \qquad (6.25)$$

Substituting Eqs. (6.21) and (6.23) into Eq. (6.25) yields

$$P_e = \frac{1}{2(N_r-1)!}\int_0^\infty \text{erfc}\left(\sqrt{\frac{1}{2}\gamma_{\text{mrc}}}\right)\frac{\gamma_{\text{mrc}}^{N_r-1}}{\gamma_{\text{av}}^{N_r}}\exp\left(-\frac{\gamma_{\text{mrc}}}{\gamma_{\text{av}}}\right)d\gamma_{\text{mrc}}$$

$$= \frac{1}{2(N_r-1)!}\int_0^\infty \text{erfc}\left(\sqrt{\frac{1}{2}\gamma_{\text{av}}x}\right)x^{N_r-1}\exp(-x)dx$$

$$(6.26)$$

where $x = \gamma_{\text{mrc}}/\gamma_{\text{av}}$ and erfc($\cdot$) denotes the *complementary error function*, discussed in Appendix E.   ∎

Unfortunately, there is no exact closed-form solution of Eq. (6.26). To proceed further, we may use numerical integration or seek an approximate solution. The reader is referred to Problem 6.25 for the derivation of an approximate formula for P_e.

6.2.3 Equal-Gain Combining

In a theoretical context, the maximal-ratio combiner is the *optimum* among linear diversity-combining techniques, in the sense that it produces the largest possible value of instantaneous output signal-to-noise ratio. However, in practical terms, there are three important issues to keep in mind:[3]

1. Significant instrumentation is needed to adjust the complex weighting parameters of the maximal-ratio combiner to their exact values, in accordance with Eq. (6.18).
2. The additional improvement in output signal-to-noise ratio gained by the maximal-ratio combiner over the selection-combiner of Section 6.2.1 is not that large, and it is quite likely that the additional improvement in receiver performance is lost in the inability to achieve the exact setting of the maximal ratio combiner.
3. So long as a linear combiner uses the diversity branch with the strongest signal, other details of the combiner may result in a minor improvement in overall receiver performance.

Issue 3 points to the formulation of the so-called equal-gain combiner, in which all the complex weighting parameters a_k have their phase angles set opposite to those of their respective multipath branches in accordance with Eq. (6.18), but, unlike the a_k in the maximal-ratio combiner, their magnitudes are set equal to some constant value— unity, for convenience of use.

We may reach a similar conclusion by examining Eq. (6.16), in which we see that the summation term $\sum_k |a_k|^2$ involving the magnitudes of the complex weighting parameters is common to both the numerator and denominator of the instantaneous output signal-to-noise ratio. That is, whether we set $a_k = ca_k e^{-j\theta_k}$, in accordance with Eq. (6.18) for the maximal-ratio combiner, or we simply set $a_k = e^{-j\theta_k}$, for all k in the equal-gain combiner, the instantaneous output signal-to-noise ratio is unchanged and hence constitutes further justification for using the equal-gain combiner in preference to the maximal-ratio combiner.

6.2.4 Square-Law Combining

Maximal-ratio combining—and, for that matter, equal-gain combining—relies on the ability to estimate the phase of the different diversity branches and to combine the signals coherently. Often, such a procedure is not practical, due to the physical separation of the diversity receivers or to hardware limitations. In this case, square-law combining offers the opportunity to obtain an advantage in diversity without requiring phase estimation.

Unlike maximal-ratio combining, square-law combining is applicable only to certain modulation techniques. In particular, it is applicable to orthogonal modulation, including modulations such as FSK or direct-sequence CDMA signals, in which different (approximately orthogonal) frequencies or sequences are used to represent different data symbols. With binary orthogonal signalling, the receiver generates the two decision variables

$$Q_{0k} = \frac{1}{\sqrt{N_o}} \int_0^T \tilde{x}_k(t) \tilde{s}_0^*(t) dt \qquad (6.27)$$

and

$$Q_{1k} = \frac{1}{\sqrt{N_o}} \int_0^T \tilde{x}_k(t) \tilde{s}_1^*(t) dt \qquad (6.28)$$

where $\tilde{s}_0(t)$ and $\tilde{s}_1(t)$ are normalized versions of the two possible binary symbols. In *orthogonal modulation*, the two signaling waveforms approximately satisfy the conditions

$$\int_0^T \tilde{s}_i(t) \tilde{s}_j^*(t) dt = \begin{cases} E_b & i = j \\ 0 & i \neq j \end{cases} \qquad (6.29)$$

where E_b is the transmitted signal energy per bit. If the binary symbol 0 is transmitted, it turns out that the two decision variables are equivalent to

$$Q_{0k} = \frac{\sqrt{E_b}\,\alpha_k e^{j\theta_k}}{\sqrt{N_0}} + \frac{w_{0k}}{\sqrt{N_0}} \qquad (6.30)$$

and

$$Q_{1k} = \frac{w_{1k}}{\sqrt{N_0}} \qquad (6.31)$$

The w_{0k} and w_{1k} are independent Gaussian random variables of zero mean and variance N_0. The square-law receiver makes a decision between the binary symbols 0 and 1 as follows:

If
$$|Q_{0k}|^2 > |Q_{1k}|^2 \qquad \text{say} \quad 0 \qquad (6.32)$$
$$\text{otherwise} \qquad \text{say} \quad 1$$

The structure of the receiver is illustrated in Fig. 6.7.

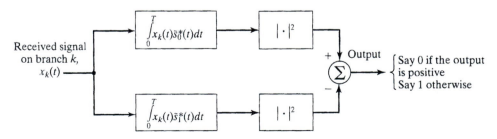

FIGURE 6.7   Illustration of receiver for orthogonal waveforms.

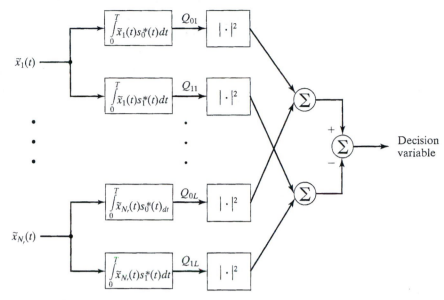

FIGURE 6.8 Block diagram of diversity receiver based on square-law combining.

With receive diversity, the detector outputs are added before being compared against the decision variables. This form of diversity, referred to as *square-law combining*, is illustrated in Fig. 6.8. With square-law combining, we form the decision variables

$$Q_0 = \sum_{k=1}^{N_r} |Q_{0k}|^2 \tag{6.33}$$

and

$$Q_1 = \sum_{k=1}^{N_r} |Q_{1k}|^2 \tag{6.34}$$

and then perform the same test as in Eq.(6.32) to decide whether a 0 or a 1 is transmitted.

Closer inspection of Eqs. (6.33) and (6.34) indicates that both Q_0 and Q_1 are the sum of squares of complex Gaussian random variables. Consequently, they both have a chi-square distribution with $2N_r$ degrees of freedom, analogous to Eq. (6.19). (The chi-square distribution is discussed in Appendix C.) The difference between the random variables Q_{0k} and Q_{1k} lies in their variance. If $s_0(t)$ was transmitted, then the variances are given by

$$
\begin{aligned}
\text{Var}(Q_{0k}) &= \frac{E_b}{N_0}\text{Var}(\alpha_k e^{j\theta_k}) + \frac{\text{Var}(w_{0k})}{N_0} \\
&= \frac{E_b}{N_0}(\text{E}[\alpha_k^2]) + \frac{\text{E}[w_{0k}^2]}{N_0} \\
&= \gamma_{\text{av}} + 1
\end{aligned}
\tag{6.35}
$$

and

$$\mathrm{Var}(Q_{1k}) = \frac{1}{N_0}\mathrm{Var}(w_{1k})$$

(6.36)

$$= 1$$

where the fading is assumed to be Rayleigh distributed with average signal-to-noise ratio γ_{av}. In this case, the probability density function for the correct symbol is given by

$$f_{Q_0}(q) = \frac{1}{(N_r-1)!}\frac{q^{N_r-1}}{(\gamma_{\mathrm{av}}+1)^{N_r}}\exp\left(-\frac{q}{\gamma_{\mathrm{av}}+1}\right)$$

(6.37)

and that for the incorrect symbol is given by

$$f_{Q_1}(q) = \frac{1}{(N_r-1)!}q^{N_r-1}\exp(-q)$$

(6.38)

Equation (6.37) follows from Eq. (6.21), with γ_{av} replaced by $\gamma_{\mathrm{av}}+1$, in accordance with Eq. (6.35), under the assumption that $s_0(t)$ is transmitted. Similarly, Eq. (6.38) follows from Eq. (6.21), with γ_{av} set equal to unity, in accordance with Eq. (6.36), under the assumption that $s_1(t)$ is transmitted.

Since Q_0 and Q_1 are independent random variables, the probability of error is the probability that $Q_0 < Q_1$ when symbol 0 is transmitted. Mathematically, this probability is obtained from the double integral

$$\mathrm{Prob}(Q_0 < Q_1) = \int_0^\infty f_{Q_0}(q_0)\left(\int_{q_o}^\infty f_{Q_1}(q_1)dq_1\right)dq_0$$

(6.39)

A similar result holds when a 1 is transmitted.

In Fig. 6.9, we compare the probability of error for coherent BFSK with square-law combining to maximal-ratio combining for diversity orders 1, 2, and 4 with binary signaling over a Rayleigh-fading channel. With BFSK, maximal-ratio combining offers approximately a 3-dB improvement over square-law combining, a performance difference that is quite small compared with the overall gains that either diversity technique usually provides.

Problem 6.2　Using Eq. (6.29) in (6.28), show that Eq. (6.31) follows, within a scaling factor. Hence, show that Q_{0k} and Q_{1k} are independent random variables.　∎

Problem 6.3　For a diversity $N_r = 2$ system and $\gamma_{\mathrm{av}} = 20$ dB, compute the fraction of the time the decision variable Q_0 drops below 10 dB when 0 has been transmitted. What is the probability that Q_1 is greater than 10 dB under the same conditions?

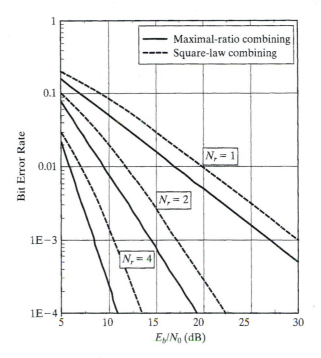

FIGURE 6.9   Comparison of diversity performance: square-law combining versus maximal-ratio combining.

Ans. Define the incomplete gamma function as

$$\Gamma(x, a) = \frac{1}{\Gamma(a)}\int_0^x s^{a-1}\exp(-s)ds$$

where

$$\Gamma(a) = \int_0^\infty s^{a-1}\exp(-s)ds$$

and $\Gamma(a) = (a - 1)!$ *if a is a positive integer. Then show*

$$P(Q_0 < 10\text{dB}) = \Gamma(10/101, 2) = 0.0046$$

and

$$P(Q_1 > 10\text{dB}) = 1 - \Gamma(10, 2) = 5 \times 10^{-4}$$ ∎

6.3 MULTIPLE-INPUT, MULTIPLE-OUTPUT ANTENNA SYSTEMS [4]

In Section 6.2, we studied space-diversity wireless communication systems employing multiple receive antennas to combat the multipath fading problem. In effect, fading was treated as a source that degrades performance, necessitating the use of space diversity on receive to mitigate it. In this section, we discuss *multiple-input, multiple-output (MIMO) wireless communications*, also referred to in the literature as *multiple-transmit, multiple-receive (MTMR) wireless communications*.

MIMO wireless communications include space diversity on receive as a special case. Most important, however, are the following three points:

1. The fading phenomenon is viewed not as a nuisance, but rather as an environmental source of possible enrichment.
2. Space diversity at both the transmit and receive ends of the wireless communications link provides the basis for a significant increase in channel capacity or spectral efficiency.
3. Unlike increasing capacity with conventional techniques, increasing channel capacity with MIMO is achieved by increasing computational complexity while maintaining the primary communication resources (i.e., total transmit power and channel bandwidth) fixed.

6.3.1 Coantenna Interference

Figure 6.10 shows the block diagram of a MIMO wireless link. The signals transmitted by the N_t transmit antennas over the wireless channel are all chosen to lie inside a common frequency band. Naturally, the transmitted signals are scattered differently by the channel. Moreover, due to multiple signal transmissions, the system experiences a spatial form of signal-dependent interference referred to as *coantenna interference (CAI)*.

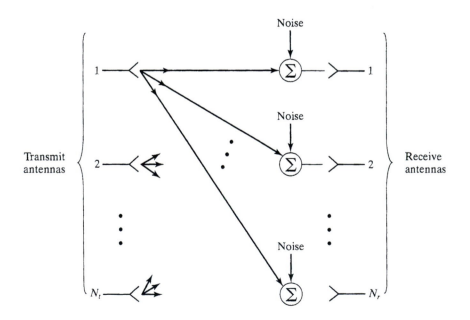

FIGURE 6.10 Block diagram of MIMO wireless link with N_t transmit antennas and N_r receive antennas.

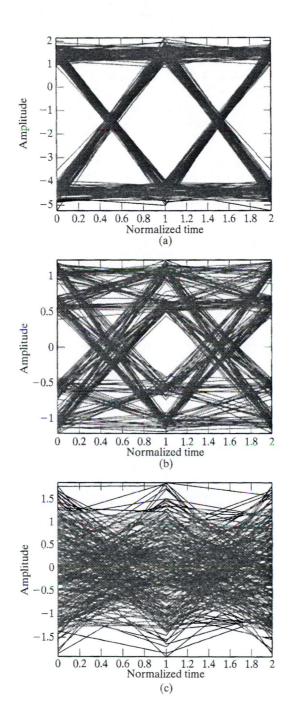

FIGURE 6.11   Effect of coantenna interference on the eye diagram for one receive antenna and different numbers of transmit antennas. (a) $N_t = 1$, (b) $N_t = 2$, (c) $N_t = 8$.

Figure 6.11 illustrates the effect of CAI for one, two, and eight simultaneous transmissions and a single receive antenna (i.e., $N_t = 1, 2, 8$ and $N_r = 1$), using BPSK; the transmitted BPSK signals are different, but they all have the same average power and occupy the same bandwidth. The figure clearly shows the difficulty that arises from CAI when the number N_t of transmit antennas is large. In particular, with eight simultaneous signal transmissions, the *eye pattern* of the received signal is practically closed. The pattern, commonly used in the study and design of digital communications, derives its name from the fact that it resembles the human eye for the transmission of binary data; the interior region of the eye pattern is called the *eye opening*. The challenge for the receiver is how to mitigate the CAI problem and thereby make it possible to provide a spectacular increase in spectral efficiency.

In a theoretical context, the spectral efficiency of a communication system is intimately linked to the channel capacity of the system. To proceed with evaluation of the channel capacity of MIMO wireless communications, we begin by formulating a baseband channel model for the system.

6.3.2 Basic Baseband Channel Model

Consider a MIMO narrowband wireless communication system built around a flat-fading channel and with N_t transmit antennas and N_r receive antennas. The antenna configuration is hereafter referred to as the pair (N_t, N_r). For a statistical analysis of the MIMO system in what follows, we use baseband representations of the transmitted and received signals, as well as of the channel. In particular, we introduce the following notation:

- The spatial parameter

$$N = \min\{N_t, N_r\} \tag{6.40}$$

  defines a *new degree of freedom* introduced into the wireless communication system by using a MIMO channel with N_t transmit antennas and N_t receive antennas.

- The N_t-by-1 vector

$$\mathbf{s}(n) = [\tilde{s}_1(n), \tilde{s}_2(n), ..., \tilde{s}_{N_t}(n)]^T \tag{6.41}$$

  denotes the complex signal vector transmitted by the N_t antennas at discrete time n. The symbols constituting the vector $\mathbf{s}(n)$ are assumed to have zero mean and common variance σ_s^2. The *total* transmit power is *fixed* at the value

$$P = N_t \sigma_s^2 \tag{6.42}$$

  For P to be maintained constant, the variance σ_s^2 (i.e., the power radiated by each transmit antenna) must be inversely proportional to N_t.

- For the flat-fading, and therefore memoryless, channel, we may use $\tilde{h}_{ik}(n)$ to denote the sampled complex gain of the channel from transmit antenna k to receive antenna i at discrete time n, where $i = 1,2,...,N_r$ and $k = 1,2,...,N_t$. We may thus express the N_r-by-N_t *complex channel matrix* as

$$\mathbf{H}(n) = \left. \begin{bmatrix} \tilde{h}_{11}(n) & \tilde{h}_{21}(n) & ... & \tilde{h}_{N_t1}(n) \\ \tilde{h}_{21}(n) & \tilde{h}_{22}(n) & ... & \tilde{h}_{N_t2}(n) \\ \vdots & \vdots & & \vdots \\ \tilde{h}_{N_r1}(n) & \tilde{h}_{N_r2} & ... & h_{N_rN_t}(n) \end{bmatrix} \right\} \begin{matrix} N_r \\ \text{receive} \\ \text{antennas} \end{matrix} \tag{6.43}$$

$$\underbrace{\phantom{\begin{bmatrix} \tilde{h}_{11}(n) & \tilde{h}_{21}(n) & ... & \tilde{h}_{N_t1}(n) \end{bmatrix}}}_{N_t \text{ transmit antennas}}$$

- The system of equations

$$\tilde{x}_i(n) = \sum_{k=1}^{N_t} \tilde{h}_{ik}(n)\tilde{s}_k(n) + \tilde{w}_i(n) \qquad \begin{matrix} i = 1, 2, ..., N_r \\ k = 1, 2, ..., N_t \end{matrix} \tag{6.44}$$

defines the complex signal received at the ith antenna due to the transmitted symbol $\tilde{s}_k(n)$ radiated by the kth antenna. The term $w_i(n)$ denotes the additive complex channel noise perturbing $x_i(n)$. Let the N_r-by-1 vector

$$\mathbf{x}(n) = [\tilde{x}_1(n), \tilde{x}_2(n), ..., \tilde{x}_{N_r}(n)] \tag{6.45}$$

denote the complex received signal vector and the N_r-by-1 vector

$$\mathbf{w}(n) = [\tilde{w}_1(n), \tilde{w}_2(n), ..., \tilde{w}_{N_r}(n)]^T \tag{6.46}$$

denote the complex channel noise vector. We may then rewrite the system of equations (6.44) in the compact matrix form

$$\mathbf{x}(n) = \mathbf{H}(n)\mathbf{s}(n) + \mathbf{w}(n) \tag{6.47}$$

Equation (6.47) describes the *basic complex channel model for MIMO wireless communications*, assuming the use of a flat-fading channel. The equation describes the input–output behavior of the channel at discrete time n. To simplify the exposition, hereafter we suppress the dependence on time n by writing

$$\mathbf{x} = \mathbf{Hs} + \mathbf{w} \tag{6.48}$$

where it is understood that all four vector–matrix terms of the equation, namely, $\mathbf{s}$, $\mathbf{H}$, $\mathbf{w}$, and $\mathbf{x}$, are in fact dependent on the discrete time n. Figure 6.12 depicts the basic channel model of Eq. (6.48).

For mathematical tractability, we assume a *Gaussian model* made up of three elements relating to the transmitter, channel, and receiver, respectively:

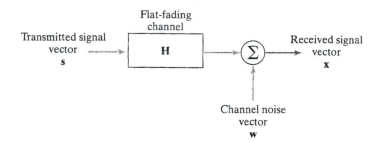

FIGURE 6.12   Depiction of the basic channel model of Eq. (6.48).

1. The N_t symbols constituting the transmitted signal vector $\mathbf{s}$ are drawn from a *white complex Gaussian codebook*; that is, the symbols $\tilde{s}_1, \tilde{s}_2, ..., \tilde{s}_{N_t}$ are independently and identically distributed (i.i.d.) complex Gaussian random variables with zero mean and common variance σ_s^2. Hence, the correlation matrix of the transmitted signal vector $\mathbf{s}$ is defined by

$$\mathbf{R_s} = \mathbf{E}[\mathbf{ss}^\dagger]$$
$$= \sigma_s^2 \mathbf{I}_{N_t} \tag{6.49}$$

where $\mathbf{I}_{N_t}$ is the N_t-by-N_t identity matrix.

2. The $N_t \times N_r$ elements of the channel matrix $\mathbf{H}$ are drawn from an ensemble of i.i.d. complex random variables with zero mean and unit variance, as shown by the complex distribution

$$h_{ik}: \quad N(0, 1/\sqrt{2}) + jN(0, 1/\sqrt{2}) \qquad \begin{matrix} i = 1, 2, ..., N_r \\ k = 1, 2, ..., N_t \end{matrix} \tag{6.50}$$

where $\mathcal{N}(.,.)$ denotes a real Gaussian distribution. On this basis, we find that the amplitude component h_{ik} is *Rayleigh* distributed, so we sometimes speak of the MIMO channel as a *rich Rayleigh scattering environment*. By the same token, we also find that the squared amplitude component, namely, $|h_{ik}|^2$, is a *chi-square random variable* with mean

$$\mathbf{E}[|h_{ik}|^2] = 1 \qquad \text{for all } i \text{ and } k \tag{6.51}$$

3. The N_r elements of the channel noise vector $\mathbf{w}$ are i.i.d. complex Gaussian random variables with zero mean and common variance σ_w^2; that is, the correlation matrix of the noise vector $\mathbf{w}$ is given by

$$\mathbf{R_w} = \mathbf{E}[\mathbf{ww}^\dagger]$$
$$= \sigma_w^2 \mathbf{I}_{N_r} \tag{6.52}$$

where $\mathbf{I}_{N_r}$ is the N_r-by-N_r identity matrix.

In light of Eq. (6.42) and the assumption that h_{ik} is a normalized random variable with zero mean and unit variance, the *average signal-to-noise ratio* (SNR) at each receiver input is given by

$$\rho = \frac{P}{\sigma_{\mathbf{w}}^2}$$

$$= \frac{N_t \sigma_{\mathbf{s}}^2}{\sigma_{\mathbf{w}}^2} \tag{6.53}$$

which, for a prescribed noise variance $\sigma_{\mathbf{w}}^2$, is fixed once the total transmit power P is fixed. Note also that (1) all the N_t transmitted signals occupy a common channel bandwidth and (2) the SNR ρ is independent of N_r.

The idealized Gaussian model described herein is applicable to indoor local area networks and other wireless environments where the mobility of the user's terminals is limited. The model, however, ignores the unavoidable ambient noise, which, as a result of experimental measurements, is known to be decidedly non-Gaussian due to the *impulse nature* of human-made electromagnetic interference as well as natural noise.[5]

6.4 MIMO CAPACITY FOR CHANNEL KNOWN AT THE RECEIVER

With the basic complex channel model at hand, we are now ready to focus the discussion on the primary issue of interest: the channel capacity of a MIMO wireless link. Two cases will be considered. The first case, discussed in Section 6.4.1, considers a link that is stationary and therefore ergodic. The second case, presented in Section 6.4.2, considers a nonergodic link, assuming quasi-stationarity from one data burst to another.

6.4.1 Ergodic Capacity

In Section 4.3, we stated that the information capacity of a *real* additive white Gaussian noise (AWGN) channel, subject to the constraint of a fixed transmit power P, is defined by

$$C = B \log_2 \left(1 + \frac{P}{\sigma_w^2} \right) \text{ bits/s} \tag{6.54}$$

where B is the channel bandwidth and σ_w^2 is the noise variance measured over B.

Given a time-invariant channel, Eq. (6.54) defines the maximum data rate that can be transmitted over the channel with an arbitrarily small probability of error incurred as a result of the transmission. With the channel used K times for the transmission of K symbols in, say, T seconds, the transmission capacity per unit time is (T/K) times the formula for C given in Eq. (6.54). Recognizing that $K = 2BT$, in accordance with the sampling theorem discussed in Chapter 4, we may express the information capacity of the AWGN channel in the equivalent form

$$C = \frac{1}{2}\log_2\left(1 + \frac{P}{\sigma_w^2}\right) \quad \text{bits/s/Hz} \tag{6.55}$$

Note that 1 bit per second per hertz corresponds to 1 bit per transmission.

With wireless communications as the medium of interest, consider next the case of a *complex*, flat-fading channel with the receiver having perfect knowledge of the channel state. The capacity of such a channel is given by[6]

$$C = \mathbf{E}\left[\log_2\left(1 + \frac{|h|^2 P}{\sigma_w^2}\right)\right] \quad \text{bits/s/Hz} \tag{6.56}$$

where the expectation is taken over the gain $h(n)$ of the channel, and the channel is assumed to be stationary and ergodic. In recognition of this assumption, C is commonly referred to as the *ergodic capacity* of the flat-fading channel, and the channel coding is applied across fading intervals (i.e., over an "ergodic" interval of channel variation with time).

It is important to note that the scaling factor of 1/2 is missing from the capacity formula of Eq. (6.56). The reason for this omission is the fact that that equation refers to a complex baseband channel, whereas Eq. (6.54) refers to a real channel. The fading channel covered by Eq. (6.56) operates on a complex signal—a signal with in-phase and quadrature components. Therefore, such a complex channel is equivalent to two real channels with equal capacities and operating in parallel—hence the result presented in that equation.

Equation (6.56) applies to the simple case of a *single-input, single-output (SISO) flat-fading channel*. Generalizing this formula to the case of a multiple-input, multiple-output (MIMO) flat-fading channel, governed by the Gaussian model described in Section 6.3.1, we find that the ergodic capacity of the MIMO channel is given by

$$C = \mathbf{E}\left[\log_2\left\{\frac{\det(\mathbf{R}_w + \mathbf{H}\mathbf{R}_s\mathbf{H}^\dagger)}{\det(\mathbf{R}_w)}\right\}\right] \quad \text{bits/s/Hz} \tag{6.57}$$

which is subject to the constraint

$$\max_{\mathbf{R}_s} \ \text{tr}(\mathbf{R}_s) \le P$$

where P is the constant transmit power. The expectation in Eq. (6.57) is over the random channel matrix $\mathbf{H}$, and the superscript $\dagger$ denotes Hermitian transposition; $\mathbf{R}_s$ and $\mathbf{R}_w$ are, respectively, the correlation matrices of the transmitted signal vector $\mathbf{s}$ and channel noise vector $\mathbf{w}$. A detailed derivation of Eq. (6.57) is presented in Appendix G.

In general, it is difficult to evaluate Eq. (6.57), except for the Gaussian model described in Section 6.5. In particular, substituting Eqs. (6.49) and (6.52) into Eq. (6.57) and simplifying yields

$$C = \mathbf{E}\left[\log_2\left\{\det\left(\mathbf{I}_{N_r} + \frac{\sigma_s^2}{\sigma_w^2}\mathbf{H}\mathbf{H}^\dagger\right)\right\}\right] \text{ bits/s/Hz} \tag{6.58}$$

Invoking the definition of the average signal-to-noise ratio ρ introduced in Eq. (6.53), we may rewrite Eq. (6.58) in the equivalent form

$$C = \mathbf{E}\left[\log_2\left\{\det\left(\mathbf{I}_{N_r} + \frac{\rho}{N_t}\mathbf{H}\mathbf{H}^\dagger\right)\right\}\right] \text{ bits/s/Hz} \tag{6.59}$$

Equation (6.59), defining the ergodic capacity of a MIMO flat-fading channel, involves the determinant of an N_r-by-N_r sum matrix followed by the logarithm to the base 2. Accordingly, we refer to this formula as the *log-det capacity formula* for a Gaussian MIMO channel.[7]

Problem 6.4   The log-det capacity formula of Eq. (6.59) assumes that $N_t \geq N_r$ for the N_r-by-N_r matrix product $\mathbf{H}\mathbf{H}^\dagger$ to be of full rank. Show that, for the alternative case, $N_t \leq N_r$, which makes the N_t-by-N_t matrix product $\mathbf{H}^\dagger\mathbf{H}$ to be of full rank, the log-det capacity formula of the MIMO link is defined by

$$C = \mathbf{E}\left[\log_2\left\{\det\left(\mathbf{I}_{N_t} + \frac{\rho}{N_r}\mathbf{H}^\dagger\mathbf{H}\right)\right\}\right] \text{ bit/s/Hz} \tag{6.60}$$

where, as before, the expectation is over the channel matrix $\mathbf{H}$.   ∎

Note that Eqs. (6.59) and (6.60) are equivalent, in that either one of them applies to all $\{N_t, N_r\}$ antenna configurations. The two formulas differentiate themselves only when the full-rank issue is of concern, as explained in Problem 6.4.

Clearly, Eq. (6.56), pertaining to a conventional flat-fading link with a single antenna at both ends, is a special case of the log-det capacity formula. Specifically, for $N_t = N_r = 1$ (i.e., no spatial diversity), $\rho = P/\sigma_w^2$, and $\mathbf{H} = h$ (with dependence on discrete time n suppressed as stated on page 361), Eq. (6.58) reduces to Eq. (6.56).

Another insightful result that follows from the log-det capacity formula is that if $N_t = N_r = N$, then as N approaches infinity, the capacity C defined in Eq. (6.58) grows asymptotically (at least) linearly with N; see Problem 6.32. That is,

$$\lim_{N \to \infty} \frac{C}{N} \geq \text{constant} \tag{6.61}$$

The asymptotic formula of Eq. (6.61) may be stated in words as follows:

> *The ergodic capacity of a MIMO flat-fading wireless link with an equal number N of transmit and receive antennas grows roughly proportionately with N.*

What this statement teaches us is that, by increasing the computational complexity resulting from the use of multiple antennas at both the transmit and receive ends of a wireless link, we are able to increase the spectral efficiency of the link in a far greater manner than is possible by conventional means (e.g., increasing the transmit signal-to-noise ratio). The potential for this very sizable increase in the spectral efficiency of a MIMO wireless communication system is attributed to the key parameter $N = \min\{N_t, N_r\}$, which, in accordance with Eq. (6.40), defines the *number of spatial degrees of freedom* provided by the system. (Later in Section 6.10, we show that N is equal to the maximal multiplexing gain.)

Problem 6.5   Show that, at high signal-to-noise ratios, the capacity gain of a MIMO wireless communication system with the channel state known to the receiver is $N = \min\{N_t, N_r\}$ bits per second per hertz for every 3-dB increase in signal-to-noise ratio.   ∎

6.4.2 Two Other Special Cases of the Log-Det Formula: Capacities of Receive and Transmit Diversity Links

Naturally, the log-det capacity formula of Eq. (6.59) for the channel capacity of an (N_t, N_r) wireless link includes the channel capacities of receive and transmit diversity links as special cases:

1. *Diversity-on-receive channel.* The log-det capacity formula of Eq. (6.60) applies to this case. Specifically, for $N_t = 1$, the channel matrix **H** reduces to a column vector, and with it, Eq. (6.60) reduces to

$$C = \mathbf{E}\left[\log_2\left\{\left(1 + \rho \sum_{i=1}^{N_r} |h_i|^2\right)\right\}\right] \text{ bits/s/Hz} \tag{6.62}$$

   Compared with the channel capacity of Eq. (6.56), for a single-input, single-output fading channel with $\rho = P/\sigma_w^2$, the squared channel gain $|h|^2$ is replaced by the sum of squared amplitudes $|h_i|^2$, $i = 1, 2, ..., N_r$. Equation (6.62) expresses the ergodic capacity due to the *linear combination* of the receive-antenna outputs, which is designed to maximize the information contained in the N_r received signals about the transmitted signal. This is simply a restatement of the maximal-ratio combining principle discussed in Section 6.2.

Problem 6.6   As pointed out previously, the selection combiner is a special case of the maximal-ratio combiner. What is the channel capacity of a wireless diversity channel using the selection combiner?

Ans. $\mathbf{E}[\log_2(1 + \rho h^2_{max})]$ *where* $h_{max} = \max\{|h_i|\}_{i=1}^{N_r}$ ∎

2. *Diversity-on-transmit channel.* The log-det capacity formula of Eq. (6.59) applies to this second case. Specifically, for $N_r = 1$, the channel matrix **H** reduces to a row

vector, and with it, Eq. (6.59) reduces to

$$C = \mathbf{E}\left[\log_2\left(1 + \frac{\rho}{N_t}\sum_{k=1}^{N_t}|h_k|^2\right)\right] \text{ bits/s/Hz} \tag{6.63}$$

where the matrix product $\mathbf{HH}^\dagger$ is replaced by the sum of squared amplitudes $|h_k|^2$, $k = 1,2,...,N_t$. Compared with Case 1 on receive diversity, the capacity of the diversity-on-transmit channel is reduced because the total transmit power is held constant, independently of the number of N_t transmit antennas.

6.4.3 Outage Capacity

To realize the log-det capacity formula of Eq. (6.59), the MIMO channel code needs to see an *ergodic process* of the random-channel processes, which, in turn, results in a hardening of the rate of reliable transmission to the $\mathbf{E}[\log_2\{\det(\cdot)\}]$ information rate (i.e., the channel capacity approaches the log-det formula). As in all information-theoretic arguments, the bit error rate would go to zero asymptotically in the block length of the code, thereby entailing a long transportation delay from the sender to the sink. In practice, however, the MIMO wireless channel is often nonergodic, and the requirement is to operate the channel under *delay constraints*. The issue of interest is then summed up as follows:

> How much information can be transmitted across a nonergodic channel, particularly if the channel code is long enough to see just one random-channel matrix?

In the situation described here, the rate of reliable information transmission (i.e., the strict Shannon-sense capacity) is zero, since, for any positive rate, the probability that the channel would not support such a rate is nonzero.

To get around this serious conceptual difficulty, the notion of *outage* is introduced into the characterization of the MIMO link. (Outage was discussed in Section 6.2 in the context of diversity-on-receive.) Specifically, *the outage probability* of a MIMO link is defined as *the probability for which the link is in a state of outage (i.e., failure) for data transmitted across the link at a certain rate R*, measured in bits per second per hertz. To proceed on this probabilistic basis, it is customary to operate the MIMO link by transmitting data in the form of bursts or frames, invoking a *quasi-static model* governed by four points:

1. The burst is *long* enough to accommodate the transmission of a large number of symbols, which, in turn, permits the use of an idealized *infinite-time horizon* that is basic to information theory.

2. Yet the burst is *short* enough that the wireless link can be treated as *quasi static* during each burst; the slow variation is used to justify the assumption that the receiver can acquire perfect knowledge of the channel state.

3. The channel matrix is permitted to change, say, from burst k to the next burst, $k + 1$, thereby accounting for statistical variations of the link.

4. The different realizations of the transmitted signal vector **s** are drawn from a *white Gaussian codebook*; that is, the correlation matrix of **s** is defined by Eq. (6.49).

Points 1 and 4 pertain to signal transmission, while points 2 and 3 pertain to the channel.

To proceed with the evaluation of outage probability, we first note that points 1 through 4 of the stochastic model just described for a nonstationary wireless link permit us to build on some of the results discussed in Section 6.4.1. In particular, in light of the log-det capacity formula of Eq. (6.59), we may view the random variable

$$C_k = \log_2\left\{\det\left(\mathbf{I}_{N_r} + \frac{\rho}{N_t}\mathbf{H}_k\mathbf{H}_k^\dagger\right)\right\} \text{ bits/s/Hz for burst } k \qquad (6.64)$$

as the expression for a sample of the wireless link. In other words, with the random-channel matrix $\mathbf{H}_k$ varying from one burst to the next, C_k will itself vary in a corresponding way. A consequence of this random behavior is that, occasionally, a draw from the cumulative distribution function of the wireless link results in a value for C_k that is inadequate to support reliable communication over the link, in which case the link is said to be in an *outage* state. Correspondingly, for a given transmission strategy, we define the *outage probability at rate R* as

$$P_{\text{outage}}(R) = \text{Prob}\{C_k < R \text{ for some burst } k\} \qquad (6.65)$$

or, equivalently,

$$P_{\text{outage}}(R) = \text{Prob}\left\{\log_2\left\{\det\left(\mathbf{I}_{N_r} + \frac{\rho}{N_t}\mathbf{H}_k\mathbf{H}_k^\dagger\right)\right\} < R \text{ for some burst } k\right\} \qquad (6.66)$$

On this basis, we may define the *outage capacity* of the MIMO link as the *maximum bit rate that can be maintained across the link for all bursts of data transmissions (i.e., all possible channel states) for a prescribed outage probability.*

Problem 6.7   To calculate the outage probability, we use the complementary cumulative distribution function of the random-channel matrix **H**, rather than the cumulative probability function itself. Explain. (By definition, the *complementary cumulative distribution function* (ccdf) is equal to unity minus the cumulative distribution function.)   ∎

EXAMPLE 6.4 Outage Capacity for Different Antenna Configurations and Varying Signal-to-Noise Ratios

In light of the random nature of the channel matrix **H**, the outage capacity is evaluated with the use of Monte Carlo simulation by computing the cumulative distribution function of the wireless link for a large number of statistically different realizations of **H**. To illustrate the simulation procedure, suppose we wish to calculate the outage capacity $C_{15\%}$ for error-free transmission for $100 - 15 = 85$ percent of the time. The calculation is performed for a (2,2) antenna configuration operating at a signal-to-noise ratio of 10 dB (i.e., $\rho = 10$).

We first obtain the *cumulative distribution function* (cdf) for this wireless link by generating a large number of Rayleigh-distributed random transfer functions under the flat-fading assumption.

According to Eq. (6.64) with $\rho = 10$ and $N_r = N_t = 2$, the capacity is given by

$$C_k = \log_2\left\{ \det\left(\mathbf{I}_2 + \frac{10}{2}\mathbf{H}_k\,\mathbf{H}_k^{\dagger}\right)\right\} \text{ bits/s/Hz}$$

for realization k of the channel transfer function. Fig. 6.13(a) plots the histogram (i.e., probability density function) of the resulting channel capacity data. This histogram is closely approximated by a *Gaussian distribution*, which should not be surprising when it is realized that an extensive amount of averaging could be involved in computing the log-det capacity formula.[8]

Integrating the probability density function curve of Fig. 6.13(a) and then subtracting the result from unity yields the complementary cumulative distribution function of the link, which is plotted in Fig. 6.13(b). Such a plot indicates the probability that a sample capacity will be greater

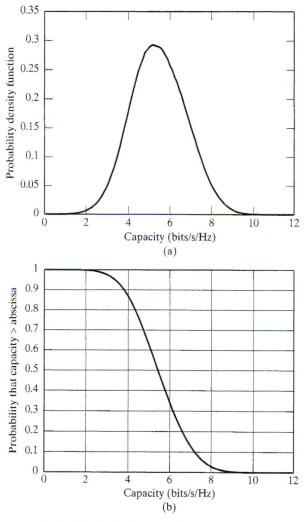

FIGURE 6.13   (a) Histogram (probability density function) of channel data for signal-to-noise ratio $\rho = 10$ dB. (b) Complementary cumulative probability distribution function corresponding to the histogram of part (a).

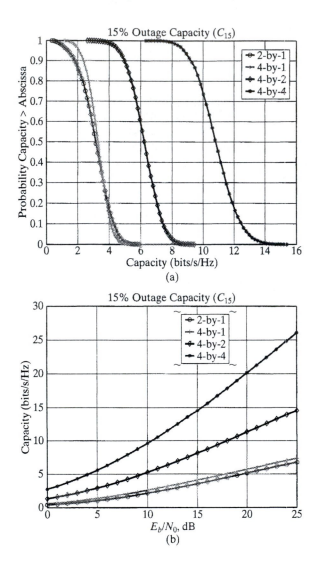

FIGURE 6.14 (a) Plots of the probability that the channel capacity is greater than the abscissa for four different antenna configurations:

$$N_t = 2, N_r = 1$$

$$N_t = 4, N_r = 1$$

$$N_t = 4, N_r = 2$$

$$N_t = 4, N_r = 4$$

(b) Plots of the outage capacity versus the signal-to-noise ratio for the four antenna configurations given in part (a).

than a threshold value (i.e., outage capacity) at a particular SNR (10 dB in this example) and for a specific antenna configuration (2 by 2 in this example). Thus, from Fig. 6.13(b), we can find the probability that the channel maintains a capacity of 4.2 bits/s/Hz for 85% of the time.

Figure 6.14 displays two different plots of the outage capacity for different signal-to-noise ratios and antenna configurations. Part (a) plots the probability that the channel capacity is greater than the abscissa versus the capacity in bits/s/Hz for a signal-to-noise ratio $\rho = 10$ dB. Part (b) of the figure portrays the picture differently by plotting the outage capacity in bits/s/Hz versus the signal-to-noise ratio in dB for varying antenna configurations.

Figure 6.14 clearly demonstrates two important points:

1. For $N_r = 1$, we see that increasing the number of transmit antennas by using $N_r = 2, 4$, results in a modest increase in the outage capacity for a fixed SNR.
2. For $N_t = 4$ and fixed SNR, there is a significant increase in the outage capacity in going from the antenna configuration (4,1) to (4,2) and a much bigger increase in the outage capacity in going from the antenna configuration (4,2) to (4,4).   ∎

Note that, at high signal-to-noise ratios, the outage probability $P_{\text{outage}}(R)$ as defined in Eq. (6.66) is approximately the same as the *frame (burst)-error probability* in terms of the signal-to-noise ratio exponent.[9] Accordingly, we may use an analysis based on the outage probability to evaluate the performance of practical space–time block coding techniques. (Space-time block codes are discussed later in Section 6.6.) That is, for a prescribed rate R, we may evaluate how the performance of a certain space–time block coding technique compares with that predicted through an outage analysis or measurement.

6.4.4 Channel Known at the Transmitter

The log-det capacity formula of Eq. (6.59) is based on the premise that the transmitter has *no* knowledge of the channel state. Knowledge of the channel state, however, can be made available to the transmitter by first estimating the channel matrix **H** at the receiver and then sending this estimate to the transmitter via a *feedback channel*.[10] In such a scenario, the capacity is optimized over the correlation matrix of the transmitted signal vector **s**, subject to the power constraint; that is, the trace of this correlation matrix is less than or equal to the constant transmit power P. Details of this optimization are presented in Appendix G.

From a practical perspective, it is important to note that the capacity gain provided by knowledge of the channel state at the transmitter over the log-det formula of Eq. (6.59) is significant only at low signal-to-noise ratios; the gain reduces to zero as the signal-to-noise ratio increases.

6.5 SINGULAR-VALUE DECOMPOSITION OF THE CHANNEL MATRIX

We may gain further insight into the behavior of a MIMO wireless communication system by applying what is known as the singular-value decomposition to the channel matrix of the system. The relationship between this algebraic decomposition of a rectangular matrix and the eigendecomposition of a Hermitian matrix formed by multiplying the matrix by its Hermitian transpose is discussed in Appendix H.

To begin the exposition, consider the matrix product $\mathbf{HH}^\dagger$ in the log-det capacity formula of Eq. (6.59). This product satisfies the Hermitian property for all $\mathbf{H}$. We may therefore diagonalize $\mathbf{HH}^\dagger$ by invoking the *eigendecomposition* of a Hermitian matrix and so write

$$\mathbf{U}^\dagger \mathbf{HH}^\dagger \mathbf{U} = \Lambda \qquad (6.67)$$

where the two new matrices $\mathbf{U}$ and Λ are described as follows:

- The matrix Λ is a *diagonal matrix* whose N_r elements are the *eigenvalues* of the matrix product $\mathbf{HH}^\dagger$.
- The matrix $\mathbf{U}$ is a *unitary matrix* whose N_r columns are the *eigenvectors* associated with the eigenvalues of $\mathbf{HH}^\dagger$.

By definition, the inverse of a unitary matrix is equal to the Hermitian transpose of the matrix, as shown by

$$\mathbf{U}^{-1} = \mathbf{U}^\dagger \qquad (6.68)$$

or, equivalently,

$$\mathbf{UU}^\dagger = \mathbf{U}^\dagger \mathbf{U} = \mathbf{I}_{N_r} \qquad (6.69)$$

where $\mathbf{I}_{N_r}$ is the N_r-by-N_r identity matrix.

Let the N_t-by-N_t matrix $\mathbf{V}$ be another unitary matrix; that is,

$$\mathbf{VV}^\dagger = \mathbf{V}^\dagger \mathbf{V} = \mathbf{I}_{N_t} \qquad (6.70)$$

where $\mathbf{I}_{N_t}$ is the N_t-by-N_t identity matrix. Since the multiplication of a matrix by the identity matrix leaves the matrix unchanged, we may inject the matrix product $\mathbf{VV}^\dagger$ into the center of the left-hand side of Eq. (6.67), thus:

$$\mathbf{U}^\dagger \mathbf{H}(\mathbf{VV}^\dagger)\mathbf{H}^\dagger \mathbf{U} = \Lambda \qquad (6.71)$$

The left-hand side of Eq. (6.71), representing a square matrix, is recognized as the product of two rectangular matrices: the N_r-by-N_t matrix $\mathbf{U}^\dagger \mathbf{HV}$ and the N_t-by-N_r matrix $\mathbf{V}^\dagger \mathbf{H}^\dagger \mathbf{U}$, which are the Hermitian of each other. Let the N_t-by-N_t matrix $\mathbf{D}$ denote a new diagonal matrix related to the N_r-by-N_r diagonal matrix Λ with $N_r \leq N_t$ by

$$\Lambda = [\mathbf{D} \ \mathbf{0}][\mathbf{D} \ \mathbf{0}]^\dagger \qquad (6.72)$$

where the null matrix $\mathbf{0}$ is added to maintain proper overall matrix dimensionality of the equation. Except for some zero elements, $\mathbf{D}$ is the *square root* of Λ. Then, examining Eqs. (6.71) and (6.72) and comparing terms, we deduce the new decomposition

$$\mathbf{U}^\dagger \mathbf{HV} = [\mathbf{D} \ \mathbf{0}] \qquad (6.73)$$

Equation (6.73) is a mathematical statement of the *singular-value decomposition (SVD) theorem*, according to which we have the following descriptions:

- The elements of the diagonal matrix

$$\mathbf{D} = \text{diag}(d_1, d_2, ..., d_{N_t}) \tag{6.74}$$

  are the *singular values* of the channel matrix $\mathbf{H}$.

- The columns of the unitary matrix

$$\mathbf{U} = [\mathbf{u}_1, \mathbf{u}_2, ..., \mathbf{u}_{N_r}] \tag{6.75}$$

  are the *left singular vectors* of matrix $\mathbf{H}$.

- The columns of the second unitary matrix

$$\mathbf{V} = [\mathbf{v}_1, \mathbf{v}_2, ..., \mathbf{v}_{N_t}] \tag{6.76}$$

  are the *right singular vectors* of matrix $\mathbf{H}$.

Problem 6.8 Applying the singular-value decomposition of Eq. (6.73) to the basic channel model of Eq. (6.48), show that for $N_r \leq N_t$:

$$\bar{\mathbf{x}} = [\mathbf{D}, \mathbf{0}]\bar{\mathbf{s}} + \bar{\mathbf{w}} \tag{6.77}$$

where

$$\bar{\mathbf{x}} = \mathbf{U}^\dagger \mathbf{x} \tag{6.78}$$

$$\bar{\mathbf{s}} = \mathbf{V}^\dagger \mathbf{s} \tag{6.79}$$

and

$$\bar{\mathbf{w}} = \mathbf{U}^\dagger \mathbf{w} \tag{6.80}$$

∎

Using the definitions of Eqs. (6.74) through (6.76), we may rewrite the decomposed channel model of Eq. (6.77) in the scalar form

$$\bar{x}_i = d_i \bar{s}_i + \bar{w}_i \qquad i = 1, 2, ..., N_r \tag{6.81}$$

According to Eq. (6.81), singular-value decomposition of the channel matrix $\mathbf{H}$ has transformed the MIMO wireless link with $N_r \leq N_t$ into N_r *virtual channels*, as illustrated in Fig. 6.15. (Note that $\bar{s}_i = 0$ for $N_r < i \leq N_t$.)The virtual channels are all *decoupled* from each other in that they constitute a parallel set of N_r *single-input, single-output (SISO) channels*, with each channel being described by the scalar input–output relation of Eq. (6.81). A comparison of the channel models of Figs. 6.12 and 6.15 immediately reveals the decoupling facilitated in the virtual model of Fig. 6.15 by the singular-value decomposition of the channel matrix $\mathbf{H}$.

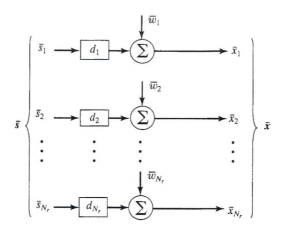

FIGURE 6.15   Set of N_r virtual decoupled channels resulting from the singular-value decomposition of the channel matrix **H**, assuming that $N_t \leq N_r$.

6.5.1 Eigendecomposition of the Log-det Capacity Formula

The log-det formula of Eq. (6.59) for the ergodic capacity of a MIMO link involves the matrix product $\mathbf{HH}^\dagger$. Substituting Eq. (6.59) into Eq. (6.67) leads to the *spectral decomposition* of $\mathbf{HH}^\dagger$ in terms of N_r *eigenmodes*, with each eigenmode corresponding to virtual data transmission using a pair of right- and left-singular vectors of the channel matrix **H** as the transmit and receive antenna weights, respectively. Thus, we may write

$$\mathbf{HH}^\dagger = \mathbf{U}\mathbf{\Lambda}\mathbf{U}^\dagger$$
$$= \sum_{i=1}^{N_r} \lambda_i \mathbf{u}_i \mathbf{u}_i^\dagger \tag{6.82}$$

where the outer product $\mathbf{u}_i \mathbf{u}_i^\dagger$ is an N_r-by-N_r matrix with a rank equal to unity. Moreover, substituting the first line of this decomposition into the determinant part of Eq. (6.59) yields

$$\det\left(\mathbf{I}_{N_r} + \frac{\rho}{N_t}\mathbf{HH}^\dagger\right) = \det\left(\mathbf{I}_{N_r} + \frac{\rho}{N_t}\mathbf{U}\mathbf{\Lambda}\mathbf{U}^\dagger\right) \tag{6.83}$$

Next, invoking the *determinant identity*

$$\det(\mathbf{I} + \mathbf{AB}) = \det(\mathbf{I} + \mathbf{BA}) \tag{6.84}$$

and then using the defining equation (6.69), we may rewrite Eq. (6.83) in the equivalent form

$$\det\left(\mathbf{I}_{N_r} + \frac{\rho}{N_t}\mathbf{HH}^\dagger\right) = \det\left(\mathbf{I}_{N_r} + \frac{\rho}{N_t}\mathbf{U}^\dagger\mathbf{U}\Lambda\right)$$

$$= \det\left(\mathbf{I}_{N_r} + \frac{\rho}{N_t}\Lambda\right) \tag{6.85}$$

$$= \prod_{i=1}^{N_r}\left(1 + \frac{\rho}{N_t}\lambda_i\right)$$

where λ_i is the ith eigenvalue of $\mathbf{HH}^\dagger$. Finally, substituting Eq. (6.85) into Eq. (6.59) yields

$$C = \mathbf{E}\left[\sum_{i=1}^{N_r}\log_2\left(1 + \frac{\rho}{N_t}\lambda_i\right)\right]\text{bits/s/Hz} \tag{6.86}$$

which is subject to constant transmit power; the expectation is over the eigenvalues of the matrix product $\mathbf{HH}^\dagger$. Equation (6.86) shows that, thanks to the properties of the logarithm, *the ergodic capacity of a MIMO wireless communication system is the sum of capacities of N_r virtual single-input, single-output channels defined by the spatial eigenmodes of the matrix product $\mathbf{HH}^\dagger$*.

According to Eq. (6.86), the channel capacity C attains its maximum value when equal signal-to-noise ratios ρ/N_t are allocated to each virtual channel in Fig. 6.15 (i.e., the N_r eigenmodes of the channel matrix $\mathbf{H}$ are all equally effective). By the same token, the capacity C is minimum when there is a single virtual channel (i.e., all the eigenmodes of the channel matrix $\mathbf{H}$ are zero except for one). The capacities of actual wireless links lie somewhere between these two extremes.

Specifically, as a result of *fading correlation* encountered in practice, it is possible for there to be a large disparity amongst the eigenvalues of $\mathbf{HH}^\dagger$; that is, one or more of the eigenvalues $\lambda_1, \lambda_2, ..., \lambda_{N_r}$ may be small. Such a disparity has quite a detrimental effect on the capacity of the wireless link, compared with the maximal condition under which all the eigenvalues of $\mathbf{HH}^\dagger$ are equal. A similar effect may also arise in a Rician fading environment when the line-of-sight (LOS) component is quite strong (i.e., the Rician factor is greater than, say, 10 dB), in which case one eigenmode of the channel is dominant. For example, when the angle spread of the incoming radio waves impinging on a linear array of receive antennas is reduced from 60° to 0.6°, the complementary cumulative distribution function (ccdf) of a MIMO wireless communication system with $(N_t, N_r) = (7,7)$ antenna configuration degenerates effectively to that of a (7,1) system.[11]

Equation (6.86), based on the log-det capacity formula of Eq. (6.59), assumes that $N_t \geq N_r$. Using the log-det capacity formula of Eq. (6.60) for the alternative case,

$N_t \leq N_r$, and following a procedure similar to that used to derive Eq. (6.86), we may show that

$$C = \mathbf{E}\left[\sum_{i=1}^{N_t} \log_2\left(1 + \frac{\rho}{N_r}\lambda_i\right)\right] \text{ bits/s/Hz} \tag{6.87}$$

where the expectation is over the λ_i, which denote the eigenvalues of the N_t-by-N_t matrix product $\mathbf{H}^\dagger\mathbf{H}$.

Problem 6.9   For the special case of $N_t = N_r = N$, show that the ergodic capacity scales linearly, rather than logarithmically, with increasing SNR as N approaches infinity.

Ans. $C \rightarrow \mathbf{E}\displaystyle\sum_{i=1}^{N}\left(\frac{\rho}{N\log_e 2}\right)\lambda_i = \left(\frac{\lambda_{\text{av}}}{\log_e 2}\right)\rho$

where

$$\lambda_{\text{av}} = \frac{1}{N}\sum_{i=1}^{N}\mathbf{E}\lambda_i$$

is the average eigenvalue of $\mathbf{HH}^\dagger = \mathbf{H}^\dagger\mathbf{H}.$ ∎

6.6   SPACE–TIME CODES FOR MIMO WIRELESS COMMUNICATIONS

Transmission techniques for MIMO wireless communications may be considered under two broadly defined categories:

1. *Unconstrained signaling techniques*, exemplified by the so-called BLAST architectures, whose aim is to increase the channel capacity by using standard channel codes.

2. *Space–time codes*, whose aim is the joint channel encoding of multiple transmit antennas.

BLAST architectures are considered under Theme Example 1 in Section 6.9. Space–time codes are discussed in this section.

As with ordinary channel codes, space–time codes employ *redundancy* for the purpose of providing protection against channel fading, noise, and interference. They may also be used to minimize the outage probability or, equivalently, maximize the outage capacity. Depending on the level of redundancy introduced into the design of space–time codes, the degree of statistical independence among the transmitted signals is correspondingly reduced.

Space–time codes may themselves be classified into two types—space–time trellis codes and space–time block codes—depending on how transmission of the signal over the wireless channel takes place.

A *space–time trellis code* permits the *serial* transmission of symbols by combining signal processing at the receiver with coding techniques that are appropriate to the use of multiple antennas at the transmitter. Space–time trellis codes, designed for two to four transmit antennas, perform extremely well in a slow-fading environment, exemplified by indoor data transmission. For decoding, a multidimensional (i.e., vector) version of the Viterbi algorithm is required. Accordingly, for a fixed number of transmit antennas, the *decoding complexity of space–time trellis codes (measured in terms of trellis states at the decoder) increases exponentially* as a function of the spectral efficiency.[12]

In a *space–time block code*, by contrast, transmission of the signal takes place in *blocks*. The code is defined by a *transmission matrix*, the formulation of which involves three parameters:

- The number of transmitted symbols, denoted by l
- The number of transmit antennas, denoted by N_t, which defines the *size* of the transmission matrix
- The number of time slots in a data block, denoted by m

With m time slots involved in the transmission of l symbols, the ratio l/m defines the *rate* of the code, which is denoted by κ.

For efficient transmission, the transmitted symbols are expressed in *complex* form. Moreover, in order to facilitate the use of *linear processing* to estimate the transmitted symbols at the receiver and thereby simplify the receiver design, orthogonality is introduced into the design of the transmission matrix. Here, we may identify two different design procedures:

- *Complex orthogonal design*, in which the transmission matrix is square, satisfying the condition for complex *orthogonality* in both the spatial and temporal sense.
- *Generalized complex orthogonal design*, in which the transmission matrix is non-square, satisfying the condition for complex orthogonality only in the temporal sense; the code rate is less than unity.

In other words, *complex orthogonality of the transmission matrix in the temporal sense is a sufficient condition for linear processing at the receiver.*

A complex orthogonal design of size N_t exists if, and only if, $N_t = 2$ (i.e., the transmitter uses two transmit antennas), and the *Alamouti code* is that code with a code rate of unity. In contrast, generalized complex orthogonal designs permit the use of more than two transmit antennas, with the result that the code rate is less than unity.

The Alamouti code is much less complex than a space–time trellis code for the same antenna configuration (i.e., two transmit antennas and a single receive antenna), but it does not perform as well as the same space–time trellis code. Nonetheless, the Alamouti code is the preferred choice, essentially because of its remarkable computational simplicity and satisfactory performance capability. Indeed, the discovery of the code by Alamouti was motivated by addressing the issue of exponentially increasing decoding complexity of space–time trellis codes, and this discovery, in turn, motivated

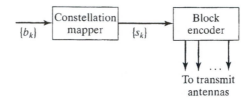

FIGURE 6.16   Block diagram of orthogonal space–time block encoder.

the search by Tarokh *et al.* for its generalization so as to accommodate the use of more than two transmit antennas.

Before proceeding to discuss the Alamouti code and generalized complex orthogonal designs of space–time block codes, we present some preliminary considerations.

6.6.1 Preliminaries

Figure 6.16 shows the baseband diagram of a *space–time block encoder*, which consists of two functional units: a mapper and the block encoder itself. The *mapper* takes the incoming binary data stream $\{b_k\}$, $b_k = \pm 1$, and generates a new *sequence of blocks*, with each block made up of multiple symbols that are complex. For example, the mapper may be in the form of an *M-ary PSK mapper* or an *M-ary QAM mapper*, which are illustrated for $M = 16$ in the signal-space diagrams of Fig. 6.17. All the symbols in a particular column of the transmission matrix are pulse shaped (in accordance with the criteria described in Section 3.4) and then modulated into a form suitable for simultaneous transmission over the channel by the transmit antennas. The pulse shaper and modulator are not shown in Fig. 6.16, as the basic issue of interest is that of baseband data transmission with an emphasis on the formulation of space–time block codes. The *block encoder* converts each block of complex symbols produced by the mapper into an l-by-N_t transmission matrix $\mathbf{S}$, where l and N_t are the *temporal* dimension and *spatial* dimension, respectively, of the transmission matrix. The individual elements of the transmission matrix $\mathbf{S}$ are made up of the complex symbols, say, $\tilde{s}_k$, generated by the mapper, their complex conjugates $\tilde{s}_k^*$, and linear combinations of $\tilde{s}_k$ and $\tilde{s}_k^*$, where the asterisk denotes complex conjugation.

EXAMPLE 6.5 Quadriphase-Shift Keying

As a simple example, consider the map portrayed by QPSK with $M = 4$. The map is described in Table 6.1, where E is the transmitted signal energy.

The input dibits (pairs of adjacent bits) are *Gray* encoded, wherein only one bit is flipped as we move from one symbol to the next. (Gray encoding was discussed in Section 3.6.) The mapped signal points lie on a circle of radius $\mathbf{E}$ centered at the origin of the signal-space diagram. ∎

Problem 6.10   Construct the map describing the *M*-ary PSK for $M = 16$ in Fig. 6.17(a).   ∎

TABLE 6.1   Gray-encoded QPSK mapper.

Dibit	Coordinates of mapped signal points
10	$\sqrt{E/2}(1, -1) = \sqrt{E}e^{j7\pi/4}$
11	$\sqrt{E/2}(-1, -1) = \sqrt{E}e^{j5\pi/4}$
01	$\sqrt{E/2}(-1, +1) = \sqrt{E}e^{j3\pi/4}$
00	$\sqrt{E/2}(+1, +1) = \sqrt{E}e^{j\pi/4}$

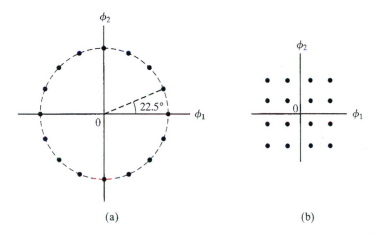

(a)   (b)

FIGURE 6.17   (a) Signal constellation of 16-PSK. (b) Signal constellation of 16-QAM.

Problem 6.11   Construct the map describing the M-ary QAM for $M = 16$ in Fig. 6.17(b).   ∎

Henceforth, the discussion of space–time block codes is confined to two-dimensional mappers exemplified by those portrayed in Fig. 6.17. That is, the output of the mapper is represented by a complex number, as illustrated in Table 6.1 for QPSK.

6.6.2 Alamouti Code[13]

The *Alamouti code* is a two-by-one orthogonal space–time block code. That is, it uses two transmit antennas and a single receive antenna, as shown in the block diagram of Fig. 6.18. Let $\tilde{s}_1$ and $\tilde{s}_2$ denote the complex symbols (signals) produced by the mapper, which are to be transmitted over the wireless channel. Signal transmission over the channel proceeds as follows:

- At some arbitrary time t, antenna 1 transmits $\tilde{s}_1$, and simultaneously, antenna 2 transmits $\tilde{s}_2$.
- At time $t + T$, where T is the symbol duration, signal transmission is switched, with $-\tilde{s}_2^*$ transmitted by antenna 1 and $\tilde{s}_1^*$ simultaneously transmitted by antenna 2.

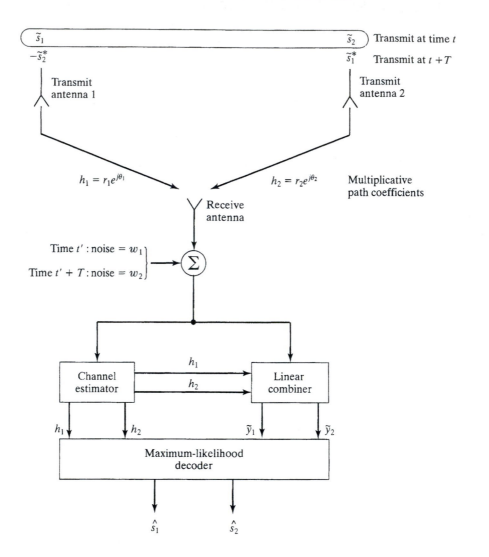

FIGURE 6.18 Block diagram of the transceiver (transmitter and receiver) for the Alamouti code. Note that $t' > t$ to allow for propagation delay.

The two-by-two space–time block code, just described, is formally written in matrix form as

$$
\mathbf{S} = \begin{bmatrix} \tilde{s}_1 & \tilde{s}_2 \\ -\tilde{s}_2^* & \tilde{s}_1^* \end{bmatrix} \implies \text{Space}
$$

$$\downarrow \text{Time}$$

$$(6.88)$$

The *transmission matrix* $\mathbf{S}$ is a *complex-orthogonal matrix* (quaternion), in that it satisfies the condition for orthogonality in both the spatial and temporal sense. To demonstrate this important property, let

$$\mathbf{S}^\dagger = \begin{bmatrix} \tilde{s}_1^* & -\tilde{s}_2 \\ \tilde{s}_2^* & \tilde{s}_1 \end{bmatrix} \implies \text{Time}$$

$$\downarrow \text{Space}$$

(6.89)

denote the *Hermitian transpose* of $\mathbf{S}$, involving both transposition and complex conjugation. To demonstrate orthogonality in a spatial sense, we multiply the code matrix $\mathbf{S}$ by its Hermitian transpose $\mathbf{S}^\dagger$ on the right, obtaining

$$\mathbf{S}\mathbf{S}^\dagger = \begin{bmatrix} \tilde{s}_1 & \tilde{s}_2 \\ -\tilde{s}_2^* & \tilde{s}_1^* \end{bmatrix} \begin{bmatrix} \tilde{s}_1^* & -\tilde{s}_2 \\ \tilde{s}_2^* & \tilde{s}_1 \end{bmatrix}$$

$$= \begin{bmatrix} |\tilde{s}_1|^2 + |\tilde{s}_2|^2 & -\tilde{s}_1\tilde{s}_2 + \tilde{s}_2\tilde{s}_1 \\ -\tilde{s}_2^*\tilde{s}_1^* + \tilde{s}_1^*\tilde{s}_2^* & |\tilde{s}_2|^2 + |\tilde{s}_1|^2 \end{bmatrix} \qquad (6.90)$$

$$= (|\tilde{s}_1|^2 + |\tilde{s}_2|^2) \begin{bmatrix} 1 & 0 \\ 0 & 1 \end{bmatrix}$$

which equals the two-by-two identity matrix, multiplied by the scaling factor $(|\tilde{s}_1|^2 + |\tilde{s}_2|^2)$.

This same result also holds for the alternative matrix product $\mathbf{S}^\dagger\mathbf{S}$, which is proof of *orthogonality* in the temporal sense. Thus, the transmission matrix of the Alamouti code satisfies the unique condition

$$\mathbf{S}\mathbf{S}^\dagger = \mathbf{S}^\dagger\mathbf{S} = (|\tilde{s}_1|^2 + |\tilde{s}_2|^2)\mathbf{I} \qquad (6.91)$$

where $\mathbf{I}$ is the two-by-two identity matrix. Note that

$$\mathbf{S}^{-1} = \frac{1}{|\tilde{s}_1|^2 + |\tilde{s}_2|^2} \mathbf{S}^\dagger \qquad (6.92)$$

In light of Eqs. (6.63), (6.90), and (6.91), we may summarize four important properties of the Alamouti code:

Property 1. Unitarity (Complex Orthogonality)

The Alamouti code is an orthogonal space–time block code in that the product of its transmission matrix with its Hermitian transpose is equal to the two-by-two identity matrix scaled by the sum of the squared amplitudes of the transmitted symbols.

Property 2. Full-Rate Complex Code

The Alamouti code (with two transmit antennas) is the only complex space–time block code with a code rate of unity. For any signal constellation, it therefore achieves full diversity at the full transmission rate.

Property 3. Linearity

The Alamouti code is linear in the transmitted symbols. That is, we may expand the code matrix **S** as a linear combination of the transmitted symbols and their complex conjugates. Doing so, we obtain

$$\mathbf{S} = \tilde{s}_1 \Gamma_{11} + \tilde{s}_1^* \Gamma_{12} + \tilde{s}_2 \Gamma_{21} + \tilde{s}_2^* \Gamma_{22} \tag{6.93}$$

where

$$\left.\begin{aligned}
\Gamma_{11} &= \begin{bmatrix} 1 & 0 \\ 0 & 0 \end{bmatrix} \\[6pt]
\Gamma_{12} &= \begin{bmatrix} 0 & 0 \\ 0 & 1 \end{bmatrix} \\[6pt]
\Gamma_{21} &= \begin{bmatrix} 0 & 1 \\ 0 & 0 \end{bmatrix} \\[6pt]
\Gamma_{22} &= \begin{bmatrix} 0 & 0 \\ -1 & 0 \end{bmatrix}
\end{aligned}\right\} \tag{6.94}$$

Property 4. Optimality of Capacity

For two transmit antennas and a single receive antenna, the Alamouti code is the only optimal space–time block code that satisfies the log-det capacity formula of Eq. (6.63).

Turning next to the design of the receiver, we assume that the channel is frequency flat and slowly varying, such that the complex multiplicative distortion introduced by the channel at time t is essentially the same as that at time $t + T$, where T is the symbol duration. This multiplicative distortion is denoted by $\alpha_k e^{j\theta_k}$ where $k = 1,2$, as indicated in Fig. 6.18. Thus, with $\tilde{s}_1$ and $\tilde{s}_2$ transmitted simultaneously at time t, the complex received signal at time $t' > t$, allowing for propagation delay, is

$$\tilde{x}_1 = \alpha_1 e^{j\theta_1} \tilde{s}_1 + \alpha_2 e^{j\theta_2} \tilde{s}_2 + w_1 \tag{6.95}$$

where w_1 is the complex channel noise at time t'. Next, with $-\tilde{s}_2^*$ and $\tilde{s}_1^*$ transmitted simultaneously at time $t + T$, the corresponding complex signal received at time $t' + T$ is

$$\tilde{x}_2 = -\alpha_1 e^{j\theta_1} \tilde{s}_2^* + \alpha_2 e^{j\theta_2} \tilde{s}_1^* + w_2 \tag{6.96}$$

where w_2 is the complex channel noise at time $t' + T$.

In the course of time, from time t' to $t' + T$, the channel estimator in the receiver produces estimates of the multiplicative distortion represented by $\alpha_k e^{j\theta_k}$, where $k = 1$, 2. Hereafter, we assume that these two estimates are accurate enough for them to be treated as essentially exact. In other words, $\alpha_1 e^{j\theta_1}$ and $\alpha_2 e^{j\theta_2}$ are both known to the receiver. Accordingly, we may reformulate the variable $\tilde{x}_1$ defined by Eq. (6.95) and the complex conjugate of the second variable $\tilde{x}_2$ defined in Eq. (6.96) in matrix form, as shown by

$$\begin{bmatrix} \tilde{x}_1 \\ \tilde{x}_2^* \end{bmatrix} = \begin{bmatrix} \alpha_1 e^{j\theta_1} & \alpha_2 e^{j\theta_2} \\ \alpha_2 e^{-j\theta_2} & -\alpha_1 e^{-j\theta_1} \end{bmatrix} \begin{bmatrix} \tilde{s}_1 \\ \tilde{s}_2 \end{bmatrix} + \begin{bmatrix} \tilde{w}_1 \\ \tilde{w}_2^* \end{bmatrix} \tag{6.97}$$

The nice thing about this equation is that the original complex signals s_1 and s_2 appear as a vector of two unknowns. It is with this goal in mind that $\tilde{x}_1$ and $\tilde{x}_2^*$ were used for the elements of the two-by-one received signal vector on the left-hand side of Eq. (6.97).

Problem 6.12   Show that the two-by-two channel matrix in Eq. (6.97) defined by the multiplicative fading factors $\alpha_1 e^{j\theta_1}$ and $\alpha_2 e^{j\theta_2}$ is an orthogonal matrix; that is, show that

$$\begin{bmatrix} \alpha_1 e^{j\theta_1} & \alpha_2 e^{j\theta_2} \\ \alpha_2 e^{-j\theta_2} & -\alpha_1 e^{-j\theta_1} \end{bmatrix}^\dagger \begin{bmatrix} \alpha_1 e^{j\theta_1} & \alpha_2 e^{j\theta_2} \\ \alpha_2 e^{-j\theta_2} & -\alpha_1 e^{-j\theta_1} \end{bmatrix} = (\alpha_1^2 + \alpha_2^2) \begin{bmatrix} 1 & 0 \\ 0 & 1 \end{bmatrix}$$

∎

Motivated by the result of Problem 6.12, we multiply both sides of Eq. (6.97) by the Hermitian transpose of the two-by-two channel matrix, namely,

$$\begin{bmatrix} \alpha_1 e^{-j\theta_1} & \alpha_2 e^{j\theta_2} \\ \alpha_2 e^{-j\theta_2} & -\alpha_1 e^{j\theta_1} \end{bmatrix}$$

and thus obtain the following pair of simultaneous equations, written in matrix form:

$$\begin{bmatrix} \alpha_1 e^{-j\theta_1} & \alpha_2 e^{j\theta_2} \\ \alpha_2 e^{-j\theta_2} & -\alpha_1 e^{j\theta_1} \end{bmatrix} \begin{bmatrix} \tilde{x}_1 \\ \tilde{x}_2^* \end{bmatrix} = (\alpha_1^2 + \alpha_2^2) \begin{bmatrix} 1 & 0 \\ 0 & 1 \end{bmatrix} \begin{bmatrix} \tilde{s}_1 \\ \tilde{s}_2 \end{bmatrix} + \begin{bmatrix} \alpha_1 e^{-j\theta_1} & \alpha_2 e^{j\theta_2} \\ \alpha_2 e^{-j\theta_2} & -\alpha_1 e^{j\theta_1} \end{bmatrix} \begin{bmatrix} \tilde{w}_1 \\ \tilde{w}_2^* \end{bmatrix}$$

$$\tag{6.98}$$

$$= (\alpha_1^2 + \alpha_2^2) \begin{bmatrix} \tilde{s}_1 \\ \tilde{s}_2 \end{bmatrix} + \begin{bmatrix} \alpha_1 e^{-j\theta_1} \tilde{w}_1 + \alpha_2 e^{j\theta_2} \tilde{w}_2^* \\ \alpha_2 e^{-j\theta_2} \tilde{w}_1 - \alpha_1 e^{j\theta_1} \tilde{w}_2^* \end{bmatrix}$$

The two-by-one *vector product* on the left-hand side of Eq. (6.98) defines the two complex outputs computed by the linear combiner in Fig. 6.18 in terms of the complex signals $\tilde{x}_1$ and $\tilde{x}_2$ received at times t' and $t' + T$. Thus, let

$$
\begin{bmatrix} \tilde{y}_1 \\ \tilde{y}_2 \end{bmatrix} = \begin{bmatrix} \alpha_1 e^{-j\theta_1} & \alpha_2 e^{j\theta_2} \\ \alpha_2 e^{-j\theta_2} & -\alpha_1 e^{j\theta_1} \end{bmatrix} \begin{bmatrix} \tilde{x}_1 \\ \tilde{x}_2^* \end{bmatrix}
$$

$$
= \begin{bmatrix} \alpha_1 e^{-j\theta_1} \tilde{x}_1 + \alpha_2 e^{j\theta_2} \tilde{x}_2^* \\ \alpha_2 e^{-j\theta_2} \tilde{x}_1 - \alpha_1 e^{j\theta_1} \tilde{x}_2^* \end{bmatrix}
$$

(6.99)

define the two-by-one vector of complex signals at the output of the combiner. Correspondingly, let

$$
\begin{bmatrix} \tilde{v}_1 \\ \tilde{v}_2 \end{bmatrix} = \begin{bmatrix} \alpha_1 e^{-j\theta_1} \tilde{w}_1 + \alpha_2 e^{j\theta_2} \tilde{w}_2^* \\ \alpha_2 e^{-j\theta_2} \tilde{w}_1 - \alpha_1 e^{j\theta_1} \tilde{w}_2^* \end{bmatrix}
$$

(6.100)

define the two-by-one vector of additive complex noise contributions in the combiner outputs. Accordingly, we may recast Eq. (6.98) in a matrix form of input–output relations describing the overall behavior of the Alamouti code, structured as in Fig. 6.18, as follows:

$$
\begin{bmatrix} \tilde{y}_1 \\ \tilde{y}_2 \end{bmatrix} = (\alpha_1^2 + \alpha_2^2) \begin{bmatrix} \tilde{s}_1 \\ \tilde{s}_2 \end{bmatrix} + \begin{bmatrix} \tilde{v}_1 \\ \tilde{v}_2 \end{bmatrix}
$$

(6.101)

In expanded form,

$$
\tilde{y}_k = (\alpha_1^2 + \alpha_2^2) \tilde{s}_k + \tilde{v}_k \qquad k = 1, 2
$$

(6.102)

Note that, due to complex orthogonality of the Alamouti code, the unwanted symbol $\tilde{s}_2$ is cancelled out in the equation for $\tilde{y}_1$ and the unwanted symbol $\tilde{s}_1$ is cancelled out in the equation for $\tilde{y}_2$. *It is these cancellations that are responsible for the simplification of the receiver.*

Note also that the scaling factor $(\alpha_1^2 + \alpha_2^2)$ can be small only if the fading coefficients α_1 and α_2 are both small. In other words, the diversity paths linking the receive antenna to the two transmit antennas must simultaneously undergo fading in order for $(\alpha_1^2 + \alpha_2^2)$ to be small. Hence, the detrimental effect of fading arises if, and only if, both diversity paths suffer from it. We may therefore state that a wireless communication system based on the Alamouti code enjoys a *two-level diversity gain.*

Problem 6.13   Ordinarily, the complex channel noise terms $\tilde{w}_1$ and $\tilde{w}_2$ are Gaussian distributed with zero mean and a common variance. Assume that $\alpha_k e^{j\theta_k}$, $k = 1, 2$, are known multiplicative constants.

(a) Show that the complex noise terms $\tilde{v}_1$ and $\tilde{v}_2$ at the combiner output are also Gaussian distributed with zero mean.

(b) Given that the real and imaginary components of $\tilde{v}_1$ and $\tilde{v}_2$ have a common variance σ_0^2, determine the common variance of $\tilde{v}_1$ and $\tilde{v}_2$.

(c) Show that the probability density function of v_1 or v_2 is

$$f_{V_k}(\tilde{v}_k) = \frac{1}{\pi\sigma_0^2}\exp\left(-\frac{|\tilde{v}_k|^2}{\sigma_0^2}\right) \qquad k = 1, 2 \tag{6.103}$$

■

Equation (6.102) is a partial description of the receiver structure depicted in Fig. 6.18 for an Alamouti-encoded system. The next and final issue to be considered is, given the noisy linear combiner output $\tilde{y}_k$ expressed in terms of the transmitted symbol $\tilde{s}_k$, how to provide an optimal estimate of $\tilde{s}_k$ for $k = 1,2$. To provide insight into this symbol-estimation problem, Fig. 6.19 illustrates the signal-space scenario for an Alamouti-encoded system based on the QPSK signal constellation. At time t, the symbols $\tilde{s}_1$ and $\tilde{s}_2$ drawn from this constellation are transmitted, followed by transmission of the modified symbols $-\tilde{s}_2^*$ and $\tilde{s}_1^*$ at time $t + T$. To account for the multipath fading phenomenon, the signal points $\tilde{s}_1$ and $\tilde{s}_2$ are weighted by the multiplying factor $(\alpha_1^2 + \alpha_2^2)$ in accordance with Eq. (6.102). The complex Gaussian noise clouds centered on the weighted signal points represent the effect of additive complex noise terms, whose intensity decreases as we move away from the weighted signal points. Figure 6.19 also includes the linear combiner outputs signified as the observations $\tilde{y}_1$ and $\tilde{y}_2$. Given the scenario pictured in this figure, what is the optimal decision for the receiver to make?

The answer to this fundamental question lies in the *maximum-likelihood decoding rule*, which assumes that the transmitted symbols $\tilde{s}_1$ and $\tilde{s}_2$ are *equally probable*. Let $\hat{s}_1$ and $\hat{s}_2$ denote the maximum-likelihood estimates of $\tilde{s}_1$ and $\tilde{s}_2$, respectively. One further item that we need to introduce is

$$d^2(\tilde{z}, \tilde{\xi}) = (\tilde{z} - \tilde{\xi})(\tilde{z} - \tilde{\xi})^*$$

which defines the squared Euclidean distance between two complex signal points denoted by $\tilde{z}$ and $\tilde{\xi}$.

To proceed then with the formulation of the maximum-likelihood decoding rule for an Alamouti-encoded system, we see from Fig. 6.18 that, given knowledge of the complex channel coefficients h_1 and h_2, the maximum-likelihood decoder performs its decision-making in response to the linear combiner outputs $\tilde{y}_1$ and $\tilde{y}_2$, which play the role of *decision statistics*. (By definition, sufficient statistics summarize the whole of the

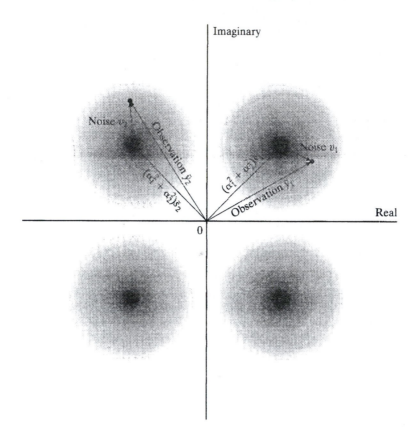

FIGURE 6.19   Signal-space diagram for Alamouti code, using the QPSK signal constellation. The signal point $\tilde{s}_1$ and $\tilde{s}_2$ are weighted by $(\alpha_1^2 + \alpha_2^2)$. The corresponding linear combiner outputs $\tilde{y}_1$ and $\tilde{y}_2$ are noisy versions of these weighted signals.

relevant information supplied by observables.) It is therefore logical that we formulate the maximum-likelihood decoding rule in terms of $\tilde{y}_1$ and $\tilde{y}_2$. To do so, we use two things: definition of the squared Euclidean distance, and the formula of Eq. (6.102) for the linear combiner outputs $\tilde{y}_k$, $k = 1,2$. In particular, we see from Eq. (6.102) that, except for the statistics of additive complex Gaussian noise component $\tilde{v}_k$, the decision statistic $\tilde{y}_k$ is uniquely determined by the transmitted symbol $\tilde{s}_k$, $k = 1,2$. Accordingly, we may simplify implementation of the maximum-likelihood decoder by decomposing it into two independent decoders, one operating on the decision statistic $\tilde{y}_1$ and the second one operating on the other decision statistic $\tilde{y}_2$, which are observed T seconds apart. On this basis, we may now formally state the maximum-likelihood decoding rule as follows:

> *Given that the receiver has knowledge of* (i) *the channel fading coefficients α_1 and α_2, and* (ii) *the set of all possible transmitted symbols in*

*the mapper's constellation denoted by S, the maximum-likelihood esti-
mates of the transmitted symbols $\tilde{s}_1$ and $\tilde{s}_2$ are respectively defined by*

$$\hat{s}_1 = \arg \min_{\varphi \in S} \left\{ d^2(\tilde{y}_1, (\alpha_1^2 + \alpha_2^2)\varphi) \right\} \tag{6.104}$$

and

$$\hat{s}_2 = \arg \min_{\varphi \in S} \left\{ d^2(\tilde{y}_2, (\alpha_1^2 + \alpha_2^2)\varphi) \right\} \tag{6.105}$$

*where the φ denote the different hypotheses for the linear com-
biner output $\tilde{y}_1$ and $\tilde{y}_2$.*

Note that the right-hand sides of Eqs. (6.104) and (6.105) do not include any references to the actual transmitted symbols $\tilde{s}_1$ and $\tilde{s}_2$, which is how it should be since
they are both unknown at the receiver. Obviously, the receiver makes the correct decisions on the transmitted symbols if both $\hat{s}_1 = \tilde{s}_1$ and $\hat{s}_2 = \tilde{s}_2$.

Returning to the scenario pictured in Fig. 6.19, we note the following

$$S = \{\tilde{s}_j\}_{j=1}^4$$

and

$$\varphi = \tilde{s}_1, \tilde{s}_2, \tilde{s}_3, \text{ or } \tilde{s}_4$$

Application of the maximum-likelihood decoding rules of Eqs. (6.104) and (6.105) to
this example scenario yields $\hat{s}_1 = \tilde{s}_1$ and $\hat{s}_2 = \tilde{s}_2$, both of which are correct.

Problem 6.14 For M-ary PSK, the mapper's signal constellation consists of M points uniformly distributed on a circle with center at the origin and radius $\sqrt{E}$, where E is the signal
energy per symbol. Show that for this particular modulation scheme, the maximum-likelihood
decoding rule of Eqs. (6.104) and (6.105) respectively reduce to

$$\hat{s}_1 = \arg \min_{\varphi_1 \in S} d^2(\tilde{y}_1, \varphi_1)$$

and

$$\hat{s}_2 = \arg \min_{\varphi_2 \in S} d^2(\tilde{y}_2, \varphi_2) \qquad \blacksquare$$

6.6.3 Performance Comparison of Diversity-on-Receive and Diversity-on-Transmit Schemes

Figure 6.20 presents a computer simulation comparing the bit error rate performance
of coherent binary PSK over an uncorrelated Rayleigh-fading channel for three different schemes:

(a) No diversity (i.e., one transmit antenna and one receive antenna)

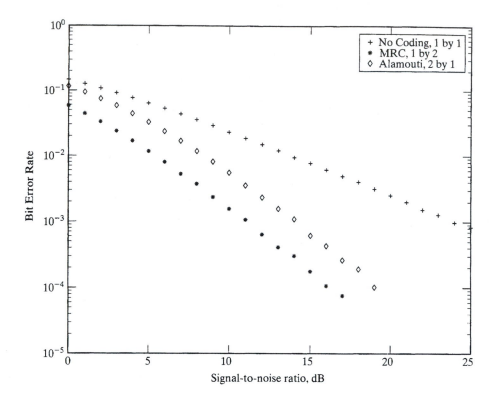

FIGURE 6.20 Comparison of the bit error rate performance of coherent BPSK over flat-fading Rayleigh channel for three configurations:
(a) No diversity.
(b) Maximal-ratio combiner ($N_t = 1, N_r = 2$).
(c) Alamouti code ($N_t = 2, N_r = 1$).

(b) The maximal-ratio combiner (i.e., one transmit antenna and two receive antennas)

(c) The Alamouti code (i.e., two transmit antennas and one receive antenna)

It is assumed that the total transmit power is the same for all three schemes, and in the case of the two diversity schemes (b) and (c), there is perfect knowledge of the channel(s) at the receiver(s).

From the figure, we see that the performance of the Alamouti code is worse by about 3 dB, compared with the maximal-ratio combiner. This 3-dB penalty is attributed to the fact that, in the space-diversity-on-transmit scheme using the Alamouti code, the transmit power in each of the two antennas is one-half of the transmit power in the space-diversity-on-receive scheme using the maximal-ratio combiner. Indeed, the diversity schemes based on the Alamouti code and maximal-ratio combiners would behave in the same way if each transmit antenna in the Alamouti code was to radiate the same power as the single transmit antenna in the maximal-ratio combiner.

Note also that, in an asymptotic sense, the plots for the MRC and Alamouti code have the same slope. This slope provides a measure of the *diversity order*. For the case at hand, the slope, and therefore the diversity order, is two, a result that confirms what we already know: the Alamouti code uses two transmit antennas, and the MRC uses two receive antennas. The issue of diversity order is considered in greater detail in Section 6.10.

6.6.4 Generalized Complex Orthogonal Space–Time Block Codes[14]

As mentioned previously, it was the remarkable computational simplicity of the Alamouti code and its capability to deliver a satisfactory performance that motivated the search for complex space–time block codes using more than two antennas. This search started with the equally pioneering work of Tarokh, Jafarkhani, and Calderbank. Building on the classic work of a number of theorists, including Radon and Hurwitz, Tarokh et al. introduced a *theory of generalized complex orthogonal designs*. This new theory applies to the construction of nonsquare orthogonal complex space–time block codes that combine coding at the transmitter with linear processing at the receiver.

Generalized complex orthogonal designs of space–time block codes distinguish themselves from the Alamouti code in three respects:

1. *A nonsquare transmission matrix*, which accommodates the use of more than two transmit antennas

2. *A fractional code rate*

3. *Orthogonality of the transmission matrix only in the temporal sense*, which is sufficient for maximum-likelihood decoding implemented in the form of a linear receiver in a manner analogous to the way the Alamouti code is implemented.

To define what we mean by a generalized complex orthogonal design, let G denote an m-by-N_t matrix, with N_t denoting the number of transmit antennas and m denoting the number of time slots. Let the entries of the matrix be designated

$$0, \pm s_1, \pm s_1^*, \pm s_2, \pm s_2^*, \ldots, \pm s_l, \pm s_l^*$$

where the number of transmitted symbols $l < m$. Then the matrix G is said to be a generalized complex orthogonalized design of size N_t and code rate $\kappa = l/m$ if it satisfies the condition for orthogonality in the temporal sense—that is, if

$$G^\dagger G = \left(\sum_{j=1}^{l} |s_j|^2 \right) I \tag{6.106}$$

where I is the N_t-by-N_t identity matrix.

The construction of space–time block codes using the generalized complex orthogonal design is exemplified by the rate-1/2 codes for transmission over a wireless channel. We have two cases:

(a) For three transmit antennas ($l = 4, m = 8$),

$$
G_3 = \begin{bmatrix}
\tilde{s}_1 & \tilde{s}_2 & \tilde{s}_3 \\
-\tilde{s}_2 & \tilde{s}_1 & -\tilde{s}_4 \\
-\tilde{s}_3 & \tilde{s}_4 & \tilde{s}_1 \\
-\tilde{s}_4 & -\tilde{s}_3 & \tilde{s}_2 \\
\tilde{s}_1^* & \tilde{s}_2^* & \tilde{s}_3^* \\
-\tilde{s}_2^* & \tilde{s}_1^* & -\tilde{s}_4^* \\
-\tilde{s}_3^* & \tilde{s}_4^* & \tilde{s}_1^* \\
-\tilde{s}_4^* & -\tilde{s}_3^* & \tilde{s}_2^*
\end{bmatrix} \implies \text{Space}
\tag{6.107}
$$

$\Downarrow$ Time

(b) For four transmit antennas ($l = 4, m = 8$),

$$
G_4 = \begin{bmatrix}
\tilde{s}_1 & \tilde{s}_2 & \tilde{s}_3 & \tilde{s}_4 \\
-\tilde{s}_2 & \tilde{s}_1 & -\tilde{s}_4 & \tilde{s}_3 \\
-\tilde{s}_3 & \tilde{s}_4 & \tilde{s}_1 & -\tilde{s}_2 \\
-\tilde{s}_4 & -\tilde{s}_3 & \tilde{s}_2 & \tilde{s}_1 \\
\tilde{s}_1^* & \tilde{s}_2^* & \tilde{s}_3^* & \tilde{s}_4^* \\
-\tilde{s}_2^* & \tilde{s}_1^* & -\tilde{s}_4^* & \tilde{s}_3^* \\
-\tilde{s}_3^* & \tilde{s}_4^* & \tilde{s}_1^* & -\tilde{s}_2^* \\
-\tilde{s}_4^* & -\tilde{s}_3^* & \tilde{s}_2^* & \tilde{s}_1^*
\end{bmatrix} \implies \text{Space}
\tag{6.108}
$$

$\Downarrow$ Time

(The subscripts in G_3 and G_4 refer to the numbers of transmit antennas.)

The symbol G used for the transmission matrices of these two codes is different from the corresponding symbol S for the Alamouti code, to emphasize the basic difference between them. For the Alamouti code, we have $SS^\dagger = S^\dagger S$, whereas the codes of Eqs. (6.107) and (6.108) do *not* satisfy the condition $GG^\dagger = G^\dagger G$.

Compared with the Alamouti code defined by the transmission matrix of Eq. (6.88), the space–time codes G_3 and G_4 are at a disadvantage in two respects:

1. The bandwidth efficiency is reduced by a factor of two.

2. The number of time slots across which the channel is required to have a constant fading envelope is increased by a factor of four.

To improve the bandwidth efficiency, we may use rate-3/4 generalized complex linear-processing orthogonal designs referred to as *sporadic codes*. The construction of these codes differs from that of G_3 and G_4—hence the use of the symbol H to denote their transmission matrices. Two cases of sporadic codes are as follows:

(a) For three transmit antennas ($l = 3, m = 4$),

$$H_3 = \begin{bmatrix} \tilde{s}_1 & \tilde{s}_2 & \tilde{s}_3/\sqrt{2} \\ -\tilde{s}_2^* & \tilde{s}_1^* & \tilde{s}_3/\sqrt{2} \\ \tilde{s}_3^*/\sqrt{2} & \tilde{s}_3^*/\sqrt{2} & (-\tilde{s}_1 - \tilde{s}_1^* + \tilde{s}_2 - \tilde{s}_2^*)/2 \\ \tilde{s}_3^*/\sqrt{2} & -\tilde{s}_3^*/\sqrt{2} & (\tilde{s}_2 + \tilde{s}_2^* + \tilde{s}_1 - \tilde{s}_1^*)/2 \end{bmatrix} \longrightarrow \text{Space} \qquad (6.109)$$

$$\big\downarrow \text{Time}$$

(b) For four transmit antennas ($l = 3, m = 4$),

$$H_4 = \begin{bmatrix} \tilde{s}_1 & \tilde{s}_2 & \tilde{s}_3/\sqrt{2} & \tilde{s}_3/\sqrt{2} \\ -\tilde{s}_2^* & \tilde{s}_1^* & \tilde{s}_3/\sqrt{2} & -\tilde{s}_3/\sqrt{2} \\ \tilde{s}_3^*/\sqrt{2} & \tilde{s}_3^*/\sqrt{2} & (-\tilde{s}_1 - \tilde{s}_1^* + \tilde{s}_2 - \tilde{s}_2^*)/2 & (-\tilde{s}_2 - \tilde{s}_2^* + \tilde{s}_1 - \tilde{s}_1^*)/2 \\ \tilde{s}_3^*/\sqrt{2} & -\tilde{s}_3^*/\sqrt{2} & (\tilde{s}_2 + \tilde{s}_2^* + \tilde{s}_1 - \tilde{s}_1^*)/2 & -(\tilde{s}_1 + \tilde{s}_1^* + \tilde{s}_2 - \tilde{s}_2^*)/2 \end{bmatrix} \longrightarrow \text{Space}$$

$$\big\downarrow \text{Time}$$

$$(6.110)$$

Table 6.2 summarizes the parameters of five space–time block codes: the Alamouti code **S** of Eq. (6.88), the generalized complex orthogonal codes of Eqs. (6.107) and (6.108), and the generalized complex orthogonal codes of the sporadic variety given in Eqs. (6.109) and (6.110).

TABLE 6.2   Summary of the Parameters of Different Space–Time Block Codes.

Space–time code	Number of transmit antennas, N_t	Number of transmitted symbols, l	Number of time slots, m	Rate, $\kappa = \dfrac{l}{m}$
S	2	2	2	1
G_3	3	4	8	1/2
G_4	4	4	8	1/2
H_3	3	3	4	3/4
H_4	4	3	4	3/4

Maximum-likelihood decoding of the space–time block codes G_3, G_4, H_3, and H_4 is achieved by using only linear processing at the receiver by virtue of the complex orthogonality of these four codes in the temporal sense. (For details of these decoding algorithms, see Problems 6.40 and 6.41; and Note 15.)

6.6.5 Performance Comparisons of Different Space–Time Block Codes Using a Single Receiver

In this section, we present computer simulation results that compare the five different space–time block codes S, G_3, G_4, H_3, and H_4, for modulation schemes that produce 3 bits/s/Hz, 2 bits/s/Hz, and 1 bit/s/Hz, assuming a single receive antenna. The incoming binary data stream is space–time block encoded in accordance with the scheme shown in Fig. 6.16.

The simulation results presented in Fig. 6.21 compare the symbol error rates for a data transmission of 3 bits/s/Hz for the following coding–modulation schemes:

(a) No coding and 8-PSK

(b) Alamouti code S, using two transmit antennas and 8-PSK

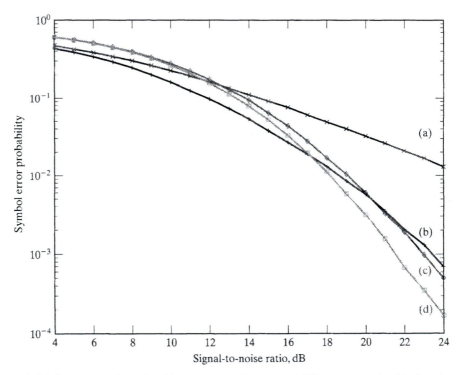

FIGURE 6.21   Comparison of the bit error rate performance of different space–time block codes:
(a) No space-time coding.
(b) Alamouti code (8-PSK).
(c) Sporadic code H_3 (16-QAM).
(d) Sporadic code H_4 (16-QAM).

(c) Sporadic space–time block code H_3, using three transmit antennas and 16-QAM

(d) Sporadic space–time block code H_4, using four transmit antennas and 16-QAM

Recognizing that the code rate for both H_3 and H_4 is 3/4, we find that the effective transmission rate is 3 bits/s/Hz. From the figure, we see that, at a symbol error rate of 10^{-3}, the 16-QAM, rate-3/4 code H_4 combination provides a gain of about 2 dB over the 8-PSK, full-rate Alamouti combination.

Figure 6.22 compares the bit error rates for a data transmission of 2 bits/s/Hz for the following coding–modulation schemes:

(a) No coding and QPSK

(b) Alamouti code **S**, using two transmit antennas and QPSK

(c) A generalized orthogonal space–time block code G_3, using three transmit antennas and 16-QAM

(d) A generalized orthogonal space–time block code G_4, using four transmit antennas and 16-QAM

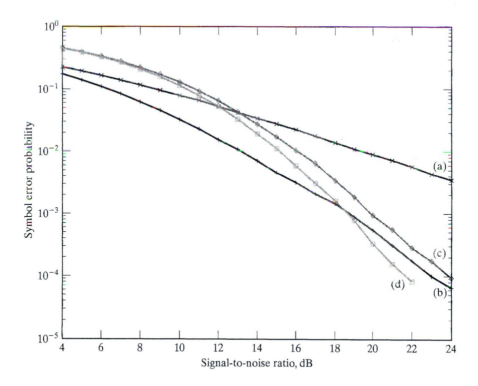

FIGURE 6.22   Comparison of bit error rate performance of different space–time block codes:
(a) No space-time coding.
(b) Alamouti code (QPSK).
(c) Code G_3 (16-QAM).
(d) Code G_4 (16-QAM).

With the chosen constellations, the effective transmit rate in each case is 2 bits/s/Hz. The figure shows that, at a symbol error rate of 10^{-4}, the 16-QAM, rate-1/2 code G_4 combination produces a gain of about 1 dB over the QPSK, full-rate Alamouti combination. At a symbol error rate of 10^{-5}, the gain advantage grows to about 5 dB (not shown in Fig.6.22).

Finally, Fig. 6.23 compares the bit error rates for a data transmission of 1 bit/s/Hz for the following coding–modulation schemes:

- No coding and BPSK
- Alamouti code **S**, using two transmit antennas and BPSK
- A generalized space–time block code G_3, using three transmit antennas and QPSK
- A generalized space–time block code G_4, using four transmit antennas and QPSK

In each case, the effective transmission rate is 1 bit/s/Hz. For a bit error rate of 10^{-4}, the figure shows that the QPSK, G_4 combination provides a gain of about 5 dB over the BPSK, **S** combination. At a symbol error rate of 10^{-5}, the gain advantage grows to about 7.5 dB (not shown in Fig. 6.23).

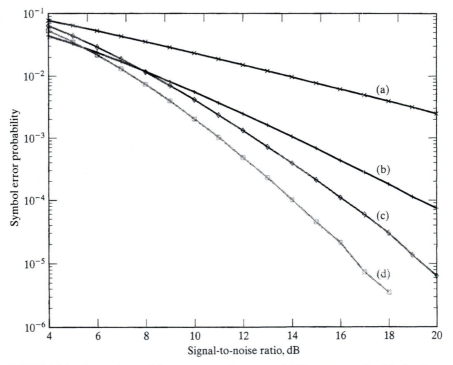

FIGURE 6.23 Comparison of bit error rate performance of different space–time block codes:
(a) No space-time coding.
(b) Alamouti code (binary PSK).
(c) Generalized code G_3 (QPSK).
(d) Generalized code G_4 (QPSK).

6.7 DIFFERENTIAL SPACE–TIME BLOCK CODES

In light of the material presented in Section 6.5 on the ergodic capacity of a MIMO wireless channel, the case for a substantial increase in spectral efficiency of the system rests on the premise that the receiver has "knowledge," or, more precisely, a "near-perfect estimate," of the channel matrix $\mathbf{H}$. Clearly, such a provision imposes a certain amount of overhead on the design and implementation of the receiver. (The topic of channel estimation was discussed in Section 3.11.) We may eliminate the need for channel estimation and thereby simplify the receiver design by using a *differential space–time block coding scheme*.[16]

The encoding system we have in mind builds on two concepts:

1. Alamouti's space–time transmit-diversity block code
2. Differential space–time block coding

The Alamouti code, involving the use of two transmit antennas and a single receive antenna in its most basic form, was discussed in Section 6.6. The space–time version of differential coding is discussed next.

6.7.1 Differential Space–Time Block Coding

In the Alamouti code, the two signals (symbols) $\tilde{s}_{1,t}$ and $\tilde{s}_{2,t}$ are transmitted on two separate antennas at time t, followed by the transmission of the related pair of signals $-\tilde{s}_{2,t+1}^*$ and $\tilde{s}_{1,t+1}^*$, at time $t+1$. Note that

-   We have appended the subscripts t and $t+1$ to the pertinent signals to distinguish a signal transmission from the one following it.
-   We have set the transmitted signal duration T equal to unity to simplify the presentation.

We assume that the total transmitted power, namely, $|\tilde{s}_{1,t}|^2 + |\tilde{s}_{2,t}|^2$, is maintained constant. To further simplify matters without loss of generality, we normalize the total transmitted power to unity. Under this assumption, it follows that the row vectors $[\tilde{s}_{1,t}, \tilde{s}_{2,t}]$ and $[-\tilde{s}_{2,t+1}^*, \tilde{s}_{1,t+1}^*]$ form an *orthonormal set*. Consequently, *any other pair of transmitted signals can be represented as a linear combination of this orthonormal set*. Specifically, given the new pair of complex signals $\tilde{s}_{3,t+2}$ and $\tilde{s}_{4,t+2}$ to be transmitted at time $t+2$, we may express them as the row vector

$$[\tilde{s}_{3,t+2}, \tilde{s}_{4,t+2}] = a_{1,t+2}[\tilde{s}_{1,t}, \tilde{s}_{2,t}] + a_{2,t+2}[-\tilde{s}_{2,t}^*, \tilde{s}_{1,t}^*] \tag{6.111}$$

where $a_{1,t+2}$ and $a_{2,t+2}$ are coefficients of the linear combination. These two coefficients are defined as the *inner products* of the row vector $[\tilde{s}_{3,t+2}, \tilde{s}_{4,t+2}]$ with the previously transmitted row vectors $[\tilde{s}_{1,t}, \tilde{s}_{2,t}]$ and $[-\tilde{s}_{2,t+1}^*, \tilde{s}_{1,t+1}^*]$, respectively;

that is,

$$[a_{1,\,t+2},\, a_{2,\,t+2}] = [\tilde{s}_{3,\,t+2},\, \tilde{s}_{4,\,t+2}]\Big[[\tilde{s}_{1,\,t},\, \tilde{s}_{2,\,t}]^\dagger,\, [-\tilde{s}_{2,\,t+1}^*,\, \tilde{s}_{1,\,t+1}^*]^\dagger\Big]$$

$$= [\tilde{s}_{3,\,t+2},\, \tilde{s}_{4,\,t+2}]\begin{bmatrix} \tilde{s}_{1,\,t}^* & -\tilde{s}_{2,\,t+1} \\ \tilde{s}_{2,\,t}^* & \tilde{s}_{1,\,t+1} \end{bmatrix} \qquad (6.112)$$

$$= [\tilde{s}_{3,\,t+2},\, \tilde{s}_{4,\,t+2}]\mathbf{S}_{t,\,t+1}^\dagger$$

where $\mathbf{S}_{t,\,t+1}^\dagger$ is the Hermitian transpose of the two-by-two transmitted signal matrix of the Alamouti code constructed at times t and $t+1$; see Eq. (6.89). Similarly, we may write

$$[-a_{2,\,t+3}^*,\, a_{1,\,t+3}^*] = [-\tilde{s}_{4,\,t+3}^*,\, \tilde{s}_{3,\,t+3}^*]\mathbf{S}_{t,\,t+1}^\dagger \qquad (6.113)$$

Combining Eqs. (6.112) and (6.113) into a single matrix relation, we may formulate the *coefficients matrix*

$$\mathbf{A}_{t+2,\,t+3} = \begin{bmatrix} a_{1,\,t+2} & a_{2,\,t+2} \\ -a_{2,\,t+3}^* & a_{1,\,t+3}^* \end{bmatrix} \qquad (6.114)$$

$$= \mathbf{S}_{t+2,\,t+3}\mathbf{S}_{t,\,t+1}^\dagger$$

where $\mathbf{S}_{t+2,\,t+3}$ is the transmitted signal matrix constructed at times $t+2$ and $t+3$; see Eq. (6.88). Equation (6.114) states that the coefficients matrix is a *product of two orthogonal Alamouti (quaternionic) matrices*.

Problem 6.15   Show that the coefficients matrix $\mathbf{A}_{t+2,\,t+3}$ is an orthonormal matrix; that is, show that the inverse of $\mathbf{A}_{t+2,\,t+3}$ equals its Hermitian transpose.   ∎

Since $\mathbf{S}_{t,\,t+1}$ is a unitary matrix by virtue of orthonormality of its two constituent row vectors, it follows that

$$\mathbf{S}_{t,\,t+1}^{-\dagger} = \mathbf{S}_{t,\,t+1}$$

We may therefore solve Eq. (6.114) for the *transmission matrix* $\mathbf{S}_{t+2,\,t+3}$ by writing

$$\mathbf{S}_{t+2,\,t+3} = \mathbf{A}_{t+2,\,t+3}\mathbf{S}_{t,\,t+1}^{-\dagger} \qquad (6.115)$$

$$= \mathbf{A}_{t+2,\,t+3}\mathbf{S}_{t,\,t+1}$$

Equation (6.115) provides the *basis for differential space–time block encoding at the transmitter*.

The corresponding basis for differential decoding at the receiver lies in Eq. (6.114), modified by the channel. To be specific, we assume that the wireless link is quasi-static with the channel matrix $\mathbf{H}$ remaining essentially constant over two consecutive blocks of signal transmission (i.e., over the time interval $(t, t + 3)$). Then, in the absence of channel noise, the received signal matrix in response to the transmitted signal matrix $\mathbf{S}_{t, t+1}$ is given by (see Problem 6.43)

$$\mathbf{X}_{t, t+1} = \mathbf{S}_{t, t+1}\mathbf{H} \tag{6.116}$$

Similarly, the received signal matrix in response to the next transmitted signal matrix $\mathbf{S}_{t+2, t+3}$ is given by

$$\mathbf{X}_{t+2, t+3} = \mathbf{S}_{t+2, t+3}\mathbf{H} \tag{6.117}$$

With the mathematical structure of Eq. (6.114) in mind, we use Eqs. (6.116) and (6.117) to construct the new two-by-two matrix

$$\begin{aligned}
\mathbf{Y}_{t+2, t+3} &= \mathbf{X}_{t+2, t+3}\mathbf{X}_{t, t+1}^{\dagger} \\
&= \mathbf{S}_{t+2, t+3}\mathbf{H}(\mathbf{S}_{t, t+1}\mathbf{H})^{\dagger} \\
&= \mathbf{S}_{t+2, t+3}\mathbf{H}\mathbf{H}^{\dagger}\mathbf{S}_{t, t+1}^{\dagger}
\end{aligned} \tag{6.118}$$

(Note that the elements of the matrix $\mathbf{Y}$ should not be confused with the y's defined in Eq. (6.99).)

From the solution to Problem 6.12, we note that

$$\mathbf{H} = \begin{bmatrix} \alpha_1 e^{j\theta_1} & \alpha_2 e^{j\theta_2} \\ \alpha_2 e^{-j\theta_2} & -\alpha_1 e^{-j\theta_1} \end{bmatrix} \tag{6.119}$$

and

$$\mathbf{H}^{\dagger}\mathbf{H} = \mathbf{H}\mathbf{H}^{\dagger} = (\alpha_1^2 + \alpha_2^2)\mathbf{I} \tag{6.120}$$

where $\mathbf{I}$ is the two-by-two identity matrix. Accordingly, Eq. (6.118) reduces to

$$\begin{aligned}
\mathbf{Y}_{t+2, t+3} &= (\alpha_1^2 + \alpha_2^2)\mathbf{S}_{t+2, t+3}\mathbf{S}_{t, t+1}^{\dagger} \\
&= (\alpha_1^2 + \alpha_2^2)\mathbf{A}_{t+2, t+3}
\end{aligned} \tag{6.121}$$

where, in the second line, we made use of Eq. (6.114).

Equation (6.121) forms the *basis for differential space–time block decoding at the receiver*. Here again, in light of the scaling factor $(\alpha_1^2 + \alpha_2^2)$, the receiver exhibits a two-level diversity gain in the same way that the coherent Alamouti receiver discussed in Section 6.6 does.

Signal transmission begins by sending an arbitrary pair of signals $\tilde{s}_{1,0}$ and $\tilde{s}_{2,0}$ at time $t = 0$, followed by sending the related pair of signals $-\tilde{s}_{2,1}^*$ and $\tilde{s}_{1,1}^*$ at time $t = 1$. No information is conveyed by sending these signals; rather, the two transmissions at times $t = 0$ and $t = 1$ provide the receiver with a known *frame of reference* for facilitating the differential space–time block decoding process. The transmission of information-bearing data begins at $t = 2$, at which time the receiver commences the differential decoding process.

In what follows, we restrict the discussion to *M*-ary PSK. This method of modulation, involving *M* signal points in a two-dimensional space, befits its use for constructing the two-by-two transmitted signal matrix characterizing the Alamouti code. With two sets of signal points involved in the formulation of Eqs. (6.115) and (6.121), we may identify two signal spaces. One, denoted by $\mathcal{A}$, is spanned by the pair of complex coefficients (a_1, a_2) constituting the matrix **A**. The second, denoted by S, is spanned by the complex signals $\tilde{s}_1, \tilde{s}_2$ constituting the matrix **S**. These two signal spaces have the following properties:

Property 1.

With *M*-ary PSK as the method of modulation used to transmit the Alamouti code, the points representing the signal space S are uniformly distributed on a circle of unit radius (assuming that the transmitted signal power is normalized to unity). Correspondingly, the points representing the signal space $\mathcal{A}$ constitute a quadrature amplitude modulation (QAM) constellation.

Property 2.

The minimum distance between the points in the signal space S is equal to the minimum distance between the points in the signal space $\mathcal{A}$.

Property 3.

The constellation of points in the input space $\mathcal{A}$ involves an expansion in the size of the alphabet, compared with that in the constellation of points in the transmitted signal space S.

To construct the matrix **A**, we need a *bijective mapping* of blocks of 2*b* bits onto the signal space $\mathcal{A}$. The mapping is bijective in the sense that it is *one-to-one and onto*, as shown in the next two examples. The first example illustrates the constellation expansion property; the second illustrates the use of Gray coding as a principled way of constructing the bijective mapping.

EXAMPLE 6.6 Constellation Expansion

Let $\tilde{s}_0$ and $\tilde{s}_1$ denote the pair of symbols transmitted at times $t = 0$ and $t = 1$, respectively, followed by the transmission of symbols $\tilde{s}_2$ and $\tilde{s}_3$ at times $t = 2$ and $t = 3$. We may thus express the corresponding matrices as

$$\mathbf{S}_{0,1} = \begin{bmatrix} \tilde{s}_0 & \tilde{s}_1 \\ -\tilde{s}_1^* & \tilde{s}_0^* \end{bmatrix}$$

and

$$\mathbf{S}_{2,3} = \begin{bmatrix} \tilde{s}_2 & \tilde{s}_3 \\ -\tilde{s}_3^* & \tilde{s}_2^* \end{bmatrix}$$

Accordingly, the use of Eq. (6.114) yields

$$\mathbf{A}_{2,3} = \mathbf{S}_{2,3}\mathbf{S}_{0,1}^{\dagger}$$

$$= \begin{bmatrix} \tilde{s}_2 & \tilde{s}_3 \\ -\tilde{s}_3^* & \tilde{s}_2^* \end{bmatrix} \begin{bmatrix} \tilde{s}_0^* & -\tilde{s}_1 \\ \tilde{s}_1^* & \tilde{s}_0 \end{bmatrix}$$

$$= \begin{bmatrix} \tilde{s}_2\tilde{s}_0^* + \tilde{s}_3\tilde{s}_1^* & -\tilde{s}_2\tilde{s}_1 + \tilde{s}_3\tilde{s}_0 \\ -\tilde{s}_3^*\tilde{s}_0^* + \tilde{s}_2^*\tilde{s}_1^* & \tilde{s}_3^*\tilde{s}_1 + \tilde{s}_2^*\tilde{s}_0 \end{bmatrix}$$

To proceed further, consider the simple example of BPSK as the method of modulation used to transmit the Alamouti code, exemplified by the matrices $\mathbf{S}_{0,1}$ and $\mathbf{S}_{2,3}$. Specifically, suppose the symbols $\tilde{s}_0$, $\tilde{s}_1$, $\tilde{s}_2$, and $\tilde{s}_3$ come from the constellation $\{-1, +1\}$ depicted in Fig. 6.24(a). Then, clearly, the four elements of the coefficients matrix $\mathbf{A}_{2,3}$ come from the constellation $\{-2, 0, +2\}$ depicted in Fig. 6.24(b). Comparing these two constellations, we immediately see that there is an expansion in the size of the alphabet, in accordance with Property 3.

The constellation expansion illustrated in Fig. 6.44 arises from the fact that, although the BPSK forms a multiplicative group (e.g., the QPSK constellation is the product of two BPSK constellations), it does not form an additive group over the integers. In contrast, the matrix $\mathbf{A}$ involves both multiplicative and additive groups—hence the constellation expansion.

It is also noteworthy that the minimum distance between the signal points in the input constellation $\mathcal{A}$ of Fig. 6.24(b) is the same as the minimum distance between the transmitted signal points in the constellation $\mathcal{S}$ of Fig. 6.24(a) in accordance with Property 2. In both cases, the minimum distance is 2.   ∎

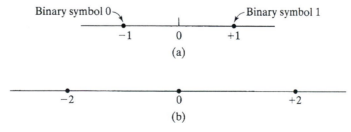

FIGURE 6.24   a) Signal constellation for BPSK.
(b) Expanded constellation for input data.

For a demonstration of Property 1, the reader is referred to Problem 6.45, in which it is requested that the reader start with a 4-QAM (i.e., QPSK) constellation for signal space $\mathcal{S}$ and arrive at a 9-QAM constellation for signal space $\mathcal{A}$. The solution to Problem 6.45 provides further confirmation of Properties 2 and 3.

EXAMPLE 6.7 Gray Coding for Bijective Mapping

To continue with the differential space–time coding using BPSK, we need a frame of reference for signal transmission at times $t = 0$ and $t = 1$. With the constellation of Fig. 6.44(a) as the polar basis for representing the BPSK, we choose the dibit 00 as the frame of reference. Then, with the symbol 1 representing the first 0 bit transmitted on antenna 1 and the symbol -1 representing the second 0 bit transmitted on antenna 2, the corresponding signal matrix is

$$S_{0,1} = \begin{bmatrix} -1 & -1 \\ 1 & -1 \end{bmatrix} \quad \text{dibit } 00$$

With this frame of reference fixed, we may now go on to determine the values assumed by the coefficients matrix $\mathbf{A}_{2,3}$ for the transmission of Gray-encoded dibits 00, 01, 11, and 10, using Eq. (6.114). The results of this calculation are summarized as follows:

1. *Transmitted data dibit* 00

$$S_{2,3} = \begin{bmatrix} -1 & -1 \\ 1 & -1 \end{bmatrix}$$

$$\mathbf{A}_{2,3} = \begin{bmatrix} 2 & 0 \\ 0 & 2 \end{bmatrix}$$

2. *Transmitted data dibit* 01

$$S_{2,3} = \begin{bmatrix} -1 & 1 \\ -1 & -1 \end{bmatrix}$$

$$\mathbf{A}_{2,3} = \begin{bmatrix} 0 & -2 \\ 2 & 0 \end{bmatrix}$$

3. *Transmitted data dibit* 11

$$S_{2,3} = \begin{bmatrix} 1 & 1 \\ -1 & 1 \end{bmatrix}$$

$$\mathbf{A}_{2,3} = \begin{bmatrix} -2 & 0 \\ 0 & -2 \end{bmatrix}$$

4. *Transmitted data dibit* 10

$$S_{2,3} = \begin{bmatrix} 1 & -1 \\ 1 & 1 \end{bmatrix}$$

$$\mathbf{A}_{2,3} = \begin{bmatrix} 0 & 2 \\ -2 & 0 \end{bmatrix}$$

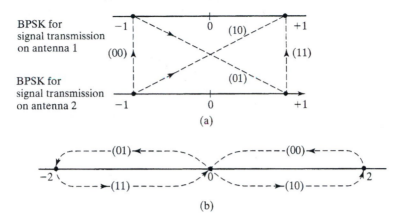

FIGURE 6.25   (a) Gray encoding of input dibits.
(b) Gray encoding of transmitted dibits.

TABLE 6.3   Signal Transmissions, Assuming 00 as the Frame of Reference.

Transmitted dibit	Coefficients at time $t = 2$		Transmitted signals at time $t = 2$	
	a_2	a_3	$\tilde{s}_2$	$\tilde{s}_3$
00	2	0	−1	−1
01	0	−2	−1	1
11	−2	0	1	1
10	0	2	1	−1

Figure 6.25 illustrates the bijective mapping of the input signal space $\mathcal{A}$ onto the transmitted signal space $\mathcal{S}$. Part (a) refers to the input dibits, and part (b) refers to the corresponding transmitted pair of bits transmitted on antennas 1 and 2.

Table 6.3 summarizes (1) the pair of coefficients a_2 and a_3 representing an incoming dibit, and (2) the corresponding pair of symbols $\tilde{s}_2$ and $\tilde{s}_3$ transmitted on antennas 1 and 2, assuming the dibit 00 as the frame of reference. ∎

Problem 6.16   Fill in the details leading to the formulation of Fig. 6.25 and Table 6.3. ∎

Problem 6.17   The results presented in Fig. 6.25 and Table 6.3 assume the frame of reference 00. Repeat Problem 6.16, this time using the frame of reference 11. ∎

6.7.2   Transmitter and Receiver Structures

As mentioned previously, Eq. (6.115) forms the mathematical basis for constructing the transmitter. On this basis, we may construct the block diagram depicted in Fig. 6.26(a), which consists of the following parts:

1. *A mapper*, which generates the entries that make up matrix $\mathbf{A}_{t+2,t+3}$ in response to the incoming pair of data bits sent by the source of information at times $t + 2$ and $t + 3$.

2. *A differential encoder*, which transforms the matrix $\mathbf{A}_{t+2,t+3}$ into the matrix $\mathbf{S}_{t+2,t+3}$. The two row vectors of matrix $\mathbf{S}_{t+2,t+3}$ are transmitted on a pair of antennas at times $t + 2$ and $t + 3$. The differential encoder itself consists of two components:

   2.1. *A delay unit* $z^{-2}\mathbf{I}$, which feeds back the matrix $\mathbf{S}_{t,t+1}$ (i.e., the previous value of $\mathbf{S}_{t+2,t+3}$) to the input of the differential encoder.

   2.2. *A multiplier*, which multiplies the matrix inputs $\mathbf{A}_{t+2,t+3}$ and $\mathbf{S}_{t,t+1}$ to provide the transmitted signal matrix $\mathbf{S}_{t,t+3}$ in accordance with Eq. (6.115).

To construct the receiver, we return to Eq. (6.121). On the basis of this equation, we may construct the block diagram of the receiver of Fig. 6.26(b), which consists of the following functional blocks:

1. *A differential decoder*, which itself consists of two components:

   1.1. *A delay unit* $z^{-2}\mathbf{I}$, which feeds forward the matrix $\mathbf{X}_{t,t+1}$ (i.e., the previous value of the received signal matrix $\mathbf{X}_{t+2,t+3}$).

   1.2. *A multiplier*, which multiplies the matrices $\mathbf{X}_{t+2,t+3}$ and $\mathbf{X}_{t,t+1}$ to produce the new matrix $\mathbf{Y}_{t+2,t+3}$.

2. *A signal estimator*, which computes the matrix $\hat{\mathbf{A}}_{t+2,t+3}$ that is closest to $\mathbf{Y}_{t+2,t+3}$ in terms of Euclidean distance; the estimate so computed is denoted by $\hat{\mathbf{A}}_{t+2,t+3}$.

3. *An inverse mapper*, which operates on the estimate $\hat{\mathbf{A}}_{t+2,t+3}$ to produce corresponding estimates of the original pair of data bits transmitted at times $t + 2$ and $t + 3$.

6.7.3 Noise Performance

In deriving Eq. (6.121), we ignored the presence of additive channel noise. It is in recognition of the unavoidable presence of channel noise that we spoke of "estimates" of transmitted data bits in describing the structure of the receiver.

How, then, does channel noise affect the performance of the receiver depicted in Fig. 6.26? To answer this question, we remind ourselves of the classical problem pertaining to the scalar case of differential phase-shift keying (DPSK). In this context, it is well known that the performance of a DPSK receiver corrupted by additive white Gaussian noise is worse than that of coherent BPSK. (See Section 3.11) So it is also for the differential space–time block coding system of Fig. 6.26, compared with the coherent version of the system discussed in Section 6.6.

Using computer simulations, Fig. 6.27 compares the noise performance of the differential space–time block coding system of Fig. 6.26 with that of the coherent version of the system. The simulations are based on the following configuration:

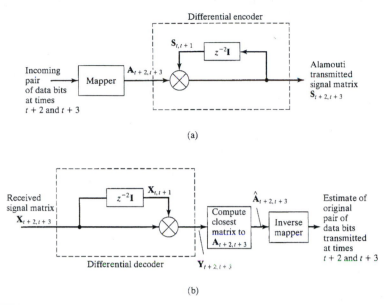

(a)

FIGURE 6.26   Block diagrams of (a) differential space–time block encoder.
(b) differential space–time block decoder, where z^{-1} = unit-delay operator and $\mathbf{I}$ = two-by-two
identity matrix.

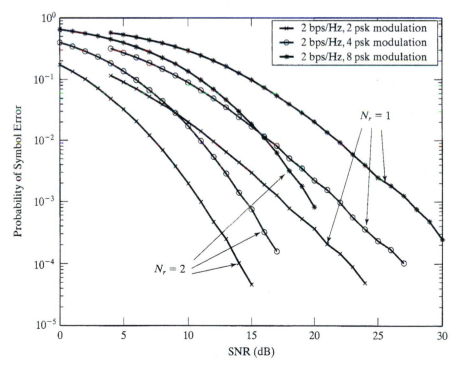

FIGURE 6.27   Noise–performance comparison of differential space–time block coding for
specific M-ary PSK modulation strategies in a Rayleigh-fading environment.

- A single bit of data sent on each of the two transmit antennas
- BPSK modulation for transmitting the complex symbols $\tilde{s}_1$ and $\tilde{s}_2$ at time t, followed by the transmission of $-\tilde{s}_2^*$ and $\tilde{s}_1^*$ at time $t + 1$

The simulation results presented in Fig. 6.27 indicate a loss of 3 dB in receiver performance that is incurred through the use of differential space–time block coding. This loss in performance may be justified by formulating the additive noise corrupting the received signal matrix **X**. The formulation is discussed in Problem 6.47, where it is shown that, under the assumption of a high signal-to-noise ratio, the average power of the additive noise is increased by about 3 dB, compared with that of the corresponding coherent receiver.

6.8   SPACE-DIVISION MULTIPLE ACCESS AND SMART ANTENNAS

The material presented up to this point has focused on the use of *space* as a basis for antenna diversity, the use of which improves spectral efficiency by counteracting the multipath-channel-fading problem. In this section, we discuss the role of space in cellular systems, the development of which has also resulted in a multifold increase in spectral efficiency by using directional antennas.

In *cellular systems*, a few channels are broadcast by the base station or shared by all users on the uplink. The majority of the traffic-bearing (and revenue-generating) channels are point-to-point, between a base station and a single user terminal, in which case we speak of the communications being *directional* in nature. The recognition that *user terminals can be spatially separated by virtue of their angular directions is the basis of space-division multiple access (SDMA)*, a technique that relies on the use of *directional antennas* to distinguish among users. A simple example is presented in Fig. 6.28, which shows a base station in a cellular system with sector antennas; that is, each antenna covers one sector—in this case, 120°— of the cell. In the illustration, each base station requires three nonoverlapping antennas, each with a field of view of 120°.

Consider the case in which there is a one-in-seven reuse pattern, as suggested by the figure. If there are N users per cell, then the power radiated on one of the sector antennas is $(N/3)P_T G_T$, where P_T is the average power per user and G_T is the antenna gain. To close the communications link, $P_T G_T$ must be the same whether the antenna is directional or omnidirectional. Consequently, we conclude that the total power radiated with a sector antenna is one-third of that produced by an omnidirectional antenna. By extension, a user terminal receiver suffers only one-third of the interference that would be produced by omnidirectional base-station antennas with the same number of users.

In the uplink, all user terminals have omnidirectional antennas, but only one-third of them are in the field of view of the base-station antenna. So the interference is reduced by two-thirds in this direction as well.

The general conclusion from this discussion is that, with 120° sector antennas at the base station, the number of user terminals can be tripled relative to the omnidirectional case and still maintain the same interference levels. There are many advantages to this approach:

- It can be applied with any of the multiple-access strategies discussed previously: FDMA, TDMA, or CDMA.
- It allows multiple users to operate on the same frequency and/or time slot in the same cell.
- It thus leads to more users in the same spectrum and improved capacity.
- The technology can be applied at the base station without affecting the mobile terminals.

Although the same idea may also be applied to the mobile terminal, at present it is more difficult to build a commercially acceptable mobile terminal that has a directional antenna. However, this is an area of active investigation.

With conventional base stations using omnidirectional antennas, when user density grows beyond the capacity of a single cell, the growth is accommodated by dividing the initial cell into a number of smaller cells in a process known as *cell splitting*. Power control is used to reduce the interference among these smaller cells. Although sector antennas are more expensive than omnidirectional antennas, it is still more economical to add sector antennas than it is to add new base stations.

The same technology can also be applied to satellite systems in which the field of view of the satellite is divided between a number of spotbeams and, depending upon the isolation between the beams, frequency and/or time slots may be reused in all or some of the beams.

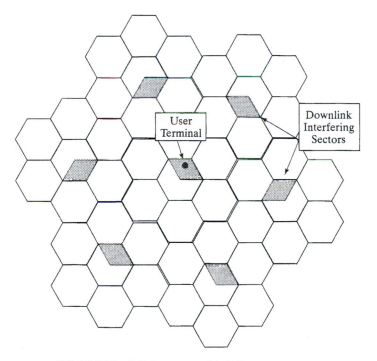

FIGURE 6.28   Cellular system with 120° sector antennas.

SDMA relies on *smart antennas*, in the sense that it takes advantage of the directional nature of radio communications. Some examples of smart antennas are the following:

1. The simplest example of a smart antenna is the *sector antenna* just described. This antenna provides significant capacity gains simply by dividing the service area of each base station into three (or more) angular sections with a significant amount of isolation between them.

2. *Switched-beam antennas* are the next step in the evolution of smart antennas. These antennas have a number of fixed beams that cover $360°$. Switched-beam antennas are typically narrower than sector antennas. The receiver selects the beam that provides the best signal and interference reduction.

3. The most advanced example is the *adaptive antenna*. This antenna dynamically adjusts its pattern to minimize the effects of noise, interference, and multipath. With adaptive antennas, there is one beam for each user.

In Section 6.8.1, we will discuss how an adaptive antenna may be implemented. There are many advantages to smart antennas for mobile applications:

- *Greater range.* Since the antennas are directional, they have larger gains and can therefore provide a stronger signal strength for the same transmit power.

- *Fewer base stations.* For areas with a low user density, fewer base stations are required, because the existing base station has a greater range. In areas with a high user density, there is less interference, due to the greater user isolation provided by the directional antennas. Hence, a single base station can serve more users.

- *Better building penetration.* This potential benefit is due to the greater signal strength and increased transmitter gain.

- *Less sensitivity to power control errors.* This additional benefit is due to the better isolation among different user signals.

- *More responsive to traffic hot spots.* In areas such as airports and conference centers, user densities can become quite high, and directional antennas allow one or a small number of base stations to service these areas better.

SDMA improves system capacity by allowing greater spectrum reuse, through (1) minimization of the effects of interference and, (2) increasing signal strength for both the user terminal and the base station.

6.8.1 Antenna Arrays[17]

In this section, we analyze the behavior of a linear array of antenna elements receiving a signal, as illustrated in Fig. 6.29. For simplicity, we assume that the transmit antenna is omni-directional, but, in general, it could be directional. The key assumption is that the distance between the transmitter and receiver is large enough for the emitted wave to be viewed as a planar wavefront for the purposes of antenna analysis.

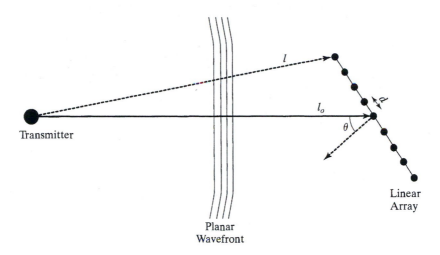

FIGURE 6.29   Plane wave incident on a linear antenna array.

The receiving antenna is positioned at an angle θ with respect to the linear array, as shown in the figure. Let the complex envelope of the transmitted signal be denoted by

$$\tilde{s}(t) = m(t)e^{j2\pi ft} \tag{6.122}$$

where $m(t)$ is the modulating signal and f is the transmission frequency. Then the received symbols at a distance l from the transmitter is given by

$$\tilde{r}(t, l) = A(l)m(t - l/c)e^{j2\pi f(l/c)} \tag{6.123}$$

where c is the speed of light and $A(l)$ is the path attenuation as a function of the distance l. (The symbol l used here to denote distance should not be confused with that used in section 6.6 to denote the number of transmitted symbols.) At this point, we make several key assumptions:

- The incident field is a plane wave. This assumption relies on the source being sufficiently distant and the array being physically small enough.
- The spacing of the antenna elements is small enough that there is no variation in amplitude among the signals received at the different elements. That is, the attenuation $A(l) \approx A(l_0) \equiv A_0$ is the same for all antenna elements.
- The bandwidth of the modulating signal is small compared with the carrier frequency f. This assumption implies that there will be little variation in the modulation over the physical dimensions of the antenna; that is,

$$m(t - l/c) \approx m(t - l_0/c) \equiv m_0(t)$$

  for each element of the array. (c denotes the speed of light.)

- There is no mutual coupling between the antenna elements; that is, they can be treated independently.

Under these assumptions, the analysis depends solely on the phase relationship of the different elements, in a manner similar to the manner in which Rayleigh fading depends on the phase relationship of the different multipath rays.

Assume that there are $N + 1$ uniformly spaced antenna elements, where N is even and d is the antenna spacing. Let the antenna elements be numbered from $-N/2$ to $N/2$. If antenna element k is at distance l_k from the transmitting antenna, then the corresponding carrier phase is given by

$$
\begin{aligned}
\phi(t) &= 2\pi f(t - l_k/c) \\
&= 2\pi f(t - l_0/c) - 2\pi f(l_k - l_0)/c \\
&\equiv \phi_0(t) - 2\pi f \Delta l_k/c \\
&\equiv \phi_0(t) - \Delta \phi_k
\end{aligned}
\tag{6.124}
$$

Since the relative distance of the kth element is $\Delta l_k = kd\sin\theta$, the corresponding phase offset at antenna element k is given by

$$
\begin{aligned}
\Delta \phi_k &= (2\pi f/c)(kd\sin\theta) \\
&= 2\pi\left(\frac{kd}{\lambda}\right)\sin\theta
\end{aligned}
\tag{6.125}
$$

where λ is the wavelength of the radio transmission. Consequently, from the assumptions underlying Eq.(6.123), the received signal at element k is given by

$$
\begin{aligned}
s_k(t) &= s_0(t)e^{j2\pi\left(\frac{kd}{\lambda}\right)\sin\theta} \\
&= a_k(\theta)s_0(t)
\end{aligned}
\tag{6.126}
$$

where

$$
s_0(t) = A_0 m_0(t)e^{j\phi_0(t)}
\tag{6.127}
$$

is the attenuated and delayed version of the transmitted signal; in Eq. (6.127),

$$
a_k(\theta) = e^{j2\pi\left(\frac{kd}{\lambda}\right)\sin\theta}
\tag{6.128}
$$

is a complex rotation. A phased array computes a linear sum of the signals received at each element shown in Fig. 6.29, yielding the received signal

$$
\begin{aligned}
r(t) &= \sum_{k=-N/2}^{N/2} w_k^* s_k(t) \\
&= (\mathbf{w}^\dagger \mathbf{a})s_0(t)
\end{aligned}
\tag{6.129}
$$

where

$$\mathbf{w} = [w_{-N/2}, ..., w_{N/2}]^T$$

is the *weighting vector* and

$$\mathbf{a}(\theta) = [a_{-N/2}(\theta),...,a_{N/2}(\theta)]^T$$

We have thus explicitly demonstrated the dependence of $\mathbf{a}$ on θ, the angle of arrival of the wavefront. The term $\mathbf{w}^\dagger\mathbf{a}$ is a *complex gain* that is applied to the received signal; it depends on the choice of the weighting vector $\mathbf{w}$ and angle θ.

EXAMPLE 6.8 Antenna Pattern with Uniform Weighting

Suppose that we select the weighting vector $\mathbf{w} = [1,1,...,1]^T$. Then, in this case, the complex gain of the receiving array antenna as a function of θ is written as

$$\begin{aligned} G(\theta) &= \mathbf{w}^\dagger\mathbf{a}(\theta) \\ &= \sum_{k=-N/2}^{N/2} e^{j2\pi\left(\frac{kd}{\lambda}\right)\sin\theta} \end{aligned} \tag{6.130}$$

If we set

$$\Theta = e^{j2\pi\left(\frac{d}{\lambda}\right)\sin\theta} \tag{6.131}$$

then we recognize that Eq. (6.130) is the sum of a geometric series, as shown by

$$\begin{aligned} G(\theta) &= \sum_{k=-N/2}^{N/2} \Theta^k \\ &= \Theta^{-N/2}\sum_{k=0}^{N} \Theta^k \\ &= \Theta^{-N/2}\left(\frac{1-\Theta^{N+1}}{1-\Theta}\right) \\ &= \frac{\Theta^{1/2}(\Theta^{-(N+1)/2} - \Theta^{(N+1)/2})}{\Theta^{1/2}(\Theta^{-1/2} - \Theta^{1/2})} \\ &= \frac{\Theta^{-(N+1)/2} - \Theta^{(N+1)/2}}{\Theta^{-1/2} - \Theta^{1/2}} \end{aligned} \tag{6.132}$$

The difference terms in the last line of Eq. (6.132) are proportional to the exponential representation of the sine function and may thus be written as

$$G(\theta) = \frac{\sin\left(\pi(N+1)\left(\frac{d}{\lambda}\right)\sin\theta\right)}{\sin\left(\pi\left(\frac{d}{\lambda}\right)\sin\theta\right)} \tag{6.133}$$

Figure 6.30 plots the gain $G(\theta)$ as a function of θ, assuming that $N = 6$ and the element spacing $d = \lambda/4$. The figure shows that the antenna has a strong gain in the direction $\theta = 0°$ and falls off quickly in other directions. ∎

If the signal is arriving from a direction θ_0, then it is desirable to have the maximum signal strength in that direction. That is, we wish to maximize $\mathbf{w}^\dagger\mathbf{a}(\theta_0)$. Accordingly, we may use the Cauchy–Schwarz inequality of Eq. (6.15) to show that

$$\mathbf{w} = c\mathbf{a}(\theta_0) \tag{6.134}$$

is the optimum weight vector, which is aligned with $\mathbf{a}(\theta_0)$; the scalar c is not to be confused with the speed of light. Without loss of generality, we choose the scalar $c = 1$. In Fig. 6.31, we plot the optimum gain,

$$\begin{aligned} G(\theta) &= \mathbf{w}^\dagger\mathbf{a}(\theta) \\ &= \mathbf{w}(\theta_0)^\dagger\mathbf{a}(\theta) \end{aligned} \tag{6.135}$$

for $\theta_0 = 45°$. As desired, we have produced an antenna that points in the direction of θ_0.

Figure 6.32 illustrates how we may produce an antenna that points in any direction simply by adjusting the weights. For a linear array, the quality of the pointing does depend somewhat on the angle θ_0. In the *end-fire* directions (i.e., large θ), however, the quality of pointing is not as good as that of other directions for small θ.

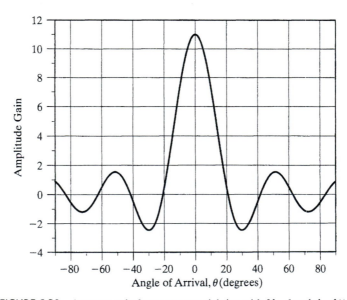

FIGURE 6.30   Antenna gain for constant weighting with $N = 6$ and $d = \lambda/4$.

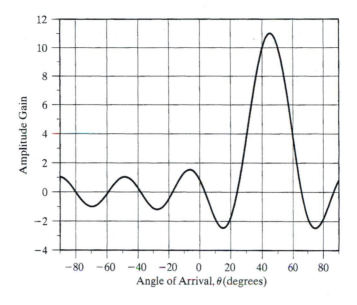

FIGURE 6.31 Gain for maximum signal strength with $\theta_0 = 45°$.

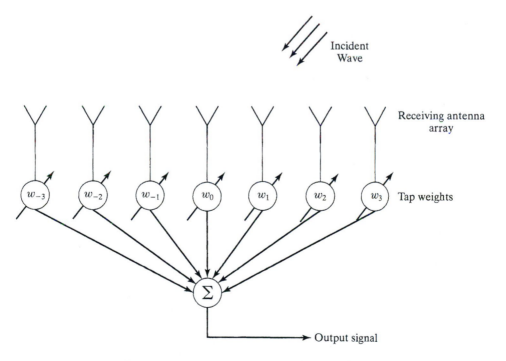

FIGURE 6.32 Antenna with adjustable elemental weights.

It is also clear, conceptually, how we could use the same array to generate multiple antennas; for example, we may use the same antenna elements with two different weighting vectors $\mathbf{w}_1$ and $\mathbf{w}_2$.

In addition, we may *adapt* the antenna to track a mobile terminal by adjusting the weights $\mathbf{w}(n)$ in discrete time n as a function of the user position. Such adaptive algorithms are discussed in Appendix I. The linear array is just one possible arrangement of antenna elements; more sophisticated arrangements (e.g., planar arrays) can provide higher gain and performance that is more uniform in all directions.

Problem 6.18 Suppose the received signal is known to consist of two strong multipath rays arriving from different directions. Describe how to design a phased array to capture the energy in both rays.   ∎

Problem 6.19 Derive an expression for the antenna gain $G(\theta)$ under the conditions of Example 6.18, assuming that the number of antenna elements, N, is even.

$$Ans.\ G(\theta) = \frac{\sin\left(\pi N\left(\frac{d}{\lambda}\right)\sin\theta\right)}{\sin\left(\pi\left(\frac{d}{\lambda}\right)\sin\theta\right)}$$ ∎

6.8.2 Multipath with Directional Antennas

For large cells, referred to as *macrocells*, the base-station antenna is usually mounted on a tall mast and is free of most significant local multipath. In such an environment, almost all multipath is generated in the vicinity of the mobile terminal for both the uplink and downlink paths. Several models have been described in the literature; one of the original ones, *Lee's model*,[18] is illustrated in Fig. 6.33. With this model, we have a number of effective scatterers surrounding the mobile terminal in a circular pattern. Reflections from these scatterers cause the multipath seen at the base station. Measurements indicate that the radius of the scatterers is typically 100 to 200 wavelengths (λ), a value consistent with other measurements which indicate that the typical angle spreads for macrocells with transmitter–receiver separations of 1 km are from 2° to 6° when measured at 800 MHz. From this observation, two rules of thumb for spacing antennas to achieve good diversity have been developed: at the mobile, it is recommended that the antenna spacing be at least 0.2λ; at the base station, an antenna separation of approximately 40λ is recommended.

With *microcells*, the separation between the base station and the user terminal is much smaller. Consequently, if we were to apply the Lee model, the circle of effective scatterers could come close to, or even encompass, the base station. In addition, microcell base station antennas are often not mounted high, so there will be scattering effects around the base station as well as the mobile. These observations motivate a geometrical model, referred to as the *single-bounce elliptical model*,[19] for multipath propagation in microcells. With this model, we begin by considering multipath that consists of a single reflection having a fixed excess delay (relative to the direct path between the base station and user terminal). From geometrical considerations, this

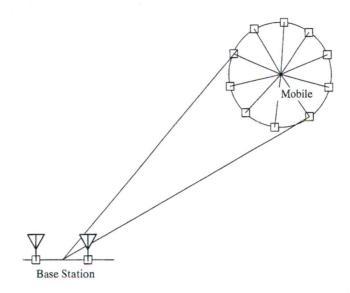

FIGURE 6.33   Lee's model for multipath propagation in macrocells.

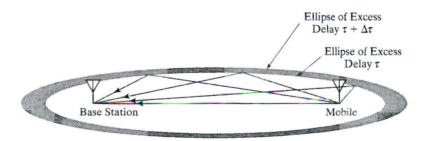

FIGURE 6.34   Single-bounce elliptical model for multipath propagation in microcells.

assumption defines an ellipse with the base station and mobile terminal situated at its foci and the reflectors located on its perimeter. We may then parameterize a series of ellipses with the excess delay parameter τ. All the sources of multipath with a delay between τ and $\tau + \Delta\tau$ are assumed to be due to reflectors in the area between the two corresponding ellipses, as shown in Fig. 6.34. By assuming the reflectors to be uniformly distributed, we can determine the distribution of the direction of arrival of multipath as a function of the time of arrival. This single-bounce elliptical model leads to some interesting observations:

- For a small number of reflectors, most multipath has a small excess delay and a small angular distribution; that is, paths with low excess delays and higher powers are clustered around the main path.

- As the number of reflectors increases, both the excess delay and the angular distribution increase.

The overall conclusion is that, with a small number of reflectors, a directional antenna may not reduce multipath as significantly as might be expected.

The previous results show that both smaller cells and directional antennas may reduce the number of multipath components that are received. A reduced number of multipath components will affect the signal's statistics. In Section 2.6.1, we considered the amplitude distribution of the received signal due to local reflections. The assumption behind that analysis was the existence of a large number of multipath rays, and the results indicated that the fading amplitude would have a Rayleigh distribution.

In this section, we will reevaluate the amplitude distribution for a small number of multipath components. Recall that the complex envelope of the received signal was modeled as (see Eq.(2.48))

$$\tilde{x}(t) = \sum_{k=1}^{N} \alpha_k e^{j\theta_k} \tilde{s}(t) \qquad (6.136)$$

where α_k is the gain or attenuation of the kth path and θ_k is the relative phase of that path. The fading amplitude is given by

$$R = \left| \sum_{k=1}^{N} \alpha_k e^{j\theta_k} \right| \qquad (6.137)$$

In Section 2.6.1, we argued that, for large N, the term on the right-hand side of Eq.(6.137) (before taking the magnitude) approaches a complex Gaussian random variable, in which case R approaches a Rayleigh-distributed random variable. With a directional antenna, N will be much smaller and a different result may follow. In Fig. 6.35, we plot the simulated distribution of R for different values of N; the Rayleigh distribution is also included in the figure for comparison. From the figure, we see that, for very small N, the probability of a deep fade is significantly larger than for large N. For $N = 6$ and larger, the distribution of R closely approximates the Rayleigh distribution. We conclude that it takes only a small number of paths to approximate a Rayleigh amplitude distribution, but for a *very* small number, the characteristics can be significantly worse.

The characteristics of the received multipath signal also depend on the bandwidth of the transmitted signal. If the transmitted signal is wideband, such as with direct-sequence spread spectrum, the delay spread is liable to be greater than a chip period. Consequently, the received signal may consist of two or more approximately independently faded signals. This is analogous to the diversity situation, and as a consequence, the variation in the received power (or envelope) is much less than is observed in the flat-fading case. The phenomenon affects directional antennas through the algorithms used for adaptation.

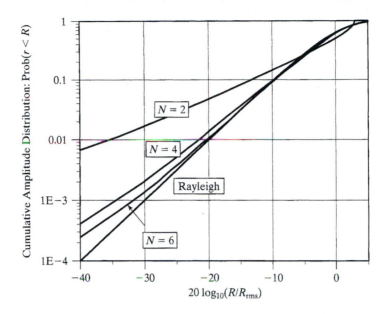

FIGURE 6.35   Amplitude distribution of multipath as a function of number of paths N.

6.9   THEME EXAMPLE 1: BLAST ARCHITECTURES

For the first theme example of the chapter, we have chosen a family of MIMO wireless communication systems popularized as *BLAST architectures*; the acronym "BLAST" stands for "Bell Laboratories Layered Space–Time." BLAST architectures use standard one-dimensional forward error-correcting codes and low-complexity interference-cancellation schemes to construct and decode powerful two-dimensional space–time codes. These MIMO systems offer spectacular increases in spectral efficiency, provided that three conditions are met:

1. The system operates in a rich Rayleigh scattering environment.

2. Appropriate coding structures are used.

3. Error-free decisions are available in the interference-cancellation schemes. This condition assumes the combined use of arbitrarily long (and therefore powerful) FEC codes and perfect decoding.

The material presented herein focuses on three specific implementations of BLAST, depending on the type of coding employed:

1. Diagonal-BLAST (D-BLAST)

2. Vertical-BLAST (V-BLAST)

3. Turbo-BLAST

6.9.1 Diagonal-BLAST Architecture[20]

In 1996, Foschini pioneered a *diagonally layered space–time architecture* known as *diagonal-BLAST,* or simply *D-BLAST.* This architecture now provides the benchmark for MIMO wireless communications. The distinctive feature of D-BLAST is its use of a diagonally layered coding structure in which the code blocks are arranged across *diagonals* in a space–time fashion, as illustrated in Fig. 6.36 for the example of $N_t = 4$ transmit antennas. The figure depicts the incoming binary data stream demultiplexed and then processed for transmission over the wireless link as four independent data substreams with equal bit rates. A distinctive feature of D-BLAST is that each encoded–modulated substream is *cycled over time.*

To illustrate the diagonal nature of the layered space-time code generated by D-BLAST, consider the example of four transmit antennas. The transmission matrix produced by this structure has the form

$$
\mathbf{S} = \begin{bmatrix}
\tilde{s}_{1,1} & \tilde{s}_{1,2} & \tilde{s}_{1,3} & \tilde{s}_{1,4} & \tilde{s}_{1,5} & \cdots & \tilde{s}_{1,K-1} & \tilde{s}_{1,K} & 0 & 0 & 0 & 0 \\
0 & \tilde{s}_{2,1} & \tilde{s}_{2,2} & \tilde{s}_{2,3} & \tilde{s}_{2,4} & \cdots & \tilde{s}_{2,K-2} & \tilde{s}_{2,K-1} & \tilde{s}_{2,K} & 0 & 0 & 0 \\
0 & 0 & \tilde{s}_{3,1} & \tilde{s}_{3,2} & \tilde{s}_{3,3} & \cdots & \tilde{s}_{3,K-3} & \tilde{s}_{3,K-2} & \tilde{s}_{3,K-1} & \tilde{s}_{3,K} & 0 & 0 \\
0 & 0 & 0 & \tilde{s}_{4,1} & \tilde{s}_{4,2} & \cdots & \tilde{s}_{4,K-4} & \tilde{s}_{4,K-3} & \tilde{s}_{4,K-2} & \tilde{s}_{4,K-1} & \tilde{s}_{4,K} & 0
\end{bmatrix}
$$

where the encoded entries in the kth row are delayed by k-1 time units with respect to those in the first row, as the entries below the diagonal are padded with zeros. The entries on the first diagonal of the transmission matrix **S** are sent to the receiver by antenna 1, the entries on the second diagonal of **S** are sent by antenna 2, and so on, with the result that the encoded symbols are transmitted by different antennas.

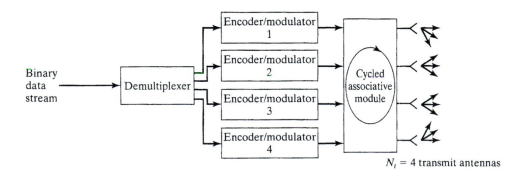

FIGURE 6.36   High-level diagram of D-BLAST transmitter for four transmit antennas.

The diagonal coding structure of D-BLAST is not only highly elegant, but also capable of achieving the channel capacity based on Eq. (6.60) for an (N_t, N_r) antenna configuration with $N_t \le N_r$, assuming that the channel is impaired by additive white Gaussian channel noise and that it operates in a quasi-static, flat-fading environment. However, a major drawback of D-BLAST lies in the need for independent coding at each diagonal layer, which implies the required use of short diagonal-layer coding schemes so as to reduce boundary space–time edge wastage; this form of wastage is illustrated in the structure of the transmission matrix **S**. Unfortunately, the use of *short* diagonal codes suffers from the deficiency of not allowing adequate time to support powerful bandwidth-efficient codes with forward error-correcting capabilities.

6.9.2 Vertical-BLAST Architecture[21]

To mitigate the computational difficulty of D-BLAST, Wolniansky et al. (1998) proposed a simplified version of BLAST known as *vertical-BLAST,* or *V-BLAST,* the first practical implementation of MIMO wireless communications to demonstrate a spectral efficiency as high as 40 bits/s/Hz in real time. In V-BLAST, the incoming binary data stream is first demultiplexed into N_t substreams, each encoded independently and mapped onto an antenna of its own by a modulator for transmission over the channel, as depicted in Fig. 6.37(a). Insofar as the transmitter is concerned, the net result is the conversion of the incoming binary data stream into a vertical vector of encoded modulated substreams—hence the name "vertical-BLAST," or "V-BLAST." Comparing this figure with Fig. 6.36 for D-BLAST, we see that in V-BLAST there is no cycling over time—hence the significant reduction in system complexity. Moreover, in the V-BLAST transmitter, every antenna transmits its own independently coded substream of data. In so doing, V-BLAST eliminates the space–time edge wastage plaguing D-BLAST, but the outage capacity achieved by V-BLAST is substantially lower than that of D-BLAST.

Turning next to the receiver depicted in Fig. 6.37(b), the signals impinging on the receive antennas are individually demodulated and then channel decoded in accordance with the corresponding operations performed in the transmitter. The detection process, leading to an estimate of the original binary data stream, is performed by the functional block labeled *ordered serial interference-cancellation (OSIC) detector,* which exploits the timing synchronism between the V-BLAST receiver and transmitter. Specifically, the detection process involves the following sequence of operations:

1. *Order determination,* in which the N_r received substreams are to be detected, in accordance with the postdetection signal-to-noise ratios of the individual substreams.
2. *Detection* of the substream, starting with the *largest signal-to-noise ratio.*
3. *Signal cancellation,* wherein the effect of the detected substream is removed from subsequent substreams.
4. *Repetition* of steps 1 through 3 until all the N_r received substreams have been individually detected.

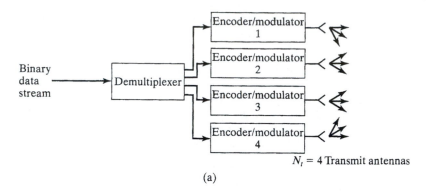

(a)

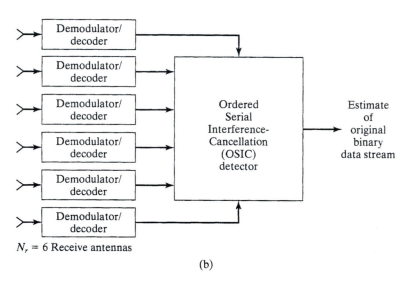

(b)

FIGURE 6.37   High-level diagram of V-BLAST.
(a) Block diagram of the transmitter four transmit antennas.
(b) Block diagram of the receiver for six receive antennas.
Note: The OSIC includes the multiplexing operation needed to restore the detected binary data stream to its original serial form.

The procedure is nonlinear, due to two factors: First, the estimated-signal-cancellation process is itself nonlinear; second, the detection step involves a quantization (slicing) operation that is appropriate to the signal constellation used in the transmitter.

Another noteworthy point is that the nulling-and-cancellation step is performed by exploiting the channel matrix **H**, which is estimated at the receiver through supervised training aided by sending a training sequence from each transmit antenna. The nulling process suppresses the interference and is followed by cancellation of the estimated signal in question.

It is important to note that for the OSIC detection process to work properly, the number of receive antennas N_r must be at least as large as the number of transmit antennas N_t, for reasons to be explained in Section 6.10.

One last comment is in order. In V-BLAST, each transmitted data stream is channel-encoded in its own way. Typically, the channel encoders in the N_t layers of the space–time system are identical. The structure of V-BLAST may be simplified by employing a single channel encoder before the demultiplexing operation in the transmitter; correspondingly, a single channel decoder is used after the multiplexing operation in the receiver.

Problem 6.20   A prototype V-BLAST system uses $N_t = 8$ transmit antennas and $N_r = 12$ receive antennas in an indoor environment with negligible delay spread. The system utilizes uncoded 16-QAM, where the term "uncoded" refers to the absence of channel coding. The system has a transmission rate of 24.3×10^3 symbols/s and a channel bandwidth of 30 kHz.

(a) Calculate the raw spectral efficiency of the system.
(b) Assuming that 80% of each burst is devoted to data transmission, calculate the payload efficiency of the system.

Ans.   (a) 25.9 bits/s/Hz
           (b) 20.7 bits/s/Hz   ∎

6.9.3 Turbo-BLAST Architecture[22]

The third BLAST architecture for high-throughput wireless communications exploits three basic ideas:

1. BLAST architecture, which uses multiple antennas on both transmit and receive.
2. *A random layered space–time (RLST) coding scheme*, which is based on the use of independent block coding and space–time interleaving.
3. *A Turbolike receiver*, also known as an *iterative detection and decoding (IDD) receiver*, which performs decoding of the RLST code and iterative estimation of the channel matrix.

In light of points 1 and 3, the new architecture is referred to as *Turbo-BLAST*. This architecture is also referred to as *Turbo-MIMO*. (The turbo coding principle was discussed in Chapter 4.)

Figure 6.38(a) gives a high-level picture of the Turbo-BLAST transmitter with $N_t = 4$ transmit antennas. The encoding process in the transmitter, responsible for generating a *serially concatenated RLST code*, involves the following steps:

- *Demultiplexing* the incoming bit stream into N_t substreams of equal bit rate
- *Independent block-encoding* of each data substream, using the same predetermined linear block FEC codes
- *Interleaving* the encoded substreams by means of a space–time permuter, which is independent of the incoming bit streams
- *Full-rate space–time encoding*, facilitated by the channel matrix; the term "full-rate" means that the symbol-constellation in the transmitter's mapper is designed to have a size equal to the spatial degree of freedom $N = \min(N_t, N_r)$.

The structure of the RLST encoder, designed on the basis of independent block coding of each transmit data substream and periodically cyclic space–time interleaving, is shown in Fig. 6.38(b). The transmitter pictured here is the multiple-antenna generalization of the serial turbo encoder discussed in Section 4.18 on joint equalization and decoding. For optimal performance of the RLST code, the receiver should use the *maximum a posteriori probability (MAP) decoding algorithm*, which is described in Appendix F. However, the computational complexity of this algorithm for the RLST code becomes increasingly unmanageable as the number of transmit or receive antennas is increased. Specifically, with K denoting the length of each layer in the RLST code, the MAP decoding algorithm needs to select one of $2^{N_t K}$ sequences, all of which increase exponentially with increasing N_t. To mitigate the computational

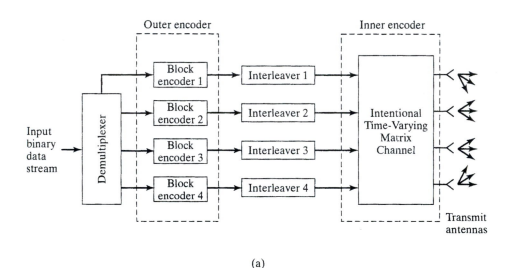

(a)

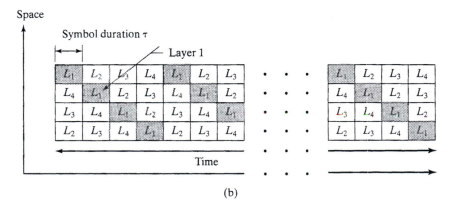

(b)

FIGURE 6.38 (a) High-level picture of the Turbo-BLAST transmitter with four transmit antennas. (b) Illustrating the structure of the random layered space-time (RLST) code generated in the transmitter.

complexity problem, we may use the near-optimal turbolike receiver depicted in Fig. 6.39. The interleavers used to design the RLST codes provide the basis for a near-optimal *iterative detection and decoding* (IDD) process with feasible computational complexity. The receiver has two stages, in accordance with the receive part of the turbo coding principle:

1. *The inner decoder*, consisting of a soft-input, soft-output (SISO) detector, which is designed to counteract the intersymbol interference (ISI) problem due to the multipath fading channel.

2. *The outer decoder*, consisting of N_t parallel SISO channel decoders, which is designed to correct symbol errors incurred during the course of transmission over the channel.

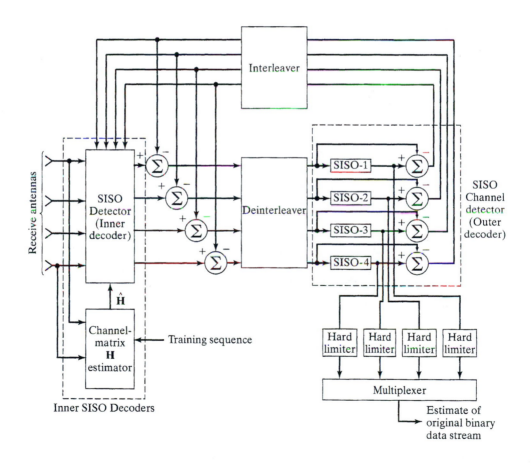

FIGURE 6.39   High-level block diagram of iterative decoder for Turbo-BLAST for four receive antennas. (Note: The abbreviation SISO stands for soft-input, soft output; it should not be confused with single-input, single-output antenna configuration.)

The detector and decoder stages of the receiver are separated by *space–time de-interleavers and interleavers*, which may be viewed as the spatial generalizations of their counterparts in the turbo decoder. The space–time deinterleavers and interleavers are used to compensate for the interleaving operations used in the transmitter and to de-correlate outputs before feeding them to the next decoding stage. The two sets of summers, where differences are computed, provide *extrinsic information* from the inner decoder to the outer decoder via the deinterleaver and from the outer decoder to the inner decoder via the interleaver, thereby forming a closed feedback loop around the two decoding stages in accordance with the turbo-coding principle. The iterative receiver produces new and better estimates at each iteration of the receiver and repeats the information-exchange process a number of times, to improve the channel decisions and channel estimates. The design of intersubstream coding at the transmitter, based on independent coding of each substream, leads to a simplification of the receiver, which now needs to select only one of 2^K sequences (where K is the length of each transmitted burst), separately for each transmitted sequence.

To reconstruct the original binary data stream, two things are done, as illustrated in Fig. 6.39:

1. When the IDD process for an information bit is terminated, the output of the SISO channel decoders are hard-limited.
2. The resulting hard-limiter outputs are multiplexed into a serial form, consistent with that of the original binary data stream.

During the *first iteration* of the receiver, a short training sequence is used to produce a preliminary estimate of the channel matrix **H**. Unfortunately, with a short training sequence, it may be difficult to achieve a good estimate of the time-varying MIMO channel, particularly in an outdoor environment. To mitigate this problem, the channel matrix is reestimated with the use of newly derived estimates of symbols at each subsequent iteration of the receiver. The *bootstrapping* technique described herein tends to extract maximal information out of each burst of data received.

6.9.4 Experimental Performance Evaluation of Turbo-BLAST versus V-BLAST[23]

We conclude Theme Example 1 by comparing the performance of QPSK-modulated Turbo-BLAST with that of a correspondingly horizontal-coded V-BLAST, using real-life indoor channel measurements on various MIMO configurations. The channel measurements were acquired by utilizing a narrowband testbed in an indoor environment with negligible delay spread. At the transmit end, each substream of 100 information bits was independently encoded, using a rate-1/2 convolutional code generator (7,5), and then interleaved by means of space–time interleavers. The space interleavers were designed with diagonal layering interleavers (Fig. 6.38); the time interleavers were chosen randomly, and no attempt was made to optimize their design. With regard to horizontal-coded V-BLAST, each of the substreams was first, independently coded via rate-1/2 convolutional code with generator (7,5) and, second, QPSK modulated. Using the measured channel characteristics, and then, using various BLAST configurations, we evaluated the performance of Turbo-BLAST over a wide range of signal-to-noise ratios. For the first two experiments, we evaluated the Turbo-BLAST system with the exact channel matrix. In the third experiment, we evaluated the performance with channel estimation, using a short training sequence and an iteratively estimated channel.

Experiment 1. Turbo-BLAST versus V-BLAST for $N_t = 5, 6, 7, 8$ and $N_r = 8$.

We first consider BLAST configurations with fewer transmit antennas than receive antennas. Figure 6.40 compares the bit-error-rate performance of Turbo-BLAST (solid traces) and coded V-BLAST (broken traces) for antenna configurations of 8 receive and 5 to 8 transmit antennas. The Turbo-BLAST system produces the best receiver performance within the first 10 iterations of its operation. The bit-error-rate performance of both V-BLAST and Turbo-BLAST improves with a decreasing number of transmitters, with Turbo-BLAST outperforming V-BLAST in all four cases. In

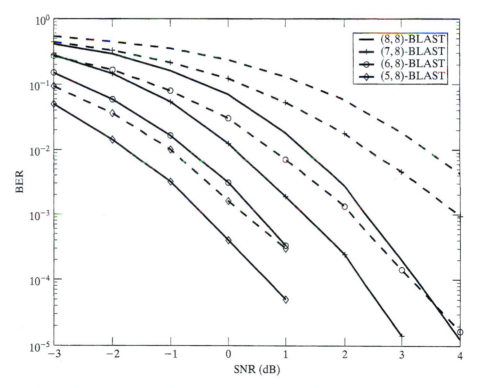

FIGURE 6.40   Bit-error-rate (BER) performance for $N_t = 5, 6, 7$, and 8 and $N_r = 8$, using convolutional code with rate $1/2$ and constraint length 3 and using QPSK modulation.
Solid traces: Turbo-BLAST.
Broken traces: V-BLAST.

terms of V-BLAST performance, a substantial gain in bit-error-rate performance is realized with fewer transmit antennas. For example, at a BER of 10^{-3}, Turbo-BLAST achieves 2–3 dB of gain over V-BLAST for $N_t = 7$ and $N_r = 8$, whereas only 0.5 dB of gain is attained when $N_t = 5$ and $N_r = 8$.

Experiment 2. Turbo-BLAST versus V-BLAST for $N_t = 8$ and $N_r = 5,6,7,8$.

We next consider BLAST configurations with fewer receive antennas than transmit antennas. Figure 6.41 compares the bit-error-rate performance of Turbo-BLAST (solid traces) with that of horizontal-coded V-BLAST (broken traces). With antenna configurations of eight transmit and five to eight receive antennas, here again we find that Turbo-BLAST gives the best overall performance within the first 10 iterations. The figure reveals a major limitation of the V-BLAST system, namely, the inability to work efficiently with fewer receive antennas than transmit antennas. In the context of Turbo-BLAST, we can make two observations from Fig. 6.41: First, the bit-error-rate performance of Turbo-BLAST improves with an increasing number of receivers, with Turbo-BLAST outperforming V-BLAST in all four cases; second, increasing the number N_r of receive antennas from 7 to 8 offers little benefit.

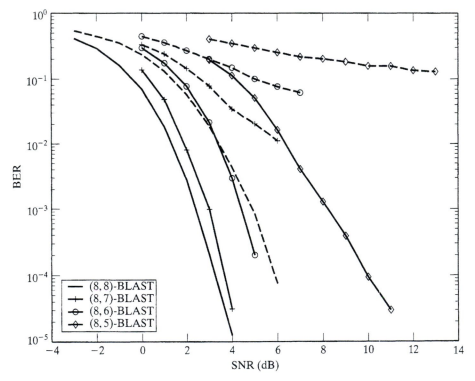

FIGURE 6.41   Bit-error-rate (BER) performance for $N_t = 8$ and $N_r = 5, 6, 7,$ and 8, using convolutional code with rate $1/2$ and constraint length 3, and using QPSK modulation.
Solid traces: Turbo-BLAST.
Broken traces: V-BLAST.

Experiment 3. Turbo-BLAST for $N_t = N_r = 8$ and iterative channel estimation.

In Fig. 6.42, we compare the performances of two decoding schemes: (1) an iterative decoder with initial channel estimation using 16 training symbols only, and (2) an iterative decoder with initial channel estimation and iterative refined channel estimation. The bit-error-rate performance results are compared for Turbo-BLAST architectures with perfect channel knowledge and with perfect channel and interference knowledge. The figure shows the convergence behavior of the IDD receiver of Fig. 6.39 at a signal-to-noise ratio of 3 dB. Although the bit-error-rate performance of the IDD receiver with iterative channel estimation is initially (i.e., at the first iteration) worse than the decoder with perfect channel knowledge, at the fifth iteration of the decoding process it comes very close in performance to the decoder with knowledge of channel state information. Moreover, both decoders converge very close to the decoder that has knowledge of both the channel and the interference (dashed trace). Because of channel estimation errors, the bit-error-rate performance of the IDD receiver with initial channel estimates is about 2–4 dB worse than the schemes with channel knowledge only and the ideal case with knowledge of the channel and interference.

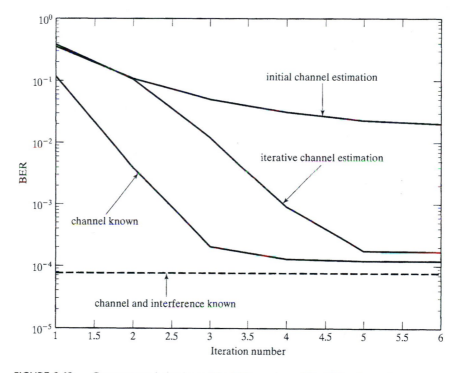

FIGURE 6.42   Convergence behaviors of the IDD receiver of Fig.6.39 under various conditions: bit-error-rate (BER) performance (at SNR = 3 dB) versus the number of iterations for $N_t = 8$ and $N_r = 8$, using convolutional code with rate 1/2, constraint length 3, and QPSK modulation.
Solid traces: Turbo-BLAST for varying channel conditions.
Broken traces: Ideal performance.

6.10 THEME EXAMPLE 2: DIVERSITY, SPACE–TIME BLOCK CODES, AND V-BLAST

Theme example 1 was devoted to three different BLAST architectures, which constitute an important class of multiple-input, multiple-output (MIMO) wireless communications. But then, MIMO wireless systems also include antenna diversity and space–time block codes as special cases. It is therefore in order that we discuss two pertinent issues:

1. Evaluation of the practical merits of a diversity-on-receive antenna system versus its diversity-on-transmit counterpart.

2. Comparison of the receiver performance of a space–time block code with that of a BLAST architecture, both involving diversity at the transmit and receive ends of the wireless link.

The purpose of this second theme example is to address these two issues, using qualitative arguments for dealing with issue 1 and computer experiments for illustrating issue 2.

6.10.1 Diversity-on-Receive versus Diversity-on-Transmit

The diversity-on-receive technique is of long standing, with its roots dating back to the 1920s.[24] In contrast, the diversity-on-transmit technique is of recent origin, with its discovery in 1998 credited to Alamouti.[13]

These two antenna diversity techniques rely on the use of multiple antennas, on the receive side of a wireless communication system in one case and on the transmit side of the system in the other case. Therein lie the realities that shape the use of the two techniques in practice.

Consider first the case of diversity-on-receive. The need for multiple antennas operating at radio frequencies and for the ancillary instruments to select the strongest diversity branch in the case of selection combining, for example, makes the use of diversity-on-receive at the terminal (mobile) units in a wireless communication system rather awkward (requiring the use of dual antennas) and expensive to implement. The picture is, however, quite different at the base station, which is equipped, in terms of both transmit power and real estate, to serve thousands of terminal units. Indeed, it is for this reason that the diversity-on-receive technique is routinely included in the construction of base stations.

In a loose sense, diversity-on-transmit may be viewed as playing a *dual* role to that of diversity-on-receive in counteracting the multipath-channel-fading problem. Accordingly, the two techniques can serve complementary purposes in their use, side by side at the base station:

*   Diversity-on-receive looks after the channel fading that affects incoming signals from terminal units operating inside the base station's coverage on the downlink.

*   Diversity-on-transmit looks after the channel fading that affects outgoing signals transmitted by the base station to the terminal units in its coverage, *without any feedback from the mobile station to the base station* on the uplink.

To implement the diversity-on-transmit, for example, exemplified by the Alamouti code requires simply (i) the use of two transmit antennas at the base station and (ii) a linear decoding receiver with a single receive antenna at the terminal unit, both of which make sense in practical terms. Indeed, it is for this reason that the Alamouti code has been adopted in third-generation (3G) wireless systems.

In short, through the use of both diversity-on-receive and diversity-on-transmit at a base station, the reception, as well as the transmission, of information-bearing signals at the base station are made *reliable*.

6.10.2 Space–Time Block Codes versus V-BLAST

Consider next the issue of comparing the receiver performance of a space–time block code (STBC) system with that of a BLAST system. For the STBC system, we have picked the expanded version of the Alamouti code; see Problem 6.37. For the BLAST system, we have picked V-BLAST. Both of these multiple-antenna configurations are the simplest schemes in their respective classes of MIMO wireless communications. Moreover, in order to highlight the underlying differences between them, both systems are uncoded (i.e., no channel coding is used in either one of them).

In what follows, the performance of each of these two MIMO systems is evaluated in the content of a two-part experiment:

Experiment 1. Receiver performances

The objective of this experiment is to evaluate the effect of increasing signal-to-noise ratio (SNR) on the symbol error rate (SER) for a prescribed spectral efficiency.

Four specific $\{N_t, N_r\}$ wireless systems are evaluated:

(i) STBC: $\{2, 2\}$
       V-BLAST: $\{2, 2\}$
(ii) STBC: $\{2, 4\}$
       V-BLAST: $\{2, 4\}$

For all four multiple-antenna systems, the spectral efficiency is maintained at 4 bits/s/Hz. The constancy of spectral efficiency is achieved by using different modulation schemes: 4-PSK (QPSK) for V-BLAST and 16-PSK for STBC. V-BLAST sends independent symbols on its two transmit antennas, and with each transmission using 4-PSK, the spectral efficiency is therefore $2 \times \log_2 4 = 4$ bits/s/Hz. As with the STBC, since the Alamouti code is a rate-1 code, the use of 16-PSK yields the spectral efficiency $\log_2 16 = 4$ bits/s/Hz, as desired.

The results for the experiment are plotted in Fig. 6.43. On the basis of these results, we may make the following observations:

(i) *Two-by-two antenna systems*

- For low signal-to-noise ratios (5 to 17.02 dB), V-BLAST outperforms STBC. This result is in perfect agreement with statements reported in Section 4.11. Specifically, therein we made the point that an uncoded system outperforms a channel encoded system when the signal-to-noise ratio is low. In the context of our present experiments, recall that V-BLAST is uncoded whereas the STBC system

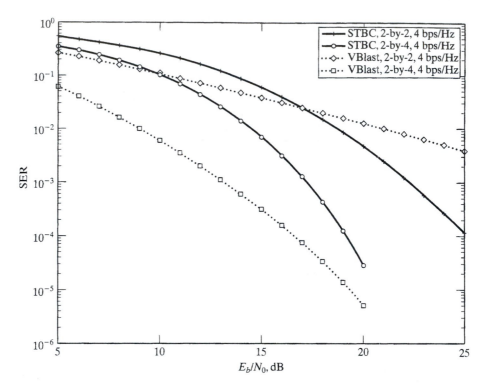

FIGURE 6.43   BER versus E_b/N_0 for four multiple-antenna systems studied under Experiment 1, assuming a fixed spectral efficiency of 4 bits/s/Hz.

uses a space-time encoder. Note, however, the SER (around 0.6 down to 0.023) is not small enough for either system to be of practical value in a wireless communications environment. But then recognize that if forward error-correction (FEC) channel codes are included in either system, then uncoded symbol error rates as high as 0.2 may be acceptable.

- For high signal-to-noise ratios (in excess of 17 dB), STBC begins to outperform V-BLAST. For example, at SNR = 25 dB, the SER produced by the STBC system is slightly larger than 10^{-4}. However, this improvement in receiver performance is attained at the cost of a significant increase in SNR.

(ii) *Two-by-four antenna systems*

- In doubling the number of receive antennas from two to four, the benefit of receive diversity is enhanced, thereby improving the performance of both STBC and V-BLAST. However, the important point to note here is the fact that, for low SNR, the improvement in receiver performance for V-BLAST is significantly better than that for STBC. For example, for SER = 10^{-3}, V-BLAST requires an SNR of 13.15 dBs. For this same SNR, the SER produced by STBC is 2.3×10^{-2}, which is more than an order of magnitude worse than the corresponding result for V-BLAST.

- The crossover SNR, at which STBC begins to outperform V-BLAST, is well in excess of 20 dB.

An important question is, for low SNR, how do we explain the significantly better receiver performance produced by V-BLAST over STBC in going from two to four receive antennas? The answer to this question resides in two factors:

1. The choice of modulation for a fixed spectral efficiency, 16-PSK for the STBC system and 4-PSK for V-BLAST, makes the former system much harder to demodulate than the latter. Asymptotically, there is a constant difference in receiver performance between these two modulation schemes, but, most important, the 16-PSK degrades more quickly than 4-PSK at low signal-to-noise ratios.

2. From Chapter 4, we recall that an uncoded system outperforms its coded counterpart at low signal-to-noise ratios. Thus, although both MIMO systems do not use channel coding, the STBC system does use the full-rate Alamouti code, which may be another possible factor for the observed difference in performance between the STBC and V-BLAST systems at low signal-to-noise ratios.

The influences of these two factors on the behavior of STBC and V-BLAST systems at low signal-to-noise ratios become more pronounced as the number of receive antennas is increased, hence the low-SNR results displayed in Fig. 6.43.

6.10.3 Diversity Order and Multiplexing Gain[25]

At various points in this chapter, we have made reference to diversity order. This subsection presents a brief discussion of the diversity order of MIMO wireless links and its experimental evaluation.

To proceed with the discussion, we first note that ultimately all the results relating to diversity order reflect in the behavior of the *average frame error rate* with respect to the signal-to-noise ratio; a frame or packet is another way of referring to a burst of data transmission across the wireless link. Consider then a space-time scheme whose average frame error rate, expressed as a function of the signal-to-noise ratio ρ, is denoted by $\text{FER}(\rho)$. On this basis, we may make the following statement:

A space-time coding scheme whose asymptotic behavior is described by

$$d_o = -\lim_{\rho \to \infty} \left\{ \frac{\log(\text{FER}(\rho))}{\log \rho} \right\} \tag{6.138}$$

is said to have a diversity order d_o.

In other words, the asymptotic slope of the average frame error rate plotted as a function of the signal-to-noise ratio, on a log-log scale, is equal to the diversity order d_o. As pointed out on page 371, the frame error rate is approximately the same as the outage probability in terms of the exponent signal-to-noise ratio. Hence, we may equally well formulate the definition of the diversity order d_o in terms of the outage probability P_{outage}.

An implication of Eq. (6.137) is that, in a Rayleigh fading environment, a space-time coding scheme with N_t transmit antennas and N_r receive antennas is capable of attaining the *maximal diversity order*

$$d_{o,\max} = N_t \times N_r \tag{6.139}$$

In this context, we may cite two simple examples of interest:

1. For the Alamouti code with $N_t = 2$ and $N_r = 1$, we have $d_o = 2$.
2. For the maximal-ratio combiner with $N_t = 1$ and $N_r = 2$, we again have $d_o = 2$.

Both of these results are intuitively satisfying.

What about BLAST systems that are designed to maximize ergodic capacity? Is there tension between maximizing capacity (i.e. transmission rate) and maximizing diversity order (i.e., reliability of communication)? To answer these two questions, we introduce the definition of multiplexing gain. Specifically, we make the following statement:

A space-time coding scheme, whose ergodic capacity expressed as a function of the signal-to-noise ratio ρ, denoted by $C(\rho)$, is said to have a multiplexing gain r if the asymptotic behavior (also see Eq. (6.40))

$$\lim_{\rho \to \infty} \frac{C(\rho)}{\log \rho} = r \tag{6.140}$$

holds.

An implication of the formula of Eq. (6.140) is that in a high signal-to-noise ratio regime, the *maximal multiplexing gain* is defined by

$$r_{\max} = \min\{N_t, N_r\} \tag{6.141}$$

Continuing the discussion of BLAST systems, consider a BLAST system that uses the ordered serial interference cancellation (OSIC) system to deal with the co-antenna interference (CAI) problem. Unlike space-time-block code systems designed to maximize diversity order, the diversity order of a BLAST system varies from one layer (antenna) to the next one by virtue of the ordered way in which the OSIC detector solves the CAI problem. Specifically, to process the first layer output, $N_t - 1$ interferences have to be cancelled. Thus, with $N_t - 1$ of the transmit antennas (i.e., degree of freedom) committed to this cancellation process, the space diversity attainable by the first layer is

$$N_r - (N_t - 1) = N_r - N_t + 1$$

Next, when the OSIC detector processes the second layer output, there will now be $N_t - 2$ interferences to be cancelled, making the space diversity attainable by the second layer assume the value

$$N_r - (N_t - 2) = N_r - N_t + 2$$

and so on for the remaining layers. The overall diversity order of the BLAST system is defined by the *minimum* of the diversity orders for all the N_r layers of the system. On this basis, we may now define the diversity order of a BLAST system using the OSIC detector as

$$d_{o, \text{BLAST}} = N_r - N_t + 1 \tag{6.142}$$

which is less than the diversity order attainable by the corresponding STBC system.

From the discussion presented herein, we see that MIMO wireless communication systems provide a *capacity-diversity trade-off*, depending on how the system is actually configured: increased diversity leads to improved reliability of the system, whereas increased capacity leads to improved data transmission across the wireless channel.

Experiment 2. Diversity Orders of STBC and V-BLAST Systems

The objective of this second experiment is to experimentally evaluate the diversity orders of STBC and V-BLAST systems in light of the material presented in Section 6.10.3. The parameters of the two classes of MIMO systems considered in the experiment are summarized in Table 6.4. The modulation strategies used in the experiment are as follows:

    STBC: 16-PSK
    V-BLAST: 4-PSK

For the experimental evaluations, we plot the symbol error rate (rather than the frame error rate) versus the signal-to-noise ratio. Although indeed the relationship between the symbol error rate and frame error rate is rather complicated, there is a strong correspondence between them. Heuristically, the two of them behave similarly in an asymptotic sense. Following a coarse argument, we may formulate the *probability of correct frame* as

$$P_{\text{correct}}(\text{Frame},\rho) = 1 - \text{FER}(\rho) \tag{6.143}$$

$$= [1 - \text{SER}(\rho)]^{K}$$

where K is the number of symbols contained in a frame. Asymptotically, in the sense of increasing signal-to-noise ratio, ρ, and therefore decreasingly small symbol error rate, we may approximate Eq. (6.142) as follows:

$$1 - \text{FER}(\rho) \approx 1 - K \times \text{SER}(\rho)$$

or, equivalently,

$$\text{FER}(\rho) \approx K \times \text{SER}(\rho) \tag{6.144}$$

Equation (6.140) suggests that, in the SNR exponent, the frame and symbol error rates (probabilities) behave similarly. In other words, their respective diversity orders (i.e., the asymptotic slopes versus the signal-to-noise ratio on log-log scales) tend to behave similarly, which is confirmed by numerical evaluations. Accordingly, in light of Eqs. (6.137) and

TABLE 6.4 MIMO Configurations for Experiment 2.

MIMO configuration	Number of Antennas		Diversity order	
	N_t	N_r	Theory	Experiment
STBC	2	1	2	1.92
	2	2	4	3.99
V-BLAST	2	2	1	0.97
	2	3	2	1.91

(6.143), we may approximately reformulate the diversity order as

$$d_o \approx - \lim_{\rho \to \infty} \left\{ \frac{\log(\mathrm{SER}(\rho))}{\log\rho} \right\} \qquad (6.145)$$

Turning now to the experimental evaluations, we present two sets of results:

1. Figure 6.44(a) plots the symbol error rates versus the signal-to-noise ratio for the two STBC configurations listed in Table 6.4(a). The last two columns of this table present the theoretical values of diversity orders determined from Eq. (6.145) and the corresponding experimental values derived from the plots of Fig. 6.44(a).

2. Figure 6.44(b) plots the symbol error rates versus the signal-to-noise ratio for the two V-BLAST configurations listed in Table 6.4(b). Here again the last two columns of this table present the theoretical values of diversity orders determined from Eq. (6.145) and the corresponding experimental values derived from the plots of Fig. 6.44(b).

On the basis of the results on diversity order presented in the two parts of Table 6.4, we may state that there is good agreement between theory and experiment for both the STBC and BLAST systems considered herein.

6.11 THEME EXAMPLE 3: KEYHOLE CHANNELS [26]

In a MIMO channel, the ability to exploit space-division multiple-access techniques for spectrally efficient wireless communications is determined by the rank of the complex channel matrix **H**. (The rank of a matrix is defined by the number of independent columns in the matrix.) For a given (N_t, N_r) antenna configuration, it is desirable that the rank of **H** equal the minimum one of N_t transmit and N_r receive antennas, for it is only then that we are able to exploit the full potential of the MIMO antenna configuration. Under special conditions, however, the rank of the channel matrix **H** is reduced to unity, in which case the scattering (fading) energy flow across the MIMO link is effectively confined to a very narrow pipe, and with it, the channel capacity is severely degraded. The third theme example of the chapter explains the origin of this physical phenomenon, which is commonly referred to as the *keyhole channel* or *pinhole channel*.

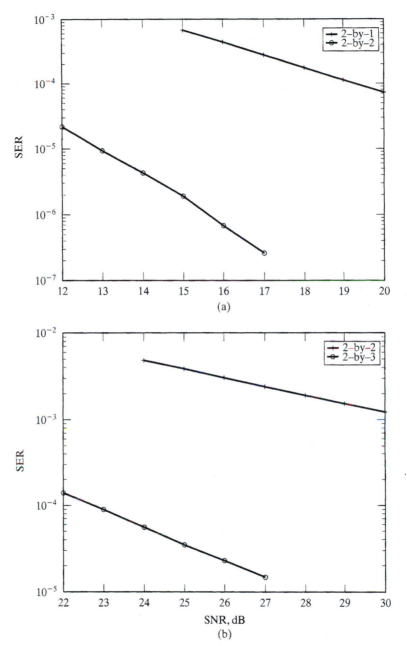

FIGURE 6.44 Experimental evaluation of diverstiy order for MIMO wireless links.
(a) Space-time block codes for two antenna configurations:
      (i) $N_t = 2, N_r = 1$. (ii) $N_t = 2, N_r = 2$ Modulation scheme: 16-PSK
(b) V-BLAST architecture for two antenna configurations:
      (ii) $N_t = 2, N_r = 1$. (ii) $N_t = 2, N_r = 3$ Modulation scheme: 4-PSK
[To achieve accurute statistics (i.e., asymptotic results), large amount of data were used in the experimental plots of Fig. 6.44.]

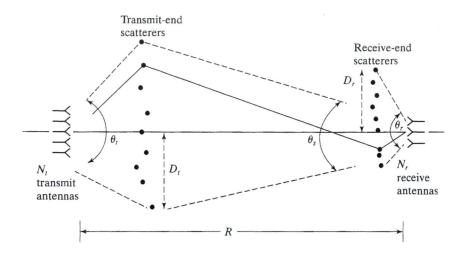

FIGURE 6.45 Propagation layout of MIMO communication link depicting two sets of scatterers, one in proximity to the transmit antennas and the other in proximity to the receive antennas.

Consider the idealized scattering environment pictured in Fig. 6.45. The environment embodies two sets of *significant scatterers*, one in proximity to the N_t transmit antennas and the other in proximity to the N_r receive antennas. Yet, the two sets of scatterers are located far enough away from their respective antenna arrays to justify the assumption of plane-wave propagation across the link. The *radial extents* of the two sets of scatterers are respectively denoted by D_t and D_r, both of which are assumed to be small compared with the range R (i.e., the distance between the transmit and receive antenna arrays). A few other assumptions pertaining to the scattering field are as follows:

- Both sets of scatterers, each numbering S, behave like *ideal reflectors*.

- The numbers of scatterers on both sides of the link are large enough to ensure the occurrence of random fading.

- The scatterers on the transmit side of the link are uniformly spaced, acting like an array of *virtual transmit antennas*, with each one of them receiving the transmitted radio signal and resending it (with no energy loss) toward the receiver.

- The scatterers on the receive side of the link are also uniformly spaced, acting like an array of *virtual receive antennas* with average spacing $2D_r/S$ and angle spread D_t/R.

Note that the scattering environment described in Fig. 6.45 is *spatially fixed and therefore time invariant.*

The *GBGP propagation model,*[27] named in recognition if its originators, Gesbert, Bölcskei, Gore, and Paulraj, is based on the scattering environment pictured in Fig. 6.45. According to this model, the complex transfer function of the MIMO link is given by

$$\mathbf{H} = \frac{1}{\sqrt{S}}\mathbf{R}_{\theta_r, d_r}^{1/2}\mathbf{G}_r\mathbf{R}_{\theta_S, 2D_r/S}^{1/2}\mathbf{G}_t\mathbf{R}_{\theta_t, d_t}^{1/2} \tag{6.146}$$

where the five matrix terms of the equation (working backwards) are defined as follows:

$\mathbf{R}_{\theta_t, d_t}$ = N_t-by-N_t correlation matrix of the transmitted signal vector for transmit angle spread θ_t and antenna spacing d_t

$\mathbf{G}_t$ = S-by-N_t matrix of i.i.d. Rayleigh-fading coefficients produced by the set of S scatterers at the transmit end of the link

$\mathbf{R}_{\theta_S, 2D_r/S}$ = S-by-S matrix of the scatterer-to-scatterer cross-correlations between the two sets of scatterers with angle spread θ_S and scatterer spacing $2D_r/S$

$\mathbf{G}_r$ = N_r-by-S matrix of i.i.d. Rayleigh-fading coefficients produced by the set of S scatterers at the receive end of the link

$\mathbf{R}_{\theta_r, d_r}$ = N_r-by-N_r correlation matrix of the received signal vector for receive angle spread θ_r and antenna spacing d_r

In light of these definitions, decomposition of the N_r-by-N_t complex transfer function $\mathbf{H}$ of the (N_r, N_t) link, as shown in Eq. (6.146), is intuitively satisfying on the following grounds:

• The individual matrix terms are arranged in reverse order, starting from the receive antennas and progressing toward the transmitter, stage by stage, exactly in the same order as that shown in Fig. 6.45.

• The overall dimensions of the multiple matrix product are N_r by N_t, as is readily shown by the fact that

$$(N_r \times N_r) \cdot (N_r \times S) \cdot (S \times S) \cdot (S \times N_t) \cdot (N_t \times N_t) = (N_r \times N_t)$$

• Each element of the square matrices $\mathbf{R}_{\theta_t, d_t}$, $\mathbf{R}_{\theta_S, 2D_r/S}$ and $\mathbf{R}_{\theta_r, d_r}$ represents an autocorrelation function, the unit of which is the square of that for a transmitted or received signal. In contrast, each element of the channel transfer matrix $\mathbf{H}$ is defined as a signal ratio—hence the need for taking the square root of these three correlation matrices.

• Each element of the rectangular matrices $\mathbf{G}_t$ and $\mathbf{G}_r$ is a dimensionless ratio.

The rank of the channel matrix $\mathbf{H}$ is controlled primarily by the middle matrix term, $\mathbf{R}_{\theta_S, 2D_r/S}^{1/2}$, which is entirely due to scatterer-to-scatterer cross-correlations. Moreover, the matrix decomposition of Eq. (6.146) shows that it is indeed possible for the Rayleigh-fading processes at both the transmit and receive ends of the MIMO link to be completely uncorrelated (i.e., the matrices $\mathbf{G}_t$ and $\mathbf{G}_r$ to be of full rank), yet the overall rank of the channel matrix $\mathbf{H}$ may collapse to unity because the matrix

$\mathbf{R}^{1/2}_{\theta_S, 2D_r/S}$ is of unit rank. In such an eventuality, the MIMO link effectively assumes a *single degree of freedom* and, accordingly, exhibits *perfect cross-correlations* between the array of receive antennas and the array of transmit antennas. This special form of MIMO link is called a *keyhole channel* or *pinhole channel*, whose capacity is equivalent to that of a single-input, single-output link operating at the same signal-to-noise ratio as that of the MIMO link.

In physical terms, the keyhole channel arises when the product of the radial extents of the transmit and receive sets of scatterers, divided by the range between the transmit and receive antennas, is small compared with the wavelength λ of the transmitted radio signal; that is,

$$\frac{D_t D_r}{R} << \lambda \qquad (6.147)$$

Under this condition, the angle spread θ_S of the scatterers at both ends of the link assumes a small value, and with it, the rank of the matrix $\mathbf{R}^{1/2}_{\theta_S, 2D_r/S}$ collapses.

The occurrence of keyhole channels has been confirmed by computer simulations and outdoor measurements. Fortunately, the loss in rank due to keyhole channels does not appear to be prevalent in most physical environments, because the condition of Eq. (6.147) is realized only infrequently.

6.12 SUMMARY AND DISCUSSION

In this chapter, we discussed different forms of space diversity, the main idea behind which is that two or more propagation paths connecting the receiver to the transmitter are better than a single propagation path. In historical terms, the first form of space diversity used to mitigate the multipath fading problem was that of *receive diversity*, involving a single transmit antenna and multiple receive antennas. Under receive diversity, we discussed the selection combiner, maximal-ratio combiner, equal-gain combiner, and square-law combiner. The selection combiner is the simplest form of receive diversity, operating on the principle that it is possible to select, among N_r receive-diversity branches, a particular diversity branch with the largest output signal-to-noise ratio; the branch so selected defines the desired received signal. The maximal-ratio combiner is more powerful than the selection combiner by virtue of the fact that it exploits the full information content of all the N_r receive-diversity branches about the transmitted signal of interest. The maximal-ratio combiner is characterized by a set of N_r receive-complex weighting factors, which are chosen to maximize the output signal-to-noise ratio of the combiner. The equal-gain combiner is a simplified version of the maximal-ratio combiner, with each weighting parameter set equal to unity. The square-law combiner differs from the selection combiner, maximal-ratio combiner, and equal-gain combiner in that it is nonlinear, but applicable only to orthogonal modulation techniques.

By far the most powerful form of space diversity is the use of multiple antennas at both the transmit and receive ends of the wireless link. The resulting configuration is referred to as a *multiple-input, multiple-output (MIMO)* wireless communication system, which includes receive diversity and transmit diversity as special cases of space diversity. The novel feature of the MIMO system is that, in a rich Rayleigh scattering environment, it can provide a high spectral efficiency, which may be explained as follows: The signals radiated simultaneously by the transmit antennas arrive at the input of each receive antenna in an uncorrelated manner due to the rich scattering mechanism of the channel. The net result is the potential for a spectacular increase in the spectral efficiency of the wireless link. Most importantly, the spectral efficiency increases roughly linearly with the number of transmit or receive antennas, whichever is less. This result assumes that the receiver has knowledge of channel state information. The spectral efficiency of the MIMO system can be further enhanced by including a feedback channel from the transmitter to the receiver, whereby the channel state is also made available to the transmitter, and with it, the transmitter is enabled to exercise control over the transmitted signal.

Increasing spectral efficiency in the face of multipath fading is one important motivation for using MIMO transmission schemes. Another important motivation is the development of *space–time codes*, whose aim is the joint coding of multiple transmit antennas so as to provide protection against channel fading, noise, and interference. In this context, of particular interest is a class of block codes referred to as *orthogonal* and *generalized complex orthogonal space–time block codes*. In this class of codes, the *Alamouti code*, characterized by a two-by-two transmission matrix, is the only full-rate complex orthogonal space–time block code. The Alamouti code satisfies the condition for complex orthogonality or unitarity in both the spatial and temporal sense. In contrast, the generalized complex orthogonal space–time codes can accommodate more than two transmit antennas; they are therefore capable of providing a larger coding gain than the Alamouti code for a prescribed bit error rate and total transmission rate at the expense of a reduced code rate and increased computational complexity. However, unlike the Alamouti code, the generalized complex orthogonal space–time codes satisfy the condition for complex orthogonality only in the temporal sense. Accordingly, the complex orthogonal space–time codes, including the Alamouti code and generalized forms, permit the use of linear receivers.

The complex orthogonal property of the Alamouti code is exploited in the development of a *differential space–time block coding scheme*, which eliminates the need for channel estimation and thereby simplifies the receiver design. This simplification is, however, attained at the expense of degradation in receiver performance, compared with the coherent version of the Alamouti code, which assumes knowledge of the channel state information at the receiver.

Space was also discussed in the context of space-division multiple-access (SDMA), the mechanization of which relies on the use of highly directional antennas. SDMA improves system capacity by allowing a greater reuse of the available spectrum through a combination of two approaches: minimization of the effects of interference and increased signal strength for both the user terminal and the base station. Advanced techniques such as phased-array antennas and adaptive antennas, which have been researched extensively under the umbrellas of signal processing and radar for more than three decades, are well suited for implementing the practical requirements of both approaches.

Under the three theme examples, we discussed three different BLAST architectures issues relating to antenna diversity, spectral efficiency, as well as keyhole channels. Each of the BLAST architectures, namely, diagonal-BLAST, vertical-BLAST, and Turbo-BLAST, offers distinct features of its own. Diagonal-BLAST (D-BLAST) makes it possible to closely approximate the ergodic channel capacity in a rich scattering environment and may therefore be viewed as the benchmark BLAST architecture. But it is impractical, as it suffers from a serious space–time edge wastage. Vertical-BLAST (V-BLAST) mitigates the computational difficulty problem of D-BLAST at the expense of a reduced channel capacity. Turbo-BLAST uses a random layered space–time code at the transmitter and incorporates the turbo coding principle in designing an iterative receiver. In so doing, Turbo-BLAST offers a significant improvement in spectral efficiency over V-BLAST, yet the computational complexity is maintained at a manageable level. In terms of performance, Turbo-BLAST outperforms V-BLAST for a prescribed (N_t, N_r) antenna configuration, but does not perform as well as D-BLAST.

The different BLAST architectures were discussed in Theme Example 1. The material presented in Theme Example 2 taught us the following:

- The two basic forms of diversity, namely, transmit diversity and receive diversity, play complementary roles, with both of them being located at the base station.

- For low SNR and fixed spectral efficiency, V-BLAST outperforms space–time block codes (STBCs) on $\{N_t, N_r\}$ antenna configurations with $N_r > N_t$.

- Assuming the use of forward error-correction channel codes, a two-by-two STBC system could provide an adequate performance for wireless communications at low SNR.

- Diversity order is determined experimentally by measuring the asymptotic slope of the average frame error rate (or average symbol error rate) plotted versus the signal-to-noise ratio on a log-log scale.

- MIMO systems provide a trade-off between outage capacity and diversity order, depending on how the system is configured.

The degenerate occurrence of keyhole channels, discussed in Theme Example 3 arises when the rank of the channel matrix is reduced to unity, in which case the capacity of the MIMO link is equivalent to that of a single-input, single-output link operating at the same signal-to-noise ratio. Fortunately, the physical occurrence of keyhole channels is a rare phenomenon.

One last comment is in order: the discussion of channel capacity presented in the chapter focused on *single-user* MIMO links. Although, indeed, wireless systems in current use cater to the needs of multiple users, the focus on single users may be justified on the following grounds:

- The derivation of MIMO channel capacity is much easier to undertake for single users than multiple users.

- Capacity formulas are known for many single-user MIMO cases, whereas the corresponding multiuser ones are unsolved.

Simply put, very little is known about the channel capacity of *multiuser* MIMO links, unless the channel state is known at both the transmitter and receiver.[28]

NOTES AND REFERENCES

[1] For detailed discussions of the receive diversity techniques of selection combining, maximal-ratio combining, and square-law combining, see Schwartz et al. (1966), Chapter 10.

[2] The term "maximal-ratio combiner" was coined in a classic paper on linear diversity combining techniques by Brennan (1959).

[3] The three-point exposition presented in Section 6.2.3 on maximal-ratio combining follows S. Stein in Schwartz (1966), pp. 653–654.

[4] For expository discussions of the many facets of MIMO wireless communications, see the papers by Gesbert et al. (2003), Diggavi et al. (2003), and Goldsmith et al. (2003). The paper by Diggavi et al. includes an exhaustive list of references to MIMO wireless communications and related issues. For books on wireless communications using multiple antennas, see Hottinen et al. (2003) and Vucetic and Yuan (2003).

[5] Impulsive noise due to human-made electromagnetic interference is discussed in Blackard and Rappaport (1993), and Wang and Poor (1999); see also Chapter 2.

[6] The formula of Eq. (6.56), defining the ergodic capacity of a flat-fading channel, is derived in Ericson (1970).

[7] The log-det capacity formula of Eq. (6.59) for MIMO wireless links operating in rich scattering environments was derived independently by Teletar (1995) and Foschini (1996); Teletar's report was published subsequently as a journal article (1999). For a detailed derivation of the log-det capacity formula, see Appendix G.

[8] The Gaussian approximation of the probability density function of the instantaneous channel capacity of a MIMO wireless link, which is governed by the log-det formula, is discussed in detail in Hochwald et al. (2003).

[9] The result that at high signal-to-noise ratios the outage probability and frame (burst) error probability are the same is derived in Zheng and Tse (2002).

[10] MIMO wireless communications systems incorporating the use of feedback channels are discussed in Vishwanath et al. (2001), Simon and Moustakas (2003), and Hochwald et al. (2003). The latter paper introduces the notion of rate feedback by quantizing the instantaneous channel capacity of the MIMO link.

[11] The effect of correlation fading on the channel capacity of MIMO wireless communications is discussed in Shiu et al. (2000) and Smith et al. (2003).

[12] Space–time trellis codes are discussed in Tarokh et al. (1998).

[13] The Alamouti code was pioneered by Siavash Alamouti (1998); the code has been adopted in third-generation (3G) wireless systems, in which it is known as space–time transmit diversity (STTD).

[14] The generalized space–time orthogonal codes were originated by Tarokh et al. (1999a,b).

[15] The decoding algorithms (written in MATLAB) for the Alamouti code **S**, and the orthogonal space–time codes $\mathbf{G}_3$, $\mathbf{G}_4$, $\mathbf{H}_3$, and $\mathbf{H}_4$ due to Tarokh et al. are presented in the Solutions Manual to this book. It should, however, be noted that there are minor errors in the original decoding algorithms for $\mathbf{H}_3$, and $\mathbf{H}_4$ listed in the Appendix to the paper by Tarokh et al. (1998). These errors have been corrected in the pertinent MATLAB codes.

[16] Differential space–time block coding, based on the Alamouti code, was first described by Tarokh and Jafarkhani (2000). See also the article by Diggavi et al. (2002), which combines this form of differential coding with orthogonal frequency-division multiplexing (OFDM) for signal transmission over fading frequency-selective channels; OFDM was discussed in Chapter 3.

[17] Chapter 3 of Liberti and Rappaport (1999) describes more general models for phased arrays other than linear and where gain in elevation angle as well as azimuth is of interest. Chapter 8 of the same book describes various algorithms for adapting the weighting vector, depending upon the direction of arrival of the signal.

[18] The circular model for effective scatterers was proposed in Lee (1982).

[19] In Chapter 7 of Liberti and Rappaport (1999), the single-bounce elliptical model is described in greater detail. Note that the model does not take into account the effects of diffraction.

[20] The D-BLAST architecture was pioneered by Foschini (1996) and discussed further in the papers by Foschini and Gans (1999) and Foschini et al. (2003).

[21] The first experimental results in the V-BLAST architecture were originally reported in the article by Golden et al. (1999); see also the paper by Foschini et al. (2003), in which this particular form of BLAST is referred to as horizontal-BLAST, or H-BLAST.

[22] The Turbo-BLAST architecture was first described by Sellathurai and Haykin (1998), with additional results reported subsequently in the papers by the same authors (2000, 2002, 2003).

[23] The experimental results presented in Figs. 6.40 through 6.42 are reproduced from the paper by Sellathurai and Haykin (2002) with permission of the IEEE.

[24] According to deHaas (1927, 1928), the possibility of using antenna diversity for mitigating short-term fading effects in radio communications was apparently first discovered in experiments with spaced receiving antennas operating in the high-frequency (HF) band. For additional historical notes, see Chapter 10 by Seymour Stein in Schwartz et al. (1966).

[25] For definitions of the diversity order and multiplexing gain of MIMO wireless communication systems and the implications of these definitions in terms of system behavior, see Digavi (2003).

[26] Keyhole channels, also dubbed pinhole channels, were described independently by Gesbert et al. (2002) and Chizhik et al. (2002).

[27] The GBGP model for MIMO wireless links is described in Gesbert et al. (2002).

[28] Multiuser MIMO wireless systems are discussed in Diggavi et al. (2003) and Gold-smith et al. (2003).

ADDITIONAL PROBLEMS

Diversity-on-receive techniques

Problem 6.21 A receive-diversity system uses a selection combiner with two diversity paths. An outage occurs when the instantaneous signal-to-noise ratio γ drops below $0.25 \gamma_{av}$, where γ_{av} is the average signal-to-noise ratio. Determine the probability of outage experienced by the receiver.

Problem 6.22 The average signal-to-noise ratio in a selection combiner is 20 dB. Compute the probability that the instantaneous signal-to-noise ratio of the device drops below $\gamma = 10$ dB for the following number of receive antennas:

    (a) $N_r = 1$
    (b) $N_r = 2$
    (c) $N_r = 3$
    (d) $N_r = 4$
        Comment on your results.

Problem 6.23 Repeat Problem 6.22 for $\gamma = 15$ dB.

Problem 6.24 In Section 6.2.2, we derived the optimum values of Eq. (6.18) for complex weighting factors of the maximal-ratio combiner using the Cauchy–Schwartz inequality. This problem addresses the same issue, using the standard maximization procedure. To simplify matters, the number N_r of diversity paths is restricted to two, with the complex weighting parameters denoted by a_1 and a_2.

    Let

$$a_k = x_k + jy_k \qquad k = 1, 2$$

Then the complex derivative with respect to a_k is defined by

$$\frac{\partial}{\partial a_k^*} = \frac{1}{2}\left(\frac{\partial}{\partial x_k} + j\frac{\partial}{\partial y_k}\right) \qquad k = 1, 2$$

Applying this formula to the combiner's output signal-to-noise ratio γ_c of Eq. (6.14), derive Eq. (6.18).

Problem 6.25 In this problem, we develop an approximate formula for the probability of error, P_e, produced by a maximal-ratio combiner for coherent FSK. We start with Eq. (6.25), and for small γ_{mrc}, we may use the following approximation for the probability density function:

$$f_\Gamma(\gamma_{mrc}) = \frac{1}{\gamma_{av}^{N_r}(N_r - 1)!}\,\gamma_{mrc}^{N_r - 1}$$

(a) Using the conditional probability of error for coherent BFSK, that is,

$$\text{Prob}(\text{error}|\gamma_{\text{mrc}}) = \frac{1}{2}\text{erfc}\left(\sqrt{\frac{1}{2}\gamma_{\text{mrc}}}\right)$$

derive the approximation

$$P_e \approx \frac{1}{2\left(\frac{1}{2}\gamma_{\text{av}}\right)^{N_r}(N_r-1)!}\int_0^\infty \text{erfc}(\sqrt{y})y^{N_r-1}dy$$

where $y = \frac{1}{2}\gamma_{\text{mrc}}$.

(b) Integrating the definite integral by parts and using the definition of the complementary error function, show that

$$P_e \approx \frac{1}{2\sqrt{\pi}\left(\frac{1}{2}\gamma_{\text{av}}\right)^{N_r}N_r!}\int_0^\infty e^{-y}y^{N_r-\frac{1}{2}}dy$$

(c) Finally, using the definite integral

$$\int_0^\infty e^{-y}y^{N_r-\frac{1}{2}}dy = \left(N_r-\frac{1}{2}\right)!$$

obtain the desired approximation

$$P_e \approx \frac{1}{2\sqrt{\pi}\left(\frac{1}{2}\gamma_{\text{av}}\right)^{N_r}}\frac{\left(N_r-\frac{1}{2}\right)!}{N_r!}$$

Problem 6.26

(a) Using the approximation for $f_\Gamma(\gamma_{\text{mrc}})$ given in Problem 6.25, determine the probability of symbol error for a maximal-ratio combiner that uses noncoherent BFSK.

(b) Compare your result of part(a) with that of Problem 6.25 for coherent BFSK.

Problem 6.27

(a) Continuing the approximation to $f_\Gamma(\gamma_{\text{mrc}})$, determine the probability of symbol error for a maximal-ratio combiner that uses coherent BPSK.

(b) Compare your result of part(a) with that of Problem 6.25 for coherent BFSK.

Problem 6.28   As discussed in Section 6.2.3, an *equal-gain combiner* is a special form of the maximal-ratio combiner for which the weighting factors are all equal. For convenience of presentation, the weighting parameters are set to unity. Assuming that the instantaneous signal-to-noise ratio γ is small compared with the average signal-to-noise ratio γ_{av}, derive an approximate formula for the probability density function of γ.

Problem 6.29   Compare the performances of the following linear diversity-on-receive techniques:

(a) Selection combiner

(b) Maximal-ratio combiner

(c) Equal-gain combiner

Base the comparison on signal-to-noise improvement, expressed in dB, for $N_r = 2, 3, 4, 5,$ and 6 diversity branches.

Problem 6.30   Show that the maximum-likelihood decision rule for the maximal-ratio combiner may be formulated in the following equivalent forms:

(a) Choose symbol s_i over s_k if and only if

$$(\alpha_1^2 + \alpha_2^2)|s_i|^2 - y_1 s_i^* - y_1^* s_i < (\alpha_1^2 + \alpha_2^2)|s_k|^2 - y_1 s_k^* - y_1^* s_k \qquad k \neq i$$

(b) Choose symbol s_i over s_k if and only if

$$(\alpha_1^2 + \alpha_2^2 - 1)|s_i|^2 + d^2(y_1, s_i) < (\alpha_1^2 + \alpha_2^2 - 1)|s_k|^2 + d^2(y_1, s_k) \qquad k \neq i$$

Here, $d^2(y_1, s_i)$ denotes the squared Euclidean distance between the received signal y_1 and constellation points s_i.

Problem 6.31   It may be argued that, in a rather loose sense, transmit-diversity and receive-diversity antenna configurations are the *dual* of each other, as illustrated in Fig. 6.46.

(a) Taking a general viewpoint, justify the mathematical basis for this duality.

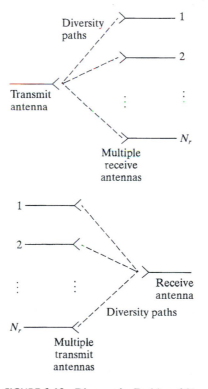

FIGURE 6.46   Diagram for Problem 6.31.

(b) However, we may cite the example of frequency-division diplexing (FDD), in which, in a strict sense, the duality depicted in Fig. 6.44 is violated. How is it possible for the violation to arise in this example?

MIMO Channel Capacity

Problem 6.32   In this problem, we continue with the solution to Problem 6.9, namely,

$$C \to \left(\frac{\lambda_{av}}{\log_e 2}\right)\rho \qquad \text{as } N \to \infty$$

where $N_t = N_r = N$ and λ_{av} is the average eigenvalue of $\mathbf{HH}^\dagger = \mathbf{H}^\dagger\mathbf{H}$,

(a) Justify the asymptotic result given in Eq. (6.61)—that is,

$$\frac{C}{N} \geq \text{constant}$$

What is the value of the constant?

(b) What conclusion can you draw from the asymptotic result?

Problem 6.33   By and large, the treatment of the ergodic capacity of a MIMO channel, as presented in Sections 6.3 and 6.5, focused on the assumption that the channel is Rayleigh distributed. In this problem, we expand on that assumption by considering the channel to be Rician distributed. In such an environment, we may express the channel matrix as

$$\mathbf{H} = a\mathbf{H}_{sp} + \mathbf{H}_{sc}$$

where $\mathbf{H}_{sp}$ and $\mathbf{H}_{sc}$ denote the specular and scattered components, respectively. To be consistent with the MIMO model described in Section 6.3, the entries of both $\mathbf{H}_{sp}$ and $\mathbf{H}_{sc}$ have unit amplitude variance, with $\mathbf{H}_{sp}$ being deterministic and $\mathbf{H}_{sc}$ consisting of iid complex Gaussian-distributed variables with zero mean. The scaling parameter a is related to the Rice K-factor by the formula

$$K = 10\log_{10} a^2 \text{dB}$$

(a) Considering the case of a pure line of sight (LOS), show that the MIMO channel has the deterministic capacity

$$C = \log_2(1 + N_r a^2 \rho) \quad \text{bits/s/Hz}$$

where N_r is the number of receive antennas and ρ is the total signal-to-noise ratio at each receiver input.

(b) Compare the result obtained in part (a) with that pertaining to the pure Rayleigh distributed MIMO channel.

(c) Explore the more general situation, involving the combined presence of both the specular and scattered components in the channel matrix $\mathbf{H}$.

Problem 6.34   Suppose that an additive, temporally stationary Gaussian interference $\mathbf{v}(t)$ corrupts the basic channel model of Eq. (6.48). The interference $\mathbf{v}(t)$ has zero mean and correlation matrix $\mathbf{R}_v$. Evaluate the effect of $\mathbf{v}(t)$ on the ergodic capacity of the MIMO link.

Problem 6.35   Consider a MIMO link for which the channel may be considered to be essentially constant for τ uses of the channel.

(a) Starting with the basic channel model of Eq. (6.48), formulate the input–output relationship of this link, with the input described by the N_r-by-τ matrix

$$\mathbf{S} = [\mathbf{s}_1, \mathbf{s}_2, \ldots, \mathbf{s}_\tau]$$

(b) How is the log-det capacity formula of the link correspondingly modified?

Orthogonal Space-Time Block Codes

Problem 6.36   The objective of this problem is to fill in the mathematical details that lie behind the formulas of Eqs. (6.104) and (6.105) for the maximum-likelihood estimates $\hat{s}_1$ and $\hat{s}_2$.

(a) Starting with Eq. (6.102) for the combiner output $\tilde{y}_k$ and using Eq. (6.103) for the probability density function of the additive complex Gaussian noise $\tilde{v}_k$, formulate the expression for the likelihood function of transmitted symbol $\tilde{s}_k$, $k = 1,2$.

(b) Hence, using the result of part (a), derive the formulas of Eqs. (6.103) and (6.104).

Problem 6.37   Figure 6.47 shows the extension of orthogonal space–time codes to the Alamouti code, using two antennas on transmit and receive. The sequence of signal encoding and transmissions is identical to that of the single-receiver case of Fig. 6.18. Table 6.5(a) defines the channels between the transmit and receive antennas. Table 6.5(b) defines the outputs of the receive antennas at times t' and $t' + T$, where T is the symbol duration.

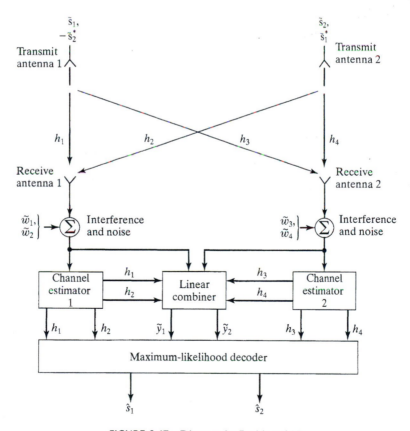

FIGURE 6.47   Diagram for Problem 6.37.

TABLE 6.5   Table for Problem 6.36.

(a)

	Receive antenna 1	Receive antenna 2
Transmit antenna 1	h_1	h_3
Transmit antenna 2	h_2	h_4

(b)

	Receive antenna 1	Receive antenna 2
Time t'	$\tilde{x}_1$	$\tilde{x}_3$
Time $t' + T$	$\tilde{x}_2$	$\tilde{x}_4$

(a) Derive expressions for the received signals $\tilde{x}_1, \tilde{x}_2, \tilde{x}_3$, and $\tilde{x}_4$, including the respective additive noise components, in terms of the transmitted symbols.

(b) Derive expressions for the line of combined outputs in terms of the received signals.

(c) Derive the maximum-likelihood decision rule for the estimates $\tilde{s}_1$ and $\tilde{s}_2$.

Problem 6.38   This problem explores a new interpretation of the Alamouti code. Let

$$\tilde{s}_i = s_i^{(1)} + js_i^{(2)} \qquad i = 1, 2$$

where $s_i^{(1)}$ and $s_i^{(2)}$ are both real numbers. The complex entry $\tilde{s}_i$ in the two-by-two Alamouti code is represented by the two-by-two real orthogonal matrix

$$\begin{bmatrix} s_i^{(1)} & s_i^{(2)} \\ -s_i^{(2)} & s_i^{(1)} \end{bmatrix} \qquad i = 1, 2$$

Likewise, the complex-conjugated entry $\tilde{s}_i^*$ is represented by the two-by-two real orthogonal matrix

$$\begin{bmatrix} s_i^{(1)} & -s_i^{(1)} \\ s_i^{(2)} & s_i^{(2)} \end{bmatrix} \qquad i = 1, 2$$

(a) Show that the two-by-two complex Alamouti code **S** is equivalent to the four-by-four *real* transmission matrix

$$\mathbf{S}_4 = \left[\begin{array}{cc|cc} s_1^{(1)} & s_1^{(2)} & s_2^{(1)} & s_2^{(2)} \\ -s_1^{(2)} & s_1^{(1)} & -s_2^{(2)} & s_2^{(1)} \\ \hline -s_2^{(1)} & s_2^{(2)} & s_1^{(1)} & -s_1^{(2)} \\ -s_2^{(2)} & -s_2^{(1)} & s_1^{(2)} & s_1^{(1)} \end{array} \right]$$

(b) Show that $\mathbf{S}_4$ is an orthogonal matrix.

(c) What is the advantage of the complex code $\mathbf{S}$ over the real code $\mathbf{S}_4$?

Problem 6.39

(a) Show that the generalized complex orthogonal space-time codes of Eqs. (6.107) and (6.108) satisfy the temporal orthogonality condition

$$\mathbf{G}^{\dagger}\mathbf{G} = \mathbf{I}$$

where the superscript † denotes Hermitian transposition and $\mathbf{I}$ denotes the identity matrix.

(b) Likewise, show that the sporadic complex orthogonal space-time codes of Eqs. (6.109) and (6.110) satisfy the temporal orthogonality condition

$$\mathbf{H}^{\dagger}\mathbf{H} = \mathbf{I}$$

Problem 6.40   Applying the maximum-likelihood decoding rule, derive the optimum receivers for the generalized complex orthogonal space–time codes of Eqs. (6.107) and (6.108).

Problem 6.41   Repeat Problem 6.40 for the sporadic complex orthogonal space–time codes of Eqs. (6.109) and (6.110).

Problem 6.42   Show that the channel capacity of the Alamouti code is equal to the sum of the channel capacities of two single-input, single-output systems.

Differential Space–Time Block Coding

Problem 6.43   Equation (6.116) defines the input–output matrix relationship of the differential space–time block coding system described in Section 6.7. Starting with Eqs. (6.98) and (6.99), derive Eq. (6.116).

Problem 6.44   The constellation expansion illustrated in Fig. 6.44 is based on the polar baseband representation {-1, +1} for BPSK transmissions of the Alamouti code on antennas 1 and 2. Explore the constellation expansion property of differential space–time coding for the following two situations:

(a) Frame of reference: dibit 00

(b) Frame of reference: dibit 11

Comment on your results.

Problem 6.45   In this problem, we investigate the use of QPSK for transmission of the Alamouti code on antennas 1 and 2. The corresponding input block of data will be in the form of quadbits (i.e., 4-bit blocks). Perform the investigation for each of the two QPSK constellations depicted in Fig. 6.48. Use 0000 as the frame of reference.

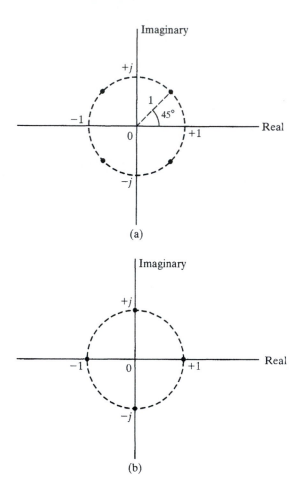

FIGURE 6.48 Diagram for Problem 6.44.

Problem 6.46 Repeat Problem 6.45 for the frame of reference 1111.

Problem 6.47 In the analytic study of differential space–time block coding presented in Section 6.7, we ignored the presence of channel noise. This problem addresses the extension of Eq. (6.116) by including the effect of channel noise.

(a) Starting with Eq. (6.101), expand the formulas of Eqs. (6.116) and (6.117) by including the effect of channel noise modeled as additive white Gaussian noise.

(b) Using the result derived in part (a), expand the formula of Eq. (6.121) by including the effect of channel noise, which consists of the following components:

   (i) Two signal-dependent noise terms

   (ii) A multiplicative noise term consisting of the product of two additive white Gaussian noise terms

(c) Show that, when the signal-to-noise ratio is high, the noise term (ii) of part (b) may be ignored, with the result that the remaining two signal-dependent noise terms (i) double the average power of noise compared with that experienced in the coherent detection of the Alamouti code.

Theme Examples

Problem 6.48   In this problem, we repeat Experiment 1 of Section 6.10, but this time we investigate the effect of increasing signal-to-noise ratio (SNR) on the symbol error rate (SER) for a prescribed modulation scheme, still operating in a Rayleigh fading environment.

(a) Using 4-PSK for both STBC and V-BLAST, plot the SER versus SNR for the following antenna configurations:

  (i) $N_t = 2, N_r = 2$
  (ii) $N_t = 2, N_r = 4$

(b) What conclusions do you draw from the experimental results of part (a)?

Problem 6.49   Continuing with Problem 6.48, suppose the STBC and V-BLAST systems use 4-PSK. This time, however, we wish to display the spectral efficiency in bits/s/Hz versus the SNR. How would you expect the performance curve of STBC to compare against that of V-BLAST? Explain.

Problem 6.50   Compare the relative merits of STBC systems versus BLAST systems in terms of the following issues:

- Capacity
- Diversity order
- Multiplexing gain
- Computational complexity

Problem 6.51   In Chapter 2, we discussed the reciprocity theorem in the context of a single-input, single-output wireless communication link. Show that the theorem also applies to Eq. (6.146); that is, show that the channel matrix **H** of the MIMO link satisfies the Hermitian property.

C H A P T E R 7

Wireless Architectures

7.1 INTRODUCTION

"Anyone, anywhere, anytime" is a popular advertising slogan for wireless services. This universal connectivity is the Holy Grail of today's communications systems. Previous chapters presented the various techniques for communicating over wireless channels and the various methods for sharing the radio spectrum with different users. In this final chapter of the book, we will present an overview of how these basic capabilities are *organized* and *managed* to provide reliable and timely services to the end user.

Section 7.2 begins with a comparison of the different multiple-access techniques. These techniques provide the baseline for what follows in the chapter. In Section 7.3, the open system interconnection (OSI) reference model for end-to-end communications is presented. The relationships among the different layers of the model are described, from physical through networking to application. The OSI model leads to the discussion in Section 7.4 of which parts of the model are affected by wireless communications and how the physical layer and multiple-access strategies integrate with higher layers. In Sections 7.5, 7.6, and 7.7, we delve into the wireless-link management functions of protocols and signaling, power control, and handover, respectively. These issues are the key to providing reliable and timely services to the user. The network layer of wireless services is discussed in Section 7.8, complemented by three theme examples that introduce and compare various wireless telephone and data networks.

7.2 COMPARISON OF MULTIPLE-ACCESS STRATEGIES

The parallel development of wireless transmission techniques and the efficient use of the radio spectrum has been highlighted in previous chapters. We have introduced four multiple-access strategies that follow the theme of increased efficiency in sharing the radio spectrum:

1. FDMA, in which users share the spectrum by dividing it into different frequency channels.

2. TDMA, in which users time-share the spectrum.

3. CDMA, in which all users use the same spectrum simultaneously, but the number of users is limited by their multiple-access interference.

4. SDMA, in which users share the spectrum in angular direction with the use of smart antennas.

Practical systems are usually a *hybrid* of two or more of these multiple-access strategies. The hybrid approach is employed for a number of reasons:

- To provide a reasonable growth strategy
- To reduce the complexity of the overall system
- To be backwardly compatible with an existing system

Although one approach may have a significant technical advantage over another, there may be other factors, such as economic considerations, that prevent the use of the basic strategy of interest. For example, in a single-user system, the use of CDMA would be difficult to justify, since it requires a very large amount of bandwidth.

In Table 7.1, various properties of the multiple-access strategies are compared; and we clarify this comparison with the following comments:

1. *Modulation.* TDMA and FDMA rely heavily on the choice of a modulation strategy to maximize spectral efficiency, as we have discussed in Chapter 3. To achieve a higher throughput in the same bandwidth, we must use higher order modulation strategies. With CDMA, the simple method of BPSK modulation is all that is required, although, for practical symmetry considerations, QPSK is often used. The choice of modulation strategy and the use of SDMA are independent.

2. *Forward error-correction (FEC) coding.* All multiple-access techniques are affected by the vagaries of the wireless channel. With FDMA and TDMA, the redundancy introduced by FEC coding requires a higher transmission rate, and thus a greater bandwidth, if the same basic throughput is to be maintained. This is the classic *trade-off between bandwidth and power efficiency.* With CDMA, FEC coding can be added without increasing the system bandwidth or harming the processing gain. The inclusion of FEC is transparent to SDMA. If transmit diversity is implemented, then there can be increased bandwidth with SDMA.

3. *Source coding.* The use of source coding improves the bandwidth efficiency of all multiple-access techniques. However, CDMA is in a position to take greater advantage of voice activation than are other techniques, since its bandwidth efficiency is determined by average interference.

4. *Diversity.* To obtain diversity with FDMA, we require either multiple transmitters or multiple receivers, or both, which is an added hardware expense. The same can be said of TDMA, except when it is used as part of a TDMA/FDMA hybrid. In that case, frequency-hopped TDMA can provide some diversity advantage. The large bandwidth of CDMA naturally provides some frequency diversity, and this can be used advantageously with a RAKE receiver. The implementation cost of a RAKE receiver is less than the dual-receiver cost of an FDMA system with frequency diversity. With the smart antennas of SDMA, there will be a reduction in space diversity due to antenna directivity, but there will also be a corresponding reduction in fading effects.

TABLE 7.1   Comparison of Different Multiple-Access Strategies.

	FDMA	TDMA	CDMA	SDMA
Modulation	- relies on bandwidth-efficient modulation	- relies on bandwidth-efficient modulation	- simple modulation	- transparent
Forward error correction	- increases power efficiency at expense of bandwidth efficiency	- increases power efficiency at expense of bandwidth efficiency	- can be implemented without affecting bandwidth efficiency	- transparent
Source coding	- improves efficiency	- improves efficiency	- improves efficiency - voice activation advantage	- transparent
Diversity	- requires multiple transmitters or receivers	- requires multiple transmitters or receivers - can be frequency-hopped	- includes frequency diversity when implemented with a RAKE receiver	- single antenna reduces space diversity - orthogonal coding improves diversity with multiple transmit antennas
User terminal complexity	- simple	- medium complexity	- more complex	- requires smart antennas
Handover	- hard	- hard	- soft	- potentially soft
System complexity	- large number of simple components	- reduced number of channel units	- large number of complex interacting components	- additional complexity related to antennas
Multiple-access interference	- limited by system planning	- limited by system planning	- dynamic power control	- limited by resolution of antennas
Fading	- flat-fading - no diversity - simple to track	- may be frequency-selective - may need equalizer	- frequency-selective diversity via RAKE receiver	- reduced multipath
Bandwidth efficiency	- hard limits - based on modulation and channel spacing	- hard limits - based on modulation and channel spacing	- soft limits	- depends on antenna resolution
Synchronization	- low resolution	- mid-resolution	- high resolution	- requires terminal location
Flexibility	- fixed data rate	- data rate variable in discrete steps	- can provide a variety of data rates without affecting signal in space	- transparent
Voice and data integration	- possible, but may require revisions to system	- straightforward using multiple slots	- multicode transmission, which may decrease efficiency of mobile terminal	- transparent
Evolution	- bandwidth to fit application	- requires medium initial bandwidth	- requires large initial bandwidth	- flexible, can be added as needed - does not affect mobile

5. *User terminal complexity.* With the progression from FDMA through TDMA to CDMA comes an evolution of terminal complexity. SDMA systems introduce a different and additional form of complexity—one related to the antennas—that is not present in any of the other systems.

6. *Handover.* With their single-receiver terminals, both FDMA and TDMA are somewhat handicapped when they must switch between base stations at a cell boundary. With CDMA, since the same frequencies are used in adjacent cells, it is easier to implement a "dual receiver" and provide a soft handover capability. In SDMA, there are additional handover requirements associated with users moving between beams.

7. *System complexity.* With an FDMA system, users can operate somewhat independently. With TDMA, the level of cooperation among users must increase to share slots. With CDMA, the system must delegate spreading codes, power control information, and synchronization information. These complexity aspects are transparent to SDMA, for which antenna pointing and terminal location are the main concerns.

8. *Multiple-access interference.* Since FDMA and TDMA tend to be limited by worst-case interference, interference is often limited in the system planning stage by the fixed assignment of frequency groups to specific cells. With CDMA, the same bandwidth is used everywhere, and performance is limited by average interference levels. However, CDMA relies heavily on accurate power control to eliminate the near–far problem. With SDMA, multiple-access interference depends upon the beamwidth and antenna sidelobes; SDMA can also reduce the sensitivity to power control accuracy.

9. *Fading-channel sensitivity.* FDMA systems are typically narrowband and therefore suffer from flat fading. If the fading is not severe, then simple channel estimation and forward error correction can often compensate for its effects. TDMA systems are typically medium-bandwidth solutions; consequently, they observe some frequency selectivity. This requires the implementation of an equalizer; indeed, the implementation of a robust tracking equalizer in wireless channels is an issue of practical importance. Due to their large bandwidth, CDMA systems face frequency-selective channels, but take advantage of this natural diversity with a RAKE receiver.

10. *Bandwidth efficiency.* For single-cell systems, FDMA and TDMA systems are generally more bandwidth efficient than CDMA systems, because they do not have to cope with multiple-access interference. However, once their frequency plan is made and the modulation selected, the maximum throughput is fixed. For multicell systems, CDMA holds an advantage because it can reuse frequencies everywhere, while FDMA and TDMA have much lower frequency reuse rates because they are limited by peak interference levels. CDMA can often add a user at the expense of a small degradation of existing users.

11. *Synchronization.* Wireless systems using FDMA, TDMA, and CDMA show a progression in synchronization resolution and a corresponding progression in complexity. The main concern of FDMA is symbol timing; by contrast, TDMA terminals must contend with both symbol timing and slot timing, and CDMA terminals must contend with chip timing. However, it should be noted that, with the increase in bandwidth, terminals become less sensitive to frequency errors. SDMA has the additional synchronization-like requirement related to determining the location of the terminal.

12. *Flexibility.* FDMA is the least flexible of the techniques; once the service is designed, any change requires a redesign. With TDMA, higher data rates can be provided by assigning more slots per user, usually with very little change to the hardware. With CDMA, different data rates can be provided by trading off the spreading rate (processing gain), making it very flexible. All of these techniques are transparent to SDMA.

13. *Voice and data integration.* The comments regarding flexibility also apply to the integration of voice and data over the same terminal. With TDMA, it is possible as well to make use of periods of voice inactivity to transmit data, thus making the system more efficient. CDMA can easily integrate voice and data, but usually it leads to multicode transmissions, which may reduce the efficiency of the user-terminal power amplifier.

14. *Evolution.* Evolving from a small system to a large system is easiest with the FDMA approach. We can easily start with a single-user system and remain relatively efficient at each step. With TDMA, start-up efficiency is related to the transmission rate; the system can evolve easily through the addition of more TDMA channels using an FDMA overlay. With CDMA, there is a large start-up cost, because a large bandwidth to serve perhaps only a few initial user terminals is needed. With SDMA, the evolution can be quite smooth, since that approach is primarily a base-station technology. If smart antennas are to be deployed on mobile terminals, they may have to be retrofitted as the antenna beams become more dense.

7.3   OSI REFERENCE MODEL[1]

Today's public communication networks are complicated systems. Networks such as the public switched telephone network (PSTN) and the Internet provide seamless connections between cities, across oceans, and between different countries and continents, languages, and cultures. Wireless is only one component of these complex systems. To illustrate how wireless can affect the design of such systems, we set forth a conceptual model of the communications process. The model was developed explicitly for computer communications over public networks, but it has been used as a paradigm for all communications.

The *open system interconnection* (OSI) reference model is a seven-layer model for the functions that occur in a communication process. Figure 7.1 illustrates the model. Each layer performs one or a number of related functions in the communications process. The same model applies to both the source and the sink of the information, and the layers are paired. The seven layers, or a subset thereof, are often referred to as a *protocol stack.* Processes (or layers) at the same level in the protocol stack are referred to as *peer* processes.

The only physical connection between the different layers of the two protocol stacks occurs at the physical layer. At each of the remaining layers, there is a *virtual* connection between peers. By this, we mean that a protocol at layer N communicates with a *peer* at the same level; it does so by nesting through all the lower layers of the protocol stack. The dashed line in the figure indicates an example in which the transport layer on one side communicates with the transport layer on the other side. The following is a simplified explanation of the functions of the different layers of this model:

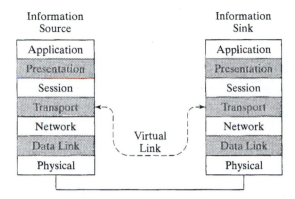

FIGURE 7.1   Seven layers of OSI model of communications.

1. *Physical layer.* This first layer provides a physical mechanism for transmitting bits between any pair of nodes. The module for performing this function is often called a *modem* (*mo*dulator and *dem*odulator).

2. *Data link layer.* This second layer performs error correction or detection in order to provide a reliable error-free link to the higher layers. Often, the data link layer will retransmit packets that are received in error, but in some implementations it discards them and relies on higher layers to do the retransmission. The data link layer is also responsible for the ordering of packets, to make sure that all packets are presented to the higher layers in the proper order. The role of the data link layer is more complicated when multiple nodes share the same media, as usually occurs in wireless systems. The component of the data link layer that controls *multiaccess communications* is the *medium access control (MAC) sublayer*, the purpose of which is to allow frames to be sent over the shared media without undue interference from other nodes.

3. *Network layer.* This third layer has many functions. One function is to determine the *routing* of the packet. A second is to determine the *quality of service (QoS)*, which is often controlled by the choice of a *connectionless* or *connection-oriented* service. A third function is *flow control*, to ensure that the network does not become congested. Note that the network layer can generate its own packets for control purposes. Often, there is a need to connect different subnetworks together. The connectivity is accomplished by adding an *Internet sublayer* to the network layer—a sublayer that provides the necessary translation facilities between the two networks.

4. *Transport layer.* This fourth layer separates messages into packets for transmission and reassembles the packets at the other end. If the network layer is unreliable, the transport layer provides reliable end-to-end communications by retransmitting incomplete or erroneous messages; it also restarts transmissions after a connection failure. In addition, the transport layer may provide a multiplexing function by combining sessions with the same source and destination; an example is parallel sessions consisting of a gaming application and a messaging application. The other peer transport layer would separate the two sessions and deliver them to the respective peer applications.

5. *Session Layer.* This fifth layer finds the right delivery service and determines access rights.

6. *Presentation Layer.* The major functions of the presentation layer are data encryption, data compression, and code conversion.

7. *Application Layer.* This final layer provides the interface to the user.

The OSI reference model was developed for computer communications, but, in the course of time, it has also been applied to other communication systems, including wireless communications. However, in many of these systems, the analogy is often qualitative, as the distinction between the layers becomes hazy. For example, in some implementations, any or all of the following may occur:

- The session, transport, and network layers are treated as a single entity.
- The error correction is performed in the modem as part of the physical layer.
- The presentation and application layers are combined into a single application.

Even when the levels are blurred in an implementation, the seven-layer division is useful for understanding the different functions that may be performed.

In most layers, a peer process at one end is paired with a single peer process at the other end. The exceptions to one-to-one pairing are the network layer and the MAC sublayer. For the network layer, all (geographically distributed) peer processes must work together to configure routing tables and ensure that the network does not become congested. Similarly, at the MAC sublayer, all peer processes must work together to ensure that the medium is shared in an efficient manner.

EXAMPLE 7.1 E-mail

As an example of the seven-layer model, consider sending an e-mail message over a wireless local area network. In this case, the seven layers at the information source can be identified as follows:

- *Application layer.* This layer is the e-mail client software on the host computer. The layer provides the interface for entering the message and the name and address of the recipient.
- *Presentation layer.* This layer, also likely to be part of the e-mail client on the host computer, performs encryption and compression of the message if they are requested.
- *Session layer.* This layer may also be part of the e-mail client. The layer determines where the e-mail account is hosted, assesses whether it is up to date, and submits passwords.
- *Transport layer.* This layer will subdivide the e-mail message into smaller packets if it is too long and will add a transport header for numbering the packets. Usually, this header will be the Transport Control Protocol (TCP) of the well-known TCP/IP protocol. The transport layer will also indicate to the network layer the source and destination of the message.
- *Network layer.* This layer will add the network header that includes source and destination addressing. It will also determine the routing path. For example, if the host computer is connected to both a wired *local area network* (LAN) and a wireless LAN, the network layer would choose which of the two options to use, based on its routing tables.

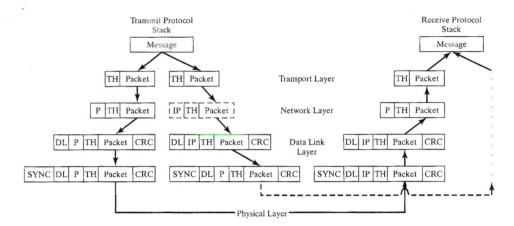

FIGURE 7.2   Addition and deletion of packet headers: TH—transport layer header; IP—Internet packet header; DL—data link header; CRC—cyclic redundancy check; SYNC—physical layer synchronization header.

- *Data link layer.* The data link layer will add its own header that determines the first hop of the communications path. It will also add *cyclic redundancy check* (CRC) bits that are used for error detection at the other end. The MAC sublayer will determine whether the radio link is busy, and when it is found to be free, the MAC sublayer will pass the data to the physical layer for transmission.
- *Physical layer.* This layer may add its own header for synchronization purposes, will add redundant bits for error correction at the other end, and then will modulate the data on a carrier for transmission over the channel.

Each layer has an associated protocol that must be understood by the peer entities at the other end. However, the protocols of each layer are designed to be independent of each other. Consequently, the protocol at one layer can be changed without affecting the protocols at the other layers. Some layers add header information to the basic message being transmitted, as shown in Fig. 7.2.

The reverse of the process will occur at the other end of the communications link: each layer will remove its peer contribution to the packet and confirm that the message is still intact. Note, however, that the other end is not the e-mail recipient, but rather the e-mail server. Getting the information to the final destination would require polling of the server by the destination machine and a repetition of the entire process just set forth.   ∎

7.4   THE OSI MODEL AND WIRELESS COMMUNICATIONS

How does the OSI reference model relate to wireless communications? The basic tenet of the OSI model is that the protocols and procedures of the various levels should be independent of the protocols and procedures of the other levels. On this basis, we would expect only the physical layer to be affected by the choice of the transmission medium. In actuality, however, wireless communications often affects three layers—the physical, data link, and network layers (as has been demonstrated throughout the book).

The key resource in wireless systems is the radio spectrum. At the physical layer, the emphasis is on modulation, source coding, channel coding, and detection techniques to maximize the use of the radio spectrum; these are issues that have occupied much of the material covered in the book.

At the link layer, the emphasis is on how the spectrum is shared, either in time, frequency, area, or angular direction. At a high level, the spectrum sharing is determined by the multiple-access strategy, as we have discussed in previous chapters. At the lower levels, sharing depends upon the link layer protocols—particularly those of the MAC sublayer—and how they are mapped to physical channels. Related issues are the *handover* between different base stations and *power control* of mobile terminals. At the network layer, the emphasis is on routing and quality of service. Routing aspects include determining the user terminal location. For example, if a mobile terminal has wandered from its "home cell," the network must contact the home cell regarding billing and authorization functions. Furthermore, with terminal movement among cells, routing patterns can be dynamic. The network quality of service is related to data rates that a particular terminal can support and data rates that the current radio link will support.

All these issues must be addressed, while at the same time addressing the service demands of the customer in terms of quality of service, response time, and so on. In the sections that follow, we will look at the protocol functions of the MAC sublayer and at several implementations of the network layer.

7.5   MAC SUBLAYER SIGNALING AND PROTOCOLS

The lower level MAC layer functions describe the rules for sharing frequencies, time slots, cells, and codes and influence how the physical layer is designed. These rules constitute the protocol that is used by all who share the radio spectrum for this service. For applications such as voice, the response times for the MAC protocol are critical. As a consequence, there are often specialized channels for supporting the protocol. Thus, the medium is shared not only with user traffic, but also with these signaling channels.

To illustrate the specialized channels, consider the next example, which describes a MAC protocol for setting up a wireless telephone call. Note that a MAC sublayer generates its own packets to control link access, in addition to dealing with the packets containing information that is provided by higher layers.

EXAMPLE 7.2 Telephone Switched Circuit Protocol

Figure 7.3 shows the ladder diagram of a MAC protocol to set up a telephone call initiated by a user terminal. In the figure, time starts at the top and proceeds downward in a step-by-step manner. In the first transaction of the figure, the user terminal (UT) MAC layer listens to system information that is continually being broadcast by the base station (BS). The *system information* informs the UT when and on what channel an access request may be made.

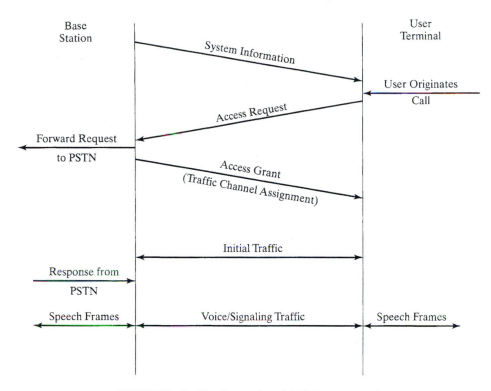

FIGURE 7.3   Ladder diagram for a MAC-layer protocol.

The UT MAC layer makes an access request when the user enters the telephone number on the user terminal. The access request includes information such as the telephone number of the called party. The BS receives the access request and replies with an access grant. The access grant informs the UT of the channel on which to transmit the voice call. If the telephone number is in the *Public Switched Telephone Network (PSTN)*, the request for making the call will also be forwarded to the networking facilities attached to the base station and then on to the PSTN. If the connection to the PSTN subscriber is successful, then the PSTN circuit is connected back to the assigned channel. Voice communication over the link so established then proceeds.

What we have just described is the ideal situation. In practice, the access request is often made over a *random-access channel*, where there is the possibility of collisions and the request being lost; so the protocol must also include a retry mechanism, to be used if it does not succeed on the first try. The access grant is typically provided over a dedicated channel, but wireless channels can be unreliable and the grant may be lost. Hence, the protocol must include a mechanism for handling the case of a lost access grant. In addition, the access grant may be a denial due to a channel not being currently available, in which case the UT usually interprets the denial as a busy signal. Alternatively, a successful connection may not be made with the PSTN, in which case the assigned channels must be released.   ∎

Example 7.2 illustrates many features of MAC protocols. There are several functions that must be performed by both the UT and the BS; there can be random components, and there can be dedicated components. For telephone calls, these must be done in a timely manner, typically less than 1 second in order for the system to function satisfactorily. This requirement places significant constraints on how the transmission medium is shared.

The MAC sublayer is an important consideration in the design of wireless systems. It is a common convention for cellular telephone systems to assign different MAC functions to *logical channels*—connections between link layer entities (i.e., peer entities) on either end of the channel. Logical channels may then be assigned to the same or different *physical channels*; that is, they will use an FDMA, TDMA, or CDMA strategy for multiplexing the different logical channels, in a manner consistent with the *system* architecture. For example, access grants and data acknowledgement are not high-bandwidth functions, so they might easily be time-division multiplexed onto the same physical channel. The types of physical channels are usually much less in number than the types of logical channels. A small number of different types of physical channels, with similar bandwidth, modulation, and frame structures, simplifies the transmitter and receiver design. In many terminal designs, there can be only one physical channel active at a time.

The different logical channels for the MAC sublayer can be divided into four types:

1. *Synchronization and broadcast channels.*   The purpose of these logical channels is to assist the mobile terminal in gaining access to the resources of the mobile telephone network. The basic resource provided by these networks is a telephone channel. Before requesting such a channel, the mobile terminal must find and synchronize to the local base station. To assist and accelerate this process, the base station may transmit an unmodulated signal or some other reference signal that the mobile can quickly find and synchronize with. Following this initial synchronization step, the mobile terminal looks for a system broadcast channel that may provide such information as the following:

- the system time
- a base station ID, affording information about neighboring BS's
- information regarding frame timing
- information regarding the channel to use for requesting telephone channels
- information regarding the channel to use to listen for pages.

2. *Paging and access channels.*   The purpose of these logical channels is to provide a short messaging service between the base station and the mobile terminal to allow the rapid setup of a telephone call. Examples of messages that might be transmitted over these logical channels are as follows:

- *Log-on and log-off messages,* wherein the mobile terminal logs onto (or logs off) of the base station and informs it that it is listening for calls. Associated with log-on is a confirmation message from the base station to the user if the user is a valid customer.

- *Pages*, or short messages from the base station to the mobile terminal informing the user terminal that there is currently an incoming telephone call. The page may contain information regarding which channel the call will be set up on. The user terminal would confirm its acceptance of the call and then tune to that channel.
- *Accesses*. When the UT is to make a call, it must first make an access request, typically over a random-access channel. The BS responds with an access grant message that may confirm or deny the request; if the former, it will usually provide the traffic channel information.

3. *Control channels.*   The purpose of these logical channels is to provide a communications link for control information that may need to be transferred between the BS and UT to maintain the quality of the link. There can be common control channels such that the control information applies to all UTs, or there can be dedicated control channels such that the information is specific to one user. Common control channels are typically used to allocate dedicated control channels. Two typical applications for dedicated control channels are the following:

- *Power control*. With this channel, one end of the communications link requests the other end of the link to increase or decrease the transmit power in order to maintain a consistent quality of communications.
- *Handover*. When a UT moves from the coverage area of one BS to that of another, there needs to be a handover of resources. The handover may involve changing the communications channel, transmit power, and signaling channels.

4. *Traffic channels.*   Traffic channels are the main resource of the network and are the logical channels that provide the exchange of information between the end users. In telephone networks, traffic channels are often dedicated for the duration of a call (except for possible cell handovers). With data applications over newer systems, some traffic channels are shared and provide a packet data service among a number of UTs and the BS. There are traffic channels in both the downlink and uplink directions.

In practice, to meet the required response times for wireless telephone systems, a portion of the transmission medium is dedicated to performing certain functions (e.g., signaling channels) on a permanent basis.

7.6   POWER CONTROL[2]

One of the system-level goals of wireless communications is to maximize utilization of the radio spectrum, which often translates into the maximization of the number of users. Many cellular systems are *interference limited*; that is, the number of users in the system is limited, not by channel parameters such as fading and receiver noise, but rather by interference. The near–far problem is a simple example of these interference limits. (See Example 7.3.) Consequently, improving spectral efficiency and increasing capacity implies minimizing the interference, which, in turn, is directly related to minimizing the transmitted power of each terminal.

EXAMPLE 7.3 The Near–Far Problem

In a given cellular system, the distance of a mobile terminal from the base station may range from 100 m to 10 km. Given a fourth-power propagation loss, what is the expected difference in received power levels at the base station if both mobile terminals transmit at the same power level?

From the statistical propagation models of Section 2.4.1, the receive power levels of the two signals are given respectively by

$$P_{100\,m} = \frac{\beta P_T}{r_{100\,m}^4} \text{ and } P_{10\,km} = \frac{\beta P_T}{r_{10\,km}^4} \tag{7.1}$$

So their relative power level is

$$\frac{P_{100\,m}}{P_{10\,km}} = \frac{(10\,km)^4}{(100\,m)^4} = 80 \text{ dB} \tag{7.2}$$

Even in a CDMA system that has a spreading rate of 512, which is equivalent to a processing gain of 27 dB, the stronger user would prevent detection of the weaker user in this scenario. ■

The goal of power control is to solve the near–far problem of communications. The goal is accomplished by *adjusting each transmitter's power so that the received signal is at the minimum acceptable level*, which will, in turn, minimize interference with other users. The conclusion is that, in cellular systems, and particularly CDMA systems, accurate, effective power control must be employed to maximize spectral efficiency. There are a number of ways of performing power control, which we discuss in the rest of this section.

7.6.1   Open Loop

Open-loop power control refers to the procedure whereby the mobile terminal measures its received signal level and adjusts its transmit power accordingly. The fundamental advantage of open-loop control is that it is fast: It does not need to wait for a round-trip delay between the base station and the mobile terminal before the correction can be applied.

The main disadvantage of open-loop power control is the limited correlation between received power levels on the uplink and the downlink. There is reasonable correlation between the average power levels of the up- and downlinks. Variations in these average effects are due to shadowing effects (slow lognormal fading) caused by terrain, vegetation, and buildings. For example, if the terminal goes into the shadow of a building, we expect both the uplink and downlink to be significantly attenuated. Open-loop power control can provide good coarse correction factors for these types of variations. That is, open-loop power control is good for tracking median power levels and slow variations such as those due to shadowing.

For fast Rayleigh fading, open-loop power control is not effective, because down- and uplink transmissions usually occur in separate frequency bands. As we have seen in Chapter 2, small changes in the carrier phase cause large differences in the propagation characteristics. Consequently, with different frequencies, there is virtually no correlation between up- and downlink fast fading.

In some cases, open-loop power control is applied in an asymmetric fashion. A high received signal level means that the transmit power should be decreased immediately, while a low received signal level means that the transmit power should be increased slowly. The purpose of the asymmetric strategy is to minimize the interference caused to other users in the event of an error in the power measurement.

7.6.2 Closed Loop

With *closed-loop power control* of the mobile terminal, the base station measures the received signal strength and then sends corresponding adjustments to the terminal. The delay between measurement and application is a critical parameter in closed-loop power control. Delay creeps into the process in a number of ways, as illustrated in Fig. 7.4:

1. To provide a reasonably accurate measurement of the received signal strength, the measurement must be averaged over several symbol periods.

2. The power control adjustment must be multiplexed with the outgoing transmission, which, in turn, implies a processing delay. To minimize this delay, power control adjustments are not forward error-correction encoded or interleaved, as such encoding would add an additional delay of 10 to 20 milliseconds, which may be unacceptable.

3. The correction incurs the transmission delay corresponding to the separation between the mobile terminal and the base station.

4. At the receiver, since the power-control adjustments are uncoded and thus less reliable, they should be averaged over several symbols, a process that incurs further delay.

Typically, power-control adjustments are sent as a single bit command—for example, a 1 to decrease and a 0 to increase the power by a predetermined amount, often about 1 dB. The delay, of the second and third components in the preceding list, is minimized by sending these power-control bits on a frequent basis. In some systems, power-control bits are sent at an 800-Hz to 1.5-kHz rate. While this improves response times, it is an added overhead for the system.

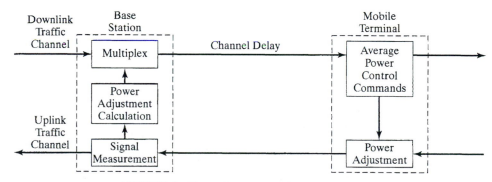

FIGURE 7.4 Illustration of closed-loop power control.

Due to the delay in the power-control loop, power control will not be able to compensate for the fast Rayleigh fading that may be expected at vehicular speeds. In fact, at high vehicle speeds, closed-loop power control may *degrade* performance. However, it is at these faster fading rates that interleaving provides a significant benefit, thus nicely compensating for the shortcomings of power control in this area.

7.6.3 Outer-Loop Power Control

Fast closed-loop power control is based on measurements of the strength of the signal at the base station and comparisons with an expected signal strength, which is based on expectations of noise and interference on the received path according to the system design. To ensure that no systematic errors enter into the comparison, there may be a slower closed-loop power control that is based on frame error rate measurements. The latter measurements are often based on a cyclic redundancy check (CRC) included with the frame and are used for error detection after error-correction decoding. Since the system or the application is designed to tolerate a specified frame error rate, the base station monitors the frame rate to ensure that it is within the expected range. Too high an error rate means that the user terminal is not getting the expected service; too low an error rate means that the user terminal is generating more interference than is acceptable. If the error rate is not in the expected range, that information is taken into account when power control adjustments are sent to the mobile terminal.

7.6.4 Other Considerations

With regard to intercell interference, power control limits the total power emitted by terminals in adjacent cells, but does not control how this interference affects individual terminals. It is important to note that fading will not be correlated between a mobile terminal and two different base stations. That is, a user terminal signal observed at two different base stations will have uncorrelated fading characteristics. Thus, while a user terminal may momentarily have a weak signal at the desired base station, it may at the same time be a relatively strong interferer at another base station.

In a practical system, the integration of the different power control strategies (open- and closed-loop) must be carefully considered, for it is important that the different compensatory measures not counteract one another. Another important point is that power control be based on a measured signal level, not the signal-to-noise ratio. Using the signal-to-noise ratio as a measure can result in an unstable situation: Since most systems that use power control are interference limited, an across-the-board increase in the signal level in these systems will also increase the noise level, resulting in the same SNR and further signal-level increases.

Transmit and receive diversity will be important components of third-generation mobile cellular systems. With multiple intentional paths between the transmitter and the receiver come more complications to the power control strategy depending upon the combining strategy employed.

7.7   HANDOVER[3]

Mobile terminals in a cellular environment will occasionally cross a cell boundary. In that event, the ideal situation is to transfer the call to the base station of the new cell in a procedure known as a *handover*. If the handover does not occur and the call is lost, it is referred to as a *dropped call*. A handover requires the identification of the new base station as well as potentially new traffic and signaling channels.

The initiation of a handover is usually based on the received signal strength. In first-generation systems, the received signal strength on the uplink is measured by the current base station. In second-generation systems, handovers are generally *mobile assisted*. That is, the mobile makes measurements of broadcast channels from the surrounding cells and reports the results to the current base station. In either case, the measurements are provided to a central processing unit, usually referred to as the *mobile switching center* (MSC) that has links to the relevant base stations and that can control the handover.

A call may be dropped for a number of reasons:

- The received signal drops below an acceptable level before the handover is completed; such an event can happen when a mobile phone is in a vehicle moving at high speed.
- There is too much processing burden at the mobile switching center, which can occur if too many handovers are being attempted simultaneously.
- There are no free channels available in the new cell to transfer the call to.

The system is usually designed to minimize the number of dropped calls. From a quality-of-service (QoS) viewpoint, a dropped call is more annoying than a *blocked call* (a call that cannot be initiated). In such a sense, cellular systems rank handovers above new calls, and often channels will be *reserved* in each cell for handovers, so as to reduce the probability that no free channels are available. Ranking handovers higher reduces the number of channels available for new calls; nevertheless, it is a trade-off made to maintain the quality of service.

7.7.1   Handover Algorithms

There are many algorithms for performing handover. One approach is to define a threshold T at a set level above the minimum acceptable signal level; that is,[4]

$$T \text{ (dBm)} = \textit{Minimum acceptable power level} \text{ (dBm)} + \delta_{th} \text{ (dB)} \qquad (7.3)$$

Whenever the received power level falls below the threshold T, a handover is initiated. There are several design trade-offs in this algorithm:

- The additive term δ_{th} cannot be too small, or else it will result in too many dropped calls due to the signal dropping below an acceptable level before the handover is completed.
- Nor can δ_{th} be too large, or else there will be too many handovers, putting a processing burden on the MSC.

- Averaging of the received signal level must be performed, to ensure that dropping below the threshold is not due just to momentary fading; however, averaging introduces a delay in processing.

These trade-offs must be made in the design of the system. In first-generation systems, the handover time is often on the order of 10 seconds and δ_{th} is typically set between 6 and 12 dB. With second-generation TDMA systems, the handover time is often 1 to 2 seconds and δ_{th} is between 0 and 6 dB.

If the received signal strength of a neighboring base station is stronger than that at the current base station, the algorithm of Eq. (7.3) does not necessarily cause a handover. For example, if the mobile terminal has a particularly good propagation path, it may stay with the current base station, well past the "cell boundary" of the adjacent cell. This unintentional extension of the cell boundary is sometimes referred to as *cell dragging*, and it can result in excessive interference by the mobile terminal in the new cell.

A second handover algorithm, which avoids this problem, is possible when the algorithm is mobile assisted. In such a case, a handover is initiated when the power of the neighboring cell base station exceeds that of the current base station by a specified amount or for a certain period. With this second algorithm, there are similar design trade-offs related to the selection of the power difference or waiting time.

The number of handovers that occur depends on the size of the cell and the velocity of the terminals. Smaller cells and higher velocities mean more handovers. Thus, both denser cell packing and terminals on vehicles place greater demands on the handover algorithm.

7.7.2 Multiple-Access Considerations

There are some special multiple-access considerations with handover, which should be taken into account in the design of wireless networks:

- With FDMA and TDMA/FDMA combination systems, the mobile terminal must change the frequencies of both its traffic channel and its signaling channels, as well as synchronize to the new base station. This process is referred to as a *hard handover*.

- With CDMA, there may be a *soft-handover* capability. Since the same frequencies are used in each cell, there is no need to change channels in order to change base stations. That is, multiple base stations receive the same signal, in a form of space diversity. While this is desirable from an individual user's perspective, it cannot be done for a large number of users simultaneously because of the burden it places on system resources.

- With SDMA, if the antenna beams are fixed, then a handover will be necessary whenever a mobile terminal crosses a beam boundary. This requirement could significantly increase the number of handovers. If the beams are not fixed, but instead track users, then handover may become necessary when two users occupying the same channel move near to one another. In either case, since the beams are all provided by the same base station, the handover processing requirements are simplified.

7.8   NETWORK LAYER

Among the reasons that a wireless physical layer may affect layers above the data link layer are the following :

- The reliability of the wireless link and the error-control measures taken by the data link layer may be less than what is expected by higher layers.
- In applications such as voice communication, there are real-time requirements that affect all of the layers.
- If the terminals are mobile, they can move between cells or be activated in areas far from their home location. Terminal mobility adds a dynamic aspect to the network that must be continually monitored.

7.8.1   Cellular Networks

Wireless communication has come to the forefront with the advent of cellular-radio telephony, the intended goal of which is to allow a user to move freely within a service area and simultaneously communicate with any telephone subscriber in the world. A block diagram of the cellular network, showing how the components fit together is presented in Fig. 7.5.

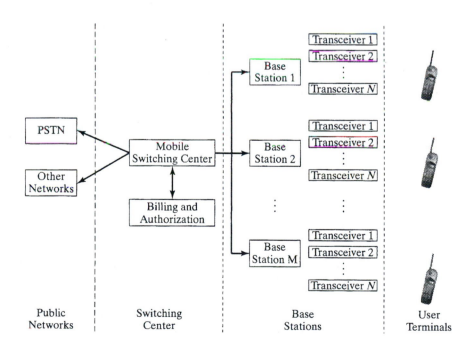

FIGURE 7.5   General interface to PSTN for a mobile telephone network.

Each cell has a base station in its center; depending on the service, the cell radius can range from under a kilometer to over 20 km. Each BS has a number of transceivers (transmitters/receivers) to service the channels in use in that cell. The transceivers perform mainly physical-layer functions and provide service to higher layers. The base station implements the *service provider* side of the physical layer and the MAC sublayer. Note that MAC functions are not simply peer to peer—that is, BS to *mobile subscribers* (UT)—but will likely have some BS-to-BS interactions as well.

The base stations are themselves connected to a *mobile switching center* (MSC) by dedicated wire lines or by fixed wireless. The mobile switching center has four important roles:

1. The MSC performs a network layer function, acting as the interface between the cellular radio system and the public switched telephone network.

2. The MSC performs a MAC sublayer function, providing overall supervision and control of the mobile communications.

3. The MSC *authorizes* the use of the system if the user has a valid account; otherwise, service can be denied. The MSC also performs a *billing* function and must keep track of system utilization by individual users.

4. Related to item 3, the MSC finds the *home location* of a user terminal if it is activated away from its usual service area. Finding the home location is necessary to perform the required authorization and billing functions.

One of the MAC functions performed by the mobile switching center is monitoring the received signal level of the call in progress. The monitoring process is used both for power control and handover functions, as described previously.

The mobile telephone can also be used to deliver data. In fact, integration of voice and data is one of the goals of third-generation systems. In these systems, the mobile switching center is augmented with a data-switching center that interfaces with other data networks. In this case, there must be a connection between a wireless and a wired network. A protocol stack for this type of connection is shown in Fig. 7.6. The source and destination nodes have full protocol stacks; intermediate points along the path have only partial protocol stacks, up to and including the network layer. There may also be an Internet sublayer if the message crosses network boundaries.

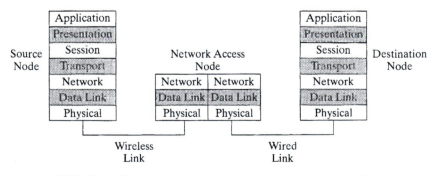

FIGURE 7.6   Protocol stacks for combined wireless/wired connection.

7.8.2 Indoor LANs[5]

Where wireless data networks are piggybacked onto existing cellular telephone networks, they use a MAC sublayer similar to the telephone system. On the other hand, those networks dedicated to data—particularly, indoor local area networks (LANs)—often use a variation of TDMA known as carrier-sense multiple-access (CSMA), described in Section 4.19. With this access method, the transmission medium is monitored, and transmission occurs only when the medium appears unoccupied. A random back-off time follows a busy medium condition. The CSMA strategy is related to the MAC (IEEE Standard 802) strategy used on wired Ethernet networks; thus, wireless networks that use this strategy are often considered a form of wireless Ethernet.

As may be expected, indoor networks have a terminology different from that of outdoor networks:

- Cells are referred to as *service sets*, with the distinction that service sets may physically overlap or even coincide.
- User terminals are referred to as *stations (STAs)*.
- All terminals are stations, but each service set has one station that acts as a base station, referred to as an *access point (AP)*.

Indoor LANs have a number of interesting features that are not usually exhibited by cellular data networks:

- Indoor LANs are often formed without preplanning. The simplest LAN may consist only of two stations, and others may be added in a spontaneous manner. This type of operation is often referred to as an *ad hoc network*.
- LANs of the same type may overlap, due to the spontaneous manner in which they are created. The MAC protocol must be robust under these circumstances.

LANs include some features that are common to many data networks:

- Long packets received from the network layer are split into smaller units for transmission over the wireless channel. This process of *fragmentation* increases reliability by increasing the probability of successful transmission. The receiving MAC layer performs *de-fragmentation*.
- All traffic intended for a single destination requires immediate acknowledgment. The sender schedules a retransmission if no acknowledgment is received.
- User terminals often have a power save mode. In this mode, the access point buffers all messages until the UT indicates that it is ready to receive data.
- Each UT must *associate* itself with an access point before communication between higher layers can occur. If a terminal moves, it may change its association, a process known as *reassociation*.
- The IEEE 802.11 MAC layer protocol uses a technique known as *virtual carrier sense* for scheduling interference-free transmissions. This is a distributed scheduling strategy whereby one station reserves the medium and all stations that can "hear" the reservation, respect it.

- Virtual carrier sense is a form of distributed control. This network also supports a form of point control wherein one station acts as a master and distributes a beacon message with scheduling information at regular intervals.

- IEEE 802.11 provides the ability to encrypt the contents of messages (privacy up to the equivalent of a wired LAN).

- In many cases, higher layer connections (i.e., network and transport layer connections) are maintained, although the terminal may be reassociated with another AP. This does not apply beyond the extended service set.

7.9 THEME EXAMPLE 1: WIRELESS TELEPHONE NETWORK STANDARDS

In Table 7.2, we compare the physical layer characteristics of a number of wireless telephone systems. All of the systems described in the table are cellular systems, and they all take advantage of the technology, the channel, and the services in a different manner.

TABLE 7.2   Comparison of Physical Layer Characteristics of Various Telephone Network Standards.

	DECT	GSM	IS-95	WCDMA
Frequency Band	1880–1900 MHz	935–960 MHz (F) 890–915 MHz (R)	869–894 MHz (F) 824–849 MHz (R)	1920–1980 MHz (F) 2110–2117 MHz (R) various for TDD
Channel BW	1.728 MHz	200 kHz	1.25 MHz	5 MHz
Modulation	GMSK	GMSK	BPSK	QPSK
Data rates	1.152 Mbps	270.8 kbps	1200, 2400, 4800, 9600 bps	up to 2 Mbps
Access strategy	FDMA/TDMA/TDD	FDMA/TDMA/FH	FDMA/CDMA	FDMA/CDMA/FDD FDMA/CDMA/TDD
Cell size	<300 m	<35 km	<35 km	< 35 km
Forward error- correction coding	none (16-bit CRC)	variable incl. rate-1/2 convol.	variable incl. rate-1/2, -1/3 convol.	Variable incl. rate-1/2, -1/3 convol.
Frame size	10 ms	4.61 ms	20 ms	10 ms
Voice encoding	ADPCM at 32 kHz	RELP at 13 kbps	CELP at 9.6 kbps and 14.4 kbps	Adaptive multirate ACELP 4.75 to 12.2 kbps
Traffic channels/ RF channel	12	8	up to 63 in theory	Depends upon data rate
Diversity	Antenna diversity at BS	Frequency hopped	Spread spectrum with RAKE receiver	Space-time block coding with transmit diversity

The standards shown in the table are as follows:

(i) *DECT.*[6] The *Digital Enhanced Cordless Telephone* (DECT) is a cordless telephone standard for high traffic density and short-range communications. The main application is to provide service to an in-building office scenario with many cordless users served by a single telephone exchange. This is a microcellular system with a base station for each cell that supports the physical layer for each call and a cluster controller that supports the MAC sublayer and network operations for a group of base stations, similar in function to the mobile switching center of Fig. 7.5.

(ii) *GSM.* This second-generation standard for cellular telephony was described in Section 4.17.

(iii) *IS-95.*[7] This second-generation standard for cellular telephony is one of the first terrestrial wireless systems to introduce spread-spectrum techniques for commercial use. It was described in Section 5.11.

(iv) *WCDMA.*[8] This third-generation standard was described in Section 5.14.

The following observations can be made from the comparisons in Table 7.2:

- All of the services operate in similar frequency bands. This is not a technology-related issue, but rather a regulatory issue, in that these frequency bands have been designated for this purpose.

- All of the techniques have an underlying FDMA strategy; TDMA and CDMA are the common multiple-access strategies used in a hybrid approach with FDMA.

- The channel bandwidth depends upon the multiple-access strategy.

- Modulation strategies vary, but they all tend to be of low complexity. Simple or low-order modulations, such as BPSK and QPSK, tend to be the most *robust* in the propagation conditions of mobile radio.

- Data rates for each mobile terminal are not high. For second-generation terminals such as GSM and IS-95, data rates per terminal are typically 10 to 30 kbps. The limitations are the available bandwidth, need, and reliability. Higher data rates need more power or shorter transmission distances.

- The cell size varies, depending on the intended application and power at the user terminal. The available power, battery life, and related issues are critical items in the design of mobile radio systems.

- Forward error-correction (FEC) coding is added to improve the robustness of transmission over radio channels. Almost all systems include forward error-correction coding.

- The frame size is small and is motivated by vocoder (voice coder) characteristics. It is somewhat coincidental and beneficial that the frame size has the same dimension as the de-correlation time of the channel, so that behavior on one frame is often uncorrelated with that of the next.

- A variety of diversity techniques are used; receive-antenna diversity at the base station is commonly used.

7.10 THEME EXAMPLE 2: WIRELESS DATA NETWORK STANDARDS

The increasing prevalence of computers has been accompanied by a growing number of wireless data networks. Some are intended for large-area cells and are implemented as extensions of mobile telephone networks. Others are intended for indoor or small LANs and provide primarily a data service that could be extended to voice by techniques such as *voice-over-Internet Protocol* (*IP*), which refers to the process of sampling and speech-encoding voice and then transmitting it as data over a network in the form of an Internet packet. Delay constraints usually limit the application of voice-over-IP to local area networks. Table 7.3 lists the physical characteristics of the following wireless data network standards:

- *GPRS*. The *General Packet Radio Service* is an extension of the GSM telephone network to provide data services to mobile users; GPRS offers a packet-switched signal that can be multiplexed with the circuit-switched voice portion of GSM.
- *WCDMA*. This is the same as the telephone network described in the previous section and represents a convergence of both data and telephone networks.

TABLE 7.3   Comparison of Physical Layer Characteristics of Various Data Network Standards.

	GPRS	WCDMA	IEEE 802.11b	Bluetooth	IEEE 802.11a
Frequency band	935–960 MHz (F) 890–915 MHz (R)	1920–1980 MHz (F) 2110–2117 MHz (R)	2.4 GHz	2.4–2.4385 GHz	5.2 GHz
Channel BW	200 kHz	5 MHz	50 MHz	80 MHz	20 MHz
Modulation	GMSK	CDMA	BPSK/QPSK FH or DS	GMSK/FH	BPSK,…,64-QAM /OFDM
Data rates	up to 116 kbps	up to 2 Mbps	up to 11 Mbps	<1Mbps	up to 54 Mbps
Access strategy	FDMA/TDMA	FDMA/CDMA	CSMA/CA	FH/TDD	FDMA/CSMA
Cell size	up to 35 km	up to 35 km	1–20 m	1–10 m	1–100m
Forward error-correction coding	variable incl. rate 1/2 convol.	variable incl. rate-1/2, -1/3 convol.	rate-1/2, -1/3 convol.	Variable repetition, Hamming, ARQ	rate-1/2 , -2/3, -3/4 convol.
Frame size	4.61 ms	10 ms	up to 20 ms	$n \times 625\mu s$ $n=1,…,5$	24 µs to 5 ms
Diversity	Frequency- hopped	Space–time block coding with Tx diversity	Dual antenna	none	Dual antenna
Network topology	point-to-multipoint/ cellular	point to multipoint/cellular	point to multipoint	point-to-point connection and connectionless	point to multipoint

- *Bluetooth*. This is a standard as a short-range cable replacement for linking components such as computers, printers, fax machines, and keyboards for very short-range communications. Bluetooth forms networks on an ad hoc basis and could serve as a wireless bridge to existing data networks. Bluetooth is described in Section 5.13.
- *IEEE 802.11b*. This is a wireless networking standard for small indoor data networks, with the intent of replacing or supplementing wired Ethernet networks. This wireless LAN standard is described in Section 5.16.
- *IEEE 802.11a*. This is similar to IEEE 802.11b in objective, except that it provides service at much higher data rates. It is described in Section 2.11.

The structure of these data networks is similar to that of telephone networks—point to multipoint, often with a cellular overlay.

In comparison to telephone networks, data networks have a much greater variation in physical characteristics. This characteristic is related to their potentially larger range of applications. In particular, we note the following points:

- The frequency bands range from 900 MHz up to 5 GHz, with the lower bands corresponding to combined mobile data-telephone networks. The other data networks listed in Table 7.3 operate in what are known as the *Industrial, Scientific, and Medical (ISM)* bands at 900 MHz, 2.4 and 5 GHz, respectively. Wireless devices can operate unlicensed in these bands, as long as their transmit powers are below certain levels and they meet certain other requirements. As a consequence, most such networks are targeting indoor LAN applications.
- The modulation techniques cover a wide spectrum of cases, moving from simple GMSK, through the linear techniques of BPSK and QPSK, to the much more complex 64-QAM and OFDM strategies.
- The channel bandwidth depends on the multiple-access strategy and data rate. It ranges from 200 kHz up to 80 MHz.
- A variety of multiple-access techniques are employed, and a convergence toward one particular technique is not evident.
- Data rates range widely and are dependent on the cell size.
- The cell size varies, ranging from very large macrocells of 35 km to picocells of about 10 m. This variation depends on the intended application and available power at the user terminal.
- All services include forward error-correction coding, which is particularly important in improving the robustness of transmission over radio channels for data applications.
- The frame size is variable, depending on the amount of data that are to be transmitted.

7.11 THEME EXAMPLE 3: IEEE 802.11 MAC

There are some differences between the MAC layer protocols for data networks and those for telephone networks. An example of the former is the IEEE 802.11 MAC

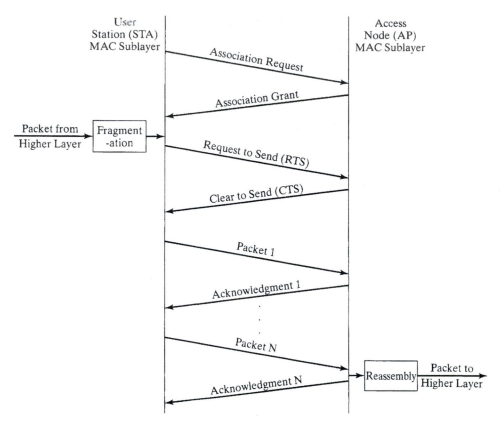

FIGURE 7.7   Illustration of IEEE 802.11 MAC protocol for transmitting a long packet.

standard, and a sample transaction of this protocol is shown in Fig. 7.7. In this example, the station executes the following steps:

1. When initially activated, the user station (STA) logs onto the local base station, referred to as an access point (AP), to inform the AP of its presence and address in the local network. This log-on procedure is referred to as an *association* with the local AP.

2. The AP *authenticates* the STA's legitimate access to the network and provides a local ID number.

3. When a packet is generated at the STA for transmission, the STA physical layer listens to the channel for activity. If the channel is not active, then a Request to Send (RTS) is transmitted, addressed to the target receiver. If the destination address is outside the local network, then the target would be the AP. The Request to Send reserves the channel for a specified period. The receiving station replies with a Clear to Send (CTS).

4. The STA then sends the packet fragments one at a time, receiving acknowledgments before sending the next, until all packet fragments have been sent.

5. All stations listen to the channel and decode all RTS messages. Once decoded, all stations, except for the one transmitting, do not transmit during the reserved time. This is the *virtual carrier sense* described in Section 7.8.2.

6. The use of RTS and CTS is optional and may not be efficient for short packets. If it is not used, then all transmissions are based on a *physical carrier sense*. In this capacity, if the medium is detected to be busy, then the STA waits until it is not busy, plus a random back-off time. The latter is intended to ensure that if a number of UTs are waiting to transmit, they do not all transmit simultaneously. If the medium is still busy after the random back-off time, then the procedure is repeated with a back-off time having a longer distribution, etc.

One variation or another of this basic MAC protocol is used in all of the IEEE 802.11 standard networks. Note that the same physical channel is used for both protocol transactions and traffic.

7.12 SUMMARY AND DISCUSSION

In this chapter, we have considered integration of the physical layer and multiple-access strategies with the network aspects of wireless systems. The topics considered were as follows:

- A comparison of the advantages and disadvantages of the different multiple-access strategies.
- The OSI reference model for communications and the aspects related to wireless communications.
- The use of specialized signaling channels as part of the MAC protocol, to meet delay requirements for voice networks.
- The need for power control in wireless cellular systems and various methods of implementing such control.
- Methods for implementing handover strategies as a terminal moves from one cell to the next.
- A comparison of wireless telephone networks and the identification of many of their common features.
- A comparison of wireless data networks and the wide range of their implementations.

Our theme examples concentrated on cellular telephone systems and local area networks. The overview encompassed by the three theme examples has been presented because they are some of the better-known wireless networks. However, it should be remembered that cellular telephone systems and wireless LANs occupy less than 3% of the radio spectrum that current technology can address. There are many other types of wireless intended applications and networks, including the following:

- *Aeronautical.* Originally intended for safety services, but recently for passenger services
- *Military.* Land, maritime, underwater, and airborne
- *Safety services.* Police, fire, ambulance

- *Broadcast.* AM, FM, television
- *Telemetry and control.* Spacecraft and remote monitoring stations

Many of these applications were originally voice oriented, but are becoming increasingly data oriented. Many of their services are provided by private networks, either satellite or terrestrial.

The chapter emphasizes the key differences between wireless and other networks that occur at the physical and multiple-access layers in the OSI model. The book, as a whole, observes that the fundamental theme of wireless evolution is to provide service with an ever-increasing spectral efficiency.

NOTES AND REFERENCES

[1] Bertsekas and Gallagher (1992) provides an in-depth look at the OSI model and data networks in general.

[2] More information on the power-control methods of IS-95 can be found in Chapter 2 of Lee and Miller (1998). The power-control methods used in WCDMA are discussed in Chapters 3 and 9 of Holma and Toskala (2000).

[3] A more detailed description of handover techniques can be found in Chapter 2 of Rappaport (1999) and Chapter 9 of Holma and Toskala (2000).

[4] Note that it is correct to say 20 dBm is 10 dB greater than 10 dBm but it is incorrect to say 20 dBm is 10 dBm greater than 10 dBm. The reason is that 20 dBm is equivalent to 100 milliwatts while 10 dBm is equivalent to 10 milliwatts; 100 milliwatts is clearly ten times larger than 10 milliwatts, not just 10 milliwatts larger.

[5] The most detailed source for information on indoor LAN networks is the standard itself (ANSI/IEEE 1999); this is an evolving document, but a good overview is provided in O'Hara and Petrick (1999).

[6] More information on the DECT network can be found in Chapter 10 of Rappaport (1999).

[7] Lee and Miller (1998) examine all aspects of the IS-95 CDMA standard; Chapter 10 of the book presents an analysis of the system capacity.

[8] WCDMA is just one of the proposed air interfaces for third-generation mobile terrestrial networks. The other major alternative is referred to as CDMA2000, a standard that is an evolution of the IS-95 standard to provide higher data rates and integrate voice and data services, similar to these functions in WCDMA. CDMA2000 is intended to be backwardly compatible with IS-95. Consequently, it has a component that has the same bandwidth as IS-95, the so-called 1x version. There is also a component that has three times the bandwidth of IS-95 that is known as the 3x version. Details of the CDMA2000 standard may be found in Chapter 15 of Garg (2000).

PROBLEMS

Problem 7.1 List the advantages of a circuit-switched operation over a store-and-forward network, and vice versa.

Problem 7.2   An aeronautical satellite service has three types of physical channels, labelled as follows:

  (i)   P-channel—a TDMA channel from base station to mobile terminal
  (ii) R-channel—a random-access channel from mobile terminal to base station
  (iii) T-channel—a TDMA channel from mobile terminal to base station.

Use a ladder diagram to illustrate a protocol allowing an aircraft to transmit a large data message.

Problem 7.3   List the four general classes of logical channels in telephone networks and the purpose of each logical channel.

Problem 7.4   Describe how GPRS could increase the utilization of a GSM network.

Problem 7.5   Is the protocol described by Fig. 7.7 robust in the case of two overlapping LANs? Explain.

Problem 7.6   Explain what it means to say that a multiple-access strategy is FDMA with a TDMA overlay. Explain how this strategy might be related to a frequency-hopping system.

Problem 7.7   Identify layers of the OSI stack in which the unreliability of the transmission link may be countered. How might it be countered?

Problem 7.8   The common size for a TCP header is 20 bytes. For an IP header, it is also 20 bytes, and for IEEE 802 MAC header, it is 14 bytes. If the channel is unreliable and frequent packet retransmissions are required, which is the best layer for initiating a retransmission and why? Assume that all lower layers simply discard erroneous packets. Suggest a method of reducing the overhead due to the TCP/IP protocol.

Problem 7.9   Describe the difficulties that may be encountered in transmitting voice over a packet data network. Under what conditions could it be successful?

Problem 7.10   A communications service is designed to have the logical channels shown in Table 7.4 and the indicated loadings. Design a multiplexing strategy for partitioning these logical channels among physical channels and determine the maximum of traffic channels. What is the relationship between the forward and return physical channels? Note that channels that are fully loaded tend to have long service delays.

TABLE 7.4   For Problem 7.10.

Logical Channels	Direction	Number	Rate
Broadcast	Forward	1	1 kbps
Paging	Forward	1	2 kbps
Access grant	Forward	1	2 kbps
Control	Forward	1	2 kbps
Traffic	Forward	—	8 kbps
Access request	Return	1	2 kbps
Acknowledgments	Return	1	1 kbps
Traffic	Return	—	8 kbps
Physical Channels			
	Forward	8	8 kbps
	Return	8	8 kbps

Problem 7.11   If the probability of a bit error over a wireless channel is p and bit errors are independent, what is the probability of packet error if a packet contains N bits? If forward error-correction coding can improve an error rate of p to an error rate of p^2, what is the improvement in the packet error rate. Forward error-correction coding requires the transmission of more channel bits, but it reduces the retransmission requirements. How would you evaluate the trade-off between the two?

Problem 7.12   Continuing with Problem 7.11, an alternative approach to improving performance is to use diversity instead of forward error-correction coding. How would this affect the trade-off?

Problem 7.13   In a LAN environment, stations may not be able to hear the transmissions of all other stations. How will this affect the performance of a CSMA strategy? How would you improve the performance of the system?

Problem 7.14   It has been observed that, during a normal telephone conversation, each speaker talks approximately 40% of the connection time. Vocoders (voice coders) often process speech in 20-millisecond blocks, producing one packet of data every 20 milliseconds when there is voice activity. If a TDMA strategy has separate forward and return bearers, with a 20-millisecond frame size and eight slots per frame, each slot capable of carrying a voice packet, what is the theoretical maximum number of voice channels this system could carry? What would prevent it from achieving the maximum? (Assume the channel data rate is eight times the vocoder data rate.)

Problem 7.15   Relate the functions of the mobile switching center to those identified in the OSI reference model of Fig. 7.2.

A P P E N D I X   A

Fourier Theory

A.1 THE FOURIER TRANSFORM[1]

Let $g(t)$ denote a *nonperiodic deterministic signal*, expressed as some function of time t. By definition, the *Fourier transform* of the signal $g(t)$ is given by the integral

$$G(f) = \int_{-\infty}^{\infty} g(t)\exp(-j2\pi ft)dt \qquad (A.1)$$

where $j = \sqrt{-1}$, and the variable f denotes *frequency*. Given the Fourier transform $G(f)$, the original signal $g(t)$ is recovered exactly using the formula for the *inverse Fourier transform*:

$$g(t) = \int_{-\infty}^{\infty} G(f)\exp(j2\pi ft)df \qquad (A.2)$$

Note that in Eqs. (A.1) and (A.2) we have used a lowercase letter to denote the time function and an uppercase letter to denote the corresponding frequency function. The functions $g(t)$ and $G(f)$ are said to constitute a *Fourier-transform pair*.

For the Fourier transform of a signal $g(t)$ to exist, it is sufficient, but not necessary, that $g(t)$ satisfy three conditions, known collectively as *Dirichlet's conditions*:

1. The function $g(t)$ is single valued, with a finite number of maxima and minima in any finite time interval.
2. The function $g(t)$ has a finite number of discontinuities in any finite time interval.
3. The function $g(t)$ is absolutely integrable; that is,

$$\int_{-\infty}^{\infty} |g(t)|dt < \infty$$

We may safely ignore the question of the existence of the Fourier transform of a time function $g(t)$ when $g(t)$ is an accurately specified description of a physically realizable signal. In other words, physical realizability is a sufficient condition for the existence of a Fourier transform. Indeed, we may go one step further and state that all finite-energy signals are Fourier transformable.

The absolute value of the Fourier transform $G(f)$, plotted as a function of frequency f, is referred to as the *amplitude spectrum* or *magnitude spectrum* of the signal $g(t)$. By the same token, the argument of the Fourier transform, plotted as a function of frequency f, is referred to as the *phase spectrum* of the signal $g(t)$. The amplitude spectrum is denoted by $|G(f)|$ and the phase spectrum is denoted by $\theta(f)$. When $g(t)$ is a real-valued function of time t, the amplitude spectrum $|G(f)|$ is symmetrical about the origin $f = 0$, whereas the phase spectrum $\theta(f)$ is antisymmetrical about $f = 0$.

Strictly speaking, the theory of the Fourier transform is applicable only to time functions that satisfy the Dirichlet conditions. (Among such functions are energy signals.) However, it would be highly desirable to extend this theory in two ways to include power signals (i.e., signals whose average power is finite). It turns out that this objective can be met through the "proper use" of the *Dirac delta function, or unit impulse.*

The Dirac delta function, denoted by $\delta(t)$, is defined as having zero amplitude everywhere except at $t = 0$, where it is infinitely large in such a way that it contains unit area under its curve; that is,

$$\delta(t) = 0 \qquad t \neq 0 \tag{A.3}$$

and

$$\int_{-\infty}^{\infty} \delta(t)dt = 1 \tag{A.4}$$

An implication of this pair of relations is that the delta function is an *even function of time*; that is, $\delta(-t) = \delta(t)$. Another important property of the Delta function is the *replication property* described by

$$\int_{-\infty}^{\infty} g(\tau)\delta(t - \tau)d\tau = g(t) \tag{A.5}$$

which states that the convolution of any function with the delta function leaves that function unchanged.

Tables A.1 and A.2 build on the formulas of Eqs. (A.1) through (A.5). In particular, Table A.1 summarizes the properties of the Fourier transform, while Table A.2 lists a set of Fourier-transform pairs.

In the time domain, a linear system (e.g., filter) is described in terms of its *impulse response*, defined as *the response of the system (with zero initial conditions) to a unit impulse or delta function $\delta(t)$ applied to the input of the system* at time $t = 0$. If the system is *time invariant*, then the shape of the impulse response is the same, no matter when the unit impulse is applied to the system. Thus, assuming that the unit impulse or delta function is applied at time $t = 0$, we may denote the impulse response of a linear time-invariant system by $h(t)$. Let this system be subjected to an arbitrary excitation $x(t)$, as in Fig. A.1(a). Then the response $y(t)$ of the system is determined by the formula

$$y(t) = \int_{-\infty}^{\infty} x(\tau)h(t - \tau)d\tau$$
$$= \int_{-\infty}^{\infty} h(\tau)x(t - \tau)d\tau \tag{A.6}$$

TABLE A.1   Summary of Properties of the Fourier Transform.

Property	Mathematical Description
1. Linearity	$ag_1(t) + bg_2(t) \rightleftharpoons aG_1(f) + bG_2(f)$ where a and b are constants
2. Time scaling	$g(at) \rightleftharpoons \dfrac{1}{\lvert a \rvert} g\left(\dfrac{f}{a}\right)$ where a is a constant
3. Duality	If $g(t) \rightleftharpoons G(f)$, then $G(t) \rightleftharpoons g(-f)$
4. Time shifting	$g(t - t_0) \rightleftharpoons G(f)\exp(-j2\pi f t_0)$
5. Frequency shifting	$\exp(j2\pi f_c t)g(t) \rightleftharpoons G(f - f_c)$
6. Area under $g(t)$	$\displaystyle\int_{-\infty}^{\infty} g(t)\,dt = G(0)$
7. Area under $G(f)$	$g(0) = \displaystyle\int_{-\infty}^{\infty} G(f)\,df$
8. Differentiation in the time domain	$\dfrac{d}{dt}g(t) \rightleftharpoons j2\pi f G(f)$
9. Integration in the time domain	$\displaystyle\int_{-\infty}^{t} g(\tau)\,d\tau \rightleftharpoons \dfrac{1}{j2\pi f}G(f) + \dfrac{G(0)}{2}\delta(f)$
10. Conjugate functions	If $g(t) \rightleftharpoons G(f)$, then $g^*(t) \rightleftharpoons G^*(-f)$
11. Multiplication in the time domain	$g_1(t)g_2(t) \rightleftharpoons \displaystyle\int_{-\infty}^{\infty} G_1(\lambda)G_2(f - \lambda)\,d\lambda$
12. Convolution in the time domain	$\displaystyle\int_{-\infty}^{\infty} g_1(\tau)g_2(t - \tau)\,d\tau \rightleftharpoons G_1(f)G_2(f)$
13. Correlation theorem	$\displaystyle\int_{-\infty}^{\infty} g_1(t)g_2^*(t - \tau)\,dt = G_1(f)G_2^*(f)$
14. Rayleigh's energy theorem	$\displaystyle\int_{-\infty}^{\infty} \lvert g(t) \rvert^2\,dt = \displaystyle\int_{-\infty}^{\infty} \lvert G(f) \rvert^2\,df$

The formula of Eq. (A.6) is called the *convolution integral.* Three different time scales are involved in it: the *excitation time* τ, *response time t, and system-memory time* $t - \tau$. Equation (A.6) is the basis of the time-domain analysis of linear time-invariant systems. It states that the present value of the response of a linear time-invariant system is the integral over the past history of the input signal, weighted according to the impulse response of the system. Thus, the impulse response acts as a *memory function* for the system.

TABLE A.2   Fourier-Transform Pairs.

Time Function	Fourier Transform
$\text{rect}\left(\dfrac{t}{T}\right)$	$T\,\text{sinc}(fT)$
$\text{sinc}(2Wt)$	$\dfrac{1}{2W}\,\text{rect}\left(\dfrac{f}{2W}\right)$
$\exp(-\pi t^2)$	$\exp(-\pi f^2)$
$\begin{cases} 1-\dfrac{\|t\|}{T}, & \|t\| < T \\ 0, & \|t\| \geq T \end{cases}$	$T\,\text{sinc}^2(fT)$
$\delta(t)$	1
1	$\delta(f)$
$g(t-t_0)$	$\exp(j2\pi f t_0)$
$\exp(j2\pi f_c t)$	$\delta(f-f_c)$
$\cos(2\pi f_c t)$	$\dfrac{1}{2}[\delta(f-f_c) + \delta(f+f_c)]$
$\sin(2\pi f_c t)$	$\dfrac{1}{2j}[\delta(f-f_c) - \delta(f+f_c)]$
$\displaystyle\sum_{i=-\infty}^{\infty} \delta(t-iT_0)$	$\dfrac{1}{T_0}\displaystyle\sum_{n=-\infty}^{\infty} \delta\left(f-\dfrac{n}{T_0}\right)$
$\displaystyle\sum_{n=-\infty}^{\infty} x(t-nT_0)$	$\dfrac{1}{T_0}\displaystyle\sum_{m=-\infty}^{\infty} X\left(\dfrac{m}{T_0}\right)\delta\left(f-\dfrac{m}{T_0}\right)$

Notes: $\delta(t)$ = delta function, or unit impulse
rect(t) = rectangular function of unit amplitude and unit duration centered on the origin
sinc(t) = sinc function

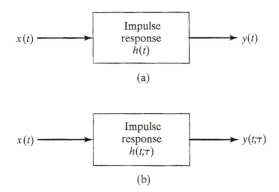

FIGURE A.1 (a) Linear system. (b) Linear time-varying system.

From Table A.1, we note that when two functions of time are convolved with each other, the operation of convolution is transformed into the multiplication of the Fourier transforms of the functions in the frequency domain. Hence, applying this property to Eq. (A.6), we may express the Fourier transform of the output signal $y(t)$ as

$$Y(f) = H(f)X(f) \tag{A.7}$$

where $X(f)$ is the Fourier transform of the input signal $x(t)$. The other quantity in Eq. (A.7), namely, $H(f)$, is called the *transfer function* of the system. It is formally defined as the Fourier transform of the impulse response $h(t)$ and is given by

$$H(f) = \int_{-\infty}^{\infty} h(t)\exp(-j2\pi ft)dt \tag{A.8}$$

Thus, the impulse response $h(t)$ provides a time-domain description of a linear time-invariant system, whereas the transfer function $H(f)$ provides an equivalent description of the system in the frequency domain.

A.2 LINEAR TIME-VARYING SYSTEMS

Consider next the case of a linear time-varying system, exemplified by a wireless communication channel. As the name implies, the impulse response of a *linear time-varying system* depends on the time at which the unit impulse is applied to the input of the system. We thus denote the impulse response of such a system by $h(t;\tau)$, where $(t - \tau)$ is the time at which the unit impulse is applied to the system and t is the time at which the resulting response is measured. (See Fig. A.1(b).) Suppose, then, that an input signal $x(t)$ is applied to a linear time-varying system with impulse response $h(t;\tau)$. Then the resulting response of the system is defined by

$$y(t) = \int_{-\infty}^{\infty} h(t;\tau)x(t - \tau)d\tau \tag{A.9}$$

where the integration is performed with respect to τ. Correspondingly, the transfer function of the system is written as $H(f;\tau)$, which is related to the impulse response $h(t;\tau)$ via the Fourier transform through the relationship

$$H(f;\tau) = \int_{-\infty}^{\infty} h(t;\tau)\exp(-j2\pi ft)\,dt \qquad (A.10)$$

Equation (A.8) is a special case of Eq. (A.10) in that, for a linear time-invariant system, we have $H(f;\tau) = H(f)$ for all τ.

A.3 SAMPLING THEOREM

In continuous-wave modulation, the carrier is typically a sinusoidal wave. In pulse modulation, by contrast, the carrier is a uniform train of pulses that are relatively short compared with the fundamental period of the carrier. The sampling theorem, described next, is basic to all the different forms of pulse modulation used in practice.

To set the stage for a statement of the sampling theorem, consider a strictly band-limited signal $x(t)$ whose frequency content is confined to a bandwidth W; that is,

$$X(f) = 0 \qquad \text{for } |f| \geq W \qquad (A.11)$$

For such a signal, the *sampling theorem* may be stated in two parts:

1. The strictly band-limited signal $x(t)$ is uniquely represented by a set of samples $x(nT_0)$, $n = 0, \pm1, \pm2, ...$, provided that the sampling rate $f_0 = 1/T_0$ is greater than twice the highest frequency component of $x(t)$; in other words, $f_0 > 2W$.
2. The original signal $x(t)$ is reconstructed from the set of samples $x(nT_0)$ for $n = 0$, $\pm1, \pm2, ...$, and $T_0 > 1/(2W)$, without loss of information, by passing this uniformly sampled signal through an ideal low-pass construction filter of bandwidth W hertz.

For a proof of the sampling theorem, we may invoke the duality property of the Fourier transform. From Table A.1, that duality property states,

If $g(t) \rightleftharpoons G(f)$, then $G(-t) \rightleftharpoons g(f)$, where the time function $G(-t)$ is obtained by substituting $-t$ for f in the Fourier transform $G(f)$ and the frequency function $g(f)$ is obtained by substituting f for t in the inverse Fourier transform $g(t)$.

From the last entry of Table A.2, we also have the Fourier-transform pair

$$\sum_{m=-\infty}^{\infty} g(t - mT_0) \rightleftharpoons f_0 \sum_{n=-\infty}^{\infty} G(nf_0)\delta(f - nf_0) \qquad (A.12)$$

where $f_0 = 1/T_0$ is the *sampling rate* and $\delta(f)$ is the Dirac delta function defined in the frequency domain. Applying the duality property to Eq. (A.12), invoking the even-function property of the delta function, and using T_0 in place of f_0 to maintain proper dimensionality in the result, we obtain

$$T_0 \sum_{n=-\infty}^{\infty} G(-nT_0)\delta(t - nT_0) \;\rightleftharpoons\; \sum_{m=-\infty}^{\infty} g(f - mf_0) \qquad (A.13)$$

To put this relation in the context of the strictly band-limited signal $x(t)$, we set $G(-t) = x(t)$ and $g(f) = X(f)$, in which case we may recast Eq. (A.13) in the desired form:

$$x_\delta(t) \;=\; T_0 \sum_{n=-\infty}^{\infty} x(nT_0)\delta(t - nT_0) \;\rightleftharpoons\; \sum_{m=-\infty}^{\infty} X(f - mf_0) \;=\; X_\delta(f) \qquad (A.14)$$

Figure A.2 presents a time–frequency description of Eq. (A.14), assuming that $X(f) = 0$ for $|f| > W$ and $f_0 > 2W$. Parts (a) and (b) of the figure depict the spectra $X(f)$ and $X_\delta(f)$, respectively, where $x(t) \rightleftharpoons X(f)$ and $x_\delta(t) \rightleftharpoons X_\delta(f)$.

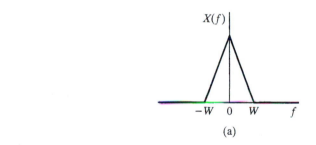

(a)

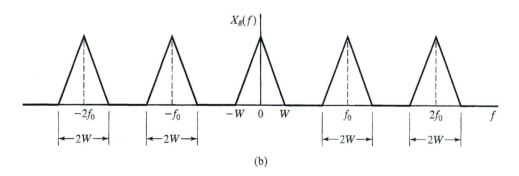

(b)

FIGURE A.2   (a) Spectrum of a signal $x(t)$ limited to the band $-W \le f < W$.
(b) Spectrum of the instantaneously sampled signal $x_\delta(t)$ for a sampling rate $f_0 > 2W$.

Given the *instantaneously sampled signal* $x_\delta(t)$ and assuming a sampling rate $f_0 > 2W$, how do we use $x_\delta(t)$ to reconstruct the original signal $x(t)$? We may do so by employing a *reconstruction system* formulated as an ideal low-pass filter of bandwidth W. To find the output of this filter in response to the sampled signal $x_\delta(t)$, we proceed in two stages:

1. Take the Fourier transform of sampled signal $x_\delta(t)$, and limit the spectrum to the frequency band $|f| \le W$ permitted by the low-pass reconstruction filter, thereby obtaining the spectrum

$$X(f) = \begin{cases} T_0 \displaystyle\sum_{n=-\infty}^{\infty} x(nT_0)\exp(-j2\pi nT_0 f) & |f| \le W \quad |f| \le W \\ 0 & |f| > W \end{cases} \qquad (A.15)$$

2. Take the inverse Fourier transform of the spectrum defined in Eq. (A.15), yielding the original signal

$$\begin{aligned} x(t) &= \int_{-\infty}^{\infty} X(f)\exp(j2\pi ft)\,dt \\ &= \int_{-W}^{W} \left(T_0 \sum_{n=-\infty}^{\infty} x(nT_0)\exp(-j2\pi nT_0 f) \right) \exp(j2\pi ft)\,dt \\ &= T_0 \sum_{n=-\infty}^{\infty} x(nT_0) \left(\frac{\sin(2\pi(t-nT_0)W)}{2\pi(t-nT_0)W} \right) \\ &= T_0 \sum_{n=-\infty}^{\infty} x(nT_0)\,\mathrm{sinc}\,(2(t-nT_0)W) \end{aligned} \qquad (A.16)$$

where the function

$$\mathrm{sinc}\,(\lambda) = \frac{\sin(\pi\lambda)}{\pi\lambda} \qquad (A.17)$$

is called the *sinc function*. Equation (A.16) states that, provided that the sampling rate f_0 satisfies the condition $f_0 > 2W$, the original signal $x(t)$ may be reconstructed as the weighted sum of the reconstruction kernel $\mathrm{sinc}(2Wt)$, where the nth component of the sum consists of the time-shifted kernel $\mathrm{sinc}(2(t-nT_0)W)$, weighted by the corresponding sample $x(nT_0)$.

Equation (A.16) verifies part (2) of the sampling theorem. Part (1) of the theorem is, in reality, merely a reformulation of part (2).

A.4   SAMPLED CONVOLUTION THEOREM

Suppose that we have two time functions, $x(t)$ and $h(t)$, with $x(t)$ limited to the frequency band $-W < f < W$. The function $x(t)$ is uniformly sampled at the rate $f_0 > 2W$ and then convolved with $h(t)$. The convolution product, denoted by $y(t)$, may be viewed as the output of a linear time-invariant system with impulse response $h(t)$, which is driven by the instantaneously sampled version of $x(t)$. The requirement is to evaluate $y(t)$.

The instantaneously sampled version of $x(t)$ is defined by (see the left-hand side of Eq. (A.14))

$$x_\delta(t) = T_0 \sum_{n=-\infty}^{\infty} x(nT_0)\delta(t - nT_0) \tag{A.18}$$

where $T_0 = 1/f_0$ is the sampling period. The convolution of $x_\delta(t)$ with $h(t)$ is defined by the integral

$$
\begin{aligned}
y(t) &= \int_{-\infty}^{\infty} h(\tau)x_\delta(t-\tau)d\tau \\[2mm]
&= \int_{-\infty}^{\infty} h(\tau)\left(T_0 \sum_{n=-\infty}^{\infty} x(nT_0)\delta(t - nT_0 - \tau) \right)d\tau \\[2mm]
&= T_0 \sum_{n=-\infty}^{\infty} x(nT_0)\int_{-\infty}^{\infty} h(\tau)\delta(t - nT_0 - \tau)d\tau
\end{aligned}
\tag{A.19}
$$

Invoking the replication property of the delta function described by Eq. (A.5), we may reduce the integral in the last line of Eq. (A.19) to

$$\int_{-\infty}^{\infty} h(\tau)\delta(t - nT_0 - \tau)d\tau = h(t - nT_0) \tag{A.20}$$

Accordingly, Eq. (A.19) simplifies to

$$y(t) = T_0 \sum_{n=-\infty}^{\infty} x(nT_0)h(t - nT_0) \tag{A.21}$$

which is the desired result. Equation (A.21) is a statement of the *sampled convolution theorem*:

> *The convolution of a continuous-time function with the instantaneously sampled version of a band-limited signal is a scaled version of the convolution sum of two time series: the original instantaneously sampled signal and the instantaneously sampled version of the continuous-time function.*

Note that Eq. (A.21) is a generalization of the expression on the left-hand side of Eq. (A.14), with the impulse response $h(t)$ taking on the role of the delta function $\delta(t)$.

A.5 OUTPUT SAMPLING OF A LINEAR TIME-VARYING CHANNEL

In the case of a linear time-invariant communication channel, it is a straightforward matter to apply the sampled convolution theorem of Eq. (A.21) to the channel output in order to proceed with the use of digital signal processing in the receiver. When, however, the channel is linear, but time varying, as, for example, in wireless communications, we have to exercise care in the selection of a suitable sampling rate for the channel output, the reason being that the impulse response of the channel, $h(t;\tau)$, depends on a second time variable, namely, τ. The problem is now more complicated, in that we have a *two-dimensional temporal situation* to handle, with the two dimensions being defined by both t and τ. For frequency analysis of the channel output and, therefore, the determination of an appropriate sampling rate, we require the use of a two-dimensional Fourier transform. Such an analysis is beyond the scope of this book; the interested reader is referred to Note 2 for further pursual of the sampling rate required for linear time-varying channels. Suffice it to say, if W_I is the bandwidth (i.e., highest frequency component) of the input signal and W_C is the bandwidth of the channel time variations, then the sampling rate for the channel output must be larger than $(W_I + W_C)$. When the channel is time invariant, W_C is zero, and this result reduces to the standard form of the sampling theorem.

A.6 CORRELATION THEOREM

Thus far, we have discussed attention on the Fourier perspectives of two basic signal-processing operations: filtering (i.e., convolution) and sampling. Another signal-processing operation basic to the study of communication systems is correlation. To be specific, consider a pair of complex-valued signals $g_1(t)$ and $g_2(t)$, which may exhibit some degree of *similarity* in their time behaviors. The similarity is quantified by the integral

$$R_{12}(\tau) = \int_{-\infty}^{\infty} g_1(t)g_2^*(t - \tau)dt \qquad (A.22)$$

where the asterisk denotes complex conjugation. The function $R_{12}(\tau)$ is called the *cross-correlation function* between $g_1(t)$ and $g_2(t)$. The *time lag* τ is introduced into one of the two signals—$g_2(t)$ in the case under consideration here—in order to explore the similarity between them. To that end, τ is made variable. Intuitively, if, on the one hand, $g_1(t)$ and $g_2(t)$ are highly similar, then we expect $R_{12}(\tau)$ to peak around some value of τ. If, on the other hand, $g_1(t)$ and $g_2(t)$ are highly dissimilar, then $R_{12}(\tau)$ would be relatively flat over a broad range of values of τ.

With Fourier analysis as the subject of interest in this appendix, it is natural that we consider the Fourier transformation of $R_{12}(\tau)$. To pursue this transformation, we first use the inverse Fourier transform that defines $g(t)$ in terms of $G(f)$ and thus rewrite Eq. (A.22) in the form of a double integral (after rearranging terms):

$$R_{12}(\tau) = \int_{-\infty}^{\infty} G_1(f) \int_{-\infty}^{\infty} g_2^*(t - \tau) e^{j2\pi ft} \, dt \, df \qquad (A.23)$$

Clearly, $R_{12}(\tau)$ is unchanged by introducing the product of the exponential $e^{j2\pi f\tau}$ and its complex conjugate $e^{-j2\pi f\tau}$ into the integral in Eq. (A.3), as is shown by

$$R_{12}(\tau) = \int_{-\infty}^{\infty} G_1(f) e^{j2\pi f\tau} \left[\int_{-\infty}^{\infty} g_2(t - \tau) e^{-j2\pi f(t - \tau)} \, d(t - \tau) \right]^* df \qquad (A.24)$$

Now, the inner integral inside the square brackets in Eq. (A.24) is recognized as the Fourier transform of $g_2(t)$, which is denoted by $G_2(f)$. Accordingly, bearing in mind the complex conjugation around the square brackets, we may finally simplify Eq. (A.24) as

$$R_{12}(\tau) = \int_{-\infty}^{\infty} G_1(f) G_2^*(f) e^{j2\pi f\tau} \, df \qquad (A.25)$$

from which we immediately infer that $G_1(f) G_2^*(f)$ is the Fourier transform of $R_{12}(\tau)$. In words, the *correlation theorem* may be stated as follows:

Given a pair of Fourier-transformable signals $g_1(t)$ and $g_2(t)$ whose cross-correlation function $R_{12}(\tau)$ is defined by Eq. (A.22), the Fourier transform of $R_{12}(\tau)$ is defined by the product $G_1(f) G_2^*(f)$, where $G_1(f)$ and $G_2(f)$ are the Fourier transforms of $g_1(t)$ and $g_2(t)$, respectively.

In applying the correlation theorem, careful attention has to be paid to the order and manner in which the functions $g_1(t)$ and $g_2(t)$ appear in Eq. (A.22) and the corresponding order of subscripts in $R_{12}(\tau)$.

Moreover, there are some similarities and basic differences between the cross-correlation and convolution integrals that should be noted:

1. In the convolution integral of Eq. (A.6), the integration is with respect to the lag variable t. By contrast, the integration in the cross-correlation integral of Eq. (A.22) is with respect to the time variable τ.

2. When both integrals are transformed into the frequency domain, the result of each transformation is expressed as a product of two Fourier transforms—but with a difference. In the case of the convolution integral, the product is simply equal to the Fourier transforms of the two signals, with the result that *convolution* is commutative. In the case of the cross-correlation integral, the Fourier transform of the particular signal delayed in the correlation process is complex

conjugated. Consequently, unlike the convolution of two integrals, the cross-correlation is *not* commutative; that is,

$$R_{12}(\tau) \rightleftharpoons G_1(f)G_2^*(f) \tag{A.26}$$

which, in general, is different from

$$R_{21}(\tau) \rightleftharpoons G_2(f)G_1^*(f) \tag{A.27}$$

A.6.1 Autocorrelation Function

When $g_1(t) = g_2(t) = g(t)$, we have the *autocorrelation function* of the signal $g(t)$, defined by

$$R_g(\tau) = \int_{-\infty}^{\infty} g(t)g^*(t - \tau)dt \tag{A.28}$$

Correspondingly, Eq. (A.26) reduces to

$$R_g(\tau) \rightleftharpoons |G(f)|^2 \tag{A.29}$$

Note that the autocorrelation function $R_g(\tau)$ is an even function of the lag τ, as is shown by

$$R_g(-\tau) = R_g(\tau) \text{ for all } \tau \tag{A.30}$$

Expanding the pair of relations summarized under Eq. (A.29), we have

$$S_g(f) = \int_{-\infty}^{\infty} R_g(\tau)e^{-j2\pi f\tau}d\tau \tag{A.31}$$

and

$$R_g(\tau) = \int_{-\infty}^{\infty} S_g(f)e^{j2\pi f\tau}df \tag{A.32}$$

where we have introduced the definition

$$S_g(f) = |G(f)|^2 \text{ for all } f \tag{A.33}$$

The new function $S_g(f)$ is called the *energy density spectrum* of the signal $g(t)$. The pair of equations (A.31) and (A.32) constitute the *Wiener–Khintchine relations* for signals with finite energy.

A.7 PARSEVAL'S RELATIONSHIPS

The energy of a complex-valued signal g(t) is defined by

$$E_g = \int_{-\infty}^{\infty} |g(t)|^2 dt \tag{A.34}$$

Putting $\tau = 0$ in Eq. (A.28) and using the definition of Eq. (A.34), we readily see that

$$E_g = R_g(0) \tag{A.35}$$

which states that the value of the autocorrelation function $R_g(\tau)$ at the origin $\tau = 0$ is equal to the energy of the signal $g(t)$.

Putting $\tau = 0$ in Eq. (A.32) and using Eqs. (A.34) and (A.35), we obtain *Parseval's energy theorem*, which states that

$$\int_{-\infty}^{\infty} |g(t)|^2 dt = \int_{-\infty}^{\infty} |G(f)|^2 df \tag{A.36}$$

In words, Parseval's energy theorem asserts that the energy of a nonperiodic signal $g(t)$ is equal to the total area under the curve of the energy density spectrum $S_g(f)$.

To deal with a periodic signal $g(t)$ of fundamental period T, we use *Parseval's power theorem*, which states that

$$\frac{1}{T}\int_0^T |g(t)|^2 dt = \sum_{k=-\infty}^{\infty} |G_k|^2 \tag{A.37}$$

where G_k are the *complex Fourier coefficients*, in terms of which the periodic signal

$$g(t) = \sum_{k=-\infty}^{\infty} G_k e^{j2\pi t/T} \tag{A.38}$$

is defined.

To deal with a periodic signal, we may use *Parseval's power theorem* to calculate the average power of the signal. To formulate this theorem, recall that a complex-valued periodic signal $g(t)$ with fundamental period T may be expanded into the *Fourier series*

$$g(t) = \sum_{k=-\infty}^{\infty} G_k e^{j2\pi k f_0 t} \tag{A.39}$$

where

$$G_k = \frac{1}{T}\int_0^T g(t) e^{-j2\pi k f_0 t} \quad k = 0, \pm 1, \pm 2, \ldots \tag{A.40}$$

are the *complex Fourier coefficients*. The fundamental frequency of the signal is itself defined by

$$f_0 = \frac{1}{T} \tag{A.41}$$

By definition, for the average power of the periodic signal $g(t)$, we have

$$P = \frac{1}{T}\int_0^T |g(t)|^2 dt \tag{A.42}$$

Accordingly, Parseval's power theorem states that we may also evaluate P by using the formula

$$P = \sum_{k=-\infty}^{\infty} |G_k|^2 \qquad \text{(A.43)}$$

where the G_k are themselves defined by Eq. (A.40).

Notes and References

[1] For an authoritative treatment of the many facets of the Fourier transform and its applications, see Bracewell (1986).

[2] For a careful discussion of the sampling rate required for linear time-varying systems, see Kailath (1959) and Médard (1995).

APPENDIX B

Bessel Functions

B.1 BESSEL FUNCTIONS OF THE FIRST KIND

Bessel functions of the first kind of integer order v are defined as the solution of the integral equation

$$J_v(z) = \frac{1}{\pi} \int_0^\pi \cos(z \sin \theta - v\theta)d\theta$$

$$= \frac{j^{-v}}{\pi} \int_0^\pi e^{jz \cos \theta} \cos(v\theta)d\theta \qquad \text{(B.1)}$$

where j is the square root of -1. The special case $v = 0$ reduces to

$$J_0(z) = \frac{1}{\pi} \int_0^\pi \cos(z \sin \theta)d\theta \qquad \text{(B.2)}$$

For a real argument z, the Bessel functions are real valued, continuously differentiable, and bounded in magnitude by unity. The even-numbered Bessel functions are symmetric and the odd-numbered Bessel functions are antisymmetric.

The Bessel function $J_v(z)$ may also be expressed as the infinite series

$$J_v(z) = \left(\frac{1}{2}z\right)^v \sum_{k=0}^\infty \frac{\left(-\frac{1}{4}z^2\right)^k}{k!\Gamma(v+k+1)} \qquad \text{(B.3)}$$

where $\Gamma(k)$ is the *gamma function*; for integer values, $\Gamma(k+1) = k!$.

We plot $J_0(z)$ and $J_1(z)$ for real-valued z in Fig. B.1. The values of these functions for a subset of z are given in Table B.1.

Problem B.1 Using the first line of Eq. (B.1), derive the second line of the equation. ∎

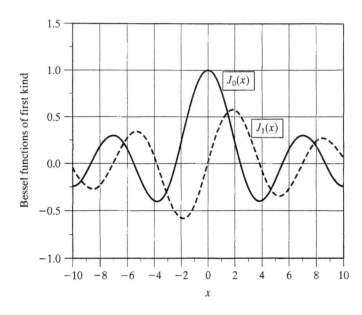

FIGURE B.1 Plots of Bessel functions of the first kind, $J_0(x)$ and $J_1(x)$.

TABLE B.1 Values of Bessel Functions and Modified Bessel Functions of the First Kind.

x	$J_0(x)$	$J_1(x)$	$I_0(x)$	$I_1(x)$
0.00	1.0000	0.0000	1.0000	0.0000
0.20	0.9900	0.0995	1.0100	0.1005
0.40	0.9604	0.1960	1.0404	0.2040
0.60	0.9120	0.2867	1.0920	0.3137
0.80	0.8463	0.3688	1.1665	0.4329
1.00	0.7652	0.4401	1.2661	0.5652
1.20	0.6711	0.4983	1.3937	0.7147
1.40	0.5669	0.5419	1.5534	0.8861
1.60	0.4554	0.5699	1.7500	1.0848
1.80	0.3400	0.5815	1.9896	1.3172
2.00	0.2239	0.5767	1.1796	1.5906
2.20	0.1104	0.5560	2.6291	1.9141
2.40	0.0025	0.5202	3.0493	2.2981
2.60	−0.0968	0.4708	3.5533	2.7554
2.80	−0.1850	0.4097	4.1573	3.3011
3.00	−0.2601	0.3391	4.8808	3.9534
3.20	−0.3202	0.2613	5.7472	4.7343
3.40	−0.3643	0.1792	6.7848	5.6701
3.60	−0.3918	0.0955	8.0277	6.7927
3.80	−0.4026	0.0128	9.5169	8.1404
4.00	−0.3971	−0.0660	11.3019	9.7595

B.2   MODIFIED BESSEL FUNCTIONS OF THE FIRST KIND

Modified Bessel functions of the first kind of integer order v are defined as the solution of the integral equation

$$I_v(z) = \frac{1}{\pi} \int_0^\pi e^{z\cos\theta} \cos(v\theta) d\theta \tag{B.4}$$

For the special case $v = 0$, Eq. (B.4) reduces to

$$I_0(z) = \frac{1}{\pi} \int_0^\pi e^{z\cos\theta} d\theta \tag{B.5}$$

For a real argument z, the modified Bessel functions are real valued, continuously differentiable, and grow exponentially as $|z|$ increases. The even-numbered modified Bessel functions are symmetric and the odd-numbered ones are antisymmetric.

The modified Bessel function may also be expressed as the infinite series

$$I_v(z) = \left(\frac{1}{2}z\right)^v \sum_{k=0}^\infty \frac{\left(\frac{1}{4}z^2\right)^k}{k!\Gamma(v+k+1)} \tag{B.6}$$

We plot $I_0(z)$ and $I_1(z)$ for real-valued z in Fig. B.2. The values of these functions for a subset of z are given in Table B.1.

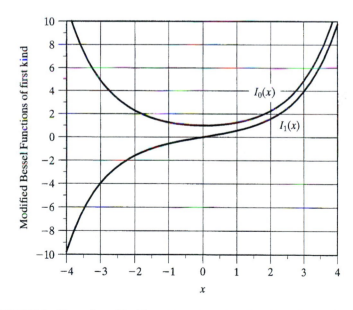

FIGURE B.2   Plots of modified Bessel functions of the first kind, $I_0(z)$ and $I_1(z)$.

APPENDIX C

Random Variables and Random Processes

C.1 SETS, EVENTS, AND PROBABILITY

Probability theory is centered on fundamental principles relating sets. In this appendix, we will consider an abstract space Ω that has elements ω. The space Ω may consist of a finite number of elements, may be countably infinite, such as the set of integers, or may be uncountable, such as the set of real numbers. We let Ξ represent all the possible subsets of Ω, including the empty set, $\varnothing$, and the complete set Ω.

Probability is a measure on any set Q in Ξ. Conceptually, if Q represents a set of elements, or *an event*, then $\text{Prob}(Q)$ is the probability of that event. Empirically, if we make N observations of this space and determine how many times n out of N trials that the observations belong to Q, then the empirical definition of the probability of the event Q is

$$\text{Prob}(Q) = \lim_{N \to \infty} (n/N) \tag{C.1}$$

This is intuitively what we think of as probability: what fraction of the time a certain event occurs.

A probability measure must satisfy three properties:

$$\begin{aligned}
&\text{Prob}(\Omega) = 1; \\
&\text{Prob}(\varnothing) = 0; \\
&\text{Prob}(A \cup B) \leq \text{Prob}(A) + \text{Prob}(B) \text{ for any } A, B \text{ in } \Xi.
\end{aligned} \tag{C.2}$$

In calculations, we are often interested in the *conditional probability* that an event A occurs, given that an event B has occurred. This is defined as

$$\text{Prob}[A|B] = \frac{\text{Prob}[A \cap B]}{\text{Prob}[B]} \tag{C.3}$$

The conditional probability is a probability measure in its own right and satisfies all of the properties of Eq. (C.2).

Bayes' theorem allows us to convert between conditioning on one event to conditioning on a different event and is given by

$$\text{Prob}(A_i|B) = \frac{\text{Prob}(B|A_i)\text{Prob}(A_i)}{\sum_{j=1}^{N}\text{Prob}(B|A_j)\text{Prob}(A_j)} \tag{C.4}$$

Bayes' theorem is often used in inferential analysis, as the expressions for conditional probability based on some events are often much simpler than those based on other events.

C.2   RANDOM VARIABLES

A *random variable* is a mapping from the abstract space Ω to the real numbers, represented as $X{:}\Omega \to \Re$, where $\Re$ is the set of real values. Conceptually, we usually consider X as the physical realization of some unknowable process. For example, X could be the voltage measured across a resistor due to thermal noise. In that case, Ω could be the state of all the electrons in the resistor.

A *discrete random variable* can take on only a discrete set of values. Sometimes these are denoted as $\{x_i\}$, where i indexes the possible values of X. For example, the number of paths in a multipath signal is a discrete random variable; it may take only the values 1,2,3,.... A *continuous random variable* can take on a continuum of values. Often, this continuum is the set of real values or the set of nonnegative reals. For example, thermal noise voltage observed at a specific instant of time is a continuous random variable.

To characterize the probabilistic behavior of random variables, we simply extend the set concepts used in the previous section. In the thermal-noise example, we may be interested in determining the probability that $X \le x$ for some value x. We would write this as $\text{Prob}(X \le x)$, but its mathematical meaning is

$$\text{Prob}(X \le x) = \text{Prob}(\{\omega \in \Omega{:}X(\omega) \le x\}) \tag{C.5}$$

That is, it is a measure on the set of those ω's such that the random variable maps the $X(\omega)$ to a value less than x. Probability is a measure on the underlying abstract set Ω. The physical realization is usually more easily understood than the abstract, but understanding the underlying concepts is often useful for resolving some probability issues.

C.3   PROBABILITY DISTRIBUTIONS AND DENSITIES

The probability that a random variable X is less than a given x is written as

$$F_X(x) = \text{Prob}(X \le x) \tag{C.6}$$

which is called the *cumulative distribution function* of the random variable X. This function is right continuous and increases monotonically, with $F(-\infty) = 0$ and $F(\infty) = 1$.

A discrete random variable will have a discrete distribution function that consists of steps at the finite or countable number of points where $\text{Prob}(X = x_i) > 0$. A continuous random variable will have a continuous distribution function. If the distribution function is a continuously differentiable function of x, then we define the *probability density function* as

$$f_X(x) = \frac{dF_X(x)}{dx} \tag{C.7}$$

Probability density functions play an important role in defining the conditional probabilities of continuous random variables. Consider the two joint events $X \leq x$ and $y \leq Y \leq y + \delta y$. We may use Bayes' rule to express the conditional probability of the first event, given the second, as

$$
\begin{aligned}
F_{X|y \leq Y \leq y + \delta y}(x) &= \frac{\text{Prob}(X \leq x, y \leq Y \leq y + \delta y)}{\text{Prob}(y \leq Y \leq y + \delta y)} \\
&= \frac{F_{XY}(x, y + \delta y) - F_{XY}(x, y)}{F_Y(y + \delta y) - F_Y(y)} \\
&= \frac{\displaystyle\int_{-\infty}^{x} \int_{y}^{y + \delta y} f_{X, Y}(u, v)\, du\, dv}{\displaystyle\int_{y}^{y + \delta y} f_Y(v)\, dv}
\end{aligned}
\tag{C.8}
$$

Differentiating Eq. (C.8) with respect to x by using Leibniz's rule, we may write

$$
\begin{aligned}
f_{X|y \leq Y \leq y + \delta y}(x) &= \frac{\displaystyle\int_{y}^{y + \delta y} f_{X, Y}(u, v)\, dv}{\displaystyle\int_{y}^{y + \delta y} f_Y(v)\, dv} \\
&\approx \frac{f_{X, Y}(x, y)\, \delta y}{f_Y(y)\, \delta y}
\end{aligned}
$$

Finally, in the limit, as δy approaches zero, assuming that $f_Y(y) \neq 0$, we have

$$f_{X|Y = y}(x) = \frac{f_{X, Y}(x, y)}{f_Y(y)} \tag{C.9}$$

It is important to note that Eq. (C.9) describes a probability density function of x for a fixed y.

C.4   EXPECTATION OF RANDOM VARIABLES

The *expected value,* or *mean value,* of a random variable X is written as $\mathbf{E}[X]$, where $\mathbf{E}$ is the *statistical expectation operator.* For a discrete random variable, the expected value of X is given by

$$\mathbf{E}[X] = \sum_{i=1}^{N} x_i \, \text{Prob}(X = x_i) \qquad \text{(C.10)}$$

For a continuous random variable that has a probability density function, the expected value of X is given by

$$\mathbf{E}[X] = \int_{-\infty}^{\infty} x f_X(x) dx \qquad \text{(C.11)}$$

If X_1 and X_2 are any two random variables, then

$$\mathbf{E}[X_1 + X_2] = \mathbf{E}[X_1] + \mathbf{E}[X_2] \qquad \text{(C.12)}$$

and

$$\mathbf{E}[\alpha X_1] = \alpha \mathbf{E}[X_1] \qquad \text{(C.13)}$$

where α is a constant. That is, expectation is a *linear* operator. In general, if $g(X)$ is any well-defined function of X, then the expected value of $g(X)$ is given by

$$\mathbf{E}[g(X)] = \sum_{i=1}^{N} g(x_i) P(X = x_i) \quad \text{or} \quad \mathbf{E}[g(X)] = \int_{-\infty}^{\infty} g(x_i) f_X(x) dx \qquad \text{(C.14)}$$

depending upon whether the random variable is discrete or continuous, respectively. Other common statistical parameters of interest are the *second moment* or *mean-square value*

$$\mathbf{E}[X^2] = \int_{-\infty}^{\infty} x^2 f_X(x) dx \qquad \text{(C.15)}$$

and the *variance*

$$\text{Var}(X) = \mathbf{E}[(X - \mathbf{E}[X])^2] = \int_{-\infty}^{\infty} (x - \mathbf{E}[X])^2 f_X(x) dx \qquad \text{(C.16)}$$

An analogous result holds for the discrete case.

C.5   COMMON PROBABILITY DISTRIBUTIONS AND THEIR PROPERTIES

Binomial distribution. Consider a discrete random variable X that can take the values 0 and 1 with probabilities $(1 - p)$ and p, respectively. Suppose N independent observations of this random variable are made and labeled X_i for $1 \leq i \leq N$. Define the new random variable

$$Y = \sum_{i=1}^{N} X_i \qquad \text{(C.17)}$$

Then Y is said to have a *binomial distribution* with parameter p; that is,

$$F_Y(y) = \sum_{j=0}^{y} \binom{N}{j} p^{N-y} (1-p)^y \qquad \text{for } 0 \le y \le N \tag{C.18}$$

The expected value and variance of Y are Np and $Np(1-p)$, respectively.

Gaussian distribution. A common continuous random variable is the Gaussian random variable. The density function of a Gaussian random variable is given by

$$f_X(x) = \frac{1}{\sqrt{2\pi}\sigma} \exp\left\{-\frac{(x-\mu)^2}{2\sigma^2}\right\} \tag{C.19}$$

where the mean of the Gaussian random variable is μ and its variance is σ^2. The distribution function of a Gaussian random variable does not have a closed-form solution, but it is usually expressed in terms of the error function as

$$F_X(x) = \int_{-\infty}^{x} f_X(s)\,ds$$

$$= \begin{cases} \dfrac{1}{2}\left(1 + \text{erf}\left(\dfrac{x-\mu}{\sqrt{2}\sigma}\right)\right) & x \ge 0 \\ 1 - F_X(-x) & x < 0 \end{cases} \tag{C.20}$$

where the *error function* is given by (see Appendix E)

$$\text{erf}(x) = \frac{2}{\sqrt{\pi}} \int_{0}^{x} e^{-z^2}\,dz \tag{C.21}$$

A linear transformation of a Gaussian random variable is also a Gaussian random variable. That is, if $X_1, X_2, ..., X_N$ are Gaussian random variables, then the composite random variable

$$Y = \sum_{i=1}^{N} b_i X_i \tag{C.22}$$

is also a Gaussian random variable. The mean of Y is given by

$$E[Y] = \sum_{i=1}^{N} b_i E[X_i] \tag{C.23}$$

If the $\{X_i\}$ are independent Gaussian random variables, then the variance of Y is given by

$$\text{Var}(Y) = \sum_{i=1}^{N} b_i^2 \text{Var}(X_i) \tag{C.24}$$

Rayleigh distribution. If X_1 and X_2 are zero-mean Gaussian random variables with a common variance σ^2, then the random variable

$$R = \sqrt{X_1^2 + X_2^2} \qquad\qquad\qquad \text{(C.25)}$$

has a *Raleigh distribution* given by (see Section C.6)

$$\text{Prob}(R < r) = 1 - e^{-r^2/2\sigma^2} \qquad\qquad r \geq 0 \qquad\qquad \text{(C.26)}$$

The corresponding probability density function is

$$f_R(r) = \frac{r}{\sigma^2} e^{-r^2/2\sigma^2} \qquad\qquad r \geq 0 \qquad\qquad \text{(C.27)}$$

The first and second moments of Y are given by

$$\mathbf{E}[R] = \sqrt{\frac{\pi}{2}}\sigma \quad \text{and} \quad \mathbf{E}[R^2] = 2\sigma^2 \qquad\qquad \text{(C.28)}$$

and the variance of Y is given by $(2 - \pi/2)\sigma^2$.

Rician distribution. If X_1 and X_2 are Gaussian random variables with means μ_1 and μ_2, respectively, and a common variance σ, then the new random variable

$$R = \sqrt{X_1^2 + X_2^2} \qquad\qquad\qquad \text{(C.29)}$$

has a *Rician distribution.* The probability density function of R is given by

$$f_R(r) = \frac{r}{\sigma^2} e^{-(r^2+s^2)/2\sigma^2} I_0\left(\frac{rs}{\sigma^2}\right) \qquad\qquad r \geq 0 \qquad\qquad \text{(C.30)}$$

where $I_0(\cdot)$ is the *modified Bessel function of order zero* (see Appendix B) and $s = \sqrt{\mu_1^2 + \mu_2^2}$. There is no known closed-form solution for the distribution function of a Rician random variable.

Chi-square distribution. If $\{X_i\}$, $i = 1, ..., N$ are zero-mean Gaussian random variables with a common variance σ^2, then the random variable

$$Y = \sum_{i=1}^{N} X_i^2 \qquad\qquad\qquad \text{(C.31)}$$

is said to have a *Chi-square distribution with N degrees of freedom.* The first two moments of Y are

$$\mathbf{E}[Y] = N\sigma^2 \qquad\qquad\qquad \text{(C.32)}$$

and

$$\mathbf{E}[Y^2] = 2N\sigma^4 + N^2\sigma^4 \qquad\qquad\qquad \text{(C.33)}$$

When $m = N/2$ is an integer, the probability density function of Y is given by

$$f_Y(y) = \frac{1}{(2\sigma^2)^m (m-1)!} y^{m-1}$$

and the cumulative distribution function of Y is given by

$$F_Y(y) = 1 - e^{-y/2\sigma^2} \sum_{k=0}^{m-1} \frac{1}{k!} \left(\frac{y}{2\sigma^2}\right)^k$$

This is one of the most common distribution functions in communications systems applications. For the case of odd N, there is no closed-form solution for $F_Y(y)$.

C.6 TRANSFORMATIONS OF RANDOM VARIABLES

Let X_1 and X_2 be continuous random variables with a joint probability density function $f_{X_1, X_2}(x_1, x_2)$, and consider the transformation defined by

$$y_1 = h_1(x_1, x_2), \text{ and } y_2 = h_2(x_1, x_2), \tag{C.34}$$

which are assumed to be one-to-one and continuously differentiable. The *Jacobian of this transformation* is defined by the matrix determinant

$$\mathbf{J}\left(\frac{y_1, y_2}{x_1, x_2}\right) = \begin{vmatrix} \dfrac{\partial y_1}{\partial x_1} & \dfrac{\partial y_1}{\partial x_2} \\[2ex] \dfrac{\partial y_2}{\partial x_1} & \dfrac{\partial y_2}{\partial x_2} \end{vmatrix} \neq 0 \tag{C.35}$$

The joint probability density function of Y_1 and Y_2 is given by

$$f_{Y_1, Y_2}(y_1, y_2) = f_{X_1, X_2}(x_1, x_2) \left| \mathbf{J}\left(\frac{y_1, y_2}{x_1, x_2}\right) \right| \tag{C.36}$$

If the transformations are not one-to-one, then other approaches must be taken. For example, consider the transformation $Y = \sqrt{X_1^2 + X_2^2}$. The cumulative distribution function of Y is given by

$$F_Y(y) = \iint_A f_{X_1, X_2}(x_1, x_2) dx_1 dx_2 \tag{C.37}$$

where A is the set of all (x_1, x_2) such that $\sqrt{x_1^2 + x_2^2} \leq y$. If X_1 and X_2 are independent, zero-mean Gaussian random variables with a variance of unity, then

$$F_Y(y) = \iint_A \frac{1}{\sqrt{2\pi}} e^{-(x_1^2 + x_2^2)/2} dx_1 dx_2 \tag{C.38}$$

If we make the transformations $x_1 = r\cos\theta$ and $x_2 = r\sin\theta$, then Eq. (C.38) becomes

$$F_Y(y) = \frac{1}{2\pi} \int_0^y \int_0^{2\pi} e^{-r^2/2} r\,dr\,d\theta$$

$$= 1 - e^{-y^2/2}$$

(C.39)

This is the Rayleigh distribution described in Section C.5.

C.7   CENTRAL-LIMIT THEOREM

Consider a sequence $\{X_n\}$ of independent and identically distributed (i.i.d.) random variables with means $\mathbf{E}[X_i] = m$ and variances $\mathbf{E}[(X_i - m)^2] = \sigma^2$. Let Y_n be a new sequence of random variables defined by the partial sums

$$Y_n = \sum_{i=1}^{n} X_i$$

(C.40)

Let μ_n be the mean of Y_n and S_n be the variance of Y_n. Define the normalized random variable

$$Z_n = \frac{Y_n - \mu_n}{S_n} = \sum_{i=1}^{n} \frac{X_i - m}{\sqrt{n}\sigma}$$

(C.41)

Then *the random variable Z_n has a distribution that is asymptotically unit normal.* That is, as n becomes large, the distribution of Z_n approaches that of a zero-mean Gaussian random variable with unit variance. This result is referred to as the *central-limit theorem*. The theorem also holds if the variables X_i are not identically distributed, but there are some restrictions.

C.8   RANDOM PROCESSES

A random process is a mapping $X:[\Omega, T] \to \Re$, where T represents a time interval such that $X(.,t)$ is a random variable for each fixed time t. To distinguish between a random variable X and a random process X, we usually write the latter as $X(t)$. If X is a discrete random variable for all t, then we say that $X(t)$ is a *discrete random process*. If X is a continuous random variable for all t, then we say that $X(t)$ is a *continuous random process*.

For each fixed value of t, we speak of the distribution function

$$F_{X(t)}(x) = \text{Prob}(X(t) < x)$$

(C.42)

We also speak of the joint distribution function

$$F_{X(t_1)X(t_2)}(x_1, x_2) = \text{Prob}(X(t_1) < x_1, X(t_2) < x_2)$$

(C.43)

For a fixed ω in Ω, the time function $X(\omega, t)$ is known as a *sample function* or *realization* of the random process.

C.9 PROPERTIES OF RANDOM PROCESSES

Let $X(t)$ be a complex random process, and define the *autocorrelation function* of that process as

$$R_X(t, s) = \mathbf{E}[X(t)X^*(s)] \tag{C.44}$$

where the asterisk denotes complex conjugation. That is, the autocorrelation function is the expectation of the product of two random variables that are parameterized by t and s. Recognizing that $R_X(0) = \mathbf{E}[|X(t)|^2]$, we see that the autocorrelation is a generalization of the second moment of a random variable. The autocorrelation is a deterministic function.

   A process whose joint distribution is invariant with time translation, that is,

$$\begin{aligned}
&\text{Prob}(X(t_1) < x_1, X(T_2) < x_2, \ldots, X(t_n) < x(n)) = \\
&\text{Prob}(X(t_1 + h) < x_1, X(t_2 + h) < x_2, \ldots, X((t_n + h) < x_n))
\end{aligned} \tag{C.45}$$

is said to be *stationary to order n* if Eq. (C.46) holds for all h and for a particular n. Many of the random processes dealt with in wireless communications are assumed to be stationary. A process is said to be *wide-sense stationary* if

$$E[X(t)] = \text{constant for all } t \text{ and } R_X(t, s) = R_X(t - s) \tag{C.46}$$

Random processes whose joint distribution functions are multivariate Gaussian are referred to as *Gaussian random processes*. If a Gaussian random process is wide-sense stationary, then it is also stationary.

C.10 SPECTRA OF RANDOM PROCESSES

In Appendix A, we defined the spectrum of a finite-energy signal $x(t)$ as the Fourier transform of that signal. However, in considering a random process $X(t)$, we have an ensemble of sample functions. To get around this difficulty, we note that the autocorrelation function of a stationary random process, namely, $R_X(\tau)$, satisfies the conditions of Fourier transformability. Accordingly, we may define the *power spectrum* or *power spectral density* of a random process $X(t)$ as the Fourier transform of its autocorrelation function $R_X(\tau)$. Denoting this new parameter by $S_X(f)$, we may thus formally write

$$S_X(f) = \int_{-\infty}^{\infty} R_X(\tau)e^{-j2\pi f\tau}d\tau \text{ and } R_X(z) = \int_{-\infty}^{\infty} S_X(f)e^{j2\pi f\tau}df \tag{C.47}$$

Note that $S_X(f)$ is measured in watt/Hz. Stated another way, the total area under the curve of $S_X(f)$, plotted as a function of frequency f, defines the average power of the process.

The autocorrelation function $R_X(\tau)$ and power spectral density $S_X(f)$ form a Fourier-transform pair, which means that the autocorrelation function $R_X(\tau)$ is the inverse Fourier transform of the power spectral density $S_X(f)$. The Fourier transform pair, linking $S_X(f)$ to $R_X(\tau)$ and vice versa, is called the *Wiener–Khintchine relation* for random processes.

An idealized example of a random process is a special form of noise commonly referred to as *white noise* $W(t)$. White noise has the property that it is uncorrelated for all nonzero time offsets. Consequently, the autocorrelation function of such a process is defined by a delta function, or

$$R_W(\tau) = \frac{N_0}{2}\delta(\tau) \tag{C.48}$$

where $N_0/2$ is the two-sided noise density in watt/Hz. The noise is referred to as "white" because the corresponding power spectrum is flat; that is,

$$S_W(f) = \frac{N_0}{2} \quad \text{for all } f \tag{C.49}$$

In other words, white noise contains all frequency components at equal strength, analogous to white light in the visible part of the spectrum. This relationship does not make any assumptions about the distribution of the random variable $n(t)$ at time t; it could be Gaussian or otherwise.

Another example of a random process is a *random binary wave*, defined by

$$x(t) = b_n \quad t_0 + nT < t < t_0 + (n+1)T \tag{C.50}$$

where t_0 is a random starting time between $[0,T]$ and $\{b_n\}$ is a sequence of independent, zero-mean random variables with values ± 1. The autocorrelation function of the process described in Eq. (C.51) is

$$
\begin{aligned}
R_X(t, t+\tau) &= \mathbf{E}[x(t)x(t+\tau)] \\[4pt]
&= \mathbf{E}_{t_0}\mathbf{E}_b[x(t)x(t+\tau)] \\[4pt]
&= \mathbf{E}_{t_0}\begin{cases} 1 & t_0 + nT < t + \tau < t_0 + (n+1)T \\ 0 & \text{otherwise} \end{cases}
\end{aligned}
\tag{C.51}
$$

where the expectation $\mathbf{E}$ has been split over the two random independent variables t_0 and $\{b_n\}$ and we have used the fact that $\mathbf{E}[b_n b_m] = 0$ if $n \neq m$. If we evaluate the expectation of Eq. (C.51) over t_0, we obtain

$$R_X(t, t + \tau) = \begin{cases} T - \tau & T > \tau > 0 \\ T + \tau & -T < \tau \leq 0 \\ 0 & \text{otherwise} \end{cases} \tag{C.52}$$

$$= \begin{cases} T - |\tau| & |\tau| < T \\ 0 & \text{otherwise} \end{cases}$$

Thus, $R_X(\tau)$ is stationary with a triangular autocorrelation function. The spectrum of the binary random wave is the Fourier transform of this autocorrelation function:

$$S_X(f) = \int_{-\infty}^{\infty} R_X(\tau) e^{-j2\pi f \tau} d\tau$$

$$= T^2 \frac{\sin^2(\pi f T)}{(\pi f T)^2} \tag{C.53}$$

$$= T^2 \text{sinc}^2(fT)$$

C.11 LINEAR FILTERING OF RANDOM PROCESSES

In communications, we filter signals for various reasons. If the signal is a random process, how do we characterize the output? Let $y(t)$ be the output resulting from applying a linear time-invariant causal filter $h(t)$ to a realization of an input random process, namely, $x(t)$, as represented by the convolution integral

$$y(t) = \int_{-\infty}^{\infty} h(\tau) x(t - \tau) d\tau = \int_{-\infty}^{\infty} x(\tau) h(t - \tau) d\tau \tag{C.54}$$

This means that the filter is applied to the particular realization $x(t) = X(\omega, t)$ of the random process, and that realization is referred as a sample path integral. A sufficient condition for $Y(t)$, the random variable composed of $\{y(\omega, t)\}$, to be a well-defined random variable for all t is that

$$\int_{-\infty}^{\infty} |h(\tau)| \mathbf{E}[|X(t - \tau)|] d\tau < \infty \tag{C.55}$$

If $Y(t)$ is well defined, then we may determine the expectation of $Y(t)$ using

$$\mathbf{E}[Y(t)] = \mathbf{E}\left[\int_{-\infty}^{\infty} h(t - \tau) X(\tau) d\tau\right] = \int_{-\infty}^{\infty} h(t - \tau) \mathbf{E}[X(\tau)] d\tau \tag{C.56}$$

and similarly for other moments of $Y(\tau)$. The interchange of the order of integration and expectation is allowed because both of these operations are linear.

If $X(t)$ is a stationary process with autocorrelation $R_X(\tau)$ and corresponding spectral density $S_X(f)$, then the spectral density of the output $Y(t)$ is given by

$$S_Y(f) = |H(f)|^2 S_X(f) \tag{C.57}$$

That is, for stationary random processes, the output power spectral density of a linear continuous-time filter is equivalent to the product of two quantities: the squared magnitude response of the filter and the input power spectral density. In addition, if the input is wide-sense stationary and the linear system is time invariant, then the output will be stationary as well. The spectral relationship of Eq. (C.57) for a stationary random process $X(t)$ may be viewed as the counterpart of Eq. (A.7) for a signal $x(t)$ with finite energy.

Analogous to the result for linear transformations of random variables, we have the result that if a linear filter has an input which is a Gaussian random process, then the output will also be a Gaussian random process.

C.12 COMPLEX RANDOM VARIABLES AND PROCESSES

In certain situations, we have to deal with the statistical characterization of complex random variables and complex random processes. (A case in point is that of the complex baseband representation of a narrowband process considered in the next section.) When we refer to a *complex random variable* $Z = X + jY$, we mean that X and Y are (real) random variables and they are described by their joint distribution function.

Similarly, if $Z(t) = X(t) + jY(t)$ is a *complex random process*, then $X(t)$ and $Y(t)$ are random processes that are characterized by their joint distributions at each time t. For example, the autocorrelation of a stationary $Z(t)$ is given by

$$
\begin{aligned}
R_Z(\tau) &= \mathbf{E}[Z(t)Z^*(t-\tau)] \\
&= \mathbf{E}[(X(t)+jY(t))(X(t-\tau)-jY(t-\tau))] \\
&= \mathbf{E}[X(t)X(t-\tau)]+\mathbf{E}[Y(t)Y(t-\tau)]+j(\mathbf{E}[Y(t)X(t-\tau)]-\mathbf{E}[X(t)Y(t-\tau)]) \\
&= R_X(\tau)+R_Y(\tau)+j(R_{YX}(\tau)-R_{XY}(\tau))
\end{aligned} \tag{C.58}
$$

where

$$
R_{XY}(\tau) = \mathbf{E}[X(t)Y(t-\tau)] \tag{C.59}
$$

C.13 COMPLEX REPRESENTATION OF NARROWBAND RANDOM PROCESSES

Let $X(t)$ be a narrowband random process centered on some frequency f_c. In a manner similar to that described in Chapter 3, we may introduce a complex baseband process $\tilde{X}(t)$ by writing

$$
X(t) = \mathrm{Re}[\tilde{X}(t)\exp(j2\pi f_c t)] \tag{C.60}
$$

where $\mathrm{Re}[\cdot]$ denotes the real part of the quantity enclosed inside the square brackets. The *complex baseband process* $\tilde{X}(t)$ is itself defined by

$$
\tilde{X}(t) = X_I(t)+jX_Q(t) \tag{C.61}
$$

where $X_I(t)$ is the *in-phase component* and $X_Q(t)$ is the *quadrature component*. Equivalently, we may express the original process $\tilde{X}(t)$ in terms of these two components as follows:

$$X(t) = X_I(t)\cos(2\pi f_c t) - X_Q(t)\sin(2\pi f_c t) \tag{C.62}$$

Correspondingly, for sample functions of $X(t,\omega)$, $X_I(t,\omega)$, and $X_Q(t,\omega)$, we may write

$$X(t, \omega) = X_I(t, \omega)\cos(2\pi f_c t) - X_Q(t, \omega)\sin(2\pi f_c t) \tag{C.63}$$

C.14 STATIONARY AND ERGODICITY

A random process is said to be *ergodic* if time averages of a sample function are equal to the corresponding ensemble average (or expectation) at a particular point in time. Mathematically, for a random process $X(t,\omega)$, this relationship can be expressed as

$$E[X(t_0, \omega)] = \lim_{T \to \infty} \frac{1}{T} \int_{-T/2}^{T/2} X(t, \omega)dt \tag{C.64}$$

where the left-hand side is the *ensemble average* (i.e., the expectation over all realizations ω at a particular point in time) and the right-hand side is the *time average* of the random process for a particular realization ω_0. In many physical applications, it is assumed that stationary processes are ergodic and that time averages and expectations can be used interchangeably.

NOTES AND REFERENCES

[1] For a detailed description of random variables and processes, see Leon-Garcia (1994).

A P P E N D I X   D

Matched Filters

D.1 MATCHED-FILTER RECEIVER

Consider a known signal $s(t)$ corrupted by additive white Gaussian noise $w(t)$, resulting in the received signal

$$x(t) = s(t) + w(t) \qquad 0 \leq t \leq T \tag{D.1}$$

What is the optimum receiver for *detecting* the known signal $s(t)$ in the received signal $x(t)$? To answer this fundamental question, we first note the following two important points:

1. The power spectral density of white noise, with sample function $w(t)$, is defined by

$$S_w(f) = \frac{N_0}{2} \quad \text{for all } f \text{ in the entire interval} \quad -\infty < f < \infty \tag{D.2}$$

The power spectral density of white noise is illustrated in Fig. D.1(a). For a stationary random process, the autocorrelation function is the inverse Fourier transform of the power spectral density. (See Appendix C.) It follows, therefore, that the autocorrelation function of white noise consists of a Dirac delta function $\delta(\tau)$, weighted by $N_0/2$, as shown in Fig. D.1(b). That is,

$$R_w(\tau) = \mathbf{E}[w(\tau)w(t-\tau)]$$
$$= \frac{N_0}{2}\delta(\tau) \tag{D.3}$$

where $\mathbf{E}$ is the statistical expectation operator. Accordingly, any two different samples of white noise are uncorrelated, no matter how closely together in time they are taken. If the white noise $w(t)$ is also Gaussian, then the two samples are statistically independent. In a sense, white Gaussian noise represents the *ultimate in randomness*.

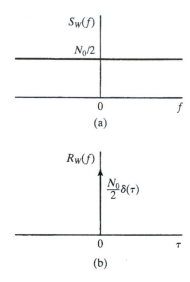

FIGURE D.1   (a) Power spectrum of the additive white noise $W(t)$.
(b) Autocorrelation function of $W(t)$.

2. Since the signal $s(t)$ is known and therefore deterministic, it follows that $s(t)$ and $w(t)$ are as uncorrelated (i.e., dissimilar) as they could ever be.

In light of point 2, we may intuitively state that, for the problem described herein, the optimum receiver consists of a *correlator* with two inputs, one being the noisy received signal $x(t)$ and the other being a locally generated replica of the known signal $s(t)$, as shown in Fig. D.2. For obvious reasons, this optimum receiver is known as the *correlation receiver*.

Another way of constructing the optimum receiver is to use a *matched filter*, defined as a linear filter whose impulse response $h(t)$ is a time-reversed, delayed version of the known signal $s(t)$; that is,

$$h(t) = \begin{cases} s(T-t) & 0 \leq t \leq T \\ 0 & \text{otherwise} \end{cases} \qquad (D.4)$$

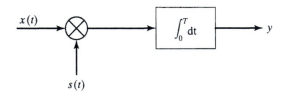

FIGURE D.2   Correlation receiver.

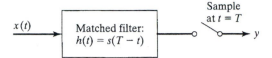

FIGURE D.3   Matched-filter receiver.

Figure D.3 shows a *matched filter receiver*, which consists of a matched filter followed by a sampler that is activated at the end of the signaling interval $t = T$. The important point to note here is that the correlation receiver of Fig. D.2 and the matched filter receiver of Fig. D.3 are equivalent insofar as their overall output samples are concerned. Specifically, for the same input signal and at the end of a signaling interval, the resulting output samples produced by these two receivers are identical.

D.2 PROBABILITY OF DETECTION

To detect a signal with a correlation receiver or a matched-filter receiver, the output sample is compared against a threshold and then a decision is made by the receiver, depending on whether the threshold is exceeded or not. In so doing, the receiver makes a decision in favor of one of two hypotheses:

Hypothesis H_1:   The known signal $s(t)$ is present in the received signal $x(t)$, a decision that is made when the threshold is exceeded.

Hypothesis H_0:   The received signal $x(t)$ consists solely of noise $w(t)$, a decision that is made when the threshold is not exceeded.

Clearly, the receiver is subject to *errors* due to the random behavior of the additive noise $w(t)$ in the received signal $x(t)$.

To calculate the *average probability of error* incurred by the receiver, we proceed by using Eq. (D.1) as the input signal applied to the correlation receiver of Fig. D.2. The resulting output sample is

$$\begin{aligned}
y &= \int_0^T x(t)s(t)dt \\
&= \int_0^T (s(t) + w(t))s(t)dt \\
&= \int_0^T s^2(t)dt + \int_0^T w(t)s(t)dt \\
&= E + \int_0^T w(t)s(t)dt
\end{aligned}$$ (D.5)

where

$$E = \int_0^T s^2(t)dt$$ (D.6)

is the *energy* of the known signal $s(t)$.

Since, by assumption, the white noise $w(t)$ is the sample function of a Gaussian process $W(t)$, it follows that the receiver output y is the sample of a Gaussian-distributed random variable Y. To complete the characterization of the receiver output, we need to determine its mean and variance.

The mean of the random variable Y is

$$
\begin{aligned}
\mu_Y &= \mathbf{E}[Y] \\
&= E + \mathbf{E}\left[\int_0^T W(t)s(t)dt\right] \\
&= E + \mathbf{E}\int_0^T [W(t)]s(t)dt \\
&= E
\end{aligned}
\tag{D.7}
$$

where we have used two facts: First, the known signal $s(t)$ is deterministic and therefore unaffected by the expectation operator $\mathbf{E}$. Second, by assumption, the mean of the white noise process $W(t)$ is zero.

The variance of the random variable Y is

$$
\begin{aligned}
\sigma_Y^2 &= \mathbf{E}[(Y-\mu_Y)^2] \\
&= \mathbf{E}\left[\int_0^T\int_0^T W(t_1)W(t_2)s(t_1)s(t_2)dt_1dt_2\right] \\
&= \int_0^T\int_0^T \mathbf{E}[W(t_1)W(t_2)]s(t_1)s(t_2)dt_1dt_2
\end{aligned}
\tag{D.8}
$$

Invoking the use of Eq. (D.2), we may write

$$
\mathbf{E}[W(t_1)W(t_2)] = \frac{N_0}{2}\delta(t_1-t_2)
\tag{D.9}
$$

Substituting Eq. (D.9) into (D.8) yields

$$
\begin{aligned}
\sigma_Y^2 &= \frac{N_0}{2}\int_0^T\int_0^T \delta(t_1-t_2)s(t_1)s(t_2)dt_1dt_2 \\
&= \frac{N_0}{2}\int_0^T s^2(t_1)dt_1 \\
&= \frac{N_0E}{2}
\end{aligned}
\tag{D.10}
$$

where E is the *signal energy*.

Putting all the pieces together, we can now say that the correlation receiver output y is the sample value of a Gaussian-distributed random variable Y with mean

$\mu_Y = E$ and variance $\sigma_Y^2 = N_0E/2$. Accordingly, we may express the probability density function of the random variable Y as

$$f_Y(y) = \frac{1}{\sqrt{2\pi}\sigma_Y}\exp\left(-\frac{(y-\mu_Y)^2}{2\sigma_Y^2}\right)$$

$$= \frac{1}{\sqrt{\pi N_0E}}\exp\left(-\frac{(y-E)^2}{N_0E}\right) \tag{D.11}$$

which is plotted in Fig. D.4.

Let λ denote the *threshold* against which the correlator output y is compared. As stated previously, when $y > \lambda$, the receiver decides in favor of hypothesis H_1; otherwise it decides in favor of hypothesis H_0. Accordingly, the conditional probability of error, given that the known signal $s(t)$ is present in the receiver input, is defined by

$$\text{Prob}(\text{say } H_0|H_1 \text{ is true}) = \int_{-\infty}^{\lambda} f_Y(y)dy \tag{D.12}$$

which is illustrated graphically in Fig. D.4. Substituting Eq. (D.11) into (D.12) yields

$$\text{Prob}(\text{say } H_0|H_1 \text{ is true}) = \frac{1}{\sqrt{\pi N_0E}}\int_{-\infty}^{\lambda}\exp\left(-\frac{(y-E)^2}{N_0E}\right)dy \tag{D.13}$$

To simplify matters, let

$$z = \frac{y-E}{\sqrt{N_0E}} \tag{D.14}$$

which means that

$$dz = \frac{dy}{\sqrt{N_0E}}$$

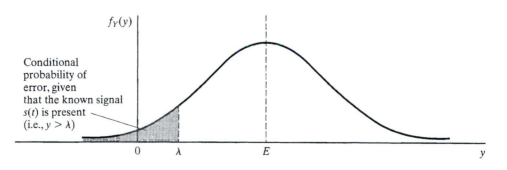

$f_Y(y)$

Conditional
probability of
error, given
that the known signal
$s(t)$ is present
(i.e., $y > \lambda$)

$0 \qquad \lambda \qquad\qquad E \qquad\qquad y$

FIGURE D.4   Probability distribution of the correlation receiver output.

Hence, we may rewrite Eq. (D.13) as

$$\text{Prob(say } H_0|H_1 \text{ is true)} = \frac{1}{\sqrt{\pi}} \int_{-\infty}^{(\lambda-E)/\sqrt{N_0 E}} \exp(-z^2)\,dz$$

$$= \frac{1}{\sqrt{\pi}} \int_{(E-\lambda)/\sqrt{N_0 E}}^{\infty} \exp(-z^2)\,dz \tag{D.15}$$

At this point in the discussion, we digress briefly to introduce a function that is closely related to the Gaussian distribution: the *error function*, defined by

$$\text{erf}(u) = \frac{2}{\sqrt{\pi}} \int_0^u \exp(-z^2)\,dz \tag{D.16}$$

Table E.1 of Appendix E gives values of the error function $\text{erf}(u)$ for the argument u in the interval $0 \le u \le 3.3$. The error function has two useful properties:

1. *Symmetry property*, described by

$$\text{erf}(-u) = -\text{erf}(u) \tag{D.17}$$

2. *Asymptote property*, which, for the argument u approaching infinity, is described by

$$\text{erf}(\infty) = \frac{2}{\sqrt{\pi}} \int_0^{\infty} \exp(-z^2)\,dz \tag{D.18}$$

$$= 1$$

Another function, the *complementary error function*, is defined by

$$\text{erfc}(u) = \frac{2}{\sqrt{\pi}} \int_u^{\infty} \exp(-z^2)\,dz \tag{D.19}$$

which is related to the error function by the formula

$$\text{erfc}(u) = 1 - \text{erf}(u) \tag{D.20}$$

We may now reformulate the conditional probability of error of Eq. (D.15) in terms of the complementary error function by writing

$$\text{Prob(say } H_0|H_1 \text{ is true)} = \frac{1}{2}\text{erfc}\left(\frac{E-\lambda}{\sqrt{N_0 E}}\right) \tag{D.21}$$

From Eq. (D.21), the following points are noteworthy:

- The signal energy E and noise spectral density N_0 have different physical interpretations, in that E is measured in joules whereas N_0 is measured in watts/hertz; yet these two units are in fact equal.

- Insofar as the signal component is concerned, the probability of error is independent of the waveform of the known signal $s(t)$, and the only parameter that matters is the signal energy E.
- The threshold λ is measured in joules.

D.3   ANOTHER PROPERTY OF THE MATCHED FILTER

Equation (D.21) sums up one important property of the matched-filter receiver in the combined presence of signal and noise at the filter input. For another important property of the matched filter, consider the case of a noiseless input. Then, with the input $x(t) = s(t)$ and the impulse response $h(t) = s(T - t)$, the resulting filter output is defined by the convolution integral

$$
\begin{aligned}
y(t) &= R_s(T - t) \\
&= R_s(t - T) \quad \text{if } w(t) = 0
\end{aligned}
\tag{D.22}
$$

The integral of Eq. (D.22) is recognized as the deterministic autocorrelation function of the signal component $s(t)$ for a lag of $T - t$, namely, $R_s(T - t)$. Accordingly, we may write

$$
\begin{aligned}
y(t) &= R_s(T - t) \\
&= R_s(T - t) \quad wt = 0
\end{aligned}
\tag{D.23}
$$

where, in the second line, we have used the fact that the autocorrelation function of a signal of finite energy is an even function of the lag (see Appendix A). In words, Eq. (D.23) states that *the output of a filter matched to an input signal is equal to the autocorrelation function of that signal, delayed by an amount equal to the duration of the signal.*

D.4   MATCHED FILTERING FOR COMPLEX SIGNALS

The material presented thus far on matched filtering applies to real-valued signals. When dealing with complex-valued signals, we make a simple modification to Eq. (D.4). Specifically, the impulse response of a filter matched to a complex-valued signal $s(t)$ is defined by

$$
h(t) = \begin{cases} s^*(T - t) & 0 \le t \le T \\ 0 & \text{otherwise} \end{cases}
\tag{D.24}
$$

where the asterisk denotes complex conjuction. Except for this minor modification, everything else presented in the Appendix remains intact.

APPENDIX E

Error Function

E.1 DEFINITIONS

The *error function*, denoted by erf(u), is defined in a number of different ways in the literature. We shall use the following definition:

$$\operatorname{erf}(u) = \frac{2}{\sqrt{\pi}} \int_0^u \exp(-z^2)dz \tag{E.1}$$

The error function has two useful properties:

1.

$$\operatorname{erf}(-u) = -\operatorname{erf}(u) \tag{E.2}$$

This is known as the *symmetry property*.

2. As u approaches infinity, erf(u) approaches unity; that is,

$$\frac{2}{\sqrt{\pi}} \int_0^\infty \exp(-z^2)dz = 1 \tag{E.3}$$

This is known as the *asymptote property*.

The *complementary error function* is defined by

$$\operatorname{erfc}(u) = \frac{2}{\sqrt{\pi}} \int_u^\infty \exp(-z^2)dz \tag{E.4}$$

The complementary error function is related to the error function as follows:

$$\operatorname{erfc}(u) = 1 - \operatorname{erf}(u) \tag{E.5}$$

Table E.1 gives values of the error function erf(u) for u in the range from 0 to 3.3.

TABLE E.1   The Error Function[a].

u	erf(u)	u	erf(u)
0.00	0.00000	1.10	0.88021
0.05	0.05637	1.15	0.89612
0.10	0.11246	1.20	0.91031
0.15	0.16800	1.25	0.92290
0.20	0.22270	1.30	0.93401
0.25	0.27633	1.35	0.94376
0.30	0.32863	1.40	0.95229
0.35	0.37938	1.45	0.95970
0.40	0.42839	1.5	0.96611
0.45	0.47548	1.55	0.97162
0.50	0.52050	1.6	0.97635
0.55	0.56332	1.65	0.98038
0.60	0.69386	1.70	0.98379
0.65	0.64203	1.75	0.98667
0.70	0.67780	1.80	0.98909
0.75	0.71116	1.85	0.99111
0.80	0.74210	1.90	0.99279
0.85	0.77067	1.95	0.99418
0.90	0.79691	2.00	0.99532
0.95	0.82089	2.50	0.99959
1.00	0.84270	3.00	0.99998
1.05	0.86244	3.30	0.999998

[a] The error function is tabulated extensively in several references; see for example, Abramowitz and Stegun (1965, pp. 297–316).

E.2   BOUNDS ON THE COMPLEMENTARY ERROR FUNCTION

Substituting $u - x$ for z in Eq. (E.4), we get

$$\text{erfc}(u) = \frac{2}{\sqrt{\pi}} \exp(-u^2) \int_{-\infty}^{0} \exp(2ux)\exp(-x^2)dx$$

For any real x, the value of $\exp(-x^2)$ lies between the successive partial sums of the power series

$$1 - \frac{x^2}{1!} + \frac{(x^2)^2}{2!} - \frac{(x^2)^3}{3!} + \dots$$

Therefore, for $u > 0$, we find, on using $(n+1)$ terms of this series, that erfc(u) lies between the values taken by

$$\frac{2}{\sqrt{\pi}} \exp(-u^2) \int_{\infty}^{0} \left(1 - x^2 + \frac{x^4}{2} - \dots \pm \frac{x^{2n}}{n!}\right) \exp(2ux)dx$$

for even n and for odd n. Putting $2ux = -v$ and using the integral

$$\int_{0}^{\infty} v^n \exp(-v)dv = n!$$

we obtain the following *asymptotic expansion* for erfc(u), assuming that $u > 0$:

$$\text{erfc}(u) \approx \frac{\exp(-u^2)}{\sqrt{\pi}u}\left[1 - \frac{1}{2u^2} + \frac{1 \cdot 3}{2^2 u^4} - \cdots \pm \frac{1 \cdot 3 \cdot 5 \cdots (2n-1)}{2^n u^{2n}}\right] \tag{E.6}$$

For large positive values of u, the successive terms of the series on the right-hand side of Eq. (E.6) decrease very rapidly. We thus deduce two simple bounds on erfc(u), one lower and the other upper, given by the inequality[1]

$$\frac{\exp(-u^2)}{\sqrt{\pi}u}\left(1 - \frac{1}{2u^2}\right) < \text{erfc}(u) < \frac{\exp(-u^2)}{\sqrt{\pi}u} \tag{E.7}$$

For large positive u, a second bound on the complementary error function erfc(u) is obtained by omitting the multiplying factor $1/u$ in the upper bound of Eq. (E.7):

$$\text{erfc}(u) < \frac{\exp(-u^2)}{\sqrt{\pi}} \tag{E.8}$$

In Fig. E.1, we have plotted erfc(u), the two bounds defined by Eq. (E.7), and the upper bound of Eq. (E.8). We see that, for $u \geq 1.5$, the bounds on erfc(u), defined by Eq. (E.7), become increasingly tight.

E.3   THE Q-FUNCTION

Consider a *standardized* Gaussian random variable X of zero mean and unit variance. The probability that an observed value of the random variable X will be greater than v is given by the *Q-function*:

$$Q(v) = \frac{1}{\sqrt{2\pi}} \int_v^\infty \exp\left(-\frac{x^2}{2}\right) \tag{E.9}$$

The *Q-function defines the area under the standardized Gaussian tail*. Inspection of Eqs. (E.4) and (E.9) reveals that the Q-function is related to the complementary error function as follows:

$$Q(v) = \frac{1}{2}\text{erfc}\left(\frac{v}{\sqrt{2}}\right) \tag{E.10}$$

Conversely, putting $u = v$, we have

$$\text{erfc}(u) = 2Q(\sqrt{2}u) \tag{E.11}$$

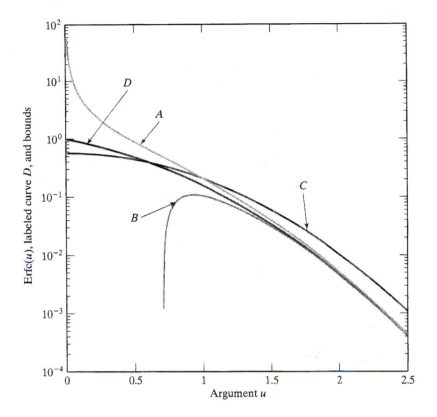

FIGURE E.1   The complementary error function erfc(u) and its bounds:

Curve A: $\dfrac{e^{-u^2}}{\sqrt{\pi}u}$

Curve B: $\dfrac{e^{-u^2}}{\sqrt{\pi}u}\left(1 - \dfrac{1}{2u^2}\right)$

Curve C: $\dfrac{e^{-u^2}}{\sqrt{\pi}}$

Curve D: erfc(u)

NOTES AND REFERENCES

[1] The derivation of Eq. (E.7) follows Blachman (1966).

APPENDIX F

MAP Algorithm

F.1 SEPARABILITY THEOREM

Following the terminology introduced in Section 4.12.4, let the vectors $\boldsymbol{\alpha}(t)$ and $\boldsymbol{\beta}(t)$ denote estimates of the state probabilities in a turbo decoder at time t that are based on past and future data, respectively. According to the *separability theorem*, the state probabilities at time t are related to $\boldsymbol{\alpha}(t)$ and $\boldsymbol{\beta}(t)$ by

$$\lambda(t) = \frac{\boldsymbol{\alpha}(t) \cdot \boldsymbol{\beta}(t)}{\|\boldsymbol{\alpha}(t) \cdot \boldsymbol{\beta}(t)\|_1} \tag{F.1}$$

where the numerator is the vector product of $\boldsymbol{\alpha}(t)$ and $\boldsymbol{\beta}(t)$ and the denominator is the L_1 norm of this product, as defined in Section 4.12.

Proof: For any m, for which $\mathrm{Prob}(s(t) = m) \neq 0$,

$$\lambda_m(t) = \mathrm{Prob}(s(t) = m | \mathbf{y})$$

$$= \frac{\mathrm{Prob}(s(t) = m, \mathbf{y})}{\mathrm{Prob}(\mathbf{y})}$$

$$= \mathrm{Prob}(\mathbf{y} | s(t) = m) \cdot \frac{\mathrm{Prob}(s(t) = m)}{\mathrm{Prob}(\mathbf{y})}$$

$$= \mathrm{Prob}(\mathbf{y}_{[1, t]} | s(t) = m) \cdot \mathrm{Prob}(\mathbf{y}_{[t+1, T]}, s(t) = m) \cdot \frac{\mathrm{Prob}(s(t) = m)}{\mathrm{Prob}(\mathbf{y})} \tag{F.2}$$

$$= \frac{\mathrm{Prob}(\mathbf{y}_{[1, t]} | s(t) = m)}{\mathrm{Prob}(s(t) = m)} \cdot \frac{\mathrm{Prob}(\mathbf{y}_{[t+1, T]}, s(t) = m)}{\mathrm{Prob}(s(t) = m)} \cdot \frac{\mathrm{Prob}(s(t) = m)}{\mathrm{Prob}(\mathbf{y})}$$

$$= \mathrm{Prob}(s(t) = m | \mathbf{y}_{[1, t]}) \cdot \mathrm{Prob}(s(t) = m | \mathbf{y}_{[t+1, T]}) \cdot \left(\frac{\mathrm{Prob}(\mathbf{y}_{[1, t]} | \mathbf{y}_{[t+1, T]})}{\mathrm{Prob}(s(t) = m) \mathrm{Prob}(\mathbf{y})} \right)$$

where the second and third lines in the development follow from Bayes' rule, the fourth line follows from the fact that the decoding process is a Markov process, and the fifth and sixth lines are further manipulations using Bayes' rule.

Often, the a priori probabilities, $\text{Prob}(s(t) = m)$, are independent of m. (In practice, some of the a priori probabilities in a turbo decoding process may be zero due to certain features of the trellis—for example, at start-up. In that case, this requirement applies only to those states whose a priori probabilities are nonzero.) This is usually the case for a time-invariant trellis, for which the inputs are equiprobable. In that case, the bracketed term in the last line on the right-hand side is independent of m. Since the summation of the left-hand side of Eq. (F.2) over m must equal unity, it follows that this bracketed term must normalize the right-hand side to sum to unity. Thus, when we identify the first and second terms in the product of the last line as $\boldsymbol{\alpha}(t)$ and $\boldsymbol{\beta}(t)$, we have

$$\lambda(t) = \frac{\boldsymbol{\alpha}(t) \cdot \boldsymbol{\beta}(t)}{\|\boldsymbol{\alpha}(t) \cdot \boldsymbol{\beta}(t)\|_1} \tag{F.3}$$

which proves the theorem.

A P P E N D I X   G

Capacity of MIMO Links

G.1   PRELIMINARIES[1]

The purpose of this appendix is to present a derivation of the log-det capacity formula of Eq. (6.59). To prepare the way for the derivation, we briefly review some basic concepts in information theory.

Consider a continuous random variable X with probability density function $f_X(x)$. The *differential entropy* of the random variable X, measured in *bits*, is defined by

$$
\begin{aligned}
h(X) &= -\int_{-\infty}^{\infty} f_X(x)\log_2 f_X(x)dx \\
&= -\mathbf{E}[\log_2 f_X(x)] \;\; \text{bits}
\end{aligned}
\tag{G.1}
$$

where $\mathbf{E}$ is the statistical expectation operator. It is important to note that the symbol X in the entropy $h(X)$ is *not* the argument of a function; rather, it merely serves the purpose of a label for the source of information.

When we have a continuous random vector $\mathbf{X}$ consisting of N random variables $X_1, X_2, ..., X_N$, we may generalize Eq. (G.1) and define the differential entropy of $\mathbf{X}$ as the N-fold integral

$$
\begin{aligned}
h(\mathbf{X}) &= -\int_{-\infty}^{\infty} f_{\mathbf{X}}(\mathbf{x})\log_2 f_{\mathbf{X}}(\mathbf{x})d\mathbf{x} \\
&= -\mathbf{E}[\log_2 f_{\mathbf{X}}(\mathbf{x})] \;\; \text{bits}
\end{aligned}
\tag{G.2}
$$

where $f_{\mathbf{X}}(\mathbf{x})$ is the joint probability density function of the random vector $\mathbf{X}$.

The logarithmic description of entropy is evident from both Eqs. (G.1) and (G.2). This particular form of description is in perfect accord with the notion of entropy in thermodynamics.

Equations (G.1) and (G.2) apply to random data, real or complex. The difference between these two forms of data manifests itself in the way in which the pertinent probability density functions are defined, as illustrated in the next example.

EXAMPLE G.1   Complex Multidimensional Gaussian Distribution

Consider an N-dimensional complex Gaussian-distributed vector $\mathbf{X}$. Each element of $\mathbf{X}$ consists of an in-phase component $X_{k,I}$ and a quadrature component $X_{k,Q}$, so

$$X_k = X_{k,I} + jX_{k,Q} \qquad k = 1,2,...,N \tag{G.3}$$

or, vectorially,

$$\mathbf{X} = \mathbf{X}_I + j\mathbf{X}_Q \tag{G.4}$$

It is assumed that $\mathbf{X}$ has zero mean. The requirement is to determine the differential entropy of $\mathbf{X}$.

If the components $\mathbf{X}_I$ and $\mathbf{X}_Q$ are orthogonal—that is, if we have

$$E[\mathbf{X}_I\mathbf{X}_Q^T] = \mathbf{O} \tag{G.5}$$

and if they are both Gaussian distributed, then they are statistically independent, or

$$f_{\mathbf{X}_I,\mathbf{X}_Q}(\mathbf{x}_I, \mathbf{x}_Q) = f_{\mathbf{X}_I}(\mathbf{x}_I)f_{\mathbf{X}_Q}(\mathbf{x}_Q) \tag{G.6}$$

The in-phase component $\mathbf{X}_I$ and quadrature component $\mathbf{X}_Q$ share the same formula for their joint probability density functions. We therefore make two important observations:

1. The components $\mathbf{X}_I$ and $\mathbf{X}_Q$ have exactly the same entropy.
2. Since the differential entropy is logarithmic in nature, it follows that the differential entropies of $\mathbf{X}_I$ and $\mathbf{X}_Q$ are additive in terms of calculating the differential entropy of $\mathbf{X}$.

Hence, we may write

$$h(\mathbf{X}_I) = h(\mathbf{X}_Q) \tag{G.7}$$

and

$$\begin{aligned} h(\mathbf{X}) &= h(\mathbf{X}_I) + h(\mathbf{X}_Q) \\ &= 2h(\mathbf{X}_I) \end{aligned} \tag{G.8}$$

The joint probability density function of the complex Gaussian vector $\mathbf{X}$ with zero mean and correlation matrix $\mathbf{R}_x$ is defined by

$$f_{\mathbf{X}}(\mathbf{x}) = \frac{1}{(2\pi)^N \det(\mathbf{R}_x)} \exp\left(-\frac{1}{2}\mathbf{x}^T\mathbf{R}_x^{-1}\mathbf{x}\right) \tag{G.9}$$

where $\mathbf{R}_x^{-1}$ is the inverse of $\mathbf{R}_x$ and $\det(\mathbf{R}_x)$ is the determinant of $\mathbf{R}_x$. Substituting Eq. (G.9) into (G.2), using the fact that the volume under $f_{\mathbf{X}}(\mathbf{x})$ is unity, and then simplifying terms, we get

$$h(\mathbf{X}) = N + N\log_2(2\pi) + \log_2\{\det(\mathbf{R}_x)\} \quad \text{bits} \tag{G.10}$$

which is uniquely defined by the correlation matrix $\mathbf{R}_x$.

For the special case of a scalar complex Gaussian random variable X, $N = 1$ and Eq. (G.10) reduces to

$$h(X) = 1 + \log_2(2\pi\sigma_X^2) \quad \text{bits} \qquad (X : \text{complex}) \tag{G.11}$$

where σ_X^2 is the variance of X. If X is real, we have

$$h(X) = \frac{1}{2}[1 + \log_2(2\pi\sigma_X^2)] \text{ bits} \qquad (X : \text{real}) \qquad \text{(G.12)}$$

For a given variance σ_X^2, the Gaussian random variable X has the largest differential entropy attainable by any random variable in its class (i.e., real or complex). A similar remark applies to a multivariate Gaussian distribution.   ∎

For the discussion at hand, we need one other notion: mutual information, which applies to a pair of related random variables or random vectors. To be specific, consider a pair of random variables X and Y with joint probability density function $f_{X,Y}(x,y)$. The *mutual information* between X and Y is defined by

$$I(X;Y) = \int_{-\infty}^{\infty} \int_{-\infty}^{\infty} f_{X,Y}(x,y)\log_2\left(\frac{f_{X|Y}(x|y)}{f_X(x)}\right)dxdy \qquad \text{(G.13)}$$

where $f_{X|Y}(x|y)$ is the conditional probability density function of X, given that $Y = y$. In words, the mutual information $I(X;Y)$ is a *measure of the uncertainty about the random variable X that is resolved by observing the second random variable Y.*

On the basis of Eq. (G.13), we may derive the following properties of mutual information that hold in general:

$$I(X;Y) \geq 0 \qquad \text{(G.14)}$$

$$I(X;Y) = I(Y;X) \qquad \text{(G.15)}$$

$$\begin{aligned} I(X;Y) &= h(X) - h(X|Y) \\ &= h(Y) - h(Y|X) \end{aligned} \qquad \text{(G.16)}$$

Here, $h(X)$ and $h(Y)$ are the differential entropies of X and Y, respectively, and

$$h(X|Y) = -\int_{-\infty}^{\infty} \int_{-\infty}^{\infty} f_{X,Y}(x,y)\log_2 f_{X|Y}(x|y)dxdy \qquad \text{(G.17)}$$

is the conditional differential entropy of X, given Y.

Formulas similar to Eqs. (G.13) through (G.17) apply to a related pair of random vectors **X** and **Y**.

With the definitions of differential entropy, conditional differential entropy, and mutual information at hand, we are ready to proceed with the derivation of the log-det capacity formula.

G.2   LOG-DET CAPACITY FORMULA OF MIMO LINK[2]

Consider a communication link with multiple antennas. Let the N_t-by-1 vector **s** denote the transmitted signal vector and the N_r-by-1 vector **x** denote the received signal vector. These two vectors are related by the *input–output relation of the channel*, namely,

$$\mathbf{x} = \mathbf{Hs} + \mathbf{w} \qquad \text{(G.18)}$$

where $\mathbf{H}$ is the *channel matrix* of the link and $\mathbf{w}$ is the additive channel noise vector. The vectors $\mathbf{s}, \mathbf{w},$ and $\mathbf{x}$ are realizations of the random vectors $\mathbf{S}, \mathbf{W},$ and $\mathbf{X},$ respectively.

In the rest of this section, the following assumptions are made:

1. The channel is stationary and ergodic.
2. The channel matrix $\mathbf{H}$ is made up of i.i.d. Gaussian elements.
3. The channel state $\mathbf{H}$ is known to the receiver, but not the transmitter.
4. The transmitted signal vector $\mathbf{s}$ has zero mean and correlation matrix $\mathbf{R_s}$.
5. The additive channel noise vector $\mathbf{w}$ has zero mean and correlation matrix $\mathbf{R_w}$.
6. Both $\mathbf{s}$ and $\mathbf{w}$ are governed by Gaussian distributions.

With both $\mathbf{H}$ and $\mathbf{x}$ unknown to the transmitter, the primary issue of interest is to determine $I(\mathbf{s}; \mathbf{x}, \mathbf{H})$, which denotes the mutual information between the transmitted signal vector $\mathbf{s}$ and both the received signal vector $\mathbf{x}$ and the channel matrix $\mathbf{H}$. Extending the definition of mutual information given in Eq. (G.13) to the problem at hand, we write

$$I(\mathbf{S};\mathbf{X},\mathbf{H}) = \iiint_{\mathcal{H}\mathcal{X}\mathcal{S}} f_{\mathbf{S},\mathbf{X},\mathbf{H}}(\mathbf{s},\mathbf{x},\mathbf{H}) \log_2\left(\frac{f_{\mathbf{S}|\mathbf{X},\mathbf{H}}(\mathbf{s}|\mathbf{x},\mathbf{H})}{f_{\mathbf{X},\mathbf{H}}(\mathbf{x},\mathbf{H})}\right) d\mathbf{s}\,d\mathbf{x}\,d\mathbf{H} \qquad \text{(G.19)}$$

where $\mathcal{S}, \mathcal{X},$ and $\mathcal{H}$ are the respective spaces pertaining to the random vectors $\mathbf{S}$ and $\mathbf{X}$ and the matrix $\mathbf{H}$. According to Bayes' rule, we have

$$f_{\mathbf{S},\mathbf{X},\mathbf{H}}(\mathbf{s},\mathbf{x},\mathbf{H}) = f_{\mathbf{S},\mathbf{X}|\mathbf{H}}(\mathbf{s},\mathbf{x}|\mathbf{H}) f_{\mathbf{H}}(\mathbf{H})$$

We may therefore rewrite Eq. (G.19) in the equivalent form

$$
\begin{aligned}
I(\mathbf{S};\mathbf{X},\mathbf{H}) &= \int_{\mathcal{H}} f_{\mathbf{H}}(\mathbf{H}) \left[\iint_{\mathcal{X}\mathcal{S}} f_{\mathbf{S},\mathbf{X}|\mathbf{H}}(\mathbf{s},\mathbf{x}|\mathbf{H}) \log_2\left(\frac{f_{\mathbf{S}|\mathbf{X},\mathbf{H}}(\mathbf{s}|\mathbf{x},\mathbf{H})}{f_{\mathbf{X},\mathbf{H}}(\mathbf{x},\mathbf{H})}\right) d\mathbf{s}\,d\mathbf{x}\right] d\mathbf{H} \\
&= \mathbf{E_H}\left[\iint_{\mathcal{X}\mathcal{S}} f_{\mathbf{S},\mathbf{X}|\mathbf{H}}(\mathbf{s},\mathbf{x}|\mathbf{H}) \log_2\left(\frac{f_{\mathbf{S}|\mathbf{X},\mathbf{H}}(\mathbf{s}|\mathbf{x},\mathbf{H})}{f_{\mathbf{X},\mathbf{H}}(\mathbf{x},\mathbf{H})}\right) d\mathbf{s}\,d\mathbf{x}\right] \\
&= \mathbf{E_H}[I(\mathbf{s};\mathbf{x}|\mathbf{H})]
\end{aligned}
\qquad \text{(G.20)}
$$

where the expectation is with respect to the channel matrix $\mathbf{H},$ and

$$I(\mathbf{s};\mathbf{x}|\mathbf{H}) = \iint_{\mathcal{X}\mathcal{S}} f_{\mathbf{S},\mathbf{X}|\mathbf{H}}(\mathbf{s},\mathbf{x}|\mathbf{H}) \log_2\left(\frac{f_{\mathbf{S}|\mathbf{X},\mathbf{H}}(\mathbf{s}|\mathbf{x},\mathbf{H})}{f_{\mathbf{X},\mathbf{H}}(\mathbf{x},\mathbf{H})}\right) d\mathbf{s}\,d\mathbf{x}$$

is the conditional mutual information between the transmitted signal vector $\mathbf{s}$ and the received signal vector $\mathbf{x}$, given the channel matrix $\mathbf{H}$. However, by assumption, the state of the channel is unknown to the transmitter. It follows, therefore, that insofar as the receiver is concerned, $I(\mathbf{s};\mathbf{x}|\mathbf{H})$ is a random variable—hence the expectation with respect to $\mathbf{H}$ in Eq. (G.20). The quantity resulting from this expectation is deterministic, defining the mutual information jointly between the transmitted signal vector **s** and both the received signal vector **x** and the channel matrix **H**. The result so obtained is indeed consistent with what we know about the notion of joint mutual information.

Next, applying the vector form of the first line in Eq. (G.16) to the mutual information $I(\mathbf{s};\mathbf{x}|\mathbf{H})$, we may write

$$I(\mathbf{s};\mathbf{x}|\mathbf{H}) = h(\mathbf{x}|\mathbf{H}) - h(\mathbf{x}|\mathbf{s},\mathbf{H}) \tag{G.21}$$

where $h(\mathbf{x}|\mathbf{H})$ is the conditional differential entropy of the input **x**, given **H**, and $h(\mathbf{x}|\mathbf{s},\mathbf{H})$ is the conditional differential entropy of the input **x**, given both **s** and **H**. Both of these entropies are random quantities, as they depend on **H**.

To proceed further, we now invoke the assumed Gaussian nature of both **s** and **H**, in which case **x** also assumes a Gaussian description. Under these assumptions, we may use Eq. (G.10) to express the entropy of the received signal **x** of dimension N_r, given **H**, as

$$h(\mathbf{x}|\mathbf{H}) = N_r + N_r\log_2(2\pi) + \log_2\{\det(\mathbf{R_x})\} \text{ bits} \tag{G.22}$$

where $\mathbf{R_x}$ is the correlation matrix of **x**. Recognizing that the transmitted signal vector **s** and the channel noise vector **w** are independent of each other, we find, from Eq. (G.18), that the correlation matrix of the received signal vector **x** is given by

$$\begin{aligned}
\mathbf{R_x} &= \mathbf{E}[\mathbf{x}\mathbf{x}^\dagger] \\
&= \mathbf{E}[(\mathbf{Hs}+\mathbf{w})(\mathbf{Hs}+\mathbf{w})^\dagger] \\
&= \mathbf{E}[(\mathbf{Hs}+\mathbf{w})(\mathbf{s}^\dagger\mathbf{H}^\dagger+\mathbf{w}^\dagger)] \\
&= \mathbf{E}[\mathbf{Hss}^\dagger\mathbf{H}^\dagger] + E[\mathbf{ww}^\dagger] \quad \text{because} \quad \mathbf{E}[\mathbf{sw}^\dagger] = \mathbf{0} \\
&= \mathbf{HE}[\mathbf{ss}^\dagger]\mathbf{H}^\dagger + \mathbf{R_w} \\
&= \mathbf{HR_sH}^\dagger + \mathbf{R_w}
\end{aligned} \tag{G.23}$$

where

$$\mathbf{R_s} = \mathbf{E}[\mathbf{ss}^\dagger] \tag{G.24}$$

and

$$\mathbf{R_w} = \mathbf{E}[\mathbf{ww}^\dagger] \tag{G.25}$$

Hence, using Eq. (G.23) in (G.22), we get

$$h(\mathbf{x}|\mathbf{H}) = N_r + N_r\log_2(2\pi) + \log_2\{\det(\mathbf{R_w}+\mathbf{HR_sH}^\dagger)\} \text{ bits} \tag{G.26}$$

Next, we note that, since the vectors **s** and **w** are independent, and since the sum of **w** plus **Hs** equals **x**, as indicated in Eq. (G.18), then the conditional differential entropy of **x**, given both **s** and **H**, is simply equal to the differential entropy of the additive channel noise vector **w**:

$$h(\mathbf{x}|\mathbf{s}, \mathbf{H}) = h(\mathbf{w}) \tag{G.27}$$

Again invoking the formula of Eq. (G.10), we have

$$h(\mathbf{w}) = N_r + N_r \log_2(2\pi) + \log_2\{\det(\mathbf{R_w})\} \text{ bits} \tag{G.28}$$

Using Eqs. (G.26), (G.27), and (G.28) in Eq. (G.21), we get

$$I(\mathbf{s};\mathbf{x}|\mathbf{H}) = \log_2\left\{\det(\mathbf{R_w} + \mathbf{H}\mathbf{R_s}\mathbf{H}^H)\right\} - \log_2\{\det(\mathbf{R_w})\}$$

$$= \log_2\left\{\frac{\det(\mathbf{R_w} + \mathbf{H}\mathbf{R_s}\mathbf{H}^H)}{\det(\mathbf{R_w})}\right\} \tag{G.29}$$

As was remarked previously, the conditional mutual information $I(\mathbf{s}; \mathbf{x}|\mathbf{H})$ is a random variable. Hence, using Eq. (G.29) in (G.20), we finally formulate the ergodic capacity of the MIMO link as the expectation

$$C = \mathbf{E_H}\left[\log_2\left\{\frac{\det(\mathbf{R_w} + \mathbf{H}\mathbf{R_s}\mathbf{H}^H)}{\det(\mathbf{R_w})}\right\}\right] \text{ bits/s/Hz} \tag{G.30}$$

which is subject to the constraint

$$\max_{\mathbf{R_s}} \text{tr}[\mathbf{R_s}] \leq P \qquad P = \text{constant transmit power}$$

where tr[.] denotes the *trace* operator, which extracts the sum of the diagonal elements of the enclosed matrix.

Equation (G.30) is the desired log-det formula for the ergodic capacity of the MIMO link. This formula is of general applicability in that correlations among the elements of the transmitted signal vector **s** and among those of the channel noise vector **w** are permitted. However, the assumptions made in its derivation involve the Gaussian aspects of **s**, **H**, and **w**.

One last comment is in order: the white Gaussian input spectrum

$$\mathbf{R_s} = \sigma_s^2 \mathbf{I}_{N_t}$$

is not necessarily optimal; nevertheless, its application does yield a lower bound to the ergodic capacity C.

G.3 MIMO CAPACITY FOR CHANNEL KNOWN AT THE TRANSMITTER[3]

The log-det formula of Eq. (G.30) for the ergodic capacity of a MIMO flat-fading channel assumes that the state of the channel is known only at the receiver. What if the state is also known perfectly at the transmitter? Then the state of the channel becomes known to the entire system, which means that we may treat the channel matrix $\mathbf{H}$ as a constant. Hence, unlike the partially known case discussed in Section G.2, there is no longer the need for invoking the expectation operator in formulating the log-det capacity. Rather, the problem becomes one of constructing the optimal $\mathbf{R_s}$ (i.e., the correlation matrix of the transmitted signal vector $\mathbf{s}$) that maximizes the ergodic capacity. To simplify the construction procedure, we consider a MIMO link with $N_r = N_t = N$. Accordingly, using the assumption of additive white Gaussian noise with variance σ_w^2 in the log-det capacity formula of Eq. (G.2), we get

$$C = \log_2\left\{ \det\left(I_N + \frac{1}{\sigma_w^2} \mathbf{H R_s H^\dagger} \right) \right\} \quad \text{bits/s/Hz} \tag{G.31}$$

We can now formally postulate the optimization problem at hand as follows:

Maximize the ergodic capacity C of Eq. (G.31) with respect to the correlation matrix $\mathbf{R_s}$, subject to two requirements:

1. *Nonnegative definite $\mathbf{R_s}$, which is a necessary requirement for a correlation matrix.*

2. *Global power constraint*

$$\text{tr}[\mathbf{R_s}] = P \tag{G.32}$$

where P is the total transmit power.

To proceed with construction of the optimal $\mathbf{R_s}$, we first use the *determinant identity*:

$$\det(\mathbf{I} + \mathbf{AB}) = \det(\mathbf{I} + \mathbf{BA}) \tag{G.33}$$

Applying this identity to Eq. (G.31) yields

$$C = \log_2\left\{ \det\left(\mathbf{I}_N + \frac{1}{\sigma_w^2} \mathbf{R_s H^\dagger H} \right) \right\} \quad \text{bits/s/Hz} \tag{G.34}$$

Diagonalizing the matrix product $\mathbf{H^\dagger H}$ by invoking the *eigendecomposition* of a Hermitian matrix, we may write

$$\mathbf{U^\dagger H^\dagger H U} = \Lambda \tag{G.35}$$

where Λ is a diagonal matrix made up of the eigenvalues of $\mathbf{H^\dagger H}$ and $\mathbf{U}$ is a unitary matrix whose columns are the associated eigenvectors. (The eigendecomposition of a Hermitian matrix is discussed in Appendix H.) We may rewrite Eq. (G.35) in the form

$$\mathbf{H^\dagger H} = \mathbf{U \Lambda U^\dagger} \tag{G.36}$$

Substituting Eq. (G.36) into Eq. (G.34), we get

$$
C = \log_2\left\{\det\left(\mathbf{I}_N + \frac{1}{\sigma_w^2}\mathbf{R_s}\mathbf{U}\boldsymbol{\Lambda}\mathbf{U}^\dagger\right)\right\} \quad \text{bits/s/Hz} \tag{G.37}
$$

Applying the determinant identity of Eq. (G.33) to Eq. (G.37) yields

$$
\begin{aligned}
C &= \log_2\left\{\det\left(\mathbf{I}_N + \frac{1}{\sigma_w^2}\boldsymbol{\Lambda}\mathbf{U}^\dagger\mathbf{R_s}\mathbf{U}\right)\right\} \\
&= \log_2\left\{\det\left(\mathbf{I}_N + \frac{1}{\sigma_w^2}\boldsymbol{\Lambda}\bar{\mathbf{R}}_\mathbf{s}\right)\right\} \quad \text{bits/s/Hz}
\end{aligned}
\tag{G.38}
$$

where

$$
\bar{\mathbf{R}}_\mathbf{s} = \mathbf{U}^\dagger\mathbf{R_s}\mathbf{U} \tag{G.39}
$$

Note that the transformed matrix $\bar{\mathbf{R}}_\mathbf{s}$ is nonnegative definite. Note also that

$$
\begin{aligned}
\operatorname{tr}[\bar{\mathbf{R}}_\mathbf{s}] &= \operatorname{tr}[\mathbf{U}^\dagger\mathbf{R_s}\mathbf{U}] \\
&= \operatorname{tr}[\mathbf{U}\mathbf{U}^\dagger\mathbf{R_s}] \\
&= \operatorname{tr}[\mathbf{R_s}]
\end{aligned}
\tag{G.40}
$$

It follows, therefore, that maximization of the capacity of Eq. (G.38) can be carried equally well over the transformed correlation matrix $\bar{\mathbf{R}}_\mathbf{s}$.

One other important point to note is that any nonnegative definite matrix $\mathbf{A}$ satisfies the *Hadamard inequality*

$$
\det(\mathbf{A}) \le \prod_k a_{kk} \tag{G.41}
$$

where the a_{kk} are the diagonal elements of the matrix $\mathbf{A}$. Hence, applying this inequality to the determinant in Eq. (G.38), we may write

$$
\det\left(\mathbf{I}_N + \frac{1}{\sigma_w^2}\boldsymbol{\Lambda}\bar{\mathbf{R}}_\mathbf{s}\right) \le \prod_{k=1}^N \left(1 + \frac{1}{\sigma_w^2}\lambda_k\bar{r}_{s,kk}\right) \tag{G.42}
$$

where λ_k is the kth eigenvalue of the matrix product $\mathbf{HH}^\dagger$ and $\bar{r}_{s,kk}$ is the kth diagonal element of the transformed matrix $\bar{\mathbf{R}}_\mathbf{s}$. The equality in Eq. (G.42) holds only when $\bar{\mathbf{R}}_\mathbf{s}$ is a diagonal matrix, which is the very condition that maximizes the ergodic capacity C.

To proceed further, we now use Eq. (G.38) and Eq. (G.42) with the equality sign to express the capacity as

$$
\begin{aligned}
C &= \log_2 \left\{ \prod_{k=1}^{N} \left(1 + \frac{1}{\sigma_w^2} \lambda_k \bar{r}_{s,kk} \right) \right\} \\
&= \sum_{k=1}^{N} \log_2 \left(1 + \frac{1}{\sigma_w^2} \lambda_k \bar{r}_{s,kk} \right) \\
&= \sum_{k=1}^{N} \log_2 \left\{ \lambda_k \left(\lambda_k^{-1} + \frac{1}{\sigma_w^2} \bar{r}_{s,kk} \right) \right\} \\
&= \sum_{k=1}^{N} \log_2 \lambda_k + \sum_{k=1}^{N} \log_2 \left(\lambda_k^{-1} + \frac{1}{\sigma_w^2} \bar{r}_{s,kk} \right)
\end{aligned}
\tag{G.43}
$$

where only the second summation is clearly adjustable. We may therefore reformulate the optimization problem at hand as follows:

Given the set of eigenvalues $\{\lambda_k\}_{k=1}^{N}$ pertaining to the matrix product $\mathbf{HH}^\dagger$, determine the optimal set of autocorrelations $\{\bar{r}_{s,kk}\}_{k=1}^{N}$ that maximizes the summation

$$
\sum_{k=1}^{N} \left(\frac{1}{\lambda_k} + \frac{1}{\sigma_w^2} \bar{r}_{s,kk} \right)
$$

subject to the constraint

$$
\sum_{k=1}^{N} \bar{r}_{s,kk} = P
\tag{G.44}
$$

The global power constraint of Eq. (G.44) follows from Eq. (G.40) and the trace definition

$$
\mathrm{tr}[\bar{\mathbf{R}}_s] = \sum_{k=1}^{N} \bar{r}_{s,kk}
\tag{G.45}
$$

The solution to this optimization problem may be determined through the *water-filling procedure*, which is well known in information theory.[3] Effectively, the solution to the water-filling problem says that, in a multiple-channel scenario, we transmit more signal power in the better channels and less signal power in the poorer channels. To be specific, imagine a vessel whose bottom is defined by the set of N dimensionless discrete levels

$$\left\{\frac{\mu - (\sigma_w^2/\lambda)}{\lambda_k}\right\}_{k=1}^{N}$$

and pour "water" into the vessel in an amount corresponding to the total transmit power P. That power is optimally divided among the N eigenmodes of the MIMO link in accordance with their corresponding "water levels" in the vessel, as illustrated in Fig. G.1 for a MIMO link with $N = 6$. The "water-fill level", denoted by the dimensionless parameter μ and indicated by the dashed line in the figure, is chosen to satisfy the constraint of Eq. (G.44). On the basis of the spatially discrete water-filling picture portrayed in Fig. G.1, we may now finally postulate the optimal $\bar{r}_{s,kk}$ to be

$$\bar{r}_{s,kk} = \left(\mu - \frac{\sigma_w^2}{\lambda_k}\right)^{+} \qquad k = 1,2,...,N \qquad \text{(G.46)}$$

where the superscript "+" signifies retaining only those terms on the right-hand side of the equation that are positive (i.e., the terms that pertain to those eigenmodes of the MIMO link for which the water levels lie below the constant μ). Correspondingly, the maximum value of the capacity of the MIMO link, in accordance with the first line of Eq. (G.43) and Eq. (G.46), is defined by

$$C = \sum_{k=1}^{N} \log_2\left(1 + \frac{1}{\sigma_w^2}\lambda_k\bar{r}_{s,kk}\right)$$

$$= \sum_{k=1}^{N} \log_2\left\{1 + \frac{1}{\sigma_w^2}\lambda_k\left(\mu - \frac{\sigma_w^2}{\lambda_k}\right)^{+}\right\} \qquad \text{(G.47)}$$

$$= \sum_{k=1}^{N} \log_2\left(\frac{\mu\lambda_k}{\sigma_w^2}\right)^{+}$$

where, as stated previously, the constant μ is chosen to satisfy the global power constraint of Eq. (G.44).

The optimal results of Eqs. (G.46) and (G.47), assuming that the channel state is known to both the transmitter and receiver, were derived by considering a MIMO link with $N_r = N_t = N$.

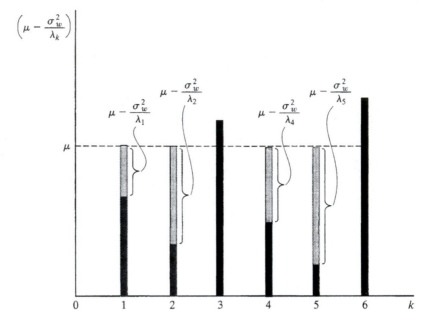

FIGURE G.1 Water-filling interpretation of the optimization procedure. For the example portrayed in the figure, we have the following source autocorrelation values

$$\bar{r}_{s,11} = \mu - \frac{\sigma_w^2}{\lambda_1}$$

$$\bar{r}_{s,22} = \mu - \frac{\sigma_w^2}{\lambda_2}$$

$$\bar{r}_{s,33} = 0$$

$$\bar{r}_{s,44} = \mu - \frac{\sigma_w^2}{\lambda_4}$$

$$\bar{r}_{s,55} = \mu - \frac{\sigma_w^2}{\lambda_5}$$

$$\bar{r}_{s,66} = 0$$

All the other nondiagonal elements of the source correlation matrix $\bar{\mathbf{R}}_s$ are zero.

NOTES AND REFERENCES

[1] For a detailed exposition of the many facets of information theory, see Cover and Thomas (1991).

[2] The first detailed derivation of the log-det capacity formula for a stationary MIMO link was presented by Telatar in an AT&T technical memorandum published in 1995 and republished as a journal paper in 1999.

[3] The waterfilling procedure is described in Cover and Thomas (1991).

APPENDIX H

Eigendecomposition

H.1 UNITARY TRANSFORMATION OF A HERMITIAN MATRIX[1]

Consider a square complex matrix $\mathbf{R}$ of dimensions M by M. The matrix $\mathbf{R}$ is assumed to be Hermitian; that is,

$$\mathbf{R}^\dagger = \mathbf{R} \qquad (H.1)$$

where the superscript † denotes Hermitian transposition. With $\mathbf{R}$ as the matrix of interest, the *eigenvalue problem* is defined by

$$\mathbf{R}\mathbf{q} = \lambda\mathbf{q} \qquad (H.2)$$

where $\mathbf{q}$ is an M-by-1 vector and λ is a scalar.

In general, there are M distinct values of the scalar λ that satisfy Eq. (H.2); these values are roots of the *characteristic equation*

$$\det(\mathbf{R} - \lambda\mathbf{I}) = 0 \qquad (H.3)$$

where $\mathbf{I}$ is the M-by-M identity matrix.

Typically, the off-diagonal elements of matrix $\mathbf{R}$ are nonzero. The *diagonalization* of $\mathbf{R}$ is achieved by expanding on the transformation described in Eq. (H.2). Specifically, we may write

$$\mathbf{Q}^\dagger\mathbf{R}\mathbf{Q} = \mathbf{\Lambda} \qquad (H.4)$$

where

$$\mathbf{\Lambda} = \text{diag}(\lambda_1, \lambda_2, ..., \lambda_M) \qquad (H.5)$$

is a *diagonal matrix* and

$$\mathbf{Q} = [\mathbf{q}_1, \mathbf{q}_2, ..., \mathbf{q}_M] \qquad (H.6)$$

is a *unitary matrix*. The scalars $\lambda_1, \lambda_2, ..., \lambda_M$ constituting the matrix $\mathbf{\Lambda}$ are called the *eigenvalues* of matrix $\mathbf{R}$, and the M-by-1 vectors $\mathbf{q}_1, \mathbf{q}_2, ..., \mathbf{q}_M$ constituting the matrix $\mathbf{Q}$ are the associated *eigenvectors* of $\mathbf{R}$.

By definition, the unitary matrix $\mathbf{Q}$ satisfies the relation

$$\mathbf{Q}\mathbf{Q}^\dagger = \mathbf{Q}^\dagger\mathbf{Q} = \mathbf{I} \tag{H.7}$$

In expanded form, we may rewrite Eq. (H.7) as

$$\mathbf{q}_i^\dagger\mathbf{q}_k = \begin{cases} 1 & \text{for } k = i \\ 0 & \text{for } k \neq i \end{cases} \tag{H.8}$$

According to Eq. (H.7), the inverse of the matrix $\mathbf{Q}$, namely, $\mathbf{Q}^{-1}$, is equal to the Hermitian transpose of $\mathbf{Q}$, or

$$\mathbf{Q}^{-1} = \mathbf{Q}^\dagger \tag{H.9}$$

In light of Eqs. (H.5) through (H.9), we may rewrite Eq. (H.4) in the equivalent form

$$\begin{aligned} \mathbf{R} &= \mathbf{Q}\mathbf{\Lambda}\mathbf{Q}^\dagger \\ &= \sum_{k=1}^{M} \lambda_k\mathbf{q}_k\mathbf{q}_k^\dagger \end{aligned} \tag{H.10}$$

Equation (H.10) is called the *spectral decomposition theorem*, which states that the *Hermitian matrix* $\mathbf{R}$ *can be expanded as the linear combination of the rank-one matrix products* $\left\{\mathbf{q}_k\mathbf{q}_k^\dagger\right\}_{k=1}^{M}$, *and the corresponding eigenvalues* $\{\lambda_k\}_{k=1}^{M}$ *are the scaling factors of the linear combination.*

H.2 RELATIONSHIP BETWEEN EIGENDECOMPOSITION AND SINGULAR-VALUE DECOMPOSITION[2]

Consider next a rectangular complex matrix $\mathbf{A}$ with dimensions L by M. Let the M-by-M matrix $\mathbf{R}$ be related to the matrix $\mathbf{A}$ as follows:

$$\mathbf{R} = \begin{cases} \mathbf{A}\mathbf{A}^\dagger & \text{for } M \geq L \\ \mathbf{A}^\dagger\mathbf{A} & \text{for } M < L \end{cases} \tag{H.11}$$

Then, according to the *singular-value decomposition (SVD) theorem*, the matrix $\mathbf{A}$ may be diagonalized as

$$\mathbf{U}^\dagger\mathbf{A}\mathbf{V} = \begin{bmatrix} \mathbf{D} & \mathbf{0} \\ \mathbf{0} & \mathbf{0} \end{bmatrix} \tag{H.12}$$

where $\mathbf{D}$ is a diagonal matrix, the $\mathbf{0}$'s are null matrices, and $\mathbf{U}$ and $\mathbf{V}$ are respectively L-by-L and M-by-M unitary matrices; that is,

$$\mathbf{U}^\dagger = \mathbf{U}^{-1} \tag{H.13}$$

and

$$\mathbf{V}^\dagger = \mathbf{V}^{-1} \tag{H.14}$$

Specifically, we may make the following statements:

- The diagonal matrix

$$\mathbf{D} = \text{diag}(d_1, d_2, ..., d_W), \quad W = \min(L,M) \tag{H.15}$$

  defines the *singular values* of matrix $\mathbf{A}$.
- The unitary matrix

$$\mathbf{U} = [\mathbf{u}_1, \mathbf{u}_2, ..., \mathbf{u}_L] \tag{H.16}$$

  defines the L *left-singular vectors* of matrix $\mathbf{A}$.
- The second unitary matrix

$$\mathbf{V} = [\mathbf{v}_1, \mathbf{v}_2, ..., \mathbf{v}_M] \tag{H.17}$$

  defines the M *right-singular vectors* of matrix $\mathbf{A}$.

Moreover, depending on whether the dimension L is greater than M or the other way around, we have two different cases in describing the relationships between singular-value decomposition and eigendecomposition:

Case 1. $L > M$
In this case, the dimension $W = M$ and the singular values $d_1, d_2,...,d_M$ are equal to the square roots of the eigenvalues of the matrix product $\mathbf{R} = \mathbf{A}^\dagger \mathbf{A}$. Correspondingly, the right-singular vectors $\mathbf{v}_1, \mathbf{v}_2,...,\mathbf{v}_M$ are the associated eigenvectors.

Case 2. $L < M$
In this second case, the dimension $W = L$ and the singular values $d_1, d_1,...,d_L$ are equal to the square roots of the eigenvalues of the alternative matrix product $\mathbf{R} = \mathbf{A}\mathbf{A}^\dagger$. Correspondingly, the left-singular vectors $\mathbf{u}_1, \mathbf{u}_2,...,\mathbf{u}_L$ are the associated eigenvectors.

NOTES AND REFERENCES

[1] The eigendecomposition of a square matrix is discussed in Chapter 5 of Strang (1980). The discussion presented therein focuses on square matrices that are real.

[2] The singular-value decomposition of a rectangular matrix is discussed in Chapter 7 of Strang (1980). Here again, the discussion focuses on real matrices. The chapter also discusses issues relating to the computation of eigenvalues.

A P P E N D I X   I

Adaptive Array Antennas

I.1 NEED FOR ADAPTIVITY

The goal of wireless communications is to allow as many users as possible to communicate reliably without regard to location and mobility. From the discussion presented in Chapter 2, we find that this goal is seriously impeded by three major channel impairments:

1. *Multipath* can cause severe fading due to phase cancellation between different propagation paths. Fading leads to a reduction in available signal power and therefore a degraded noise performance at the receiver.

2. *Delay spread* results from differences in propagation delays among the multiple propagation paths. When the delay spread exceeds about 10% of the symbol duration, the intersymbol interference experienced by the received signal reaches a significant level, thereby causing a reduction in the attainable data rate.

3. *Cochannel interference* arises in cellular systems in which the available frequency channels are divided into different sets, each of which is assigned to a specific cell and with several cells in the system using the same set of frequencies. Cochannel interference limits the *system capacity* (i.e., the largest possible number of users that can be reliably served by the system).

Typically, cellular systems use 120° sectorization at each base station, and only one user accesses a sector of a base station at a given frequency. We may combat the effects of multipath fading and cochannel interference at the base station by using three identical, but separate, *antenna arrays*, one for each sector of the base station. (The compensation of delay spread is considered later in the section.) Figure I.1 shows the block diagram of an *array signal processor*; it is assumed that there are N users whose signals are received at a particular sector of the base station and that the array for that sector consists of K identical antenna elements. A particular user is treated as the one of interest, and the remaining $N - 1$ users give rise to cochannel interference. In addition to the cochannel interference, each component of the array signal processor's input is corrupted by additive white Gaussian noise (AWGN). The analysis presented herein is

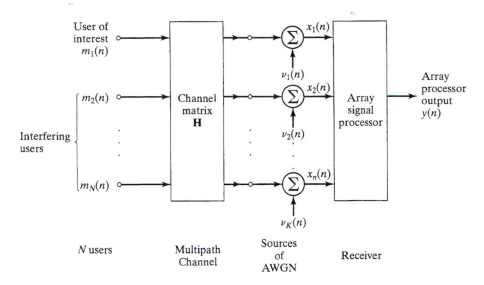

FIGURE I.1   Block diagram of array signal processor that involves K antenna elements and that is being driven by a multipath channel.

for baseband signals, which, in general, are complex valued. This means that both the channel and the array signal processor require complex characterizations of their own. The structure depicted in Fig. I.1 is drawn for one output pertaining to the user of interest. The array signal processor is duplicated for users at other frequencies at the base station.(Figure I.1 refers to a situation different from that considered in Chapter 6 and Appendix G, hence the difference in notation.)

The multipath channel is characterized by the channel matrix, which is denoted by $\mathbf{H}$. The matrix $\mathbf{H}$ has dimensions K-by-N and may therefore be expanded into N column vectors, as shown by:

$$\mathbf{H} = [\mathbf{h}_1, \mathbf{h}_2, ..., \mathbf{h}_N] \tag{I.1}$$

Column vector $\mathbf{h}_i$, $i = 1, 2, ..., N$ is of dimension K, and represents the multipath component pertaining to user i.

Given the configuration described in Fig. I.1, the goal is to design a *linear array signal processor* for the receiver that satisfies two requirements:

1. The cochannel interference produced by the $N-1$ interfering users is cancelled.
2. The output signal-to-noise ratio (SNR) for the user of interest is maximized.

Hereafter, these two requirements are referred to as design requirements 1 and 2.

To proceed with this design task, it is assumed that the multipath channel is described by flat Rayleigh fading. Then, in light of the material presented in Example 6.2, we find that the use of diversity permits the treatment of the column vectors $\mathbf{h}_2, \mathbf{h}_3, ..., \mathbf{h}_N$ as *linearly independent*, which is justified, provided that the spacing between antenna elements of the array is large enough (10 to 20 times the wavelength) for independent fading. To simplify the presentation, we suppose that user 1 is the user of interest and the remaining $N - 1$ users are responsible for co-channel interference, as indicated in Fig. I.1. The key design issue is how to find the *weight vector*, denoted by **w**, that characterizes the array signal processor. Toward that end, we may proceed as follows:

1. We choose the K-dimensional weight vector **w** to be orthogonal to the vectors $\mathbf{h}_2, \mathbf{h}_3, ..., \mathbf{h}_N$, which are associated with the interfering users. This choice fulfills design requirement 1 (i.e., the cancellation of cochannel interference).

2. To satisfy design requirement 2 (i.e., maximization of the SNR), we will briefly digress from the issue at hand to introduce the notion of a subspace. Given a *vector space*, or just a *space*, formed by a set of linearly independent vectors, a *subspace* of the space is a subset that satisfies two conditions:

   (i) If we add any two vectors $\mathbf{z}_1$ and $\mathbf{z}_2$ in the subspace, their sum $\mathbf{z}_1$ and $\mathbf{z}_2$ is still in the subspace.

   (ii) If we multiply any vector **z** in the subspace by any scalar a, the multiple $a\mathbf{z}$ is still in the subspace.

Define the subspace $\mathcal{W} \perp \{\mathbf{h}_2, \mathbf{h}_3, ..., \mathbf{h}_N\}$. Then, returning to the issue of how to maximize the output SNR for user 1, we first construct a subspace denoted by $\mathcal{W}$ whose dimension is equal to the difference between the number of antenna elements and the number of interfering users—that is, $K - (N - 1) = K - N + 1$. Next, we project the complex conjugate of the channel vector $\mathbf{h}_1$ (pertaining to user 1) onto the subspace $\mathcal{W}$. The projection so computed defines the weight vector **w**.

EXAMPLE I.1 Subspace Method for Determining the Weight Vector

To illustrate the two-step subspace method for determining the weight vector **w**, consider the simple example of a system involving two users characterized by the channel vectors $\mathbf{h}_1$ and $\mathbf{h}_2$, and an antenna array consisting of three elements; that is, $N = 2$ and $K = 3$. Then, for this example, the subspace $\mathcal{W}$ is two-dimensional, since

$$K - N + 1 = 3 - 2 + 1 = 2$$

With user 1 viewed as the user of interest and user 2 viewed as the interferer, we may construct the signal-space diagram shown in Fig. I.2. The subspace $\mathcal{W}$, shown shaded in this figure, is orthogonal to channel vector $\mathbf{h}_2$. The weight vector **w** of the array signal processor is determined by the projection of the complex-conjugated channel vector of user 1 (i.e., $\mathbf{h}_1{}^*$) onto the subspace $\mathcal{W}$, as depicted in the figure.

The important conclusion drawn from this discussion is that a linear receiver using optimum combining with K antenna elements and involving $N - 1$ interfering users has the same performance as a linear receiver with $K - N + 1$ antenna elements, without interference, independent of the multipath environment. For this equivalence to be realized, we of course require that $K > N - 1$. Provided that this condition is satisfied, the receiver cancels the cochannel interference with a diversity improvement equal to $K - N + 1$, which represents an N-fold increase in system capacity.

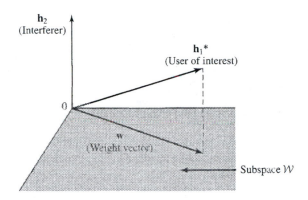

FIGURE I.2   Signal-space diagram for Example I.1, involving a user of interest, a single interferer, and an antenna array of three elements. The subspace $\mathcal{W}$, shown shaded, is two dimensional in this example.

The design of an array signal processor in accordance with the two-step subspace procedure described herein is of the *zero-forcing* kind. We say this because, given K antenna elements, the array has enough degrees of freedom to *force* the output due to the $N-1$ interfering users, represented by the linearly independent channel vectors $\mathbf{h}_2, \ldots, \mathbf{h}_K$, to zero so long as K is greater than $N-1$. Note also that this procedure includes $N=1$ (i.e., a single user with no interfering users) as a special case. In this case, the channel matrix consists of vector $\mathbf{h}_1$ that lies in the subspace $\mathcal{W}$, and the zero-forcing solution $\mathbf{w}$ equals $\mathbf{h}_1{}^*$.   ∎

The analysis presented thus far has been entirely of a *spatial* kind that ignores the effect of delay spread. What if the delay spread is significant compared with the symbol duration, and cannot, therefore, be ignored? Recognizing that delay-spread is responsible for intersymbol interference, we may incorporate a *linear* equalizer in each antenna branch of the array to compensate for delay spread. The resulting array signal processor takes the form shown in Fig. I.3, which combines temporal and spatial processing. Spatial processing is provided by the antenna array, and the temporal processing is provided by a bank of finite-duration impulse response (FIR) filters. For obvious reasons, this structure is called a *space–time processor*.

I.1.1   Adaptive Antenna Arrays[1]

The subspace design procedure for the array signal processor in Fig. I.1 assumes that the channel impairments are stationary and that we have knowledge of the channel matrix **H**. In reality, however, multipath fading, delay spread, and cochannel interference are all *nonstationary* in their own individual ways. Also, the channel characterization may be unknown. To deal with these practical issues, we need to make the receiving array signal processor in Fig. I.1 *adaptive*. Bearing in mind the scope of this book, we confine the discussion to adaptive spatial processing, assuming that the delay spread is negligible. We further assume that the multipath fading phenomenon is slow enough to justify the *least-mean-square (LMS) algorithm* to perform the adaptation.

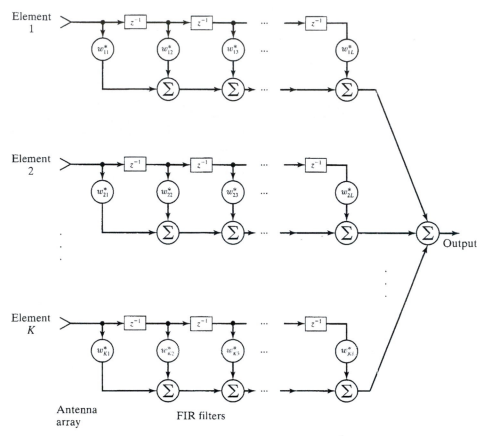

FIGURE I.3   Baseband space–time processor. The blocks labeled z^{-1} are unit-delay elements, with each delay being equal to the symbol period. The filter coefficients are complex valued. The FIR filters are all assumed to be of length L.

I.1.2   Least-Mean-Square (LMS) Algorithm

Figure I.4 shows the structure of an *adaptive antenna array*, in which the output of each antenna element is multiplied by an adjustable (controllable) weight w_k, $k = 1,2,...,K$, and then the weighted elemental outputs of the array are summed to produce the array output signal, denoted by y. The adaptive antenna array does not require knowledge of the direction of arrival of the desired signal originating from a user of interest, so long as the system is supplied with a *reference signal*, which is *correlated* with the desired signal. For example, the reference signal could correspond to a training sequence that is transmitted on a periodic basis. The output signal of the array is subtracted from the reference signal, denoted by d, to generate an *error signal* e, which is used to apply the appropriate adjustments to the elemental weights of the array. In this way, a feedback system to control the elemental weights is built into the operation of the antenna array, thereby making it adaptive to changes in the environment. Note that the block diagram is drawn for baseband processing. In a practical system, a quadrature hybrid is

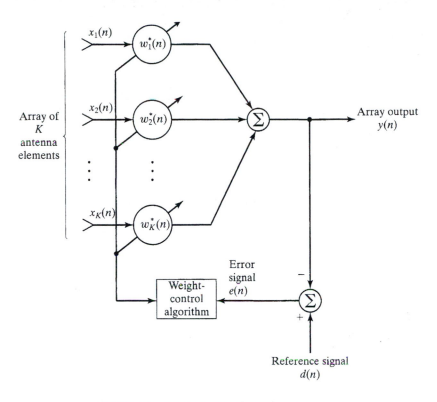

FIGURE I.4   Block diagram of adaptive antenna array.

used for each antenna element of the array to split the complex-valued received signal at each element into two components: one real and the other imaginary. The use of a hybrid has been omitted from the figure, to simplify the diagram.

Let $x_k(n)$ denote the output of the kth element in the array at discrete time n, and let $w_k(n)$ denote the corresponding value of the weight connected to this element. Then the output signal of the array (consisting of K antenna elements) is

$$y(n) = \sum_{k=1}^{K} w_k^*(n)x_k(n) \tag{I.2}$$

where $w_k^*(n)x_k(n)$ is the inner product of the complex-valued quantities $w_k(n)$ and $x_k(n)$. Denoting the reference signal as $d(n)$, we may evaluate the *error signal* as

$$e(n) = d(n) - y(n) \tag{I.3}$$

To optimize the performance of the adaptive antenna array, it is customary to use the *mean-square error*

$$J = E[|e(n)|^2] \tag{I.4}$$

as the cost function to be *minimized*. Minimization of the cost function J tends to suppress the interfering signals and thereby enhance the desired signal in the array output. The LMS algorithm minimizes the instantaneous value of the cost function J and, through successive iterations, approaches the minimum mean-square error (MMSE) (i.e., the optimum solution for the elemental weights) ever more closely. An adaptive antenna array based on the minimum mean-square error criterion is highly likely to provide a better solution than one based on the zero-forcing criterion embodied in the two-step subspace method.

The adjustment applied to the kth elemental weight is

$$\Delta w_k(n) = \mu e^*(n)x_k(n) \qquad k = 1, 2, \cdots, K \tag{I.5}$$

where μ is the *step-size parameter*. The updated value of this weight is

$$w_k(n + 1) = w_k(n) + \Delta w_k(n) \qquad k = 1, 2, \cdots, K \tag{I.6}$$

Equations (I.2), (I.3), (I.5) and (I.6), in that order, constitute the *complex LMS algorithm*.[2] The algorithm is initiated by setting $w_k(0) = 0$ for all k.

The advantages of an adaptive antenna array using the complex LMS algorithm are threefold:

1. Simplicity of implementation
2. Only a linear growth in complexity with the number of antenna elements
3. Robust performance with respect to disturbances.

However, the system suffers from the following drawbacks:

- A slow rate of convergence, which is typically 10 times the number of adjustable weights. This limits the use of the complex LMS algorithm to a slow-fading environment, for which the Doppler spread is small compared with the reciprocal of the duration of the observation interval.
- Sensitivity of the convergence behavior to variations in the reference signal and cochannel interference powers.

These limitations of the complex LMS algorithm can be overcome by using an algorithm known as *direct matrix inversion* (DMI).[3] Unlike the LMS algorithm, the DMI algorithm operates in *batch* mode, in that the computation of the elemental weights is based on a batch of L snapshots. The batch size L is chosen as a compromise between two conflicting requirements:

- The size L should be small enough for the batch of snapshots used in the computation to be justifiably treated as pseudostationary.
- The size L should be large enough for the computed values of the elemental weights to approach the MMSE solution.

The DMI algorithm is the optimum combining technique for array antennas deployed in many base stations today. The algorithm may be reformulated for recursive computation if desired.[4]

When the teletraffic is high, the base stations are ordinarily configured as micro-cells, which are small cells such as an office floor or a station deployed along a highway with directional antennas. In such a configuration, there are many inexpensive base stations in close proximity to each other. The use of adaptive antenna arrays provides the means for an alternative configuration in which there are fewer (but more expensive) base stations, further apart from each other than in the corresponding microcellular system.

NOTES AND REFERENCES

[1] For a discussion of adaptive antenna arrays and their theory, design, and applications, see Compton (1988).

[2] The least-mean-square (LMS) algorithm is discussed Haykin (2002) and Widrow and Stearns (1985).

[3] The direct matrix inversion (DMI) algorithm, also referred to as the sample matrix inversion method, is discussed in Compton (1988); see pp. 331–332.

[4] The recursive least-squares (RLS) algorithm provides an iterative method for implementing the method of least squares, which lies behind the DMI; for details, see Haykin (2002).

Bibliography

Abramowitz, M. and I. Stegun, eds. *Handbook of Mathematical Functions.* New York: Dover, 1964.

Abramson, N. "The ALOHA system—Another alternative for computer communications," in 1970 Fall Joint Compt. Conf., *AFIPS Conf. Proc.*, vol. 37. Montvale, NJ: AFIPS Press, 1970, pp. 281–285.

Abramson, N. (ed.). *Multiple Access Communications,* New York: IEEE Press, 1993.

Alamouti, S. "A simple transmitter diversity scheme for wireless communications." *IEEE J. Selected Areas in Communications,* Vol. 16, pp. 1451–1458, 1998.

Anderson, J.B., T. Aulin, and C.E. Sundberg. *Digital Phase Modulation.* New York: Plenum Press, 1986.

Anderson, J.B., *Digital Transmission Engineering.* IEEE Press, 1999.

ANSI/IEEE Std 802.11, Part 11: Wireless LAN Medium Access Control (MAC) and Physical Layer (PHY) Specifications, 1999 edition.

Bahai, A.R.S. and Saltzberg, B.R. *Multi-Carrier Digital Communications: Theory and Applications of OFDM,* Kluwer Academic/Plenum Publishers, 1999.

Bahl, L., J. Cocke, F. Jelinek, and J. Raviv. "Optimal decoding of linear codes for minimizing symbol error rate." *IEEE Trans. Inform. Theory,* Vol. IT-20, 284–287, 1974.

Barker, R.H. "Group synchronization of binary digital systems." pp. 273–2876, in W. Jackson (ed.), *Communication Theory.* New York: Academic Press, 1953.

Benedetto, J.S. and G. Montrosi. "Unveiling turbo codes: Some results on parallel concatenated coding schemes." *IEEE Trans. Inform. Theory,* Vol. 42, pp. 409–428, March 1996.

Benedetto, S. and E. Biglieri. *Principles of Digital Transmission with Wireless Applications.* New York: Kluwer Academic/Plenum Publishers, 1999.

Berrou, C. "The ten-year old turbo codes are entering into service." *IEEE Communications Magazine,* pp. 110–116, August 2003.

Berrou, C. and A. Glavieux. "Near optimum error correcting coding and decoding: Turbo-codes." *IEEE Trans. Commun.,* Vol. 44, pp. 1261–1271, 1996.

Berrou, C., A. Glavieux, and P. Thitimajshima. "Near Shannon-limit error correction coding and decoding: Turbo codes," in *Proc. 1993 International Conference on Communications,* pp. 1064–1070, Geneva, Switzerland, May 1993.

Bertsekas, D. and R. Gallager. *Data Networks,* 2nd ed. Englewood Cliffs, NJ: Prentice Hall, 1992.

Blachman, N.M. *Noise and its Effect on Communication*, McGraw-Hill, 1966.

Blackard, K.L. and T.S. Rappaport. "Measurements and models of radio frequency impulse noise for indoor wireless communications." *IEEE J. Selected Areas in Communications*, Vol. 11, pp. 991–1001, Sept. 1993.

Boutros, J., N. Gresset, and L. Brunel. "Turbo coding and decoding for multiple antenna channels." *International Symposium on Turbo Codes, Proceedings*, pp. 185–86, Brest, France, Sept. 2003.

Bracewell, R.N. *The Fourier Transform and Its Applications*, 2nd ed., rev. New York: McGraw-Hill, 1986.

Brennan, D.G. "Linear diversity combining techniques." *Proceedings of the IRE*, Vol. 47, pp. 1075–1102, June 1959.

Caire, G., G. Taricco, and E. Biglieri. "Bit-interleaved coded modulation." *IEEE Trans. Inform. Theory*, Vol. 44, pp. 927–946, 1998.

Carlson, A.B. *Communication Systems*, 2nd ed. New York: McGraw-Hill, 1975.

Cassioli, D., M.Z. Win, and A.F. Molisch. "The Ultra-Wide Bandwidth Indoor Channel: From Statistical Model to Simulations," *IEEE J. Selected Areas in Commun.*, Vol. 20, pp. 1247–1257, 2002.

Chennakeshu, S. and G.J. Saulnier. "Differential detection of $\pi/4$-shifted-DQPSK for digital cellular radio." *IEEE Trans. Vehicular Technology*, Vol. 42, pp. 46–57, 1993.

Cheung, K.W., J.H.M. Sau, and R.D. Murch. "A new empirical model for indoor propagation prediction." *IEEE Trans. Vehicular Technology*, 47(3), pp. 996–1001, August 1998.

Chizhik, D., G. Foschini, and R.A. Valenzuali. "Capacities of multi-element transmit and receive antennas: Correlations and keyholes." *Electronics Letters*, Vol. 36, pp. 1099–1100, 2000.

Chugg, K., A. Anastasopoulos, and X. Chen. *Iterative Detection: Adaptivity, complexity, and applications*. Boston: Kluwer, 2001.

Clarke, R.H. "A statistical theory of mobile radio reception." *Bell System Tech. J.*, Vol. 47, pp. 957–1000, 1968.

Clarke, G.C., and Cain, J.B., *Error-Correction Coding for Digital Communications*. New York: Plenum, 1981.

Compton, R.T. *Adaptive Antennas: Concepts and Performance*. Englewood Cliffs, NJ: Prentice-Hall, 1988.

Cover, T.M. and J.A. Thomas. *Elements of Information Theory*. New York: Wiley, 1991.

deBuda, R. "Coherent demodulation of frequency-shift keying with low deviation ratio." *IEEE Trans. Communications*, Vol. COM-20, pp. 429–435, 1972.

deHaas, A. *Radio-Nieuws*, Vol. 10, pp. 357-364, December 1927, and Vol. 11, pp. 80-88, February 1928; portions of these two papers were translated by B.B. Barrow, under the title "Translation of a historic paper on diversity reception." *Proc. IRE*, Vol. 49, pp. 367–369, January 1961.

deJager, F. and C.B. Dekker. "Tamed frequency modulation—A novel method to achieve spectrum economy in digital transmission." *IEEE Trans. Communications*, Vol. 26, pp. 534–542, May 1978.

Diggavi, S.N., N. Al-Dhahir, A. Stanoulis, and A.R. Calderbank. "Differential space–time coding for frequency on selective channels." *IEEE Communications Letters*, Vol. 6, pp. 253–255, 2002.

Diggavi, S.N., N. Al-Dhahir, A. Stanoulis, and A.R. Calderbank. "Great expectations: The value of spatial diversity in wireless networks." *Proceedings of the IEEE*, Vol. 91, 2003.

Diggavi, S.N. "Role of spatial diversity in wireless networks," Workshop on New Directions for Statistical Signal Processing in the 21st Century," Lake Louise, Alberta, Canada, 5–10 October, 2003.

Divsalar, D., and F. Pollara. "Turbo codes for PCS applications." *Proceedings of International Communications Conference, Seattle, Washington*, pp. 54-59, June 1995.

Dixon, R.C. *Spread Spectrum Systems with Commercial Applications*, 3rd ed. New York: Wiley, 1994.

Doelz, M.I. and E.H. Heald. Minimum shift data communication system, U.S. Patent 2977417, March 1961.

Elias, P. "Coding for noisy channels." *IRE Convention Record, Part 4*, pp. 37–46, March 1955.

Ericson, T. "A Gaussian channel with slow fading." *IEEE Trans. Inform. Theory*, Vol. 47, pp. 2321–2334, 2001.

Forney, G.D. Jr. "The Viterbi algorithm." *Proceedings of the IEEE*, Vol. 61, pp. 268–278, 1973.

Foschini, G.J. "Layered space–time architecture for wireless communication in a fading environment when using multi-element antennas." *Bell Labs Technical J.*, Vol. 1, No. 2, pp. 41–59, 1996.

Foschini, G.J., and M.J. Gans. "On limits of wireless communications in a fading environment when using multiple antennas." *Wireless Personal Communications*, Vol. 6, pp. 311–335, 1998.

Foschini, G.J., D. Chizkik, M.J. Gans, C. Papadies, and R.A. Valenzueli. "Analysis and performance of some basic space–time architectures." *IEEE J. Select. Areas Communications*, Vol. 21, pp. 303–320, 2003.

Foschini, G.J., G.D. Golden, R.A. Valenzuela, and P.W. Wolniansky. "Simplified processing for wireless communication at high spectral efficiency." *IEEE J. Selected Areas in Communications*, Vol. 17, pp. 1841–1852, 1999.

Gallager, R.G. *Information Theory and Reliable Communications*. New York: John Wiley and Sons, 1968.

Garg, V.K. *IS-95 CDMA and cdma 2000*. Upper Saddle River, NJ: Prentice Hall, 2000.

Gesbert, D., H. Bolseskei, D. Gore, and A. Paulraj. "Outdoor MIMO wireless channels: Models and performance prediction." *IEEE Trans. Communications*, Vol. 50, pp. 225–234, 2002.

Gesbert, D., M. Shafi, D.S. Shiu, P. Smith, and A. Naguib. "From theory to practice: An overview of MIMO space–time coded wireless systems." *IEEE J. Selected Areas in Communications*, Vol. 21, pp. 281–302, 2003.

Gold, R. "Optimal binary sequences for spread spectrum multiplexing." *IEEE Trans. Inform. Theory*, Vol. IT-14, pp. 154–156, 1968.

Golden, G.D., J.G. Forschini, R.A. Valenzuela, and P.W. Woliansky. "Detection algorithm and initial laboratory results using V-BLAST space–time communication architecture." *Electronics Letters*, Vol. 35, pp. 14–15, 1999.

Goldsmith, A., S.A. Jafar, N. Jindal, and S. Vishwaneth. "Fundamental capacity of MIMO channels." *IEEE J. Selected Areas in Communications*, Vol. 21, pp. 684–702, 2003.

Hagenauer, J. "The turbo principle: Tutorial introduction and state of the art," International Symposium Turbo Codes. Brest, France, Sept. 1997.

Hanzo, L., T. Liew, and B. Yeap. *Turbo Coding, Turbo Equalization and Space–Time Coding*. Hoboken, NJ: Wiley, 2002.

Hata M., "Empirical formal for propagation loss in land mobile radio services," *IEEE Trans. Vehic. Technol.*, Vol. 29, August 1980.

Haykin, S. *Adaptive Filter Theory*, 4th ed., Upper Saddle River, NJ: Prentice Hall, 2002.

Haykin, S. *Communication Systems*. 4th ed. New York: John Wiley, 2001.

Heegard, C. and S.B. Wicker. *Turbo Coding*. Boston: Kluwer Academic Publishers, 1999.

Holma, H. and A. Toskala (eds.). *WCDMA for UMTS*. New York: Wiley, 2000.

Holmes, J.K. *Coherent Spread Spectrum Systems.* Melbourne, FL: 1990.

Hochwald, B.M., T.L. Marzetta, and V. Tarokh. "Multi-antenna channel-hardening and its implications for rate feedback and scheduling," submitted to *IEEE Trans. Inform. Theory.*

Hottinen, A., O. Tirkkonen, and R. Wichman. *Multi-antenna Transceiver Techniques for 3G and Beyond.* Hoboken, NJ: Wiley, 2003.

ITU Recommendations. The ITU publishes a variety of recommendations regarding propagation models, interference models, and other topics of interest to radio engineers.

Jakes, W.C. (ed.). *Microwave Mobile Communications.* New York: Wiley, 1974. Reprinted by IEEE Press, 1994.

Kailath, T. *Sampling models for linear time-variant filters,* Technical Report 352, MIT RLE, Cambridge, MA, May 25, 1959.

Kaplan, E.D. (ed.). *Understanding GPS: Principles and Applications.* Boston: Artech House, 1996.

Lee, J.S. and L.E. Miller. *CDMA Systems Engineering Handbook.* Boston: Artech House, 1998.

Lee, W.C.Y. *Mobile Communications Engineering.* New York: McGraw-Hill, 1982.

Leon-Garcia, A. *Probability and Random Processes,* 2nd ed., Reading, MA: Addison-Wesley, 1994.

Liberti, J.C., Jr. and T.S Rappaport. *Smart Antennas for Wireless Communications.* Upper Saddle River, NJ: Prentice Hall, 1999.

Liew, T.H. and L. Hanzo. "Space–time codes and concatenated channel codes for wireless communications." *Proceedings of the IEEE,* Vol. 90, pp. 187–219, 2002.

Lin, S., and D.J. Costello, Jr. *Error Control Coding: Fundamentals and Applications.* Englewood Cliffs, NJ: Prentice-Hall, 1983.

Lodge, J.H. and M.L. Moher. "Maximum likelihood sequence estimation of CPM signals transmitted over Rayleigh flat-fading channels." *IEEE Trans. Communications,* Vol. 38, pp. 787–794, 1990.

McEliece, R. *The Theory of Information and Coding.* Reading, MA: Addison-Wesley, 1977.

Médard, M. *The Capacity of Time Varying Multiple User Channels in Wireless Communications.* Doctor-of-Science Thesis in Electrical Engineering, MIT, Cambridge, MA, Sept. 1995.

Mehrotra, A. *GSM System Engineering.* Boston: Artech House, 1997.

Michelson, A.M. and A.H. Levesque. *Error-control Techniques for Digital Communication.* New York: Wiley, 1985.

Moher, M. "An iterative multiuser decoder for near capacity communications." *IEEE Trans. on Communications,* Vol. 46, pp. 870–880, July 1998.

Moher, M.L. and J.H. Lodge. "TCMP–a modulation and coding strategy for Rician fading channels." *IEEE J. Selected Areas Communications,* Vol.7, pp. 1347–1355, 1989.

Monzingo, R.A. and T.W. Miller. *Introduction to Adaptive Arrays.* New York: Wiley, 1980.

Nakahima, N., et al. "A system design for TDMA mobile radios." *Proc. 40th IEEE Vehicular Technology Conference,* pp. 295–298, May 1990.

Noble, D.E. "The history of land-mobile radio communications." *Proceedings of the IRE,* Section 28 Vehicular Communications, pp. 1405–1414, May 1962.

Nyquist, H. "Thermal agitation of electric charge in conductors." *Physical Review,* second series, Vol. 32, pp. 110–113, 1928.

O'Hara, B. and A. Petrick. *IEEE 802.11 Handbook: A Designer's Companion.* New York: IEEE Press, 1999.

Okumura, Y., Ohmori, E., and Fukuda, K., "Field strength and its variability in VHF and UHF land mobile radio services," *Review of the Electrical Communications Laboratory*, Vol. 16, Sept–Oct 1968.

Oppenheim, A.V., R.W. Schafer, and J.R. Buck. *Discrete-time Signal Processing*, 2nd. ed. Upper Saddle River, NJ: Prentice Hall, 1999.

Pahlavan, K., and Levesque, A., *Wireless Information Networks*, New York: Wiley-Interscience, 1995.

Papadias, C. and G. Foschini. "On the capacity of certain space–time coding schemes." *EURASIP Journal on Applied Signal Processing* 2002:5, pp. 447–458, Hindawi Publishing Corporation, 2002.

Parsons, D. *The Mobile Propagation Channel*. New York: Wiley, 1992.

Paulraj, A. "Diversity Techniques." chapter in *CRC Handbook on Communications*, ed. J. Gibson. Boca Raton, FL: CRC Press, 11, pp. 213–223, Dec. 1996.

Peterson, R.L., R.E. Ziemer, and D.E. Borth. *Introduction to Spread Spectrum Communications*. Upper Saddle River, NJ: Prentice Hall, 1995.

Proakis, J.G. *Digital Communications*, 3rd ed. New York: McGraw-Hill, 1995.

Ramo, S., J.R. Whinnery, and T. Van Duzer. *Fields and Waves in Communications Electronics*. New York: Wiley, 1965.

Rappaport, T.S. *Wireless Communications: Principles and Practice*. Upper Saddle River, N.J.: Prentice Hall, 1999.

Rappaport, T.S. *Wireless Communications: Principles and Practice*, 2nd. ed. Upper Saddle River, N.J.: Prentice Hall, 2002.

Rice, S.O. "Mathematical analysis of random noise," *Bell System Technical J.*, Vol. 23, pp. 282–332, 1944; Vol. 24, pp. 46–156, 1945. Reprinted in N. Wax (ed.), *Selected papers on Noise and Stochastic Processes*, New York: Dover, 1954.

Roddy, W. *Satellite Communications*. 2nd ed. New York: McGraw-Hill, 1996.

Saleh, A.A.M. and R.A. Valenzuela. "A statistical model for indoor multipath propagation." *IEEE J. Selected Areas Commun.*, 5, pp. 128–137, February 1987.

Scholtz, R.A. "The origins of spread spectrum communications." *IEEE Trans. Communications*, Vol. 30, pp. 822–854, 1982.

Schwartz, M., W.R. Bennett, and S. Stein. *Communications Systems and Techniques*. New York: McGraw-Hill, 1966. Reprinted by IEEE Press, 1995.

Sellathurai, M. and S. Haykin. "Joint beamformer estimation and co-antenna interference cancelation for turbo-BLAST," in *Proc. IEEE Int. Conf. Acoustics, Speech, Signal Processing*, Salt Lake City, UT, May 2001.

Sellathurai, M. and S. Haykin. "Turbo-BLAST: A novel technique for multi-transmit multi-reccive wireless communications." *Multiaccess, Mobility, and Teletraffic for Wireless Communications*, Vol. 5, pp. 13–24, Kluwer Academic Publishers.

Sellathurai, M. and S. Haykin. "Turbo-BLAST for high speed wireless communications." *Proceedings of wireless communications and network conf.*, Chicago, Sept. 2000.

Sellathurai, M. and S. Haykin. "Turbo-BLAST for Wireless Communications: Theory and Experiments." *IEEE Trans. Signal Processing*, Special Issue on MIMO Wireless Communications, Vol. 50, pp. 2538–2546, 2002.

Sellathurai, M. and S. Haykin. "Turbo-BLAST for Wireless Communications: first experimental results," *IEEE Trans. on Vehicular Technology*, Vol. 52 pp. 530–535, 2003.

Sellathurai, M. and S. Haykin. "Turbo-BLAST: Performance Evaluation in Correlated Rayleigh Fading Environment." *IEEE J. Selected Areas of Communications*, Special issue, Vol. 21 pp. 340–349, 2003.

Shankar, P.M. *Introduction to Wireless Systems*, Hoboken, NJ: Wiley, 2002.

Shannon, C. "A mathematical theory of communication." *Bell System Technical J.*, Vol. 27, pp. 379–423, 623–656, 1948.

Shiu D.-S., G.J. Foschini, M.J. Gans, and J.M. Kahn. "Fading correlation and the effect on the capacity of multielement antenna systems." *IEEE Trans. Communications*, Vol. 48, pp. 502–513, 2000.

Simon, M.K., J.K. Omura, R.A. Scholtz, and B.K. Levitt. *Spread Spectrum Communications*, 3 Vols., Rockville, MD: Computer Science Press, 1985.

Simon, M.K., J.K. Omura, R.A. Scholtz, and B.K. Levitt. *Spread Spectrum Communications Handbook*, New York: McGraw-Hill, 1994.

Simon, S. and A. Moustakas. "Optimizing MIMO antenna systems with channel covariance feedback." *IEEE J. Selected Areas Communications*, Special Issue on MIMO Systems, Vol. 21, Iss. 3, pp. 406–417, 2003.

Sklar, B. *Digital Communications, Fundamentals and Applications*, 2nd ed., Upper Saddle River, NJ: Prentice Hall, 2001.

Smith, P.J., M. Shafi, and G. Lebrun. "MIMO capacity in Rician fading channels: An exact characterization," under preparation, 2003.

Spilker, J.J., Jr. *Digital Communications by Satellite*. Englewood Cliffs, NJ: Prentice-Hall, 1977.

Starr, T., J.M. Cioffi, and P.J. Silverman. *Understanding Digital Subscriber Line Technology*. Upper Saddle River, NJ: Prentice Hall, 1999.

Steele, R. and L. Hanzo. *Mobile Radio Communications*, 2nd ed. New York: Wiley/IEEE Press, 1999.

Stein, S. "Linear diversity combining techniques," Chapter 10 in M. Schwartz, W.R. Bennett, and S. Stein, Communication Systems and Techniques, McGraw-Hill, 1966.

Strang, G. *Linear Algebra and Its Applications*, 2nd ed. New York: Academic Press, 1980.

Stüber, G.L. *Principles of Mobile Communication*, 2nd ed. Boston: Kluwer Academic, 2001.

Stutzman, W.L. and G.A. Thiele. *Antenna Theory and Design*, 2nd ed. New York: Wiley, 1998.

Tabagi, F.A. and L. Kleinroke. "Packet switching in radio channels." *IEEE Trans. Communications*, Part I, Vol. 23, pp. 1400–1416, Part II, Vol. 23, pp. 1417–1433, 1975.

Tabagi, F.A. and L. Kleinroke. "Packet switching in radio channels." *IEEE Trans. Communications*, Part III, Vol. 24, pp. 832–844, 1976.

Tarokh, V. and H. Jafarkhani. "A differential detection scheme for transmit diversity." *IEEE J. Selected Areas in Communications*, Vol. 18, pp. 1169–1174, 2000.

Tarokh, V., H. Jafarkhani, and A. Calderbank. "Space–time block codes from orthogonal designs." *IEEE Trans. Inform. Theory*, Vol. 45, pp. 1456–1467, 1999.

Tarokh, V., H. Jafarkhani, and A.R. Calderbank. "Space–time block coding for wireless communications: Performance results." *IEEE J. Select. Areas Communications*, Vol. 17, pp. 451–460, Mar. 1999.

Tarokh, V., A. Naguib, N. Seshadri, and A. Calderbank. "Combined array processing and space–time coding." *IEEE Trans. Inform. Theory*, Vol. 45, pp. 1121–1128, 1999.

Tarokh, V., A. Naguib, N. Seshadri, and A. Calderbank. "Space–time codes for high data rate wireless communication: Performance criteria in the presence of channel estimation errors, mobility, and multiple paths." *IEEE Trans. Communications*, Vol. 47, pp. 199–207, 1999.

Tarokh, V., N. Seshadri, and A.R. Calderbank, "Space–time codes for high data rate wireless communication: Performance criterion and code construction." *IEEE Trans. Information Theory*, Vol. 44, No. 2, pp. 744–765, 1998.

Technology Industry Association (TIA), Project No. PN-2759, "Cellular system dual-mode mobile station - Base station compatibility standard," (IS-54, Revision B) Jan. 9, 1992.

Telatar, I.E. "Capacity of multi-antenna Gaussian channels." *European Trans. Telecommunications*, Vol. 10, pp. 585–595, 1999. (Originally published as AT&T Technical Memorandum, 1995).

ten Brink, S. "Convergence of iterative decoding." *Electronic Letters*, Vol. 53, No. 10, pp. 806–808, May 1999.

Tobagi, F., and Kleinrock, L., "Packet Switching in Radio Channels: Part II - The Hidden Terminal Problem in Carrier Sense Multiple-Access and the Busy-Tone Solution", *IEEE Trans. on Communications*, Vol. 23, pp. 1417–1433, December 1975.

Ungerboeck, G. "Channel coding with multilevel/phase signals." *IEEE Trans. Inform. Theory*, Vol. IT-28, pp. 55–67, 1982.

Ungerboeck, G. "Trellis-coded modulation with redundant signal sets." *Parts 1 and 2, IEEE Communications Magazine*, Vol. 25, No. 2, pp. 5–21, 1987.

Verdú, S. "Minimum probability of error for asynchronous Gaussian multiple-access channels." *IEEE Trans. Inform. Theory*, Vol. 32, pp. 85–96, 1986.

Verdú, S. *Multiuser Detection*. Cambridge, UK: Cambridge University Press, 1998.

Vishwanath, S., S. Jafar, and A. Goldsmith. "Channel capacity and beamforming for multiple transmit and receive antennas with covariance feedback." *International Communications Conference*, 2001, Helsinki, Finland.

Viterbi, A.J. *CDMA: Principles of Spread Spectrum Communications*. Reading, MA: Addison-Wesley, 1995.

Viterbi, A.J. "Error bounds for convolutional codes and an asymptotically optimum decoding algorithm." *IEEE Trans. Inform. Theory*, Vol. IT-13, pp. 260–269, 1967.

Viterbi, A.J. and J.K. Omura. *Principles of Digital Communication and Coding*. New York: McGraw-Hill, 1979.

Wang, X., and H.V. Poor. "Robust multiuser detection in non-Gaussian channels." *IEEE Trans. Signal Processing*, Vol. 47, pp. 289–305, 1999.

Widrow, B. and S.D. Stearns. *Adaptive Signal Processing*. Englewood Cliffs, NJ: Prentice-Hall, 1985.

Win, M.Z. and R.A. Scholtz. "Impulse radio: how it works," *IEEE Commun. Letters*, Vol. 2, pp. 36–38, 1998.

Wolniansky, P.W., J.G. Foschini, G.D. Golden, and R.A. Valenzuela. "V-BLAST: An architecture for realizing very high data rates over the rich-scattering wireless channel." *Proceedings of ISSSE*, Pisa, Italy, Sept. 1998.

Wozencraft, J.M. and I.M. Jacobs. *Principles of Communication Engineering*. New York: Wiley, 1965.

Zehavi, E. "8-PSK trellis codes for a Rayleigh fading channel." *IEEE Trans. Communications*, Vol. 40, pp. 873–883, May 1992.

Zheng, L. and D.N.C. Tse. "Communication on the Grassmann manifold: A geometric approach to the noncoherent multiple-antenna channel." *IEEE Trans. Inform. Theory*, Vol. 48, pp. 359–383, 2002.

Index